Geometry

Geometry

includes plane, analytic, and transformational geometries

Sixth Edition

Barnett Rich, PhD

Former Chairman, Department of Mathematics
Brooklyn Technical High School, New York City

Christopher Thomas, PhD

Professor and Chair, Department of Mathematics
Massachusetts College of Liberal Arts, North Adams, MA

Schaum's Outline Series

New York Chicago San Francisco Athens London Madrid
Mexico City Milan New Delhi Singapore Sydney Toronto

BARNETT RICH held a doctor of philosophy degree (PhD) from Columbia University and a doctor of jurisprudence (JD) from New York University. He began his professional career at Townsend Harris Hall High School of New York City and was one of the prominent organizers of the High School of Music and Art where he served as the Administrative Assistant. Later he taught at CUNY and Columbia University and held the post of chairman of mathematics at Brooklyn Technical High School for 14 years. Among his many achievements are the 6 degrees that he earned and the 23 books that he wrote, among them Schaum's Outlines of Elementary Algebra, Modern Elementary Algebra, and Review of Elementary Algebra.

CHRISTOPHER THOMAS has a BS from University of Massachusetts at Amherst and a PhD from Tufts University, both in mathematics. He first taught as a Peace Corps volunteer at the Mozano Senior Secondary School in Ghana. Since then he has taught at Tufts University, Texas A&M University, and the Massachusetts College of Liberal Arts. He has written Schaum's Outline of Math for the Liberal Arts as well as other books on calculus and trigonometry.

1 2 3 4 5 6 7 8 9 10 LOV 22 21 20 19 18 17

ISBN 978-1-260-01057-2
MHID 1-260-01057-0

e-ISBN 978-1-260-01058-9 (basic e-book)
e-MHID 1-260-01058-9

Preface to the First Edition

The central purpose of this book is to provide maximum help for the student and maximum service for the teacher.

Providing Help for the Student

This book has been designed to improve the learning of geometry far beyond that of the typical and traditional book in the subject. Students will find this text useful for these reasons:

(1) ***Learning Each Rule, Formula, and Principle***

Each rule, formula, and principle is stated in simple language, is made to stand out in distinctive type, is kept together with those related to it, and is clearly illustrated by examples.

(2) ***Learning Each Set of Solved Problems***

Each set of solved problems is used to clarify and apply the more important rules and principles. The character of each set is indicated by a title.

(3) ***Learning Each Set of Supplementary Problems***

Each set of supplementary problems provides further application of rules and principles. A guide number for each set refers the student to the set of related solved problems. There are more than 2000 additional related supplementary problems. Answers for the supplementary problems have been placed in the back of the book.

(4) ***Integrating the Learning of Plane Geometry***

The book integrates plane geometry with arithmetic, algebra, numerical trigonometry, analytic geometry, and simple logic. To carry out this integration:

(a) A separate chapter is devoted to analytic geometry.
(b) A separate chapter includes the complete proofs of the most important theorems together with the plan for each.
(c) A separate chapter fully explains 23 basic geometric constructions. Underlying geometric principles are provided for the constructions, as needed.
(d) Two separate chapters on methods of proof and improvement of reasoning present the simple and basic ideas of formal logic suitable for students at this stage.
(e) Throughout the book, algebra is emphasized as the major means of solving geometric problems through algebraic symbolism, algebraic equations, and algebraic proof.

(5) ***Learning Geometry Through Self-study***

The method of presentation in the book makes it ideal as a means of self-study. For able students, this book will enable then to accomplish the work of the standard course of study in much less time. For the less able, the presentation of numerous illustrations and solutions provides the help needed to remedy weaknesses and overcome difficulties, and in this way keep up with the class and at the same time gain a measure of confidence and security.

(6) *Extending Plane Geometry into Solid Geometry*

A separate chapter is devoted to the extension of two-dimensional plane geometry into three-dimensional solid geometry. It is especially important in this day and age that the student understand how the basic ideas of space are outgrowths of principles learned in plane geometry.

Providing Service for the Teacher

Teachers of geometry will find this text useful for these reasons:

(1) *Teaching Each Chapter*

Each chapter has a central unifying theme. Each chapter is divided into two to ten major subdivisions which support its central theme. In turn, these chapter subdivisions are arranged in graded sequence for greater teaching effectiveness.

(2) *Teaching Each Chapter Subdivision*

Each of the chapter subdivisions contains the problems and materials needed for a complete lesson developing the related principles.

(3) *Making Teaching More Effective Through Solved Problems*

Through proper use of the solved problems, students gain greater understanding of the way in which principles are applied in varied situations. By solving problems, mathematics is learned as it should be learned—by doing mathematics. To ensure effective learning, solutions should be reproduced on paper. Students should seek the why as well as the how of each step. Once students sees how a principle is applied to a solved problem, they are then ready to extend the principle to a related supplementary problem. Geometry is not learned through the reading of a textbook and the memorizing of a set of formulas. Until an adequate variety of suitable problems has been solved, a student will gain little more than a vague impression of plane geometry.

(4) *Making Teaching More Effective Through Problem Assignment*

The preparation of homework assignments and class assignments of problems is facilitated because the supplementary problems in this book are related to the sets of solved problems. Greatest attention should be given to the underlying principle and the major steps in the solution of the solved problems. After this, the student can reproduce the solved problems and then proceed to do those supplementary problems which are related to the solved ones.

Others Who Will Find This Text Advantageous

This book can be used profitably by others besides students and teachers. In this group we include: (1) the parents of geometry students who wish to help their children through the use of the book's self-study materials, or who may wish to refresh their own memory of geometry in order to properly help their children; (2) the supervisor who wishes to provide enrichment materials in geometry, or who seeks to improve teaching effectiveness in geometry; (3) the person who seeks to review geometry or to learn it through independent self-study.

BARNETT RICH
Brooklyn Technical High School
April, 1963

Introduction

Requirements

To fully appreciate this geometry book, you must have a basic understanding of algebra. If that is what you have really come to learn, then may I suggest you get a copy of Schaum's Outline of *College Algebra*. You will learn everything you need and more (things you don't need to know!)

If you have come to learn geometry, it begins at Chapter one.

As for algebra, you must understand that we can talk about numbers we do not know by assigning them variables like x, y, and A.

You must understand that variables can be combined when they are exactly the same, like $x + x = 2x$ and $3x^2 + 11x^2 = 14x^2$, but not when there is any difference, like $3x^2y - 9xy = 3x^2y - 9xy$.

You should understand the deep importance of the equals sign, which indicates that two things that appear different are actually exactly the same. If $3x = 15$, then this means that $3x$ is just another name for 15. If we do the same thing to both sides of an equation (add the same thing, divide both sides by something, take a square root, etc.), then the result will still be equal.

You must know how to solve an equation like $3x + 8 = 23$ by subtracting eight from both sides, $3x + 8 - 8 = 23 - 8 = 15$, and then dividing both sides by 3 to get $3x/3 = 15/3 = 5$. In this case, the variable was *constrained*; there was only one possible value and so x would have to be 5.

You must know how to add these sorts of things together, such as $(3x + 8) + (9 - x) = (3x - x) + (8 + 9) = 2x + 17$. You don't need to know that the ability to rearrange the parentheses is called *associativity* and the ability to change the order is called *commutativity*.

You must also know how to multiply them: $(3x + 8)\cdot(9 - x) = 27x - 3x^2 + 72 - 8x = -3x^2 + 19x + 72$

Actually, you might not even need to know that.

You must also be comfortable using more than one variable at a time, such as taking an equation in terms of y like $y = x^2 + 3$ and rearranging the equation to put it in terms of x like $y - 3 = x^2$. so $\sqrt{y - 3} = \sqrt{x^2}$ and thus $\sqrt{y - 3} = \pm x$, so $x = \pm\sqrt{y - 3}$.

You should know about square roots, how $\sqrt{20} = \sqrt{2 \cdot 2 \cdot 5} = 2\sqrt{5}$. It is useful to keep in mind that there are many *irrational numbers*, like $\sqrt{2}$, which could never be written as a neat ratio or fraction, but only approximated with a number of decimals.

You shouldn't be scared when there are lots of variables, either, such as $F = \frac{gM_1M_2}{r^2}$; thus, $Fr^2 = gM_1M_2$ by cross-multiplication, so $r = \pm\sqrt{\frac{gM_1M_2}{F}}$.

Most important of all, you should know how to take a formula like $V = \frac{1}{3}\pi r^2 h$ and replace values and simplify. If $r = 5$ cm and $h = 8$ cm, then

$$V = \frac{1}{3}\pi(5\text{ cm})^2(8\text{ cm}) = \frac{200\pi}{3}\text{ cm}^3.$$

Contents

Lines, Angles, and Triangles

1.1 Historical Background of Geometry

The word *geometry* is derived from the Greek words *geos* (meaning *earth*) and *metron* (meaning *measure*). The ancient Egyptians, Chinese, Babylonians, Romans, and Greeks used geometry for surveying, navigation, astronomy, and other practical occupations.

The Greeks sought to systematize the geometric facts they knew by establishing logical reasons for them and relationships among them. The work of men such as Thales (600 B.C.), Pythagoras (540 B.C.), Plato (390 B.C.), and Aristotle (350 B.C.) in systematizing geometric facts and principles culminated in the geometry text *Elements*, written in approximately 325 B.C. by Euclid. This most remarkable text has been in use for over 2000 years.

1.2 Undefined Terms of Geometry: Point, Line, and Plane

1.2A Point, Line, and Plane are Undefined Terms

These undefined terms underlie the definitions of all geometric terms. They can be given meanings by way of descriptions. However, these descriptions, which follow, are not to be thought of as definitions.

1.2B Point

A *point* has position only. It has no length, width, or thickness.

A point is represented by a dot. Keep in mind, however, that the dot *represents* a point but *is not* a point, just as a dot on a map may represent a locality but is not the locality. A dot, unlike a point, has size.

A point is designated by a capital letter next to the dot, thus point A is represented: *A*.

1.2C Line

A line has length but has no width or thickness.

A line may be represented by the path of a piece of chalk on the blackboard or by a stretched rubber band.

A line is designated by the capital letters of any two of its points or by a small letter, thus:

A B, C D, a, or $\overleftrightarrow{AB}$.

A *line* may be straight, curved, or a combination of these. To understand how lines differ, think of a line as being generated by a moving point. A *straight line*, such as ⟷, is generated by a point moving always in the same direction. A *curved line*, such as ⌒, is generated by a point moving in a continuously changing direction.

Two lines intersect in a point.

A straight line is unlimited in extent. It may be extended in either direction indefinitely.

A *ray* is the part of a straight line beginning at a given point and extending limitlessly in one direction: $\overrightarrow{AB}$ and A B designate rays.

In this book, the word *line* will mean "straight line" unless otherwise stated.

1.2D Surface

A *surface* has length and width but no thickness. It may be represented by a blackboard, a side of a box, or the outside of a sphere; remember, however, that these are representations of a surface but are not surfaces.

A plane surface (or *plane*) is a surface such that a straight line connecting any two of its points lies entirely in it. A plane is a flat surface.

Plane geometry is the geometry of plane figures—those that may be drawn on a plane. Unless otherwise stated, the word *figure* will mean "plane figure" in this book.

SOLVED PROBLEMS

1.1 Illustrating undefined terms

Point, line, and plane are undefined terms. State which of these terms is illustrated by (a) the top of a desk; (b) a projection screen; (c) a ruler's edge; (d) a stretched thread; (e) the tip of a pin.

Solutions

(a) surface; (b) surface; (c) line; (d) line; (e) point.

1.3 Line Segments

A straight line segment is the part of a straight line between two of its points, including the two points, called *endpoints*. It is designated by the capital letters of these points with a bar over them or by a small letter. Thus, $\overline{AB}$ or r represents the straight line segment $A \overset{r}{—} B$ between A and B.

The expression *straight line segment* may be shortened to *line segment* or to *segment*, if the meaning is clear. Thus, $\overline{AB}$ and *segment AB* both mean "the straight line segment AB."

1.3A Dividing a Line Segment into Parts

If a line segment is divided into parts:

1. The length of the whole line segment equals the sum of the lengths of its parts. Note that the length of $\overline{AB}$ is designated AB. A number written beside a line segment designates its length.
2. The length of the whole line segment is greater than the length of any part.
   Suppose $\overline{AB}$ is divided into three parts of lengths a, b, and c; thus $A \overset{a}{—}\bullet\overset{b}{—}\bullet\overset{c}{—} B$. Then $AB = a + b + c$. Also, AB is greater than a; this may be written as $AB > a$.

If a line segment is divided into two equal parts:

1. The point of division is the *midpoint* of the line segment.

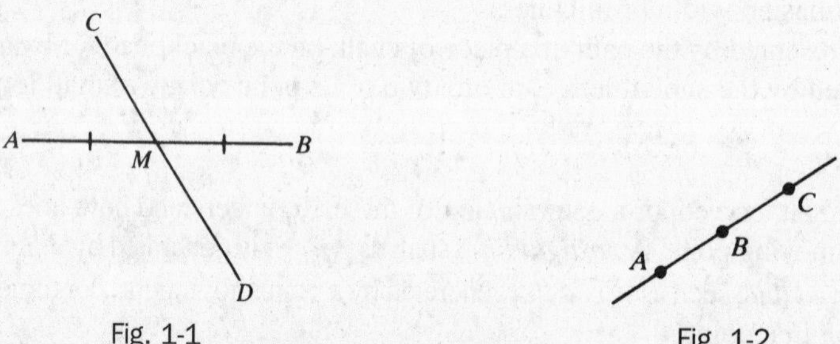

Fig. 1-1 Fig. 1-2

2. A line that crosses at the midpoint is said to *bisect* the segment.
   Because $AM = MB$ in Fig. 1-1, M is the midpoint of $\overline{AB}$, and $\overline{CD}$ bisects $\overline{AB}$. Equal line segments may be shown by crossing them with the same number of strokes. Note that $\overline{AM}$ and $\overline{MB}$ are crossed with a single stroke.
3. If three points A, B, and C lie on a line, then we say they are *collinear*. If A, B, and C are collinear and $AB + BC = AC$, then B is between A and C (see Fig. 1-2).

1.3B Congruent Segments

Two line segments having the same length are said to be *congruent*. Thus, if $AB = CD$, then $\overline{AB}$ is congruent to $\overline{CD}$, written $\overline{AB} \cong \overline{CD}$.

SOLVED PROBLEMS

1.2 Naming line segments and points

See Fig. 1-3.

(a) Name each line segment shown.

(b) Name the line segments that intersect at *A*.

(c) What other line segment can be drawn using points *A*, *B*, *C*, and *D*?

(d) Name the point of intersection of $\overline{CD}$ and $\overline{AD}$.

(e) Name the point of intersection of $\overline{BC}$, $\overline{AC}$, and $\overline{CD}$.

Obtuse

Fig. 1-3

Solutions

(a) $\overline{AB}$, $\overline{BC}$, $\overline{CD}$, $\overline{AC}$, and $\overline{AD}$. These segments may also be named by interchanging the letters; thus, $\overline{BA}$, $\overline{CB}$, $\overline{DC}$, $\overline{CA}$, and $\overline{DA}$ are also correct.

(b) $\overline{AB}$, $\overline{AC}$, and $\overline{AD}$

(c) $\overline{BD}$

(d) *D*

(e) *C*

1.3 Finding lengths and points of line segments

See Fig. 1-4.

(a) State the lengths of $\overline{AB}$, $\overline{AC}$, and $\overline{AF}$.

(b) Name two midpoints.

(c) Name two bisectors.

(d) Name all congruent segments.

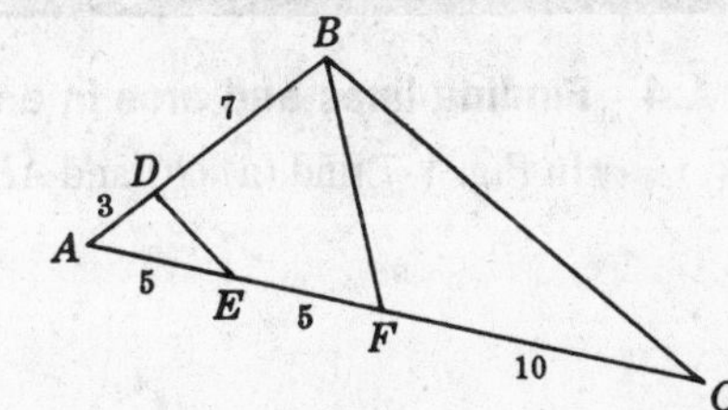

Fig. 1-4

Solutions

(a) $AB = 3 + 7 = 10$; $AC = 5 + 5 + 10 = 20$; $AF = 5 + 5 = 10$.

(b) *E* is midpoint of $\overline{AF}$; *F* is midpoint of $\overline{AC}$.

(c) $\overline{DE}$ is bisector of $\overline{AF}$; $\overline{BF}$ is bisector of $\overline{AC}$.

(d) $\overline{AB}$, $\overline{AF}$, and $\overline{FC}$ (all have length 10); $\overline{AE}$ and $\overline{EF}$ (both have length 5).

1.4 Circles

A *circle* is the set of all points in a plane that are the same distance from the *center*. The symbol for circle is ⊙; for circles, Ⓢ. Thus, ⊙*O* stands for the circle whose center is *O*.

The *circumference* of a circle is the distance around the circle. It contains 360 *degrees* (360°).

A *radius* is a segment joining the center of a circle to a point on the circle (see Fig. 1-5). From the definition of a circle, it follows that the radii of a circle are congruent. Thus, $\overline{OA}$, $\overline{OB}$, and $\overline{OC}$ of Fig. 1-5 are radii of ⊙*O* and $\overline{OA} \cong \overline{OB} \cong \overline{OC}$.

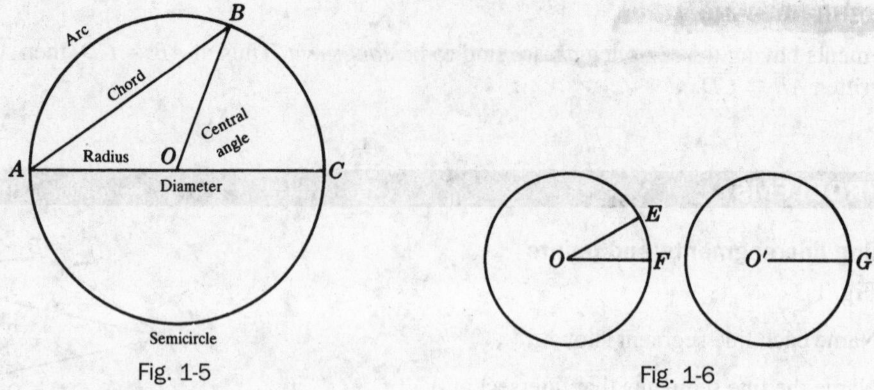

Fig. 1-5 Fig. 1-6

A *chord* is a segment joining any two points on a circle. Thus, $\overline{AB}$ and $\overline{AC}$ are chords of $\odot O$.

A *diameter* is a chord through the center of the circle; it is the longest chord and is twice the length of a radius. $\overline{AC}$ is a diameter of $\odot O$.

An *arc* is a continuous part of a circle. The symbol for arc is $\frown$, so that $\overset{\frown}{AB}$ stands for arc AB. An arc of measure 1° is 1/360th of a circumference.

A *semicircle* is an arc measuring one-half of the circumference of a circle and thus contains 180°. A diameter divides a circle into two semicircles. For example, diameter $\overline{AC}$ cuts $\odot O$ of Fig. 1-5 into two semicircles.

A *central angle* is an angle formed by two radii. Thus, the angle between radii $\overline{OB}$ and $\overline{OC}$ is a central angle. A central angle measuring 1° cuts off an arc of 1°; thus, if the central angle between $\overline{OE}$ and $\overline{OF}$ in Fig. 1-6 is 1°, then $\overset{\frown}{EF}$ measures 1°.

Congruent circles are circles having congruent radii. Thus, if $\overline{OE} \cong \overline{O'G}$, then $\odot O \cong \odot O'$.

SOLVED PROBLEMS

1.4 Finding lines and arcs in a circle

In Fig. 1-7 find (a) OC and AB; (b) the number of degrees in $\overset{\frown}{AD}$; (c) the number of degrees in $\overset{\frown}{BC}$.

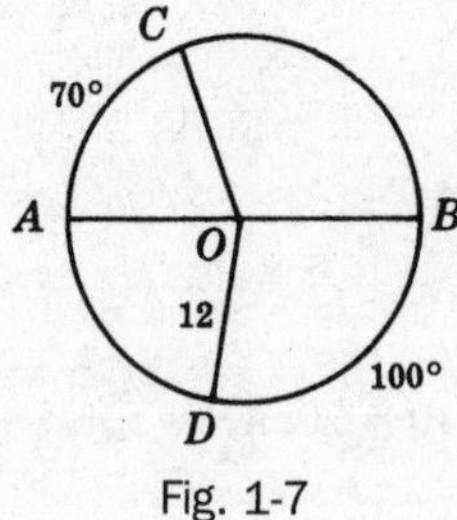

Fig. 1-7

Solutions

(a) Radius OC = radius OD = 12. Diameter AB = 24.

(b) Since semicircle ADB contains 180°, $\overset{\frown}{AD}$ contains $180° - 100° = 80°$.

(c) Since semicircle ACB contains 180°, $\overset{\frown}{BC}$ contains $180° - 70° = 110°$.

1.5 Angles

An *angle* is the figure formed by two rays with a common end point. The rays are the *sides* of the angle, while the end point is its *vertex*. The symbol for angle is ∠ or ∡; the plural is ∡s.

Thus, $\overrightarrow{AB}$ and $\overrightarrow{AC}$ are the sides of the angle shown in Fig. 1-8(a), and A is its vertex.

1.5A Naming an Angle

An angle may be named in any of the following ways:

1. With the vertex letter, if there is only one angle having this vertex, as $\angle B$ in Fig. 1-8(b).
2. With a small letter or a number placed between the sides of the angle and near the vertex, as $\angle a$ or $\angle 1$ in Fig. 1-8(c).
3. With three capital letters, such that the vertex letter is between two others, one from each side of the angle. In Fig. 1-8(d), $\angle E$ may be named $\angle DEG$ or $\angle GED$.

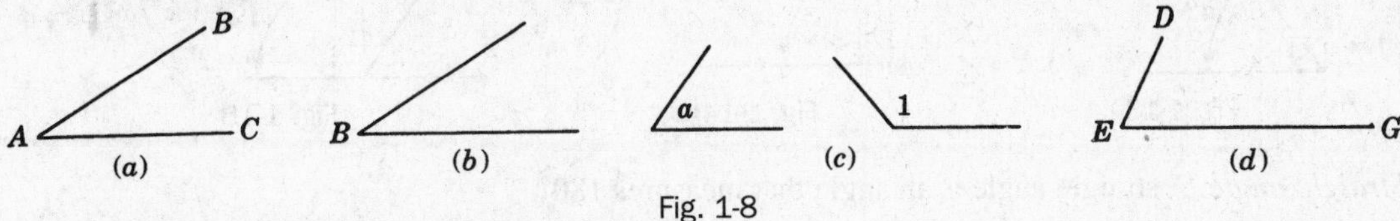

Fig. 1-8

1.5B Measuring the Size of an Angle

The size of an angle depends on the extent to which one side of the angle must be rotated, or turned about the vertex, until it meets the other side. We choose degrees to be the unit of measure for angles. The measure of an angle is the number of degrees it contains. We will write $m\angle A = 60°$ to denote that "angle A measures 60°."

The protractor in Fig. 1-9 shows that $\angle A$ measures of 60°. If $\overrightarrow{AC}$ were rotated about the vertex A until it met $\overrightarrow{AB}$, the amount of turn would be 60°.

In using a protractor, be sure that the vertex of the angle is at the center and that one side is along the 0°–180° diameter.

The size of an angle *does not* depend on the lengths of the sides of the angle.

Fig. 1-9 Fig. 1-10

The size of $\angle B$ in Fig. 1-10 would not be changed if its sides $\overrightarrow{AB}$ and $\overrightarrow{BC}$ were made larger or smaller.

No matter how large or small a clock is, the angle formed by its hands at 3 o'clock measures 90°, as shown in Figs. 1-11 and 1-12.

Fig. 1-11 Fig. 1-12

Angles that measure less than 1° are usually represented as fractions or decimals. For example, one-thousandth of the way around a circle is either $\frac{360°}{1000}$ or 0.36°.

In some fields, such as navigation and astronomy, small angles are measured in *minutes* and *seconds*. One degree is comprised of 60 minutes, written $1° = 60'$. A minute is 60 seconds, written $1' = 60''$. In this notation, one-thousandth of a circle is $21'36''$ because $\frac{21}{60} + \frac{36}{3600} = \frac{1296}{3600} = \frac{360}{1000}$.

1.5C Kinds of Angles

1. *Acute angle*: An acute angle is an angle whose measure is less than 90°.

   Thus, in Fig. 1-13 $a°$ is less than 90°; this is symbolized as $a° < 90°$.

2. *Right angle*: A right angle is an angle that measures 90°.

   Thus, in Fig. 1-14, $m(\text{rt. } \angle A) = 90°$. The square corner denotes a right angle.

3. *Obtuse angle*: An obtuse angle is an angle whose measure is more than 90° and less than 180°.

   Thus, in Fig. 1-15, 90° is less than $b°$ and $b°$ is less than 180°; this is denoted by $90° < b° < 180°$.

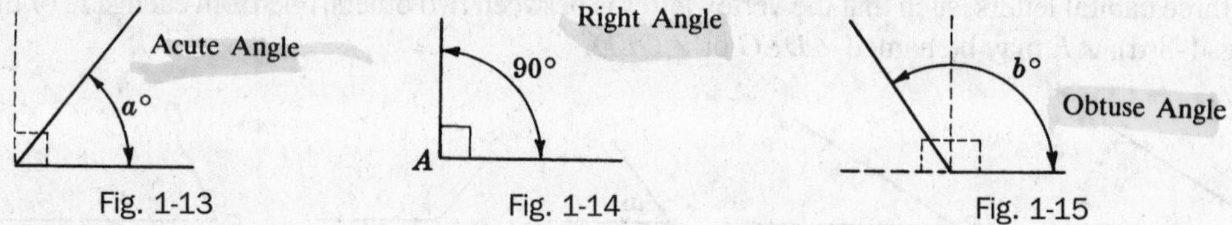

Fig. 1-13 Fig. 1-14 Fig. 1-15

4. *Straight angle*: A straight angle is an angle that measures 180°.

   Thus, in Fig. 1-16, $m(\text{st. } \angle B) = 180°$. Note that the sides of a straight angle lie in the same straight line. But do not confuse a straight angle with a straight line!

Fig. 1-16 Fig. 1-17

5. *Reflex angle*: A reflex angle is an angle whose measure is more than 180° and less than 360°.

   Thus, in Fig. 1-17, 180° is less than $c°$ and $c°$ is less than 360°; this is symbolized as $180° < c° < 360°$.

1.5D Additional Angle Facts

1. *Congruent angles* are angles that have the same number of degrees. In other words, if $m\angle A = m\angle B$, then $\angle A \cong \angle B$.

   Thus, in Fig. 1-18, rt. $\angle A \cong$ rt. $\angle B$ since each measures 90°.

Fig. 1-18 Fig. 1-19

2. A line that *bisects* an angle divides it into two congruent parts.

   Thus, in Fig. 1-19, if $\overline{AD}$ bisects $\angle A$, then $\angle 1 \cong \angle 2$. (Congruent angles may be shown by crossing their arcs with the same number of strokes. Here the arcs of $\measuredangle$s 1 and 2 are crossed by a single stroke.)

3. *Perpendiculars* are lines or rays or segments that meet at right angles.

   The symbol for perpendicular is $\perp$; for perpendiculars, $\perp$s. In Fig. 1-20, $\overline{CD} \perp \overline{AB}$, so right angles 1 and 2 are formed.

4. A *perpendicular bisector* of a given segment is perpendicular to the segment and bisects it.

   In Fig. 1-21, $\overleftrightarrow{GH}$ is the bisector of $\overline{EF}$; thus, $\angle 1$ and $\angle 2$ are right angles and M is the midpoint of $\overline{EF}$.

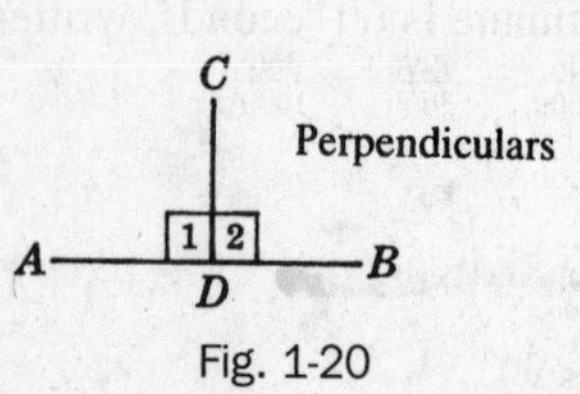

Fig. 1-20

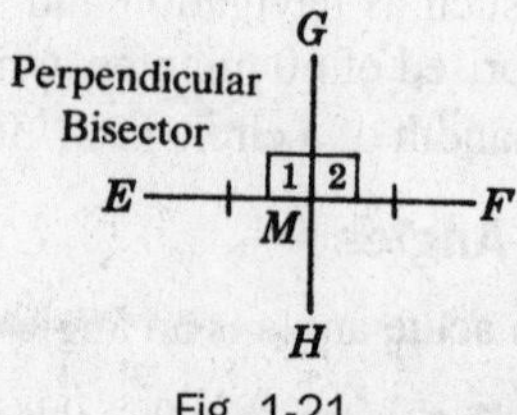

Fig. 1-21

SOLVED PROBLEMS

1.5 Naming an angle

Name the following angles in Fig. 1-22: (a) two obtuse angles; (b) a right angle; (c) a straight angle; (d) an acute angle at D; (e) an acute angle at B.

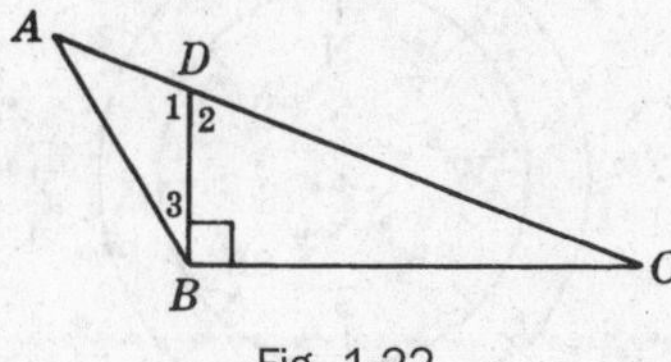

Fig. 1-22

Solutions

(a) $\angle ABC$ and $\angle ADB$ (or $\angle 1$). The angles may also be named by reversing the order of the letters: $\angle CBA$ and $\angle BDA$.

(b) $\angle DBC$

(c) $\angle ADC$

(d) $\angle 2$ or $\angle BDC$

(e) $\angle 3$ or $\angle ABD$

1.6 Adding and subtracting angles

In Fig. 1-23, find (a) $m\angle AOC$; (b) $m\angle BOE$; (c) the measure of obtuse $\angle AOE$.

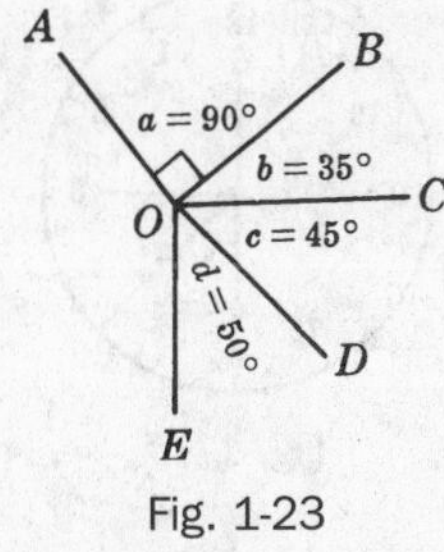

Fig. 1-23

Solutions

(a) $m\angle AOC = m\angle a + m\angle b = 90° + 35° = 125°$

(b) $m\angle BOE = m\angle b + m\angle c + m\angle d = 35° + 45° + 50° = 130°$

(c) $m\angle AOE = 360° - (m\angle a + m\angle b + m\angle c + m\angle d) = 360° - 220° = 140°$

1.7 Finding parts of angles

Find (a) $\frac{2}{5}$ of the measure of a rt. $\angle$; (b) $\frac{2}{3}$ of the measure of a st. $\angle$; (c) $\frac{1}{2}$ of 31°; (d) $\frac{1}{10}$ of 70°20′.

Solutions

(a) $\frac{2}{5}(90°) = 36°$

(b) $\frac{2}{3}(180°) = 120°$

(c) $\frac{1}{2}(31°) = 15\frac{1}{2}° = 15°30'$

(d) $\frac{1}{10}(70°20') = \frac{1}{10}(70°) + \frac{1}{10}(20') = 7°2'$

1.8 Finding rotations

In a half hour, what turn or rotation is made (a) by the minute hand, and (b) by the hour hand of a clock? What rotation is needed to turn (c) from north to southeast in a clockwise direction, and (d) from northwest to southwest in a counterclockwise direction (see Fig. 1-24)?

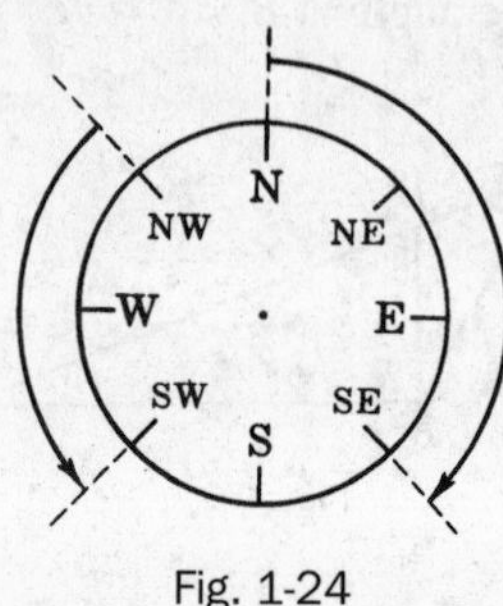

Fig. 1-24

Solutions

(a) In 1 hour, a minute hand completes a full circle of 360°. Hence, in a half hour it turns 180°.

(b) In 1 hour, an hour hand turns $\frac{1}{12}$ of 360° or 30°. Hence, in a half hour it turns 15°.

(c) Add a turn of 90° from north to east and a turn of 45° from east to southeast to get $90° + 45° = 135°$.

(d) The turn from northwest to southwest is $\frac{1}{4}(360°) = 90°$.

1.9 Finding angles

Find the measure of the angle formed by the hands of the clock in Fig. 1-25, (a) at 8 o'clock; (b) at 4:30.

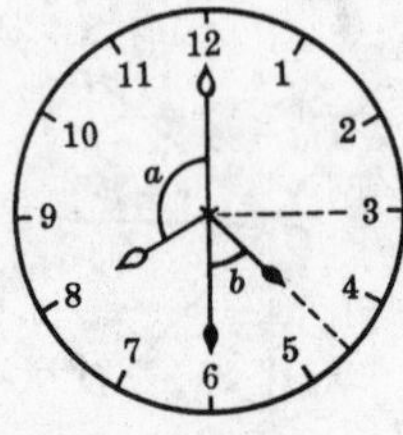

Fig. 1-25

Solutions

(a) At 8 o'clock, $m\angle a = \frac{1}{3}(360°) = 120°$.

(b) At 4:30, $m\angle b = \frac{1}{2}(90°) = 45°$.

1.10 Applying angle facts

In Fig. 1-26, (a) name two pairs of perpendicular segments; (b) find $m\angle a$ if $m\angle b = 42°$; (c) find $m\angle AEB$ and $m\angle CED$.

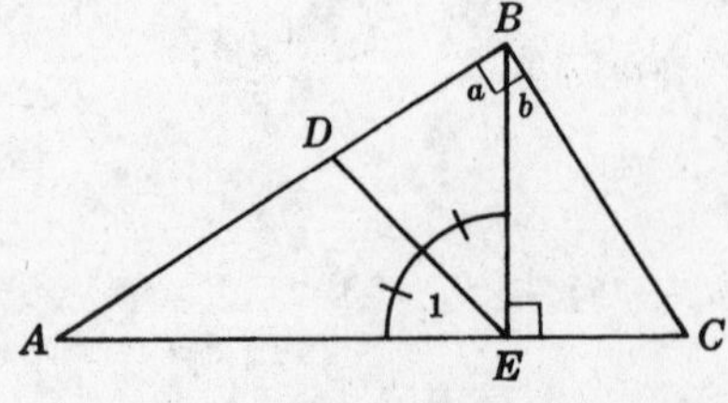

Fig. 1-26

Solutions

(a) Since $\angle ABC$ is a right angle, $\overline{AB} \perp \overline{BC}$. Since $\angle BEC$ is a right angle, $\overline{BE} \perp \overline{AC}$.

(b) $m\angle a = 90° - m\angle b = 90° - 42° = 48°$.

(c) $m\angle AEB = 180° - m\angle BEC = 180° - 90° = 90°$. $m\angle CED = 180° - m\angle 1 = 180° - 45° = 135°$.

1.6 Triangles

A *polygon* is a closed plane figure bounded by straight line segments as sides. Thus, Fig. 1-27 is a polygon of five sides, called a *pentagon*; it is named pentagon *ABCDE*, using its letters in order.

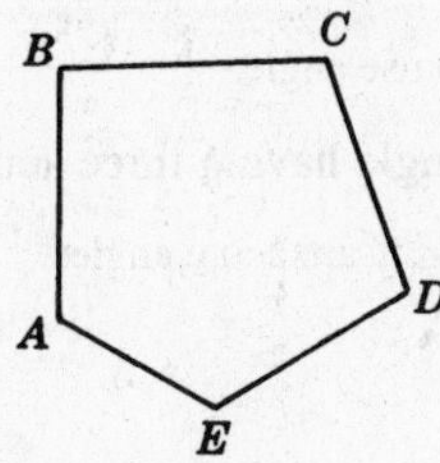

Fig. 1-27

A *quadrilateral* is a polygon having four sides.

A *triangle* is a polygon having three sides. A *vertex* of a triangle is a point at which two of the sides meet. (*Vertices* is the plural of vertex.) The symbol for triangle is $\triangle$; for triangles, $\triangle$s.

A triangle may be named with its three letters in any order or with a Roman numeral placed inside of it. Thus, the triangle shown in Fig. 1-28 is $\triangle ABC$ or $\triangle$I; its sides are $\overline{AB}$, $\overline{AC}$, and $\overline{BC}$; its vertices are *A, B,* and *C*; its angles are $\angle A$, $\angle B$, and $\angle C$.

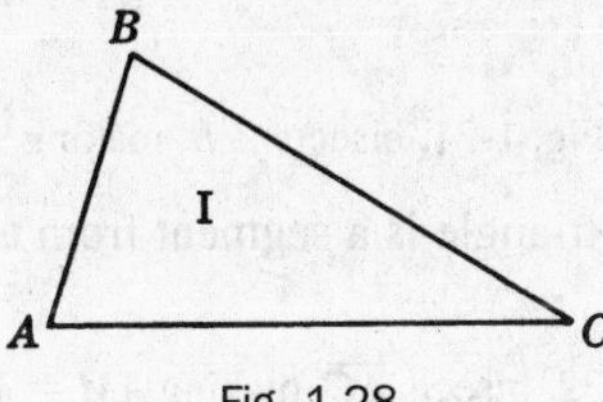

Fig. 1-28

1.6A Classifying Triangles

Triangles are classified according to the equality of the lengths of their sides or according to the kind of angles they have.

Triangles According to the Equality of the Lengths of their Sides (Fig. 1-29)

1. *Scalene triangle*: A scalene triangle is a triangle having no congruent sides.

   Thus in scalene triangle *ABC*, $a \neq b \neq c$. The small letter used for the length of each side agrees with the capital letter of the angle opposite it. Also, $\neq$ means "is not equal to."

2. *Isosceles triangle*: An isosceles triangle is a triangle having at least two congruent sides.

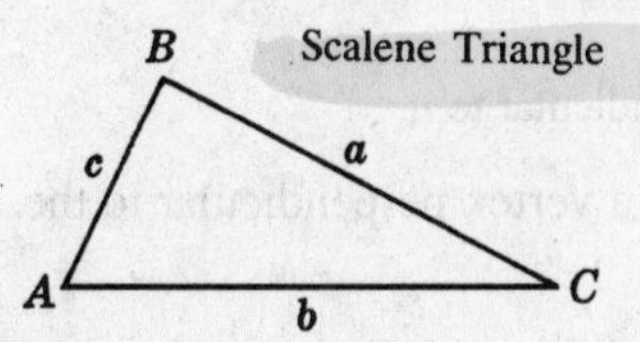

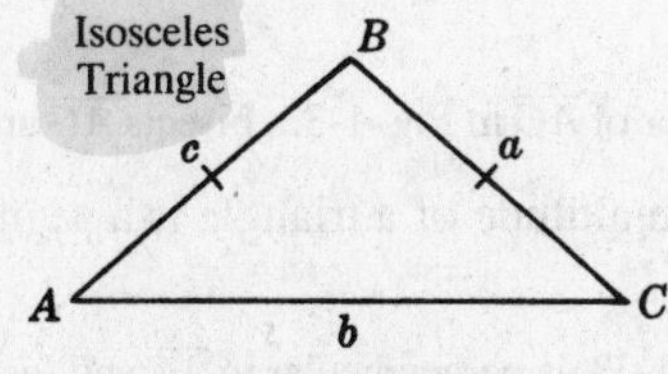

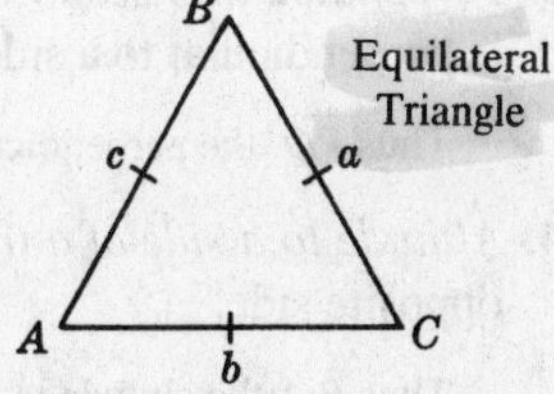

Fig. 1-29

Thus in isosceles triangle ABC, $a = c$. These equal sides are called the *legs* of the isosceles triangle; the remaining side is the *base*, b. The angles on either side of the base are the *base angles*; the angle opposite the base is the *vertex angle*.

3. *Equilateral triangle*: An equilateral triangle is a triangle having three congruent sides.

   Thus in equilateral triangle ABC, $a = b = c$. Note that an equilateral triangle is also an isosceles triangle.

Triangles According to the Kind of Angles (Fig. 1-30)

1. *Right triangle*: A right triangle is a triangle having a right angle.

   Thus in right triangle ABC, $\angle C$ is the right angle. Side c opposite the right angle is the *hypotenuse*. The perpendicular sides, a and b, are the *legs* or *arms* of the right triangle.

2. *Obtuse triangle*: An obtuse triangle is a triangle having an obtuse angle.

   Thus in obtuse triangle DEF, $\angle D$ is the obtuse angle.

3. *Acute triangle*: An acute triangle is a triangle having three acute angles.

   Thus in acute triangle HJK, $\angle H$, $\angle J$, and $\angle K$ are acute angles.

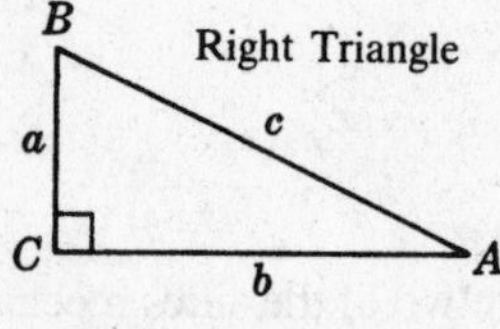

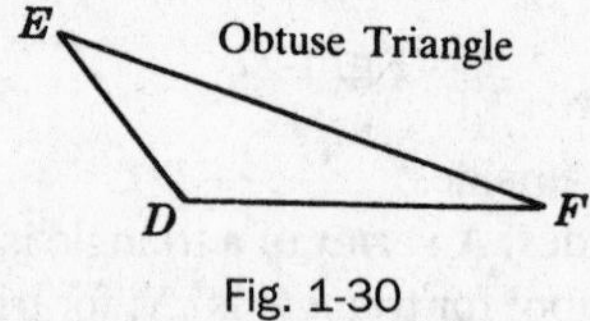

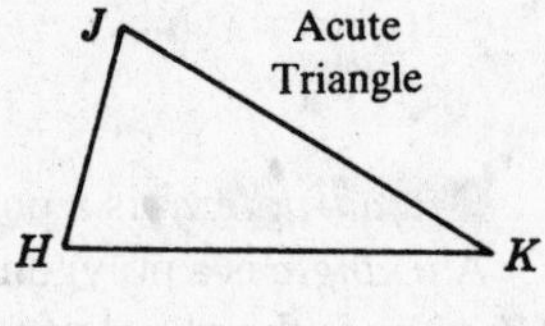

Fig. 1-30

1.6B Special Lines in a Triangle

1. *Angle bisector of a triangle*: An angle bisector of a triangle is a segment or ray that bisects an angle and extends to the opposite side.

   Thus $\overline{BD}$, the angle bisector of $\angle B$ in Fig. 1-31, bisects $\angle B$, making $\angle 1 \cong \angle 2$.

2. *Median of a triangle*: A median of a triangle is a segment from a vertex to the midpoint of the opposite side.

   Thus $\overline{BM}$, the median to $\overline{AC}$, in Fig. 1-32, bisects $\overline{AC}$, making $AM = MC$.

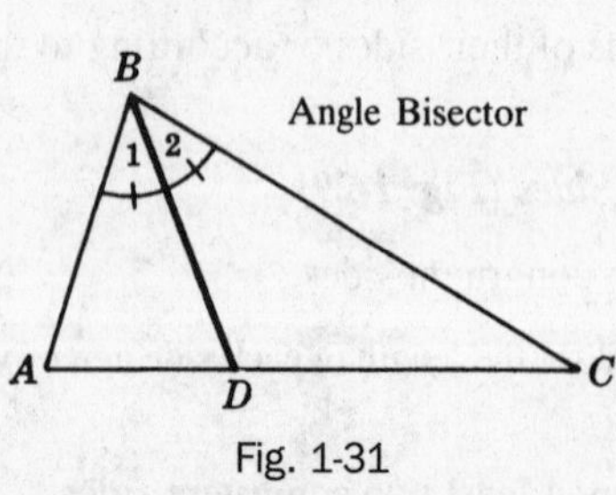

Fig. 1-31

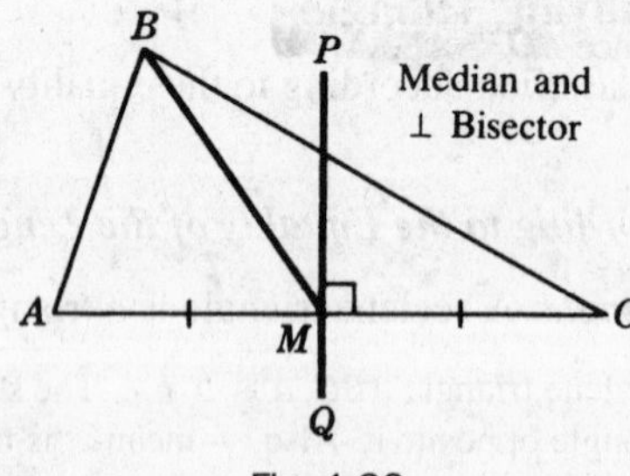

Fig. 1-32

3. *Perpendicular bisector of a side*: A perpendicular bisector of a side of a triangle is a line that bisects and is perpendicular to a side.

   Thus $\overleftrightarrow{PQ}$, the perpendicular bisector of $\overline{AC}$ in Fig. 1-32, bisects $\overline{AC}$ and is perpendicular to it.

4. *Altitude to a side of a triangle*: An altitude of a triangle is a segment from a vertex perpendicular to the opposite side.

   Thus $\overline{BD}$, the altitude to $\overline{AC}$ in Fig. 1-33, is perpendicular to $\overline{AC}$ and forms right angles 1 and 2. Each angle bisector, median, and altitude of a triangle extends from a vertex to the opposite side.

5. *Altitudes of obtuse triangle*: In an obtuse triangle, the altitude drawn to either side of the obtuse angle falls outside the triangle.

Thus in obtuse triangle *ABC* (shaded) in Fig. 1-34, altitudes $\overline{BD}$ and $\overline{CE}$ fall outside the triangle. In each case, a side of the obtuse angle must be extended.

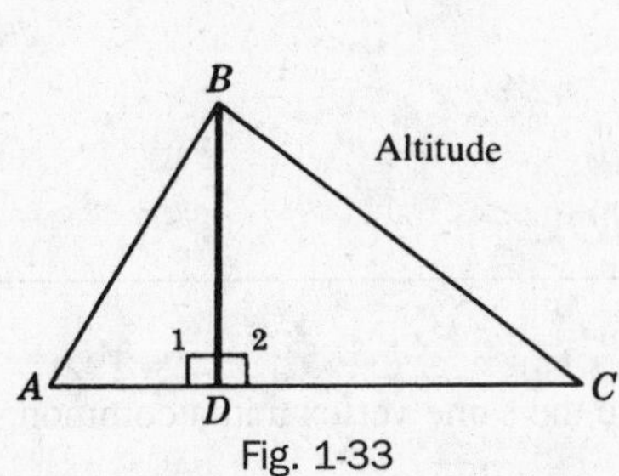

Fig. 1-33

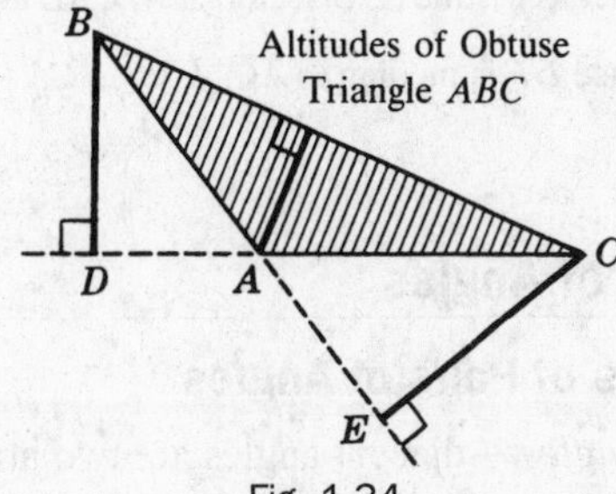

Fig. 1-34

SOLVED PROBLEMS

1.11 Naming a triangle and its parts

In Fig. 1-35, name (a) an obtuse triangle, and (b) two right triangles and the hypotenuse and legs of each. (c) In Fig. 1-36, name two isosceles triangles; also name the legs, base, and vertex angle of each.

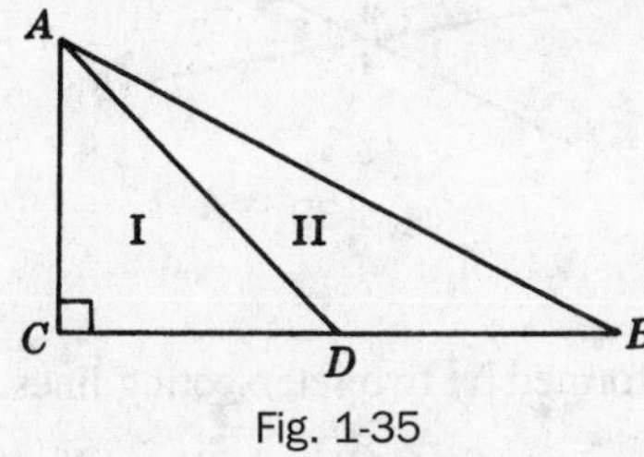

Fig. 1-35

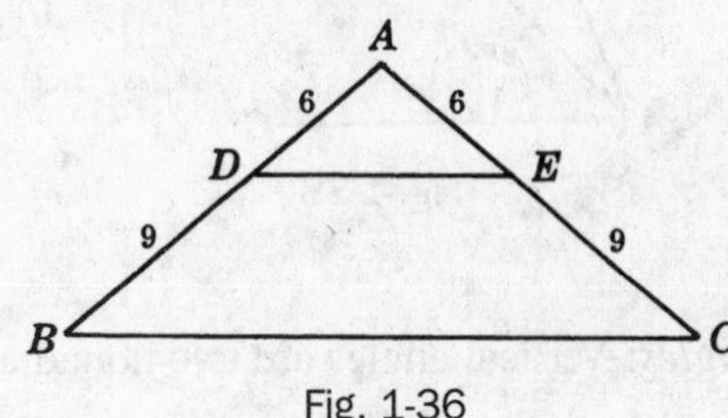

Fig. 1-36

Solutions

(a) Since $\angle ADB$ is an obtuse angle, $\angle ADB$ *or* $\triangle$II is obtuse.

(b) Since $\angle C$ is a right angle, $\triangle$I and $\triangle ABC$ are right triangles. In $\triangle$I, $\overline{AD}$ is the hypotenuse and $\overline{AC}$ and $\overline{CD}$ are the legs. In $\triangle ABC$, *AB* is the hypotenuse and $\overline{AC}$ and $\overline{BC}$ are the legs.

(c) Since $AD = AE$, $\triangle ADE$ is an isosceles triangle. In $\triangle ADE$, $\overline{AD}$ and $\overline{AE}$ are the legs, $\overline{DE}$ is the base, and $\angle A$ is the vertex angle.

Since $AB = AC$, $\triangle ABC$ is an isosceles triangle. In $\triangle ABC$, $\overline{AB}$ and $\overline{AC}$ are the legs, $\overline{BC}$ is the base, and $\angle A$ is the vertex angle.

1.12 Special lines in a triangle

Name the equal segments and congruent angles in Fig. 1-37, (a) if $\overline{AE}$ is the altitude to $\overline{BC}$; (b) if $\overline{CG}$ bisects $\angle ACB$; (c) if $\overline{KL}$ is the perpendicular bisector of $\overline{AD}$; (d) if $\overline{DF}$ is the median to $\overline{AC}$.

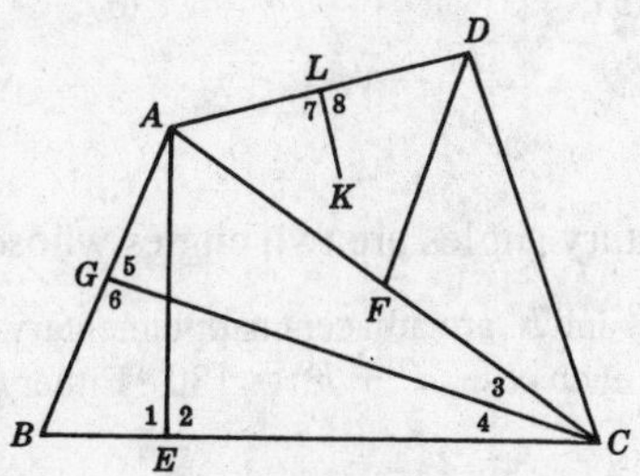

Fig. 1-37

Solutions

(a) Since $\overline{AE} \perp \overline{BC}$, $\angle 1 \cong \angle 2$.

(b) Since $\overline{CG}$ bisects $\angle ACB$, $\angle 3 \cong \angle 4$.

(c) Since $\overline{LK}$ is the $\perp$ bisector of $\overline{AD}$, $AL = LD$ and $\angle 7 \cong \angle 8$.

(d) Since $\overline{DF}$ is median to $\overline{AC}$, $AF = FC$.

1.7 Pairs of Angles

1.7A Kinds of Pairs of Angles

1. *Adjacent angles*: Adjacent angles are two angles that have the same vertex and a common side between them.

Thus, the entire angle of $c°$ in Fig. 1-38 has been cut into two adjacent angles of $a°$ and $b°$. These adjacent angles have the same vertex A, and a common side $\overrightarrow{AD}$ between them. Here, $a° + b° = c°$.

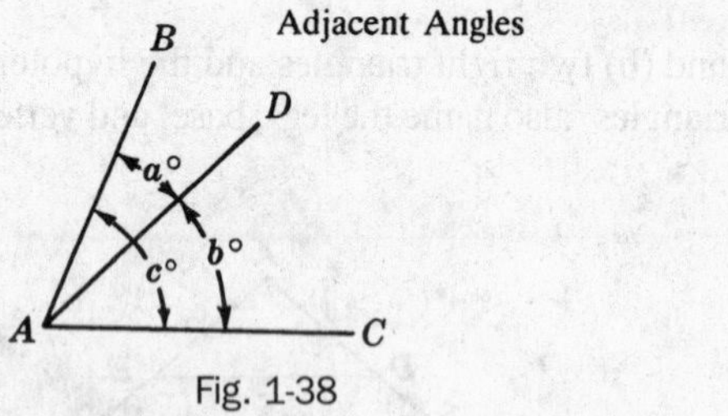

Fig. 1-38

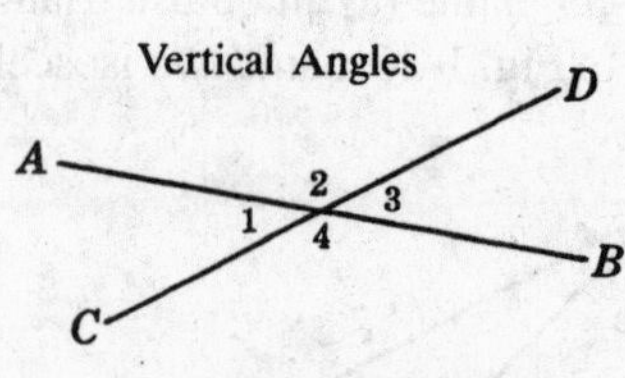

Fig. 1-39

2. *Vertical angles*: Vertical angles are two nonadjacent angles formed by two intersecting lines.

Thus, $\angle 1$ and $\angle 3$ in Fig. 1-39 are vertical angles formed by intersecting lines $\overleftrightarrow{AB}$ and $\overleftrightarrow{CD}$. Also, $\angle 2$ and $\angle 4$ are another pair of vertical angles formed by the same lines.

3. *Complementary angles*: Complementary angles are two angles whose measures total 90°.

Thus, in Fig. 1-40(a) the angles of $a°$ and $b°$ are adjacent complementary angles. However, in (b) the complementary angles are nonadjacent. In each case, $a° + b° = 90°$. Either of two complementary angles is said to be the *complement* of the other.

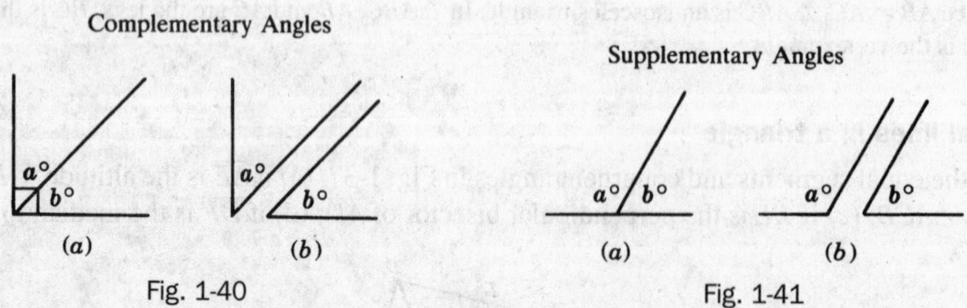

Fig. 1-40

Fig. 1-41

4. *Supplementary angles*: Supplementary angles are two angles whose measures total 180°.

Thus, in Fig. 1-41(a) the angles of $a°$ and $b°$ are adjacent supplementary angles. However, in Fig. 1-41(b) the supplementary angles are nonadjacent. In each case, $a° + b° = 180°$. Either of two supplementary angles is said to be the *supplement* of the other.

1.7B Principles of Pairs of Angles

PRINCIPLE 1: *If an angle of $c°$ is cut into two adjacent angles of $a°$ and $b°$, then $a° + b° = c°$.*

Thus if $a° = 25°$ and $b° = 35°$ in Fig. 1-42, then $c° + 25° + 35° = 60°$.

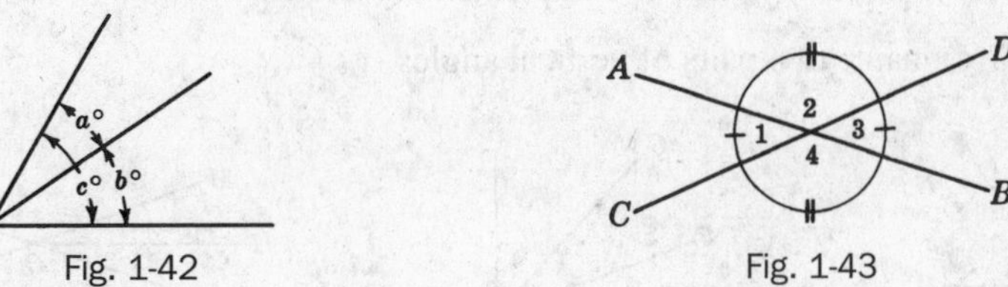

Fig. 1-42 Fig. 1-43

PRINCIPLE 2: *Vertical angles are congruent.*

Thus if $\overleftrightarrow{AB}$ and $\overleftrightarrow{CD}$ are straight lines in Fig. 1-43, then $\angle 1 \cong \angle 3$ and $\angle 2 \cong \angle 4$. Hence, if $m\angle 1 = 40°$, then $m\angle 3 = 40°$; in such a case, $m\angle 2 = m\angle 4 = 140°$.

PRINCIPLE 3: *If two complementary angles contain $a°$ and $b°$, then $a° + b° = 90°$.*

Thus if angles of $a°$ and $b°$ are complementary and $a° = 40°$, then $b° = 50°$ [Fig. 1-44(a) or (b)].

PRINCIPLE 4: *Adjacent angles are complementary if their exterior sides are perpendicular to each other.*

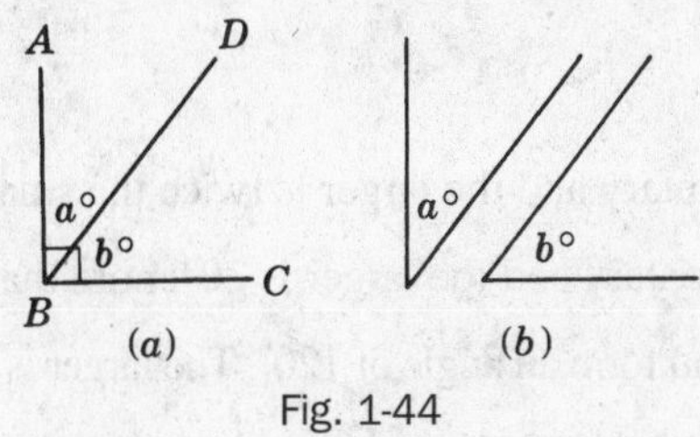

Fig. 1-44

Thus in Fig. 1-44(a), $a°$ and $b°$ are complementary since their exterior sides $\overline{AB}$ and $\overline{BC}$ are perpendicular to each other.

PRINCIPLE 5: *If two supplementary angles contain $a°$ and $b°$, then $a° + b° = 180°$.*

Thus if angles of $a°$ and $b°$ are supplementary and $a° = 140°$, then $b° = 40°$ [Fig. 1-45(a) or (b)].

PRINCIPLE 6: *Adjacent angles are supplementary if their exterior sides lie in the same straight line.*

Thus in Fig. 1-45(a) $a°$ and $b°$ are supplementary angles since their exterior sides $\overrightarrow{AB}$ and $\overrightarrow{BC}$ lie in the same straight line $\overrightarrow{AC}$.

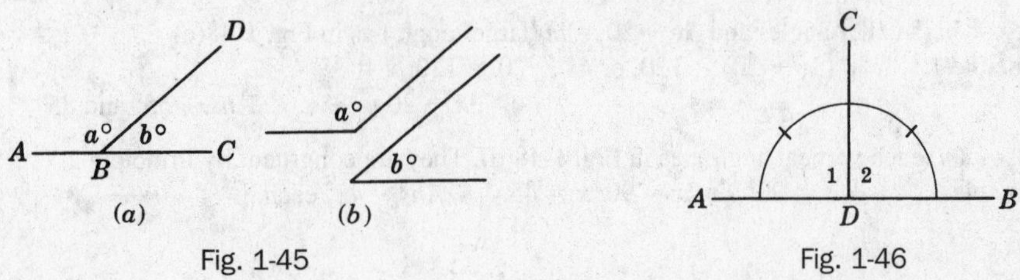

Fig. 1-45 Fig. 1-46

PRINCIPLE 7: *If supplementary angles are congruent, each of them is a right angle. (Equal supplementary angles are right angles.)*

Thus if $\angle 1$ and $\angle 2$ in Fig. 1-46 are both congruent and supplementary, then each of them is a right angle.

SOLVED PROBLEMS

1.13 Naming pairs of angles

(a) In Fig. 1-47(a), name two pairs of supplementary angles.

(b) In Fig. 1-47(b), name two pairs of complementary angles.

(c) In Fig. 1-47(c), name two pairs of vertical angles.

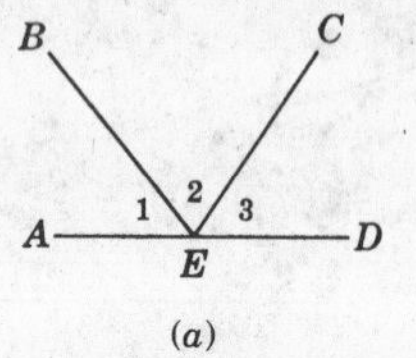

(a)

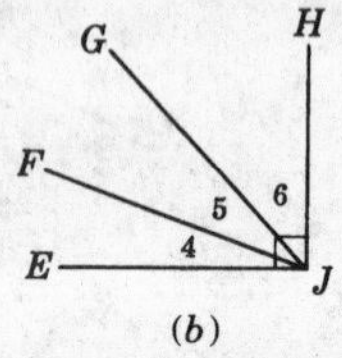

(b)

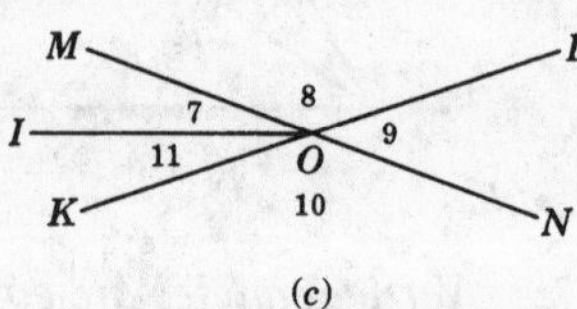

(c)

Fig. 1-47

Solutions

(a) Since their sum is 180°, the supplementary angles are (1) $\angle 1$ and $\angle BED$; (2) $\angle 3$ and $\angle AEC$.

(b) Since their sum is 90°, the complementary angles are (1) $\angle 4$ and $\angle FJH$; (2) $\angle 6$ and $\angle EJG$.

(c) Since $\overleftrightarrow{KL}$ and $\overleftrightarrow{MN}$ are intersecting lines, the vertical angles are (1) $\angle 8$ and $\angle 10$; (2) $\angle 9$ and $\angle MOK$.

1.14 Finding pairs of angles

Find two angles such that:

(a) The angles are supplementary and the larger is twice the smaller.

(b) The angles are complementary and the larger is 20° more than the smaller.

(c) The angles are adjacent and form an angle of 120°. The larger is 20° less than three times the smaller.

(d) The angles are vertical and complementary.

Solutions

In each solution, x is a number only. This number indicates the number of degrees contained in the angle. Hence, if $x = 60$, the angle measures 60°.

(a) Let $x = m$ (smaller angle) and $2x = m$ (larger angle), as in Fig. 1-48(a).
Principle 5: $x + 2x = 180$, so $3x = 180$; $x = 60$.
$2x = 120$. *Ans.* 60° and 120°

(b) Let $x = m$ (smaller angle) and $x + 20 = m$ (larger angle), as in Fig. 1-48(b).
Principle 3: $x + (x + 20) = 90$, or $2x + 20 = 90$; $x = 35$.
$x + 20 = 55$. *Ans.* 35° and 55°

(c) Let $x = m$ (smaller angle) and $3x - 20 = m$ (larger angle) as in Fig. 1-48(c).
Principle 1: $x + (3x - 20) = 120$, or $4x - 20 = 120$; $x = 35$.
$3x - 20 = 85$. *Ans.* 35° and 85°

(d) Let $x = m$ (each vertical angle), as in Fig. 1-48(d). They are congruent by Principle 2.
Principle 3: $x + x = 90°$, or $2x = 90$; $x = 45$. *Ans.* 45° each.

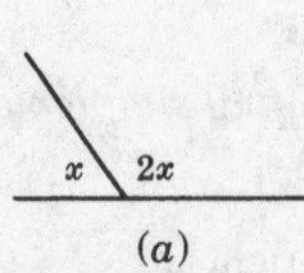

(a)

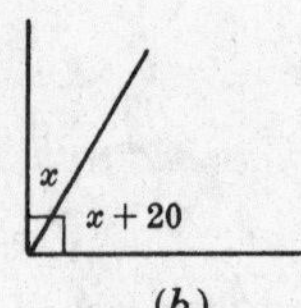

(b)

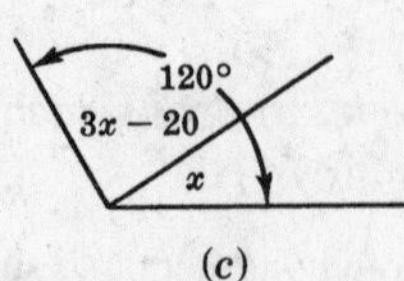

(c)

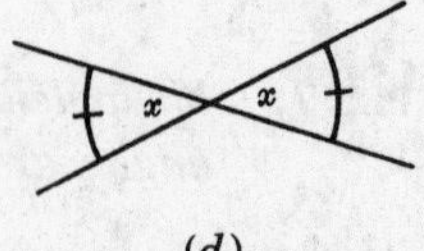

(d)

Fig. 1-48

1.15 Finding a pair of angles using two unknowns

For each of the following, be represented by a and b. Obtain two equations for each case, and then find the angles.

(a) The angles are adjacent, forming an angle of 88°. One is 36° more than the other.

(b) The angles are complementary. One is twice as large as the other.

(c) The angles are supplementary. One is 60° less than twice the other.

(d) The angles are supplementary. The difference of the angles is 24°.

Solutions

(a) $a + b = 88$
$a = b + 36$ *Ans.* 62° and 26°

(b) $a + b = 90$
$a = 2b$ *Ans.* 60° and 30°

(c) $a + b = 180$
$a = 2b - 60$ *Ans.* 100° and 80°

(d) $a + b = 180°$
$a - b = 24°$ *Ans.* 78° and 102°

SUPPLEMENTARY PROBLEMS

1.1. Point, line, and plane are undefined terms. Which of these is illustrated by (a) the tip of a sharpened pencil; (b) the shaving edge of a blade; (c) a sheet of paper; (d) a side of a box; (e) the crease of a folded paper; (f) the junction of two roads on a map? (1.1)

1.2. (a) Name the line segments that intersect at E in Fig. 1-49. (1.2)

(b) Name the line segments that intersect at D.

(c) What other line segments can be drawn using points A, B, C, D, E, and F?

(d) Name the point of intersection of $\overline{AC}$ and $\overline{BD}$.

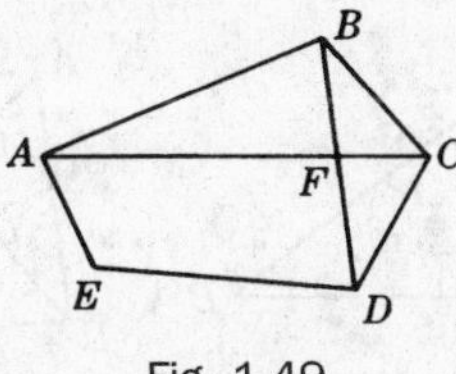

Fig. 1-49

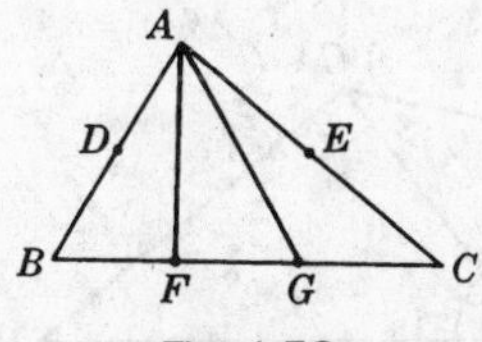

Fig. 1-50

1.3. (a) Find the length of $\overline{AB}$ in Fig. 1-50 if AD is 8 and D is the midpoint of $\overline{AB}$ (1.3)

(b) Find the length of $\overline{AE}$ if AC is 21 and E is the midpoint of $\overline{AC}$.

1.4. (a) Find OB in Fig. 1-51 if diameter $AD = 36$. (1.4)

(b) Find the number of degrees in $\overset{\frown}{AE}$ if E is the midpoint of semicircle $\overset{\frown}{AED}$. Find the number of degrees in (c) $\overset{\frown}{CD}$; (d) $\overset{\frown}{AC}$; (e) $\overset{\frown}{AEC}$.

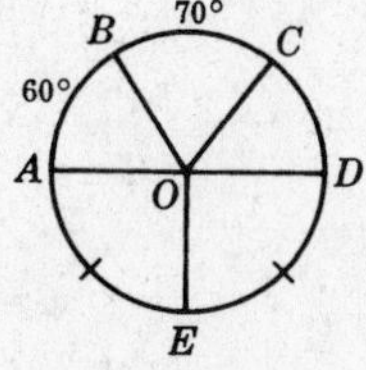

Fig. 1-51

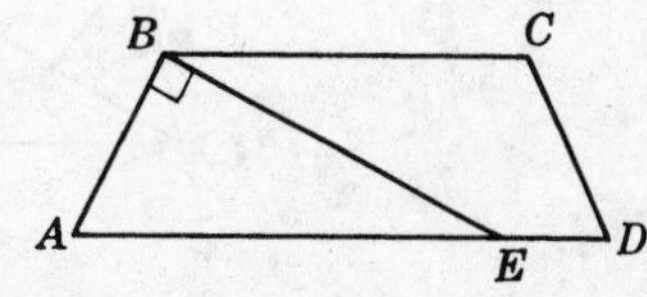

Fig. 1-52

1.5. Name the following angles in Fig. 1-52 (a) an acute angle at B; (b) an acute angle at E; (c) a right angle; (d) three obtuse angles; (e) a straight angle. (1.5)

1.6. (a) Find $m\angle ADC$ if $m\angle c = 45°$ and $m\angle d = 85°$ in Fig. 1-53. (1.6)

(b) Find $m\angle AEB$ if $m\angle e = 60°$.

(c) Find $m\angle EBD$ if $m\angle a = 15°$.

(d) Find $m\angle ABC$ if $m\angle b = 42°$.

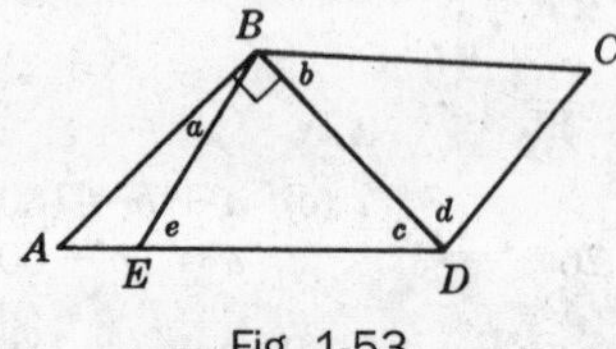

Fig. 1-53

1.7. Find (a) $\frac{5}{6}$ of a rt. ∠; (b) $\frac{2}{9}$ of a st. ∠; (c) $\frac{1}{3}$ of 31°; (d) $\frac{1}{5}$ of 45°55′. (1.7)

1.8. What turn or rotation is made (a) by an hour hand in 3 hours; (b) by the minute hand in $\frac{1}{3}$ of an hour? What rotation is needed to turn from (c) west to northeast in a clockwise direction; (d) east to south in a counterclockwise direction; (e) southwest to northeast in either direction? (1.8)

1.9. Find the angle formed by the hand of a clock (a) at 3 o'clock; (b) at 10 o'clock; (c) at 5:30 AM; (d) at 11:30 PM. (1.9)

1.10. In Fig. 1-54: (1.10)

(a) Name two pairs of perpendicular lines.

(b) Find $m\angle BCD$ if $m\angle 4$ is 39°.

If $m\angle 1 = 78°$, find (c) $m\angle BAD$; (d) $m\angle 2$; (e) $m\angle CAE$.

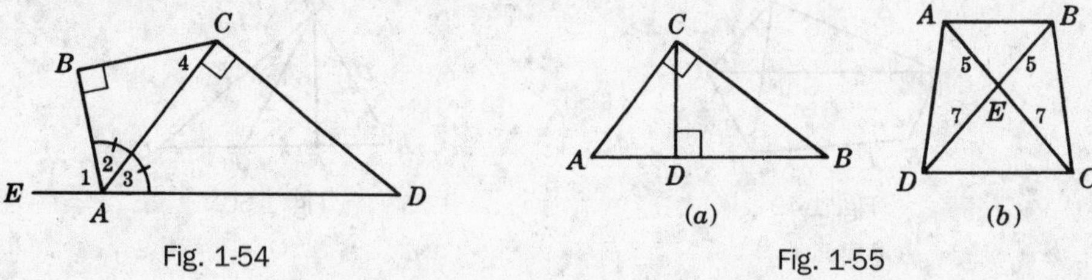

Fig. 1-54 Fig. 1-55

1.11. (a) In Fig. 1-55(a), name three right triangles and the hypotenuse and legs of each. (1.11)

In Fig. 1-55(b), (b) name two obtuse triangles and (c) name two isosceles triangles, also naming the legs, base, and vertex angle of each.

1.12. In Fig. 1-56, name the congruent lines and angles (a) if $\overline{PR}$ is a ⊥ bisector of $\overline{AB}$; (b) if $\overline{BF}$ bisects ∠ABC; (c) if $\overline{CG}$ is an altitude to $\overline{AD}$; (d) if $\overline{EM}$ is a median to $\overline{AD}$. (1.12)

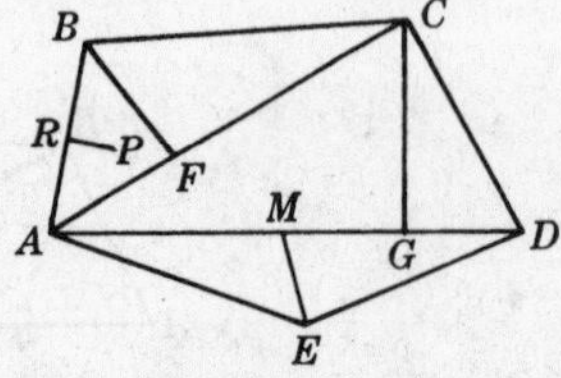

Fig. 1-56

1.13. In Fig. 1-57, state the relationship between: (1.13)

(a) ∠1 and ∠4 (d) ∠4 and ∠5

(b) ∠3 and ∠4 (e) ∠1 and ∠3

(c) ∠1 and ∠2 (f) ∠*AOD* and ∠5

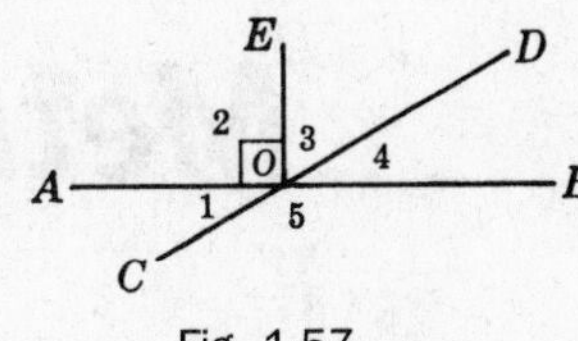

Fig. 1-57

1.14. Find two angles such that: (1.14)

(a) The angles are complementary and the measure of the smaller is 40° less than the measure of the larger.

(b) The angles are complementary and the measure of the larger is four times the measure of the smaller.

(c) The angles are supplementary and the measure of the smaller is one-half the measure of the larger.

(d) The angles are supplementary and the measure of the larger is 58° more than the measure of the smaller.

(e) The angles are supplementary and the measure of the larger is 20° less than three times the measure of the smaller.

(f) The angles are adjacent and form an angle measuring 140°. The measure of the smaller is 28° less than the measure of the larger.

(g) The angles are vertical and supplementary.

1.15. For each of the following, let the two angles be represented by a and b. Obtain two equations for each case, and then find the angles. (1.15)

(a) The angles are adjacent and form an angle measuring 75°. Their difference is 21°.

(b) The angles are complementary. One measures 10° less than three times the other.

(c) The angles are supplementary. One measures 20° more than four times the other.

CHAPTER 2

Methods of Proof

2.1 Proof By Deductive Reasoning

2.1A Deductive Reasoning is Proof

Deductive reasoning enables us to derive true or acceptably true conclusions from statements which are true or accepted as true. It consists of three steps as follows:

1. Making a *general statement* referring to a whole set or class of things, such as the class of dogs: *All dogs are quadrupeds (have four feet).*
2. Making a *particular statement* about one or some of the members of the set or class referred to in the general statement: *All greyhounds are dogs.*
3. Making a *deduction* that follows logically when the general statement is applied to the particular statement: *All greyhounds are quadrupeds.*

Deductive reasoning is called *syllogistic reasoning* because the three statements together constitute a syllogism. In a syllogism the general statement is called the major premise, the particular statement is the minor premise, and the deduction is the conclusion. Thus, in the above syllogism:

1. The major premise is: *All dogs are quadrupeds.*
2. The minor premise is: *All greyhounds are dogs.*
3. The conclusion is: *All greyhounds are quadrupeds.*

Using a circle, as in Fig. 2-1, to represent each set or class will help you understand the relationships involved in deductive reasoning.

1. Since the major premise or general statement states that all dogs are quadrupeds, the circle representing dogs must be inside that for quadrupeds.
2. Since the minor premise or particular statement states that all greyhounds are dogs, the circle representing greyhounds must be inside that for dogs.

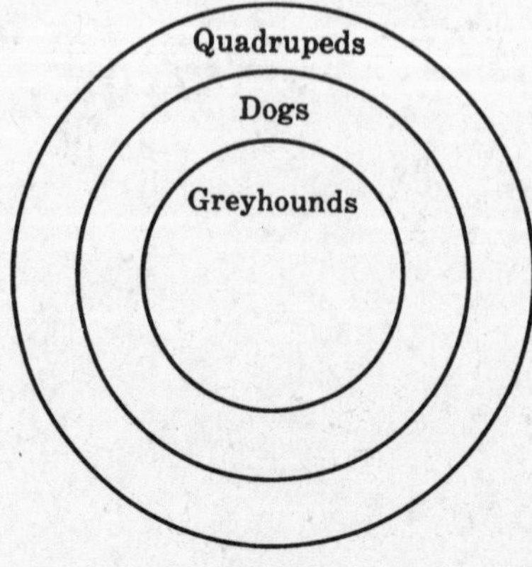

Fig. 2-1

3. The conclusion is obvious. Since the circle of greyhounds must be inside the circle of quadrupeds, the only possible conclusion is that greyhounds are quadrupeds.

2.1B Observation, Measurement, and Experimentation are not Proof

Observation cannot serve as proof. Eyesight, as in the case of a color-blind person, may be defective. Appearances may be misleading. Thus, in each part of Fig. 2-2, AB does not seem to equal CD although it actually does.

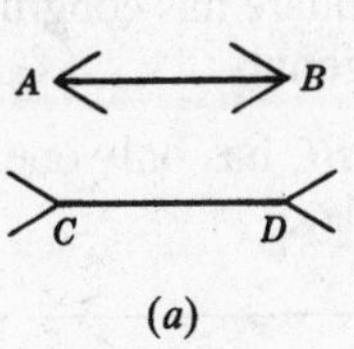

(a)

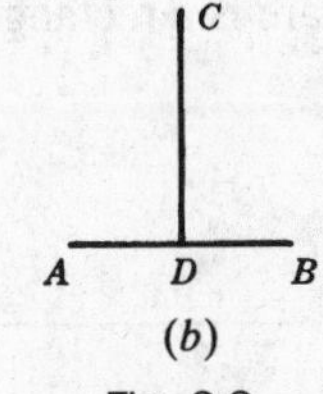

(b)

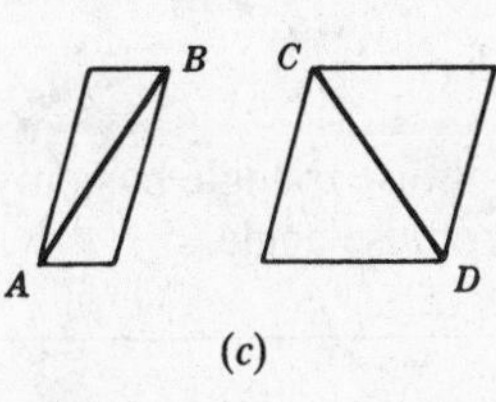

(c)

Fig. 2-2

Measurement cannot serve as proof. Measurement applies only to the limited number of cases involved. The conclusion it provides is not exact but approximate, depending on the precision of the measuring instrument and the care of the observer. In measurement, allowance should be made for possible error equal to half the smallest unit of measurement used. Thus if an angle is measured to the nearest degree, an allowance of half a degree of error should be made.

Experiment cannot serve as proof. Its conclusions are only probable ones. The degree of probability depends on the particular situations or instances examined in the process of experimentation. Thus, it is probable that a pair of dice are loaded if ten successive 7s are rolled with the pair, and the probability is much greater if twenty successive 7s are rolled; however, neither probability is a certainty.

SOLVED PROBLEMS

2.1 Using circles to determine group relationships

In (a) to (e) each letter, such as A, B, and R, represents a set or group. Complete each statement. Show how circles may be used to represent the sets or groups.

(a) If A is B and B is C, then _?_.

(b) If A is B and B is E and E is R, then _?_.

(c) If X is Y and _?_, then X is M.

(d) If C is D and E is C, then _?_.

(e) If squares (S) are rectangles (R) and rectangles are parallelograms (P), then _?_.

Solutions

(a) A is C (b) A is R (c) Y is M (d) E is D (e) Squares are parallelograms

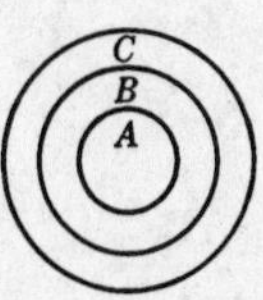

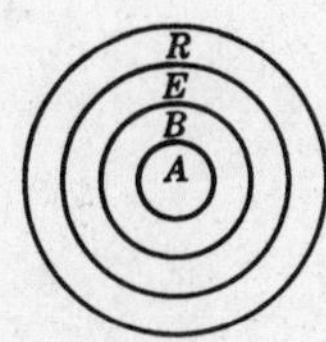

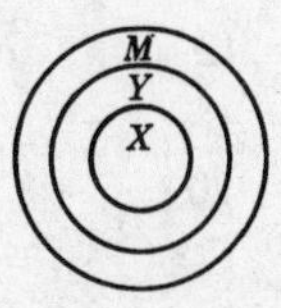

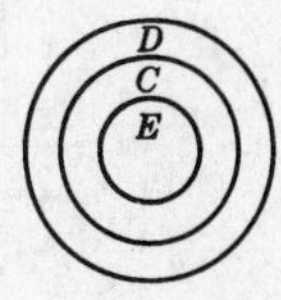

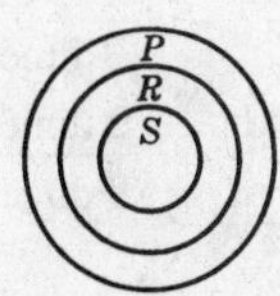

2.2 Completing a syllogism

Write the statement needed to complete each syllogism:

Major Premise (General Statement)	Minor Premise (Particular Statement)	Conclusion (Deducted Statement)
(a) A cat is a domestic animal.	Fluffy is a cat.	?
(b) All people must die.	?	Jan must die.
(c) Vertical angles are congruent.	$\angle c$ and $\angle d$ are vertical angles.	?
(d) ?	A square is a rectangle.	A square has congruent diagonals.
(e) An obtuse triangle has only one obtuse angle.	?	$\triangle ABC$ has only one obtuse angle.

Solutions

(a) Fluffy is a domestic animal.

(b) Jan is a person.

(c) $\angle c \cong \angle d$.

(d) A rectangle has congruent diagonals.

(e) $\triangle ABC$ is an obtuse triangle.

2.2 Postulates (Assumptions)

The entire structure of proof in geometry rests upon, or begins with, some unproved general statements called *postulates*. These are statements which we must willingly assume or accept as true so as to be able to deduce other statements.

2.2A Algebraic Postulates

POSTULATE 1: *Things equal to the same or equal things are equal to each other; if $a = b$ and $c = b$, then $a = c$.* (Transitive Postulate)

Thus the total value of a dime is equal to the value of two nickels because each is equal to the value of ten pennies.

POSTULATE 2: *A quantity may be substituted for its equal in any expression or equation.* (Substitution Postulate)

Thus if $x = 5$ and $y = x + 3$, we may substitute 5 for x and find $y = 5 + 3 = 8$.

POSTULATE 3: *The whole equals the sum of its parts.* (Partition Postulate)

Thus the total value of a dime, a nickel, and a penny is 16 cents.

POSTULATE 4: *Any quantity equals itself.* (Reflexive Postulate or Identity Postulate)

Thus $x = x$, $m\angle A = m\angle A$, and $AB = AB$.

POSTULATE 5: *If equals are added to equals, the sums are equal; if $a = b$ and $c = d$, then $a + c = b + d$.* (Addition Postulate)

If	7 dimes = 70 cents	If	$x + y = 12$
and	2 dimes = 20 cents	and	$x - y = 8$
then	9 dimes = 90 cents	then	$2x = 20$

POSTULATE 6: *If equals are subtracted from equals, the differences are equal; if* $a = b$ *and* $c = d$, *then* $a - c = b - d$. (Subtraction Postulate)

If	7 dimes = 70 cents	If	$x + y = 12$
and	2 dimes = 20 cents	and	$x - y = 8$
then	5 dimes = 50 cents	then	$2y = 4$

POSTULATE 7: *If equals are multiplied by equals, the products are equal; if* $a = b$ *and* $c = d$, *then* $ac = bd$. (Multiplication Postulate)

Thus if the price of one book is \$2, the price of three books is \$6.

Special multiplication axiom: Doubles of equals are equal.

POSTULATE 8: *If equals are divided by equals, the quotients are equal; if* $a = b$ *and* $c = d$, *then* $a/c = b/d$, *where* $c, d \neq 0$. (Division Postulate)

Thus if the price of 1 lb of butter is 80 cents then, at the same rate, the price of $\frac{1}{4}$ lb is 20 cents.

POSTULATE 9: *Like powers of equals are equal; if* $a = b$, *then* $a^n = b^n$. (Powers Postulate)

Thus if $x = 5$, then $x^2 = 5^2$ or $x^2 = 25$.

POSTULATE 10: *Like roots of equals are equal; if* $a = b$ *then* $\sqrt[n]{a} = \sqrt[n]{b}$.

Thus if $y^3 = 27$, then $y = \sqrt[3]{27} = 3$.

2.2B Geometric Postulates

POSTULATE 11: *One and only one straight line can be drawn through any two points.*

Thus, $\overleftrightarrow{AB}$ is the only line that can be drawn between A and B in Fig. 2-3.

Fig. 2-3

Fig. 2-4

POSTULATE 12: *Two lines can intersect in one and only one point.*

Thus, only P is the point of intersection of $\overleftrightarrow{AB}$ and $\overleftrightarrow{CD}$ in Fig. 2-4.

POSTULATE 13: *The length of a segment is the shortest distance between two points.*

Thus, $\overline{AB}$ is shorter than the curved or broken line segment between A and B in Fig. 2-5.

Fig. 2-5

Fig. 2-6

POSTULATE 14: *One and only one circle can be drawn with any given point as center and a given line segment as a radius.*

Thus, only circle A in Fig. 2-6 can be drawn with A as center and $\overline{AB}$ as a radius.

POSTULATE 15: *Any geometric figure can be moved without change in size or shape.*

Thus, $\triangle$I in Fig. 2-7 can be moved to a new position without changing its size or shape.

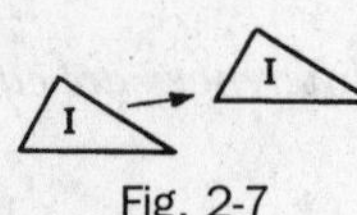

Fig. 2-7

Fig. 2-8

POSTULATE 16: *A segment has one and only one midpoint.*

Thus, only M is the midpoint of $\overline{AB}$ in Fig. 2-8.

POSTULATE 17: *An angle has one and only one bisector.*

Thus, only $\overrightarrow{AD}$ is the bisector of $\angle A$ in Fig. 2-9.

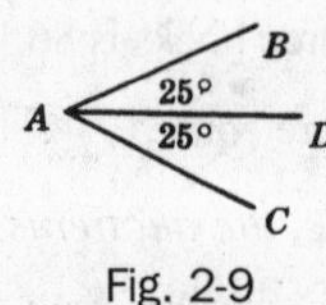

Fig. 2-9

POSTULATE 18: *Through any point on a line, one and only one perpendicular can be drawn to the line.*

Thus, only $\overrightarrow{PC} \perp \overleftrightarrow{AB}$ at point P on $\overleftrightarrow{AB}$ in Fig. 2-10.

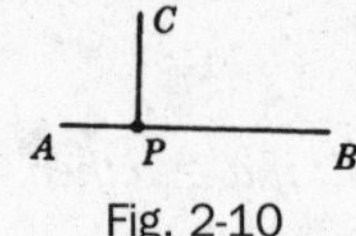

Fig. 2-10

POSTULATE 19: *Through any point outside a line, one and only one perpendicular can be drawn to the given line.*

Thus, only $\overline{PC}$ can be drawn $\perp \overleftrightarrow{AB}$ from point P outside $\overleftrightarrow{AB}$ in Fig. 2-11.

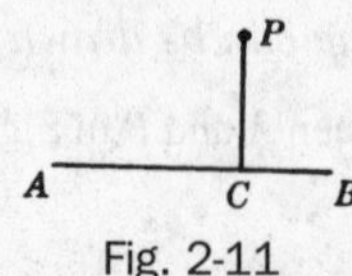

Fig. 2-11

SOLVED PROBLEMS

2.3 Applying postulate 1

In each part, what conclusion follows when Postulate 1 is applied to the given data from Figs. 2-12 and 2-13?

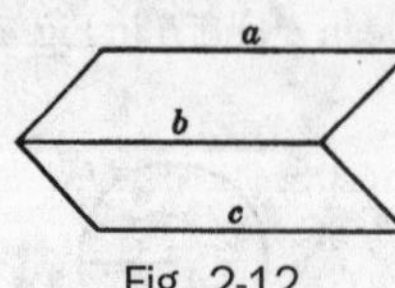

Fig. 2-12

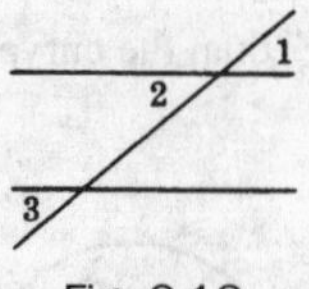

Fig. 2-13

(a) Given: $a = 10, b = 10, c = 10$

(b) Given: $a = 25, a = c$

(c) Given: $a = b, c = b$

(d) Given: $m\angle 1 = 40°, m\angle 2 = 40°, m\angle 3 = 40°$

(e) Given: $m\angle 1 = m\angle 2, m\angle 3 = m\angle 1$

(f) Given: $m\angle 3 = m\angle 1, m\angle 2 = m\angle 3$

Solutions

(a) Since a, b, and c each equal 10, $a = b = c$.

(b) Since c and 25 each equal a, $c = 25$.

(c) Since a and c each equal b, $a = c$.

(d) Since $\angle 1$, $\angle 2$, and $\angle 3$ each measures 40°, $\angle 1 \cong \angle 2 \cong \angle 3$.

(e) Since $\angle 2$ and $\angle 3$ each $\cong \angle 1$, $\angle 2 \cong \angle 3$.

(f) Since $\angle 1$ and $\angle 2$ each $\cong \angle 3$, $\angle 1 \cong \angle 2$.

2.4 Applying postulate 2

In each part, what conclusion follows when Postulate 2 is applied to the given data?

(a) Evaluate $2a + 2b$ when $a = 4$ and $b = 8$.

(b) Find x if $3x + 4y = 35$ and $y = 5$.

(c) Given: $m\angle 1 + m\angle B + m\angle 2 = 180°$, $\angle 1 \cong \angle A$, and $\angle 2 \cong \angle C$ in Fig. 2-14.

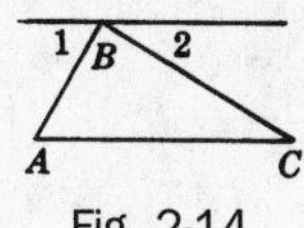

Fig. 2-14

Solutions

(a) Substitute 4 for a and 8 for b:

$$2a + 2b$$
$$2(4) + 2(8)$$
$$8 + 16 = 24 \quad \textit{Ans.}$$

(b) Substitute 5 for y:

$$3x + 4y = 35$$
$$3x + 4(5) = 35$$
$$3x + 20 = 35$$
$$3x = 15, \quad x = 5 \quad \textit{Ans.}$$

(c) Substitute $\angle A$ for $\angle 1$ and $\angle C$ for $\angle 2$:

$$m\angle 1 + m\angle B + m\angle 2 = 180°$$
$$m\angle A + m\angle B + m\angle C = 180° \quad \textit{Ans.}$$

2.5 Applying postulate 3

State the conclusions that follow when Postulate 3 is applied to the data in (a) Fig. 2.15(a) and (b) Fig. 2-15(b).

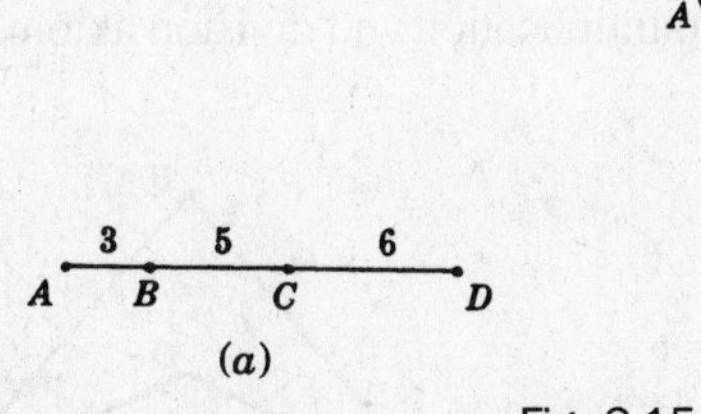

Fig. 2-15

Solutions

(a) $AC = 3 + 5 = 8$
$BD = 5 + 6 = 11$
$AD = 3 + 5 + 6 = 14$

(b) $m\angle AEC = 60° + 40° = 100°$
$m\angle BED = 40° + 30° = 70°$
$m\angle AED = 60° + 40° + 30° = 130°$

2.6 Applying postulates 4, 5, and 6

In each part, state a conclusion that follows when Postulates 4, 5, and 6 are applied to the given data.

(a) Given: $a = e$ (Fig. 2-16)

(b) Given: $a = c$, $b = d$ (Fig. 2-16)

(c) Given: $m\angle BAC = m\angle DAE$ (Fig. 2-17)

(d) Given: $m\angle BAC = m\angle BCA$, $m\angle 1 = m\angle 3$ (Fig. 2-17)

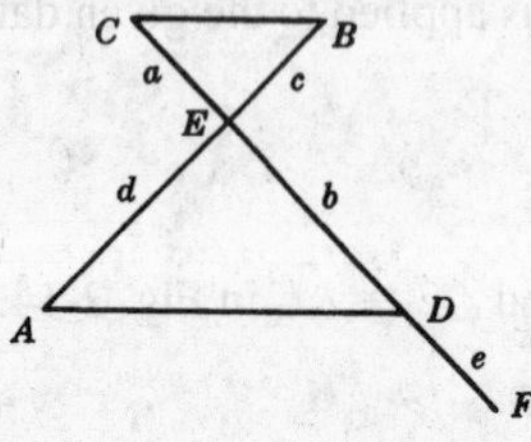

Fig. 2-16

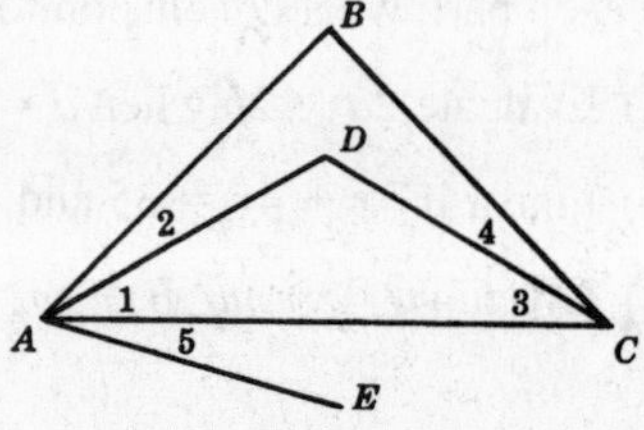

Fig. 2-17

Solutions

(a)

$a = e$	Given
$b = b$	Identity
$a + b = b + e$	Add. Post
$CD = EF$	Subst.

(b)

$a = c$	Given
$b = d$	Given
$a + b = c + d$	Add. Post.
$CD = AB$	Subst.

(c)

$m\angle BAC = m\angle DAE$	Given
$m\angle 1 = m\angle 1$	Given
$m\angle BAC - m\angle 1 = m\angle DAE - m\angle 1$	Subt. Post.
$m\angle 2 = m\angle 5$	Subst.

(d)

$m\angle BAC = m\angle BCA$	Given
$m\angle 1 = m\angle 3$	Given
$m\angle BAC - m\angle 1 = m\angle BCA - m\angle 3$	Subt. Post.
$m\angle 2 = m\angle 4$	Subst.

2.7 Applying postulates 7 and 8

State the conclusions that follow when the multiplication and division axioms are applied to the data in (a) Fig. 2-18 and (b) Fig. 2-19.

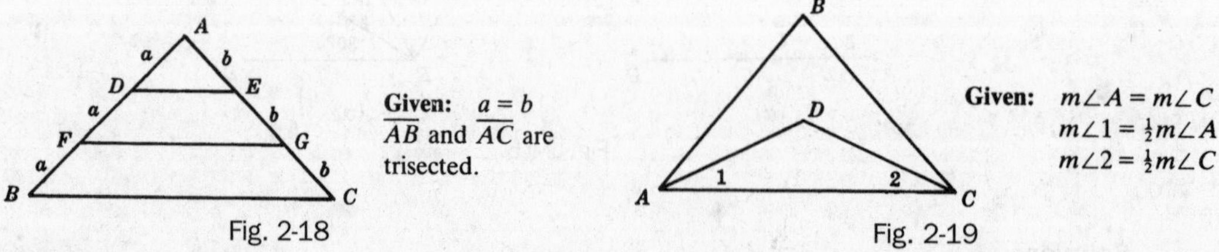

Fig. 2-18

Fig. 2-19

Solutions

(a) If $a = b$, then $2a = 2b$ since doubles of equals are equal. Hence, $AF = DB = AG = EC$. Also, $3a = 3b$, using the Multiplication Postulate. Hence, $AB = AC$.

(b) If $m\angle A = m\angle C$, then $\frac{1}{2}m\angle A = \frac{1}{2}m\angle C$ since halves of equals are equal. Hence, $m\angle 1 = m\angle 2$.

2.8 Applying postulates to statements

Complete each sentence and state the postulate that applies.

(a) If Harry and Alice are the same age today, then in 10 years __?__.

(b) Since 32°F and 0°C both name the temperature at which water freezes, we know that __?__.

(c) If Henry and John are the same weight now and each loses 20 lb, then __?__.

(d) If two stocks of equal value both triple in value, then __?__.

(e) If two ribbons of equal size are cut into five equal parts, then __?__.

(f) If Joan and Agnes are the same height as Anne, then __?__.

(g) If two air conditioners of the same price are each discounted 10 percent, then __?__.

Solutions

(a) They will be the same age. (Add. Post.)

(b) 32°F = 0°C. (Trans. Post.)

(c) They will be the same weight. (Subt. Post.)

(d) They will have the same value. (Mult. Post.)

(e) Their parts will be of the same size. (Div. Post.)

(f) Joan and Agnes are of the same height. (Trans. Post.)

(g) They will have the same price. (Subt. Post.)

2.9 Applying geometric postulates

State the postulate needed to correct each diagram and accompanying statement in Fig. 2-20.

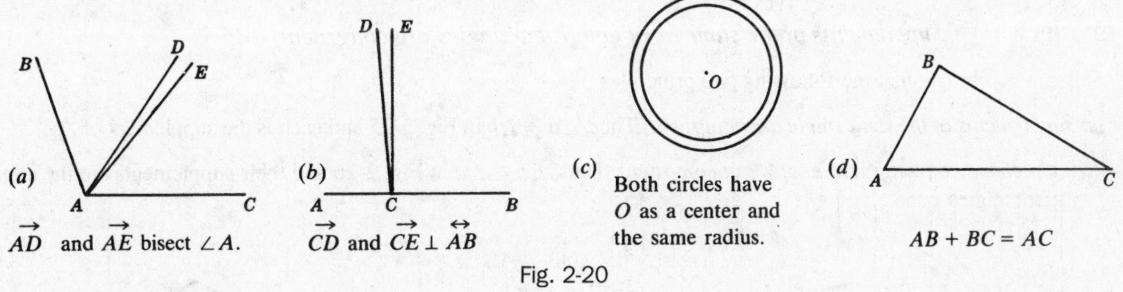

Fig. 2-20

Solutions

(a) Postulate 17. (b) Postulate 18. (c) Postulate 14. (d) Postulate 13. (*AC* is less than the sum of *AB* and *BC*.)

2.3 Basic Angle Theorems

A *theorem* is a statement, which, when proved, can be used to prove other statements or derive other results. Each of the following basic theorems requires the use of definitions and postulates for its proof.

Note: We shall use the term *principle* to include important geometric statements such as theorems, postulates, and definitions.

PRINCIPLE 1: *All right angles are congruent.*

Thus, $\angle A \cong \angle B$ in Fig. 2-21.

Fig. 2-21

PRINCIPLE 2: *All straight angles are congruent.*

Thus, $\angle C \cong \angle D$ in Fig. 2-22.

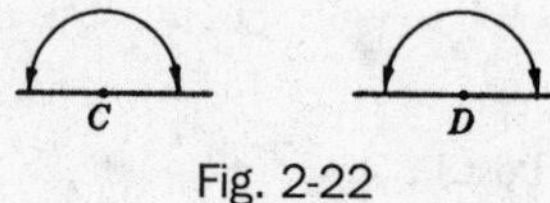

Fig. 2-22

PRINCIPLE 3: *Complements of the same or of congruent angles are congruent.*

This is a combination of the following two principles:

1. *Complements of the same angle are congruent.* Thus, $\angle a \cong \angle b$ in Fig. 2.23 and each is the complement of $\angle x$.
2. *Complements of congruent angles are congruent.* Thus, $\angle c \cong \angle d$ in Fig. 2-24 and their complements are the congruent $\angle$s x and y.

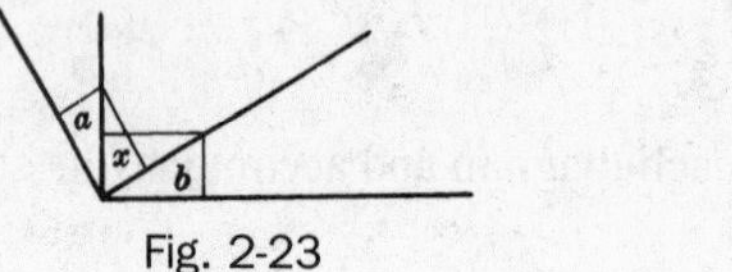

Fig. 2-23

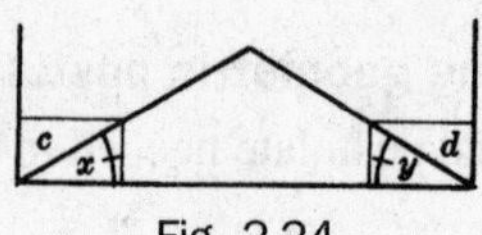

Fig. 2-24

PRINCIPLE 4: *Supplements of the same or of congruent angles are congruent.*

This is a combination of the following two principles:

1. *Supplements of the same angle are congruent.* Thus, $\angle a \cong \angle b$ in Fig. 2-25 and each is the supplement of $\angle x$.
2. *Supplements of congruent angles are congruent.* Thus, $\angle c \cong \angle d$ in Fig. 2-26 and their supplements are the congruent angles x and y.

Fig. 2-25

Fig. 2-26

PRINCIPLE 5: *Vertical angles are congruent.*

Thus, in Fig. 2-27, $\angle a \cong \angle b$; this follows from Principle 4, since $\angle a$ and $\angle b$ are supplements of the same angle, $\angle c$.

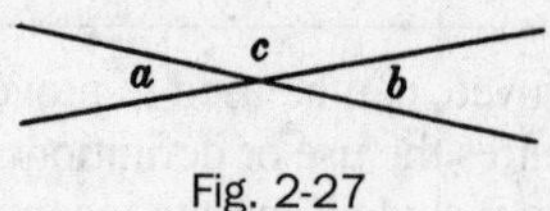

Fig. 2-27

SOLVED PROBLEMS

2.10 Applying basic theorems: principles 1 to 5

State the basic angle theorem needed to prove $\angle a \cong \angle b$ in each part of Fig. 2-28.

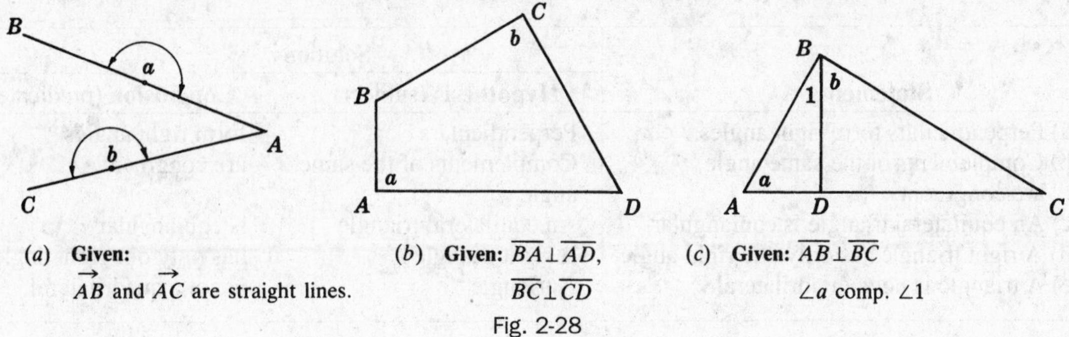

Fig. 2-28

Solutions

(a) Since $\overleftrightarrow{AB}$ and $\overleftrightarrow{AC}$ are straight lines, $\angle a$ and $\angle b$ are straight $\angle$s. Hence, $\angle a \cong \angle b$. *Ans.* All straight angles are congruent.

(b) Since $\overline{BA} \perp \overline{AD}$ and $\overline{BC} \perp \overline{CD}$, $\angle a$ and $\angle b$ are rt. $\angle$s. Hence, $\angle a \cong \angle b$. *Ans.* All right angles are congruent.

(c) Since $\overline{AB} \perp \overline{BC}$, $\angle B$ is a rt. $\angle$, making $\angle b$ the complement of $\angle 1$. Since $\angle a$ is the complement of $\angle 1$, $\angle a \cong \angle b$. *Ans.* Complements of the same angle are congruent.

2.4 Determining the Hypothesis and Conclusion

2.4A Statement Forms: Subject-Predicate Form and If-Then Form

The statements "A heated metal expands" and "If a metal is heated, then it expands" are two forms of the same idea. The following table shows how each form may be divided into its two important parts, the *hypothesis,* which tells *what is given,* and the *conclusion,* which tells *what is to be proved.* Note that in the if-then form, the word *then* may be omitted.

Form	Hypothesis (What is given)	Conclusion (What is to be proved)
Subject-predicate form: *A heated metal expands.*	**Hypothesis is subject:** *A heated metal*	**Conclusion is predicate:** *expands*
If-then form: *If a metal is heated, then it expands.*	**Hypothesis is if clause:** *If a metal is heated*	**Conclusion is then clause:** *then is expands*

2.4B Converse of a Statement

The converse of a statement is formed by interchanging the hypothesis and conclusion. Hence to form the converse of an if-then statement, interchange the if and then clauses. In the case of the subject-predicate form, interchange the subject and the predicate.

Thus, the converse of "triangles are polygons" is "polygons are triangles." Also, the converse of "if a metal is heated, then it expands" is "if a metal expands, then it is heated." Note in each of these cases that the statement is true but its converse need not necessarily be true.

PRINCIPLE 1: *The converse of a true statement is not necessarily true.*

Thus, the statement "triangles are polygons" is true. Its converse need not be true.

PRINCIPLE 2: *The converse of a definition is always true.*

Thus, the converse of the definition "a triangle is a polygon of three sides" is "a polygon of three sides is a triangle." Both the definition and its converse are true.

SOLVED PROBLEMS

2.11 Determining the hypothesis and conclusion in subject-predicate form

Determine the hypothesis and conclusion of each statement.

Statements	Solutions	
	Hypothesis (subject)	**Conclusion** (predicate)
(a) Perpendiculars form right angles.	Perpendiculars	form right angles
(b) Complements of the same angle are congruent.	Complements of the same angle	are congruent
(c) An equilateral triangle is equiangular.	An equilateral triangle	is equiangular
(d) A right triangle has only one right angle.	A right triangle	has only one right angle
(e) A triangle is not a quadrilateral.	A triangle	is not a quadrilateral

2.12 Determining the hypothesis and conclusion in if-then form

Determine the hypothesis and conclusion of each statement.

Statements	Solutions	
	Hypothesis (if clause)	**Conclusion** (then clause)
(a) If a line bisects an angle, then it divides the angle into two congruent parts.	If a line bisects an angle	then it divides the angle into two congruent parts
(b) A triangle has an obtuse angle if it is an obtuse triangle.	If it is an obtuse triangle	(then) a triangle has an obtuse angle
(c) If a student is sick, she should not go to school.	If a student is sick	(then) she should not go to school
(d) A student, if he wishes to pass, must study regularly.	If he wishes to pass	(then) a student must study regularly

2.13 Forming converses and determining their truth

State whether the given statement is true. Then form its converse and state whether this is necessarily true.

(a) A quadrilateral is a polygon.

(b) An obtuse angle has greater measure than a right angle.

(c) Florida is a state of the United States.

(d) If you are my pupil, then I am your teacher.

(e) An equilateral triangle is a triangle that has all congruent sides.

Solutions

(a) Statement is true. Its converse, "a polygon is a quadrilateral," is not necessarily true; it might be a triangle.

(b) Statement is true. Its converse, "an angle with greater measure than a right angle is an obtuse angle," is not necessarily true; it might be a straight angle.

(c) Statement is true. Its converse, "a state of the United States is Florida," is not necessarily true; it might be any one of the other 49 states.

(d) Statement is true. Its converse, "if I am your teacher, then you are my pupil," is also true.

(e) The statement, a definition, is true. Its converse, "a triangle that has all congruent sides is an equilateral triangle," is also true.

2.5 Proving a Theorem

Theorems should be proved using the following step-by-step procedure. The form of the proof is shown in the example that follows the procedure. Note that accepted symbols and abbreviations may be used.

1. Divide the theorem into its hypothesis (what is given) and its conclusion (what is to be proved). Underline the hypothesis with a single line, and the conclusion with a double line.
2. On one side, make a marked diagram. Markings on the diagram should include such helpful symbols as square corners for right angles, cross marks for equal parts, and question marks for parts to be proved equal.
3. On the other side, next to the diagram, state what is given and what is to be proved. The "Given" and "To Prove" must refer to the parts of the diagram.
4. Present a plan. Although not essential, a plan is very advisable. It should state the major methods of proof to be used.
5. On the left, present statements in successively numbered steps. The last statement must be the one to be proved. All the statements must refer to parts of the diagram.
6. On the right, next to the statements, provide a reason for each statement. Acceptable reasons in the proof of a theorem are given facts, definitions, postulates, assumed theorems, and previously proven theorems.

Step 1: **Prove:** All right angles are equal in measure.

Steps 2 and 3: **Given:** $\angle A$ and $\angle B$ are rt. $\angle$s

To Prove: $m\angle A = m\angle B$

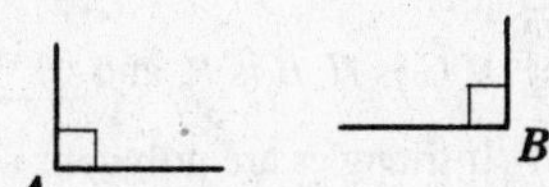

Step 4: **Plan:** Since each angle equals 90°, the angles are equal in measure, using Post. 1: Things equal to the same thing are equal to each other.

Steps 5 and 6:

Statements	Reasons
1. $\angle A$ and $\angle B$ are rt. $\angle$s.	1. Given
2. $m\angle A$ and $m\angle B$ each $= 90°$.	2. m(rt. $\angle$) $= 90°$
3. $m\angle A = m\angle B$	3. Things = to same thing = each other.

SOLVED PROBLEM

2.14 Proving a theorem

Use the proof procedure to prove that supplements of angles of equal measure have equal measure.

Step 1: **Prove:** Supplements of angles of equal measure have equal measure.

Steps 2 and 3: **Given:** $\angle a$ sup. $\angle 1$, $\angle b$ sup. $\angle 2$

$m\angle 1 = m\angle 2$

To Prove: $m\angle a = m\angle b$

Step 4: **Plan:** Using the subtraction postulate, the equal angle measures may be subtracted from the equal sums of measures of pairs of supplementary angles. The equal remainders are the measures of the supplements.

Steps 5 and 6:

Statements	Reasons
1. $\angle a$ sup. $\angle 1$, $\angle b$ sup. $\angle 2$	1. Given
2. $m\angle a + m\angle 1 = 180°$ $m\angle b + m\angle 2 = 180°$	2. Sup. $\angle$s are $\angle$s the sum of whose measures = 180°.
3. $m\angle a + m\angle 1 = m\angle b + m\angle 2$	3. Things = to the same thing = each other.
4. $m\angle 1 = m\angle 2$	4. Given
5. $m\angle a = m\angle b$	5. If =s are subtracted from =s, the differences are =.

SUPPLEMENTARY PROBLEMS

2.1. Complete each statement. In (a) to (e), each letter, such as *C*, *D*, or *R*, represents a set or group. (2.2)

(a) If *A* is *B* and *B* is *H*, then __?__.

(b) If *C* is *D* and *P* is *C*, then __?__.

(c) If __?__ and *B* is *R*, then *B* is *S*.

(d) If *E* is *F*, *F* is *G*, and *G* is *K*, then __?__.

(e) If *G* is *H*, *H* is *R*, and __?__, then *A* is *R*.

(f) If triangles are polygons and polygons are geometric figures, then __?__.

(g) If a rectangle is a parallelogram and a parallelogram is a quadrilateral, then __?__.

2.2. State the conclusions which follow when Postulate 1 is applied to the given data, which refer to Fig. 2-29. (2.3)

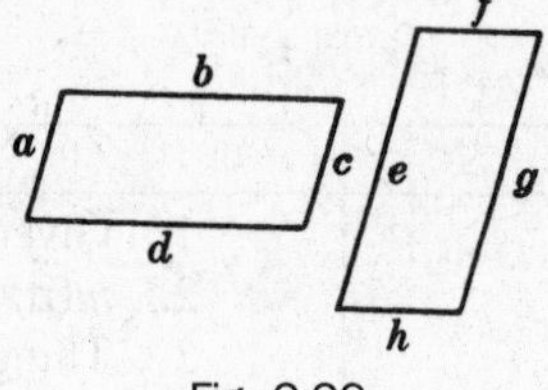

Fig. 2-29

(a) $a = 7, c = 7, f = 7$

(b) $b = 15, b = g$

(c) $f = h, h = a$

(d) $a = c, c = f, f = h$

(e) $b = d, d = g, g = e$

2.3. State the conclusions which follow when Postulate 2 is applied in each case. (2.4)

(a) Evaluate $a^2 + 3a$ when $a = 10$.

(b) Evaluate $x^2 - 4y$ when $x = 4$ and $y = 3$.

(c) Does $b^2 - 8 = 17$ when $b = 5$?

(d) Find x if $x + y = 20$ and $y = x + 3$.

(e) Find y if $x + y = 20$ and $y = 3x$.

(f) Find x if $5x - 2y = 24$ and $y = 3$.

(g) Find x if $x^2 + 3y = 45$ and $y = 3$.

2.4. State the conclusions that follow when Postulate 3 is applied to the data in Fig. 2-30(a) and (b). (2.5)

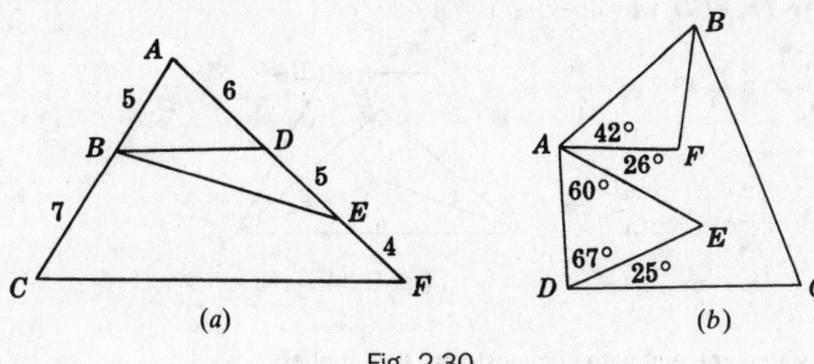

Fig. 2-30

2.5. State a conclusion involving two new equals that follows when Postulate 4, 5, or 6 is applied to the given data. (2.6)

(a) Given: $b = e$ (Fig. 2-31).

(b) Given: $b = c$, $a = d$ (Fig. 2-31).

(c) Given: $\angle 4 \cong \angle 5$ (Fig. 2-32).

(d) Given: $\angle 1 \cong \angle 3$, $\angle 2 \cong \angle 4$ (Fig. 2-32).

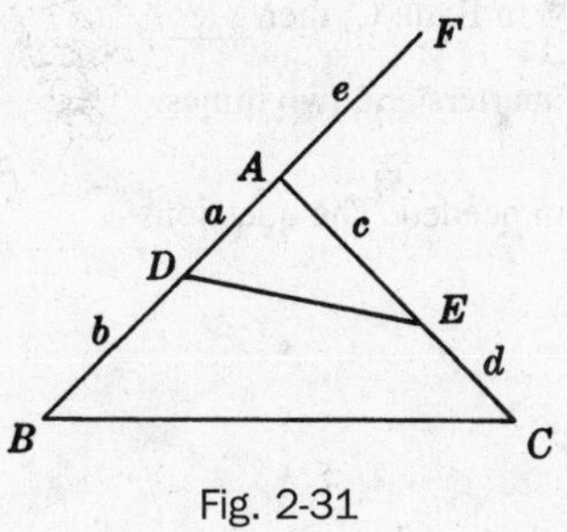

Fig. 2-31

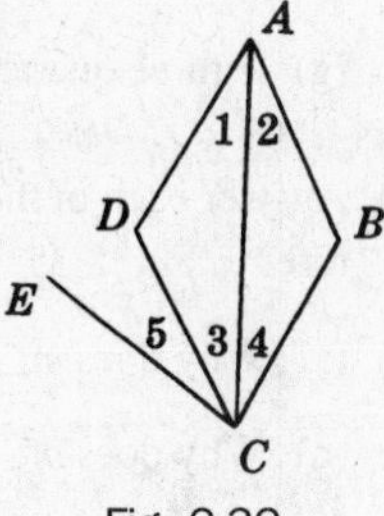

Fig. 2-32

2.6. In Fig. 2-33 $\overline{AD}$ and $\overline{BC}$ are trisected (divided into 3 equal parts). (2.7)

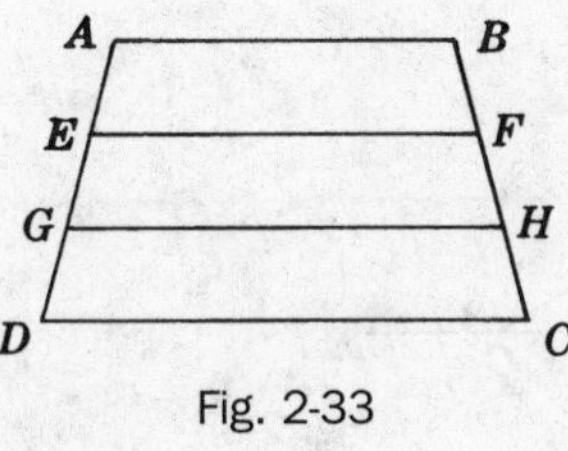

Fig. 2-33

(a) If $\overline{AD} \cong \overline{BC}$, why is $\overline{AE} \cong \overline{BF}$?

(b) If $\overline{EG} \cong \overline{FH}$, why is $\overline{AG} \cong \overline{BH}$?

(c) If $\overline{GD} \cong \overline{HC}$, why is $\overline{AD} \cong \overline{BC}$?

(d) If $\overline{ED} \cong \overline{FC}$, why is $\overline{EG} \cong \overline{FH}$?

2.7. In Fig. 2-34 $\angle BCD$ and $\angle ADC$ are trisected.

(a) If $m\angle BCD = m\angle ADC$, why does $m\angle FCD = m\angle FDC$?

(b) If $m\angle 1 = m\angle 2$, why does $m\angle BCD = m\angle ADC$?

(c) If $m\angle 1 = m\angle 2$, why does $m\angle ADF = m\angle BCF$?

(d) If $m\angle EDC = m\angle ECD$, why does $m\angle 1 = m\angle 2$?

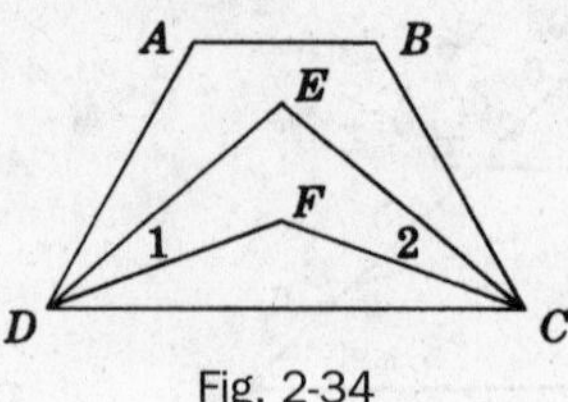

Fig. 2-34

2.8. Complete each statement, and name the postulate that applies. (2.8)

(a) If Bill and Helen earn the same amount of money each hour and their rate of pay is increased by the same amount, then __?__.

(b) In the past year, those stocks have tripled in value. If they had the same value last year, then __?__.

(c) A week ago, there were two classes that had the same register. If the same number of pupils were dropped in each, then __?__.

(d) Since 100°C and 212°F are the boiling temperatures of water, then __?__.

(e) If two boards have the same length and each is cut into four equal parts, then __?__.

(f) Since he has \$2000 in Bank A, \$3000 in Bank B and \$5000 in Bank C, then __?__.

(g) If three quarters and four nickels are compared with three quarters and two dimes, __?__.

2.9. Answer each of the following by stating the basic angle theorem needed. The questions refer to Fig. 2-35. (2.10)

(a) Why does $m\angle 1 = m\angle 2$?

(b) Why does $m\angle DBC = m\angle ECB$?

(c) If $m\angle 3 = m\angle 4$, why does $m\angle 5 = m\angle 6$?

(d) If $\overrightarrow{AF} \perp \overline{DE}$ and $\overrightarrow{GC} \perp \overline{DE}$, why does $m\angle 7 = m\angle 8$?

(e) If $\overrightarrow{AF} \perp \overline{DE}$, $\overrightarrow{GC} \perp \overline{DE}$, and $m\angle 11 = m\angle 12$, why does $m\angle 9 = m\angle 10$?

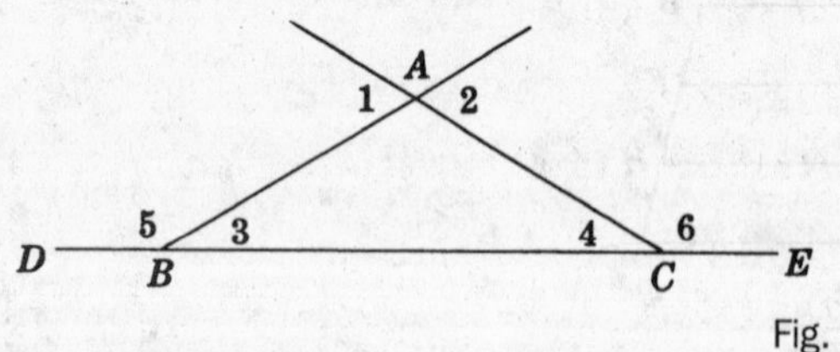

Fig. 2-35

2.10. Determine the hypothesis and conclusion of each statement. (2.11 and 2.12)

(a) Stars twinkle.

(b) Jet planes are the speediest.

(c) Water boils at 212°F.

(d) If it is the American flag, its colors are red, white, and blue.

(e) You cannot learn geometry if you fail to do homework in the subject.

(f) A batter goes to first base if the umpire calls a fourth ball.

(g) If A is B's brother and C is B's son, then A is C's uncle.

(h) An angle bisector divides the angle into two equal parts.

(i) A segment is trisected if it is divided into three congruent parts.

(j) A pentagon has five sides and five angles.

(k) Some rectangles are squares.

(l) Angles do not become larger if their sides are made longer.

(m) Angles, if they are congruent and supplementary, are right angles.

(n) The figure cannot be a polygon if one of its sides is not a straight line segment.

2.11. State the converse of each of the following true statements. State whether the converse is necessarily true. (2.13)

(a) Half a right angle is an acute angle.

(b) An obtuse triangle is a triangle having one obtuse angle.

(c) If the umpire called a third strike, then the batter is out.

(d) If I am taller than you, then you are shorter than I.

(e) If I am heavier than you, then our weights are unequal.

2.12. Prove each of the following. (2.14)

(a) Straight angles are congruent.

(b) Complements of congruent angles are congruent.

(c) Vertical angles are congruent.

Congruent Triangles

3.1 Congruent Triangles

Congruent figures are figures that have the same size and the same shape; they are the exact duplicates of each other. Such figures can be moved on top of one another so that their corresponding parts line up exactly. For example, two circles having the same radius are congruent circles.

Congruent triangles are triangles that have the same size and the same shape.

If two triangles are congruent, their corresponding sides and angles must be congruent. Thus, congruent triangles ABC and $A'B'C'$ in Fig. 3-1 have congruent corresponding sides ($\overline{AB} \cong \overline{A'C'}$, $\overline{BC} \cong \overline{B'C'}$, and $\overline{AC} \cong \overline{A'C'}$) and congruent corresponding angles ($\angle A \cong \angle A'$, $\angle B \cong \angle B'$, and $\angle C \cong \angle C'$).

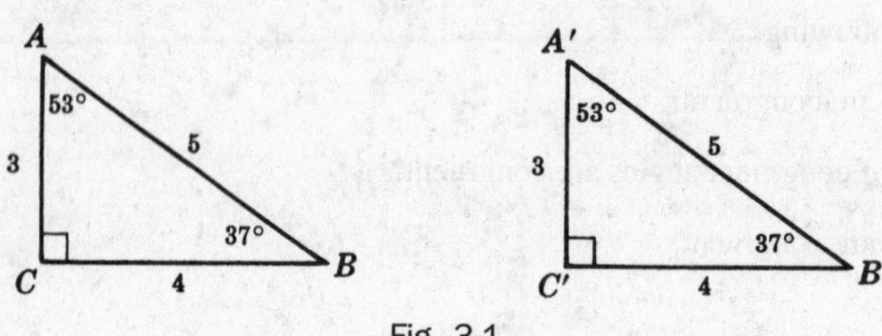

Fig. 3-1

(Read $\triangle ABC \cong \triangle A'B'C'$ as "Triangle ABC is congruent to triangle A-prime, B-prime, C-prime.")

Note in the congruent triangles how corresponding equal parts may be located. Corresponding sides lie opposite congruent angles, and corresponding angles lie opposite congruent sides.

3.1A Basic Principles of Congruent Triangles

PRINCIPLE 1: *If two triangles are congruent, then their corresponding parts are congruent.* (Corresponding parts of congruent triangles are congruent.)

Thus if $\triangle ABC \cong \triangle A'B'C'$ in Fig. 3-2, then $\angle A \cong \angle A'$, $\angle B \cong \angle B'$, $\angle C \cong \angle C'$, $a = a'$, $b = b'$, and $c = c'$.

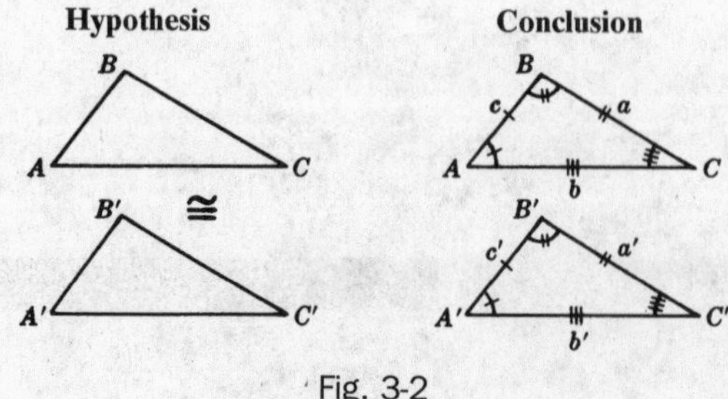

Fig. 3-2

Methods of Proving that Triangles are Congruent

PRINCIPLE 2: (Side-Angle-Side, SAS) *If two sides and the included angle of one triangle are congruent to the corresponding parts of another, then the triangles are congruent.*

Thus if $b = b'$, $c = c'$, and $\angle A \cong \angle A'$ in Fig. 3-3, then $\triangle ABC \cong \triangle A'B'C'$.

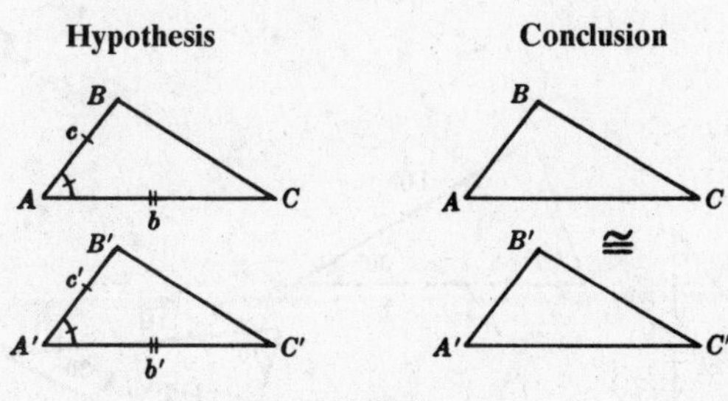

Fig. 3-3

PRINCIPLE 3: (Angle-Side-Angle, ASA) *If two angles and the included side of one triangle are congruent to the corresponding parts of another, then the triangles are congruent.*

Thus if $\angle A \cong \angle A'$, $\angle C \cong \angle C'$, and $b = b'$ in Fig. 3-4, then $\triangle ABC \cong \triangle A'B'C'$.

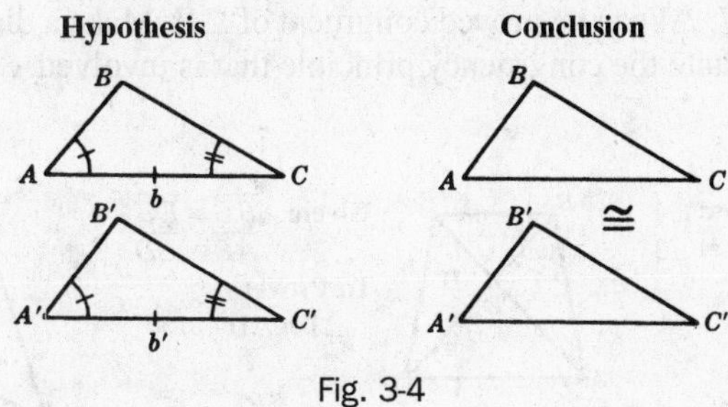

Fig. 3-4

PRINCIPLE 4: (Side-Side-Side, SSS) *If three sides of one triangle are congruent to three sides of another, then the triangles are congruent.*

Thus if $a = a'$, $b = b'$, and $c = c'$ in Fig. 3-5, then $\triangle ABC \cong \triangle A'B'C'$.

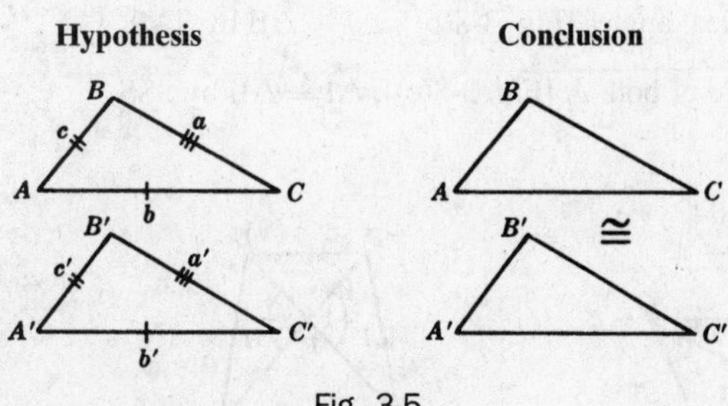

Fig. 3-5

SOLVED PROBLEMS

3.1 Selecting congruent triangles

From each set of three triangles in Fig. 3-6 select the congruent triangles and state the congruency principle that is involved.

Solutions

(a) $\triangle I \cong \triangle II$, by SAS. In $\triangle III$, the right angle is not between 3 and 4.

(b) $\triangle II \cong \triangle III$, by ASA. In $\triangle I$, side 10 is not between 70° and 30°.

(c) $\triangle I \cong \triangle II \cong \triangle III$ by SSS.

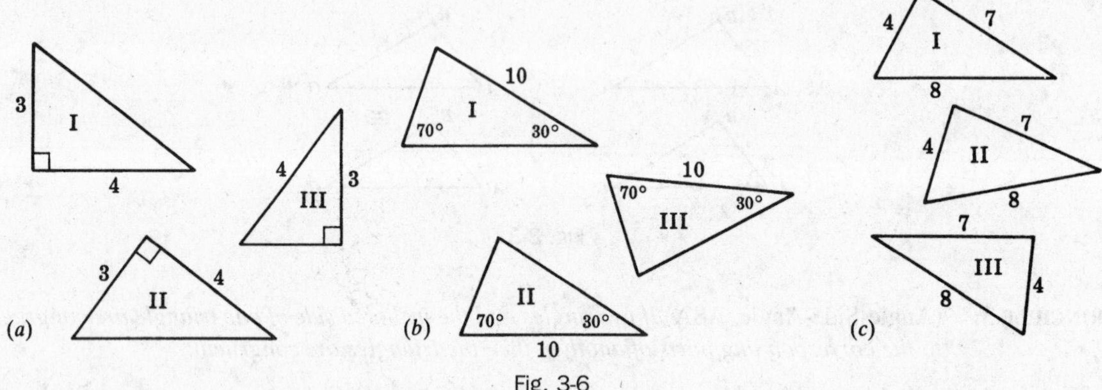

Fig. 3-6

3.2 Determining the reason for congruency of triangles

In each part of Fig. 3-7, $\triangle I$ can be proved congruent of $\triangle II$. Make a diagram showing the equal parts of both triangles and state the congruency principle that is involved.

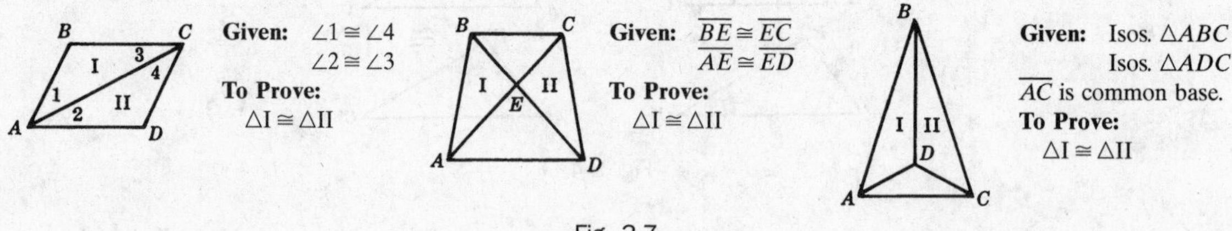

Fig. 3-7

Solutions

(a) AC is a common side of both ⚠ [Fig. 3-8(a)]. $\triangle I \cong \triangle II$ by ASA.

(b) $\angle 1$ and $\angle 2$ are vertical angles [Fig. 3-8(b)]. $\triangle I \cong \triangle II$ by SAS.

(c) BD is a common side of both ⚠ [Fig. 3-8(c)]. $\triangle I \cong \triangle II$ by SSS.

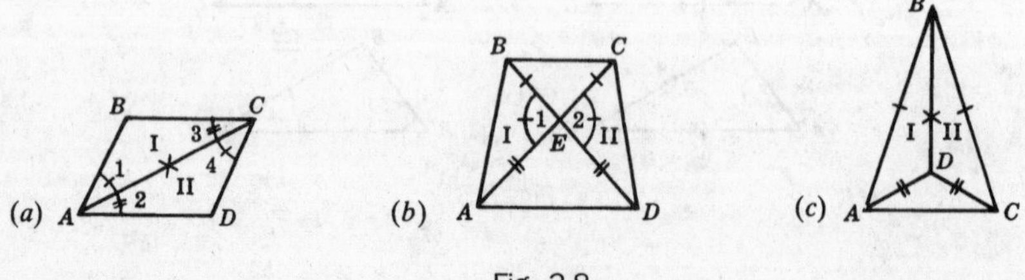

Fig. 3-8

3.3 Finding parts needed to prove triangles congruent

State the additional parts needed to prove $\triangle I \cong \triangle II$ in the given figure by the given congruency principle.

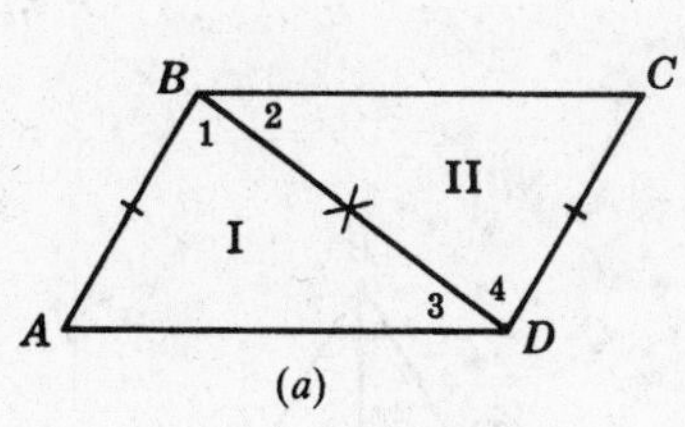

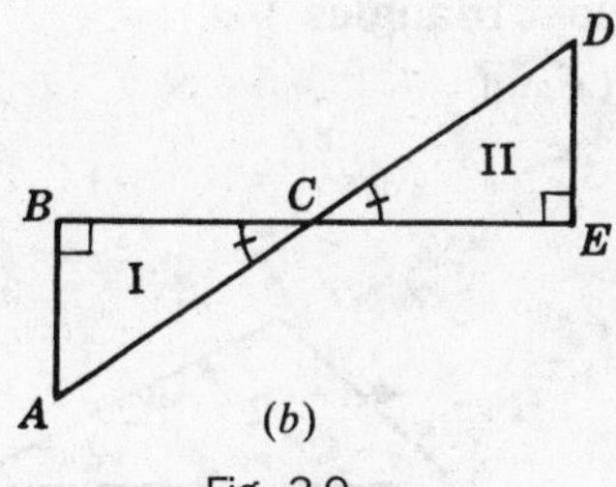

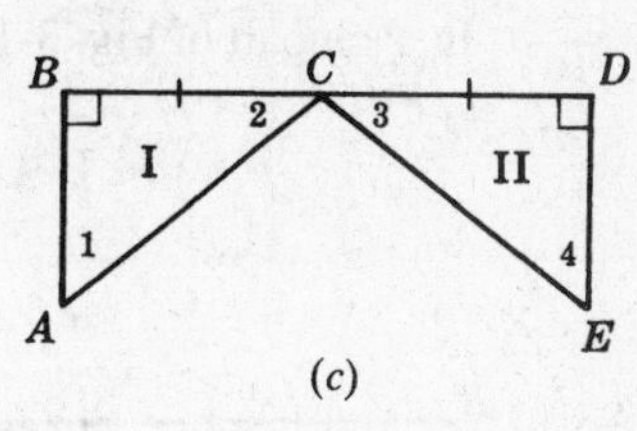

Fig. 3-9

(a) In Fig. 3-9(a) by SSS.

(b) In Fig. 3-9(a) by SAS.

(c) In Fig. 3-9(b) by ASA.

(d) In Fig. 3-9(c) by ASA.

(e) In Fig. 3-9(c) by SAS.

Solutions

(a) If $\overline{AD} \cong \overline{BC}$, then $\triangle I \cong \triangle II$ by SSS.

(b) If $\angle 1 \cong \angle 4$, then $\triangle I \cong \triangle II$ by SAS.

(c) If $\overline{BC} \cong \overline{CE}$, then $\triangle I \cong \triangle II$ by ASA.

(d) If $\angle 2 \cong \angle 3$, then $\triangle I \cong \triangle II$ by ASA.

(e) If $\overline{AB} \cong \overline{DE}$, then $\triangle I \cong \triangle II$ by SAS.

3.4 Selecting corresponding parts of congruent triangles

In each part of Fig. 3-10, the equal parts needed to prove $\triangle I \cong \triangle II$ are marked. List the remaining parts that are congruent.

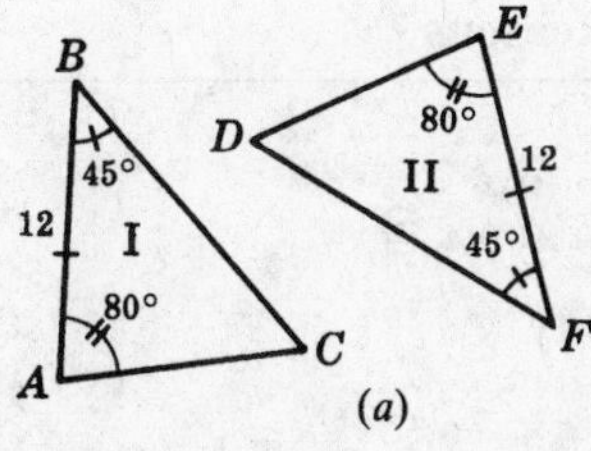

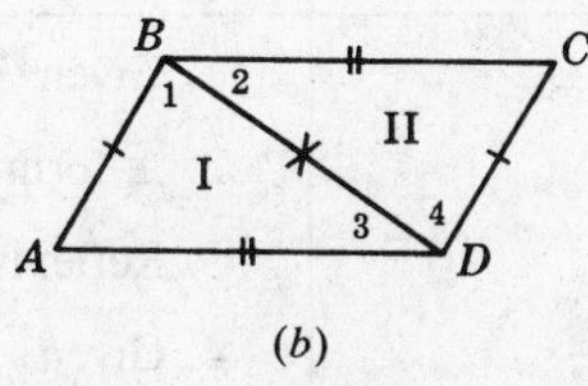

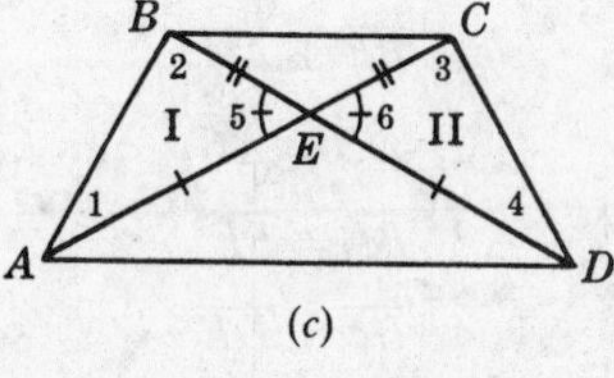

Fig. 3-10

Solutions

Congruent corresponding sides lie opposite congruent angles. Congruent corresponding angles lie opposite congruent sides.

(a) Opposite 45°, $\overline{AC} \cong \overline{DE}$, Opposite 80°, $\overline{BC} \cong \overline{DF}$. Opposite the side of length 12; $\angle C \cong \angle D$.

(b) Opposite $\overline{AB}$ and $\overline{CD}$, $\angle 3 \cong \angle 2$. Opposite $\overline{BC}$ and $\overline{AD}$, $\angle 1 \cong \angle 4$. Opposite common side $\overline{BD}$, $\angle A \cong \angle C$.

(c) Opposite $\overline{AE}$ and $\overline{ED}$, $\angle 2 \cong \angle 3$. Opposite $\overline{BE}$ and $\overline{EC}$, $\angle 1 \cong \angle 4$. Opposite $\angle 5$ and $\angle 6$, $\overline{AB} \cong \overline{CD}$.

3.5 Applying algebra to congruent triangles

In each part of Fig. 3-11, find x and y.

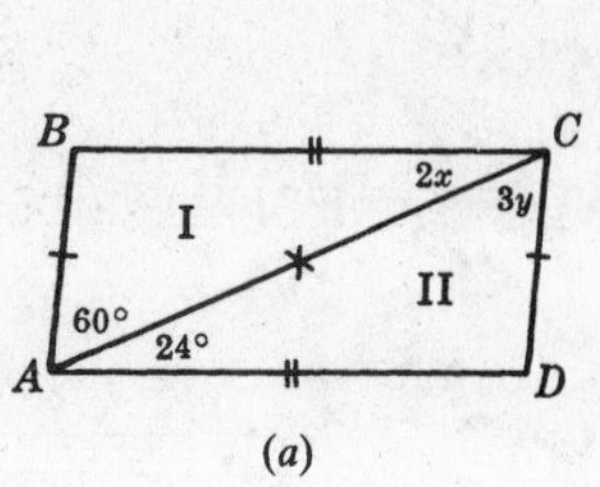

(a)

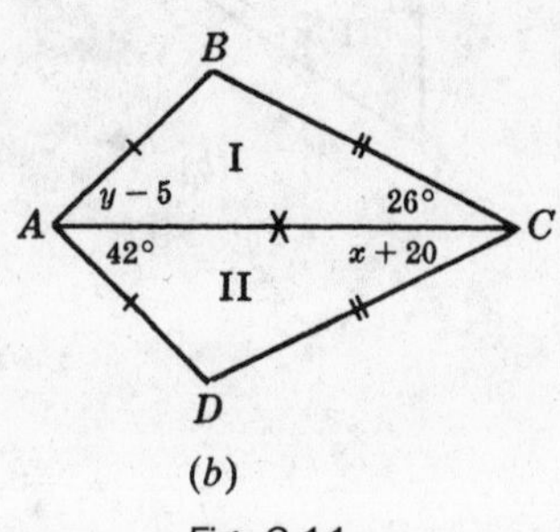

(b)

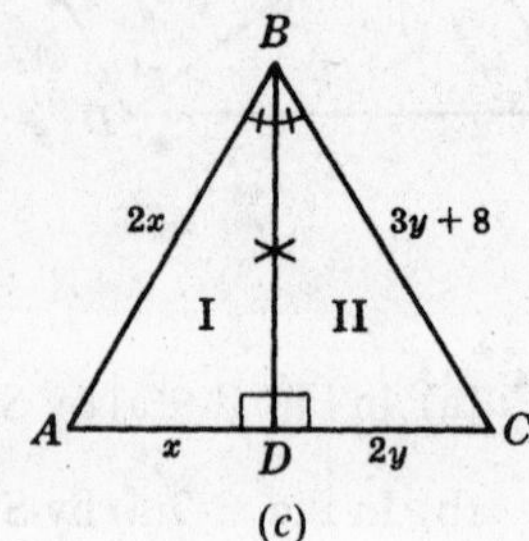

(c)

Fig. 3-11

Solutions

(a) Since $\Delta\text{I} \cong \Delta\text{II}$, by SSS, corresponding angles are congruent. Hence, $2x = 24$ or $x = 12$, and $3y = 60$ or $y = 20$.

(b) Since $\Delta\text{I} \cong \Delta\text{II}$, by SSS, corresponding angles are congruent. Hence, $x + 20 = 26$ or $x = 6$, and $y - 5 = 42$ or $y = 47$.

(c) Since $\Delta\text{I} \cong \Delta\text{II}$, by ASA, corresponding sides are congruent. Then $2x = 3y + 8$ and $x = 2y$. Substituting $2y$ for x in the first of these equations, we obtain $2(2y) = 3y + 8$ or $y = 8$. Then $x = 2y = 16$.

3.6 Proving a congruency problem

Given: $\overline{BF} \perp \overline{DE}$
$\overline{BF} \perp \overline{AC}$
$\angle 3 \cong \angle 4$

To Prove: $\overline{AF} \cong \overline{FC}$

Prove: Prove $\Delta\text{I} \cong \Delta\text{II}$

PROOF:

Statements	Reasons
1. $\overline{BF} \perp \overline{AC}$	1. Given
2. $\angle 5 \cong \angle 6$	2. ⊥s form rt. ∠s; rt. ∠s are ≅
3. $\overline{BF} \cong \overline{BF}$	3. Reflexive property
4. $\overline{BF} \perp \overline{DE}$	4. Given
5. $\angle 1$ is the complement of $\angle 3$. $\angle 2$ is the complement of $\angle 4$.	5. Adjacent angles are complementary if exterior sides are ⊥ to each other.
6. $\angle 3 \cong \angle 4$	6. Given
7. $\angle 1 \cong \angle 2$	7. Complements of ≅ ∠s are =
8. $\Delta\text{I} \cong \Delta\text{II}$	8. ASA
9. $\overline{AF} \cong \overline{FC}$	9. Corresponding parts of congruent △s are ≅

3.7 Proving a congruency problem stated in words

Prove that if the opposite sides of a quadrilateral are equal and a diagonal is drawn, equal angles are formed between the diagonal and the sides.

Solution

If the opposite sides of a quadrilateral are congruent and a diagonal is drawn, congruent angles are formed between the diagonal and the sides.

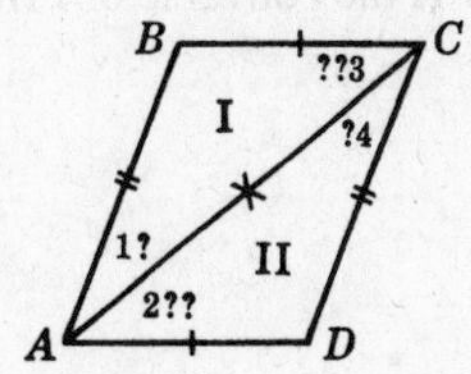

Given: Quadrilateral $ABCD$
$\overline{AB} \cong \overline{CD}$, $\overline{BC} \cong \overline{AD}$
$\overline{AC}$ is a diagonal.

To Prove: $\angle 1 \cong \angle 4$, $\angle 2 \cong \angle 3$

Plan: Prove $\Delta I \cong \Delta II$

PROOF:

Statements	Reasons
1. $\overline{AB} \cong \overline{CD}$, $\overline{BC} \cong \overline{AD}$	1. Given
2. $\overline{AC} \cong \overline{AC}$	2. Reflexive property
3. $\Delta I \cong \Delta II$	3. SSS
4. $\angle 1 \cong \angle 4$, $\angle 2 \cong \angle 3$	4. Corresponding parts of $\cong$ ⚠ are $\cong$.

3.2 Isosceles and Equilateral Triangles

3.2A Principles of Isosceles and Equilateral Triangles

PRINCIPLE 1: *If two sides of a triangle are congruent, the angles opposite these sides are congruent.* (Base angles of an isosceles triangle are congruent.)

Thus in $\triangle ABC$ in Fig. 3-12, if $\overline{AB} \cong \overline{BC}$, then $\angle A \cong \angle C$.

A proof of Principle 1 is given in Chapter 16.

Hypothesis

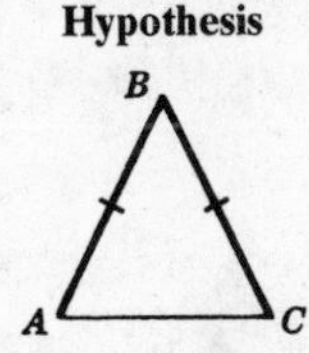

Conclusion

Fig. 3-12

Hypothesis

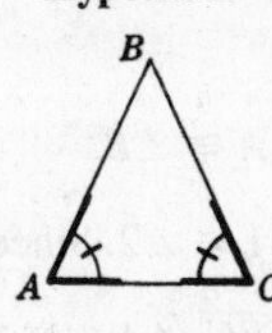

Conclusion

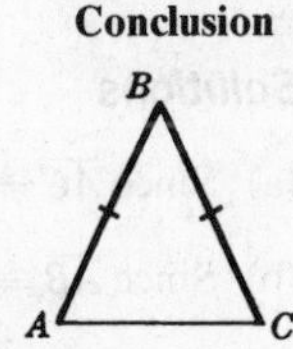

Fig. 3-13

PRINCIPLE 2: *If two angles of a triangle are congruent, the sides opposite these angles are congruent.*

Thus in $\triangle ABC$ in Fig. 3-13, if $\angle A \cong \angle C$, then $\overline{AB} \cong \overline{BC}$.

Principle 2 is the converse of Principle 1. A proof of Principle 2 is given in Chapter 16.

PRINCIPLE 3: *An equilateral triangle is equiangular.*

Thus in $\triangle ABC$ in Fig. 3-14, if $\overline{AB} \cong \overline{BC} \cong \overline{CA}$, then $\angle A \cong \angle B \cong \angle C$.

Principle 3 is a corollary of Principle 1. A *corollary* of a theorem is another theorem whose statement and proof follow readily from the theorem.

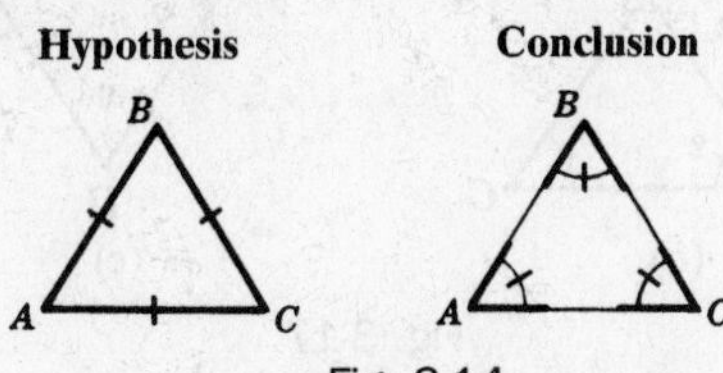

Fig. 3-14

PRINCIPLE 4: *An equiangular triangle is equilateral.*

Thus in $\triangle ABC$ in Fig. 3-15, if $\angle A \cong \angle B \cong \angle C$, then $\overline{AB} \cong \overline{BC} \cong \overline{CA}$.

Principle 4 is the converse of Principle 3 and a corollary of Principle 2.

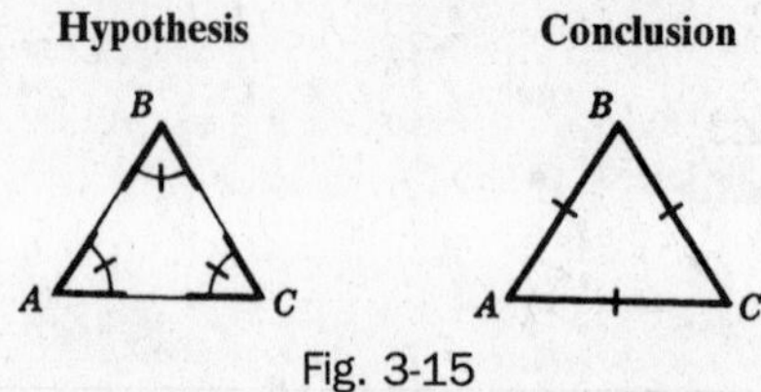

Fig. 3-15

SOLVED PROBLEMS

3.8 Applying principles 1 and 3

In each part of Fig. 3-16, name the congruent angles that are opposite congruent sides of a triangle.

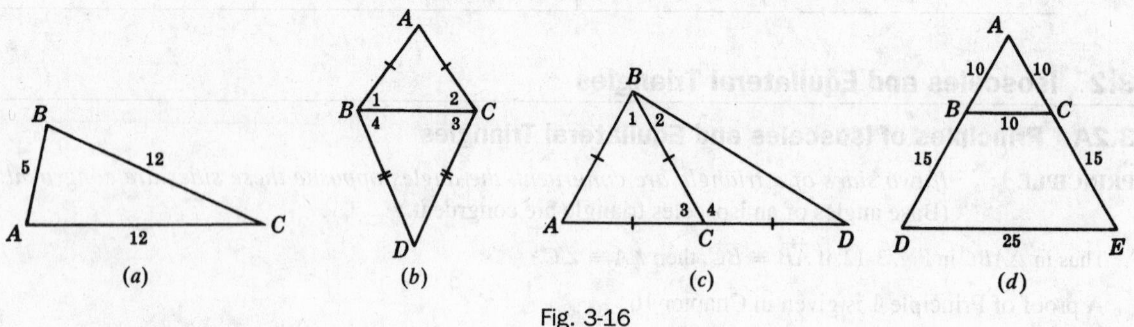

Fig. 3-16

Solutions

(a) Since $\overline{AC} \cong \overline{BC}$, $\angle A \cong \angle B$.

(b) Since $\overline{AB} \cong \overline{AC}$, $\angle 1 \cong \angle 2$. Since $\overline{BD} \cong \overline{CD}$, $\angle 3 \cong \angle 4$.

(c) Since $\overline{AB} \cong \overline{AC} \cong \overline{BC}$, $\angle A \cong \angle 1 \cong \angle 3$. Since $\overline{BC} \cong \overline{CD}$, $\angle 2 \cong \angle D$.

(d) Since $\overline{AB} \cong \overline{BC} \cong \overline{AC}$, $\angle A \cong \angle ACB \cong \angle ABC$. Since $\overline{AE} \cong \overline{AD} \cong \overline{DE}$, $\angle A \cong \angle D \cong \angle E$.

3.9 Applying principles 2 and 4

In each part of Fig. 3-17, name the congruent sides that are opposite congruent angles of a triangle.

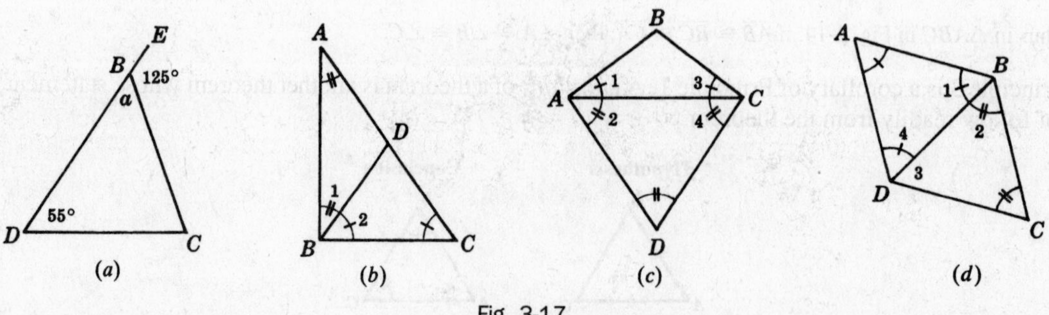

Fig. 3-17

Solutions

(a) Since $m\angle a = 55°$, $\angle a \cong \angle D$. Hence, $\overline{BC} \cong \overline{CD}$.

(b) Since $\angle A \cong \angle 1$, $\overline{AD} \cong \overline{BD}$. Since $\angle 2 \cong \angle C$, $\overline{BD} \cong \overline{CD}$.

(c) Since $\angle 1 \cong \angle 3$, $\overline{AB} \cong \overline{BC}$. Since $\angle 2 \cong \angle 4 \cong \angle D$, $\overline{CD} \cong \overline{AD} \cong \overline{AC}$.

(d) Since $\angle A \cong \angle 1 \cong \angle 4$, $\overline{AB} \cong \overline{BD} \cong \overline{AD}$. Since $\angle 2 \cong \angle C$, $\overline{BD} \cong \overline{CD}$.

3.10 Applying isosceles triangle principles

In each of Fig. 3-18(a) and (b), $\triangle$I can be proved congruent to $\triangle$II. Make a diagram showing the congruent parts of both triangles and state the congruency principle involved.

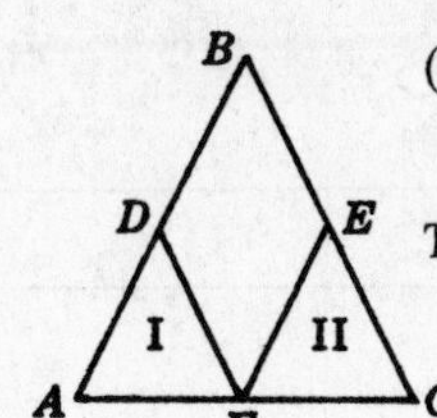

(*a*) **Given:** $\overline{AB} \cong \overline{BC}$
$\overline{AD} \cong \overline{EC}$
F is midpoint of $\overline{AC}$.
To Prove: $\triangle I \cong \triangle II$

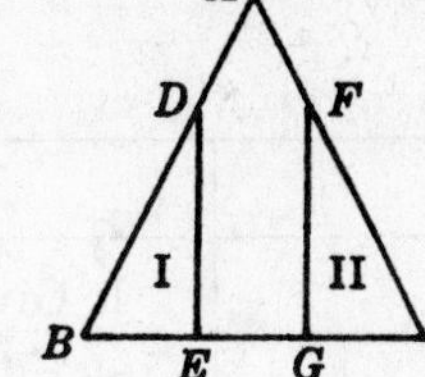

(*b*) **Given:** $\overline{AB} \cong \overline{AC}$
$\overline{BC}$ is trisected at E and G.
$\overline{DE} \perp \overline{BC}$
$\overline{FG} \perp \overline{BC}$
To Prove: $\triangle I \cong \triangle II$

Fig. 3-18

Solutions

(a) Since $\overline{AB} \cong \overline{BC}$, $\angle A \cong \angle C$. $\triangle I \cong \triangle II$ by SAS [see Fig. 3-19(a)].

(b) Since $\overline{AB} \cong \overline{AC}$, $\angle B \cong \angle C$. $\triangle I \cong \triangle II$ by ASA [see Fig. 3-19(b)].

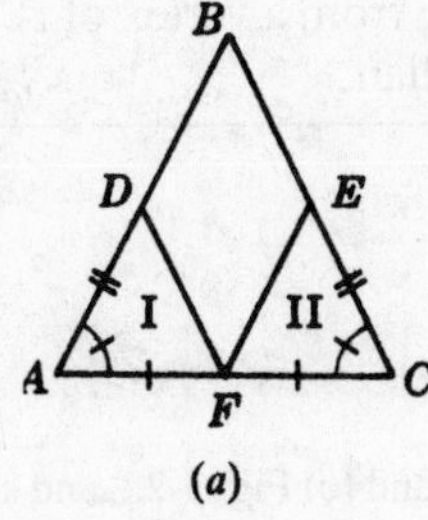

(*a*)

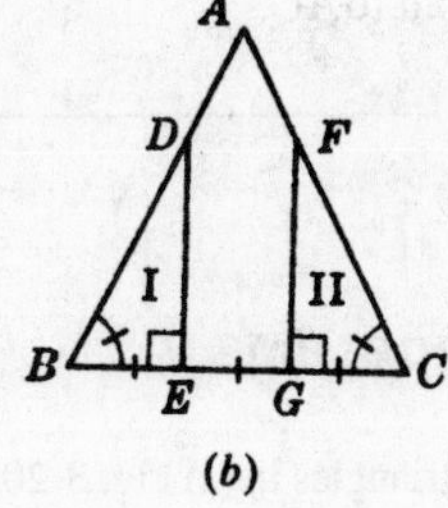

(*b*)

Fig. 3-19

3.11 Proving an isosceles triangle problem

Given: $\overline{AB} \cong \overline{BC}$
$\overline{AC}$ is trisected at D and E.
To Prove: $\angle 1 \cong \angle 2$
Plan: Prove $\triangle I \cong \triangle II$ to obtain $\overline{BD} \cong \overline{BE}$.

PROOF:

Statements	Reasons
1. AC is trisected at D and E	1. Given
2. $\overline{AD} \cong \overline{EC}$	2. To trisect is to divide into three congruent parts
3. $\overline{AB} \cong \overline{BC}$	3. Given
4. $\angle A \cong \angle C$	4. In a $\triangle$, $\angle$s opposite $\cong$ sides are $\cong$
5. $\triangle I \cong \triangle II$	5. SAS
6. $\overline{BD} \cong \overline{BE}$	6. Corresponding parts of $\triangle$s are $\cong$
7. $\angle 1 \cong \angle 2$	7. Same as 4

3.12 Proving an isosceles triangle problem stated in words

Prove that the bisector of the vertex angle of an isosceles triangle is a median to the base.

Solution

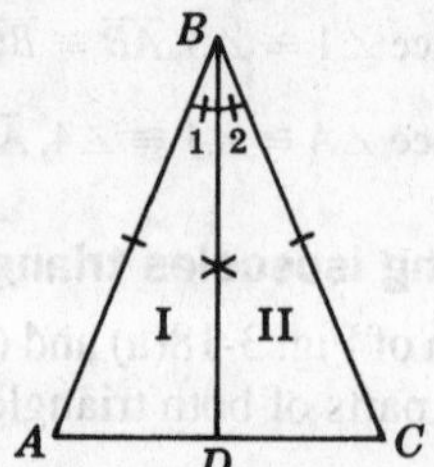

<u>The bisector of the vertex angle of an isosceles triangle</u> <u>is a median to the base.</u>

Given: Isosceles $\triangle ABC$ ($\overline{AB} \cong \overline{BC}$)
$\overline{BD}$ bisects $\angle B$

To Prove: $\overline{BD}$ is a median to $\overline{AC}$

Plan: Prove $\triangle I \cong \triangle II$ to obtain $\overline{AD} \cong \overline{DC}$.

PROOF:

Statements	Reasons
1. $\overline{AB} \cong \overline{BC}$	1. Given
2. $\overline{BD}$ bisects $\angle B$.	2. Given
3. $\angle 1 \cong \angle 2$	3. To bisect is to divide into two congruent parts.
4. $\overline{BD} \cong \overline{BD}$	4. Reflexive property.
5. $\triangle I \cong \triangle II$	5. SAS
6. $\overline{AD} \cong \overline{DC}$	6. Corresponding parts of $\cong$ $\triangle$s are $\cong$.
7. $\overline{BD}$ is a median to $\overline{AC}$.	7. A line from a vertex of $\triangle$ bisecting opposite side is a median.

SUPPLEMENTARY PROBLEMS

3.1. Select the congruent triangles in (a) Fig. 3-20, (b) Fig. 3-21, and (c) Fig. 3-22, and state the congruency principle in each case. (3.1)

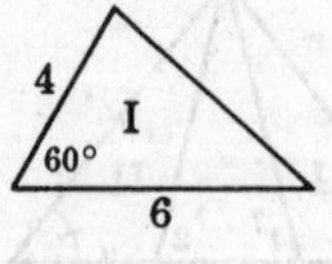

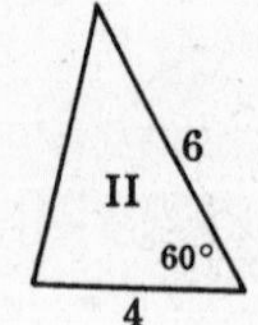

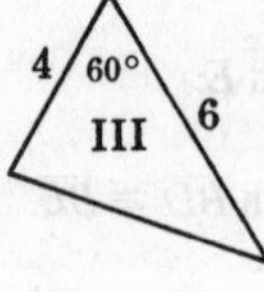

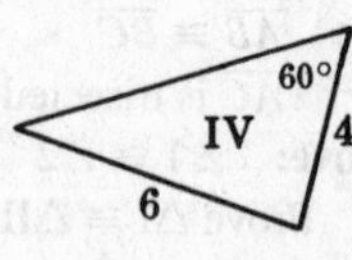

Fig. 3-20

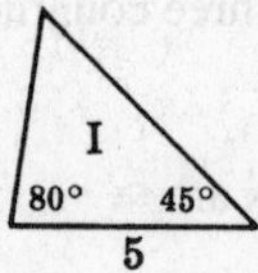

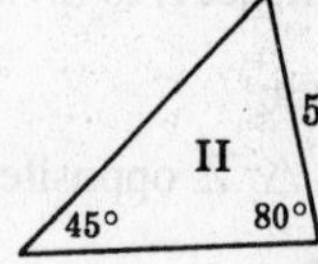

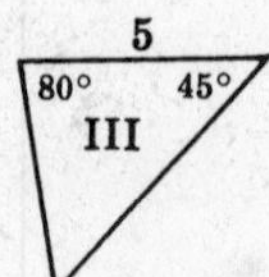

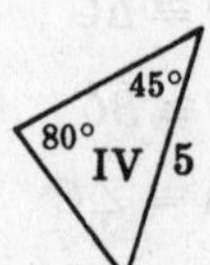

Fig. 3-21

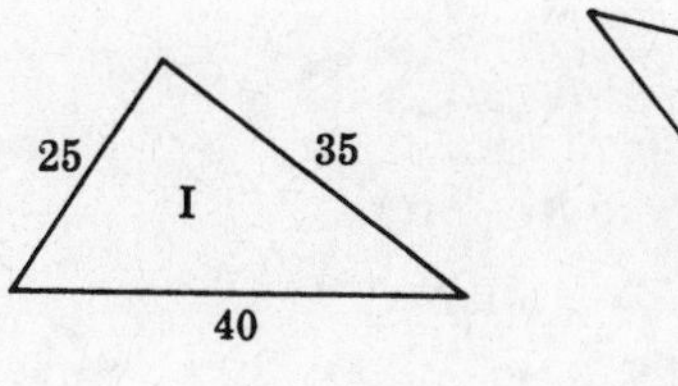

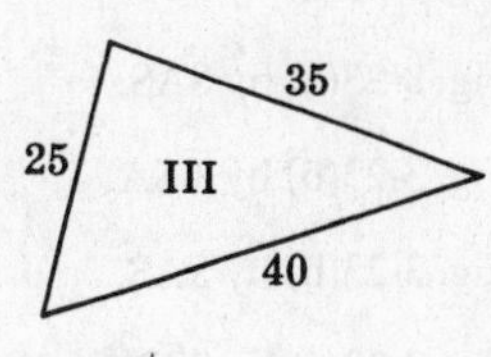

Fig. 3-22

3.2. In each figure below, $\triangle$I can be proved congruent to $\triangle$II. State the congruency principle involved. (3.2)

(*a*)

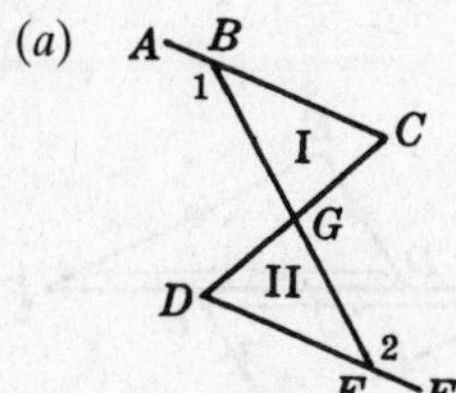

Given: $\angle 1 \cong \angle 2$
G is midpoint of $\overline{BF}$.
To Prove: $\triangle I \cong \triangle II$

(*b*)

Given: $\overline{AB} \perp \overline{BE}$
$\overline{EF} \perp \overline{BE}$
$\overline{BC} \cong \overline{DE}$
$\overline{AB} \cong \overline{EF}$
To Prove: $\triangle I \cong \triangle II$

(*c*)

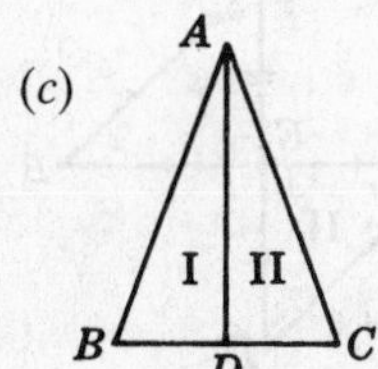

Given: $\overline{AB} \cong \overline{AC}$
$\overline{AD}$ is median to $\overline{BC}$.
To Prove: $\triangle I \cong \triangle II$

(*d*)

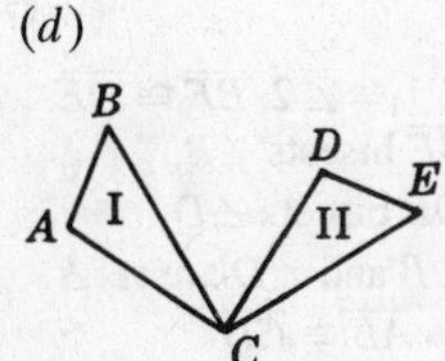

Given: $\overline{BC} \perp \overline{CE}$
$\overline{AC} \perp \overline{CD}$
$\overline{AC} \cong \overline{CD}$
$\overline{BC} \cong \overline{CE}$
To Prove: $\triangle I \cong \triangle II$

(*e*)

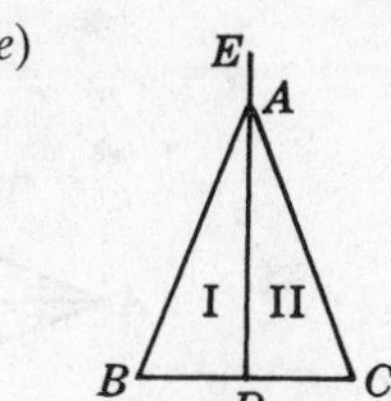

Given: $\angle EAB \cong \angle EAC$
$\overline{AD} \perp \overline{BC}$
To Prove: $\triangle I \cong \triangle II$

(*f*)

Given: $\overline{AD}$ and $\overline{BE}$ bisect each other.
To Prove: $\triangle I \cong \triangle II$

(*g*)

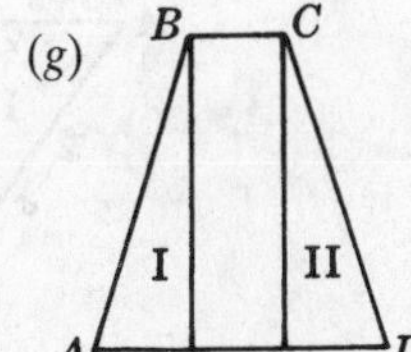

Given: $\overline{BE} \perp \overline{AD}$
$\overline{CF} \perp \overline{AD}$
$\overline{BE} \cong \overline{CF}$
$\overline{AD}$ is trisected.
To Prove: $\triangle I \cong \triangle II$

(*h*)

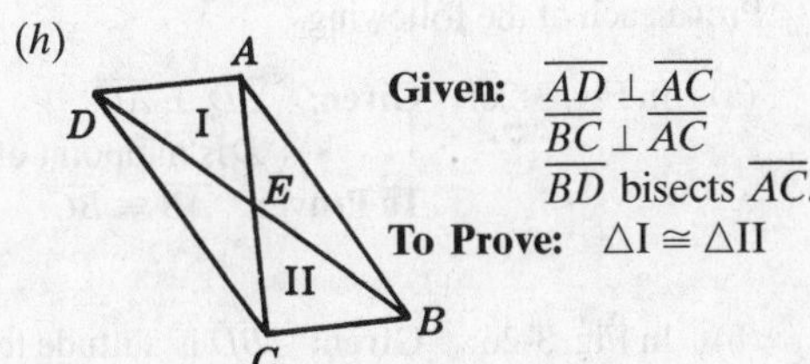

Given: $\overline{AD} \perp \overline{AC}$
$\overline{BC} \perp \overline{AC}$
$\overline{BD}$ bisects $\overline{AC}$.
To Prove: $\triangle I \cong \triangle II$

3.3. State the additional parts needed to prove $\triangle I \cong \triangle II$ in the given figure by the given congruency principle. (3.3)

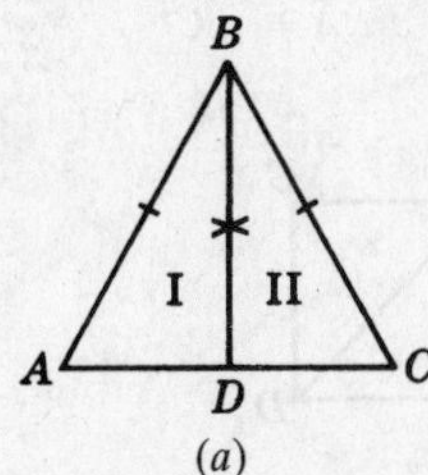

(*a*)

(*b*)

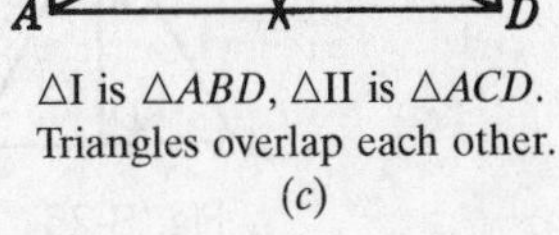

$\triangle$I is $\triangle ABD$, $\triangle$II is $\triangle ACD$.
Triangles overlap each other.

(*c*)

Fig. 3-23

(a) In Fig. 3-23(a) by SSS.

(b) In Fig. 3-23(a) by SAS.

(c) In Fig. 3-23(b) by ASA.

(d) In Fig. 3-23(b) by SAS.

(e) In Fig. 3-23(c) by SSS.

(f) In Fig. 3-23(c) by SAS.

3.4. In each part of Fig. 3-24, the congruent parts needed to prove $\Delta I \cong \Delta II$ are marked. Name the remaining parts that are congruent. (3.4)

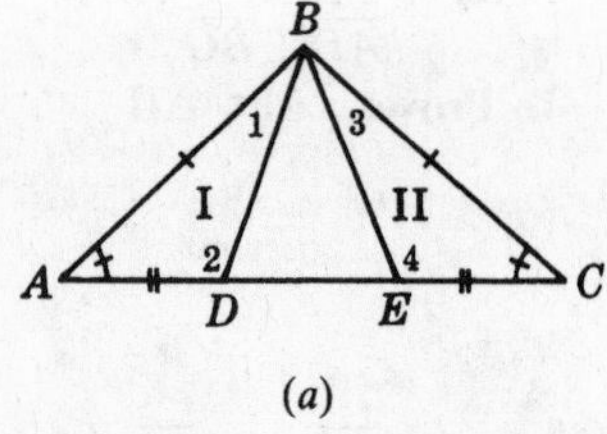

(*a*)

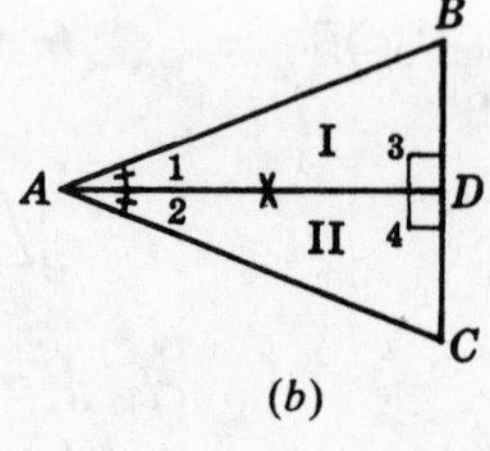

(*b*)

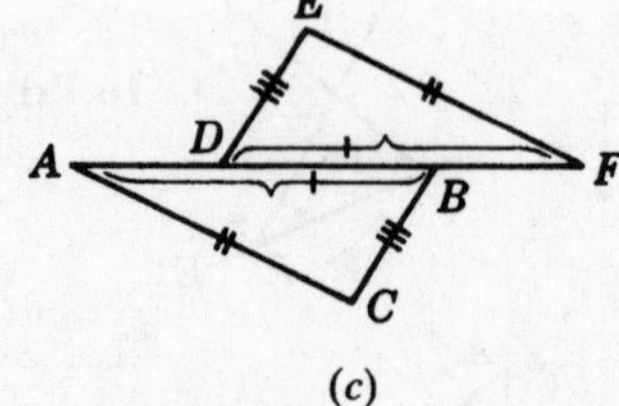

(*c*)

Fig. 3-24

3.5. In each part of Fig. 3-25, find *x* and *y*. (3.5)

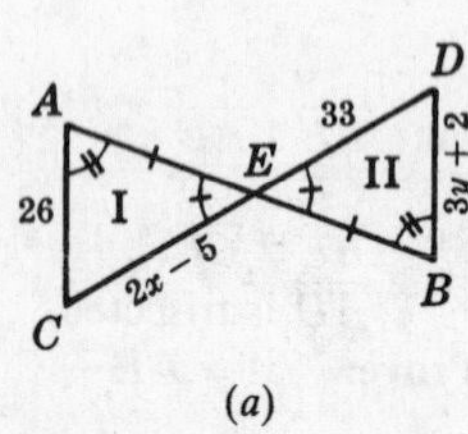

(*a*)

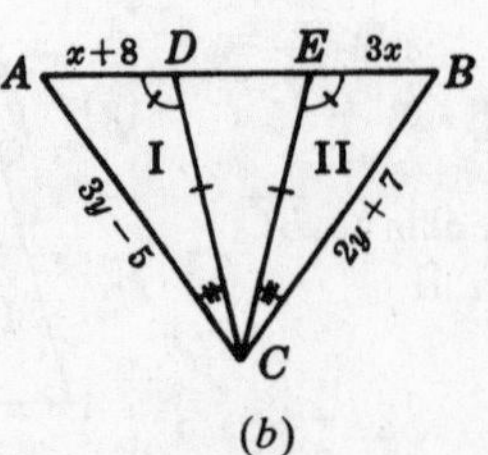

(*b*)

(*c*)

Fig. 3-25

3.6. Prove each of the following. (3.6)

(a) In Fig. 3-26: **Given:** $\overline{BD} \perp \overline{AC}$
D is midpoint of $\overline{AC}$.
To Prove: $\overline{AB} \cong \overline{BC}$

(b) In Fig. 3-26: **Given:** $\overline{BD}$ is altitude to $\overline{AC}$.
$\overline{BD}$ bisects $\angle B$.
To Prove: $\angle A \cong \angle C$

(c) In Fig. 3-27: **Given:** $\angle 1 \cong \angle 2$, $\overline{BF} \cong \overline{DE}$
$\overline{BF}$ bisects $\angle B$.
$\overline{DE}$ bisects $\angle D$.
$\angle B$ and $\angle D$ are rt. $\angle$s.
To Prove: $\overline{AB} \cong \overline{CD}$

(d) In Fig. 3-27: **Given:** $\overline{BC} \cong \overline{AD}$
E is midpoint of $\overline{BC}$.
F is midpoint of $\overline{AD}$.
$\overline{AB} \cong \overline{CD}$, $\overline{BF} \cong \overline{DE}$
To Prove: $\angle A \cong \angle C$

Fig. 3-26

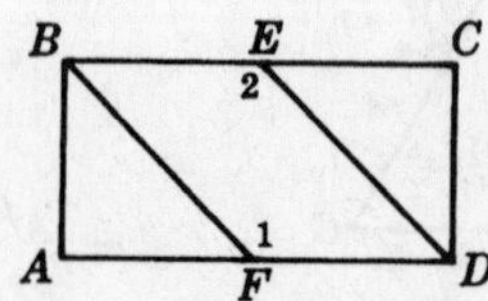

Fig. 3-27

(e) In Fig. 3-28: **Given:** $\angle 1 \cong \angle 2$
$\overline{CE}$ bisects $\overline{BF}$.
To Prove: $\angle C \cong \angle E$

(f) In Fig. 3-28: **Given:** $\overline{BF}$ and $\overline{CE}$ bisect each other.
To Prove: $\overline{BC} \cong \overline{EF}$

(g) In Fig. 3-29: **Given:** $\overline{CD} \cong \overline{C'D'}$, $\overline{AD} \cong \overline{A'D'}$
$\overline{CD}$ is altitude to $\overline{AB}$.
$\overline{C'D'}$ is altitude to $\overline{A'B'}$.
To Prove: $\angle A \cong \angle A'$

(h) In Fig. 3-29: **Given:** $\overline{CD}$ bisects $\angle C$.
$\overline{C'D'}$ bisects $\angle C'$
$\angle C \cong \angle C'$,
$\angle B \cong \angle B'$,
$BC \cong \overline{B'C'}$.
To Prove: $\overline{CD} \cong \overline{C'D'}$

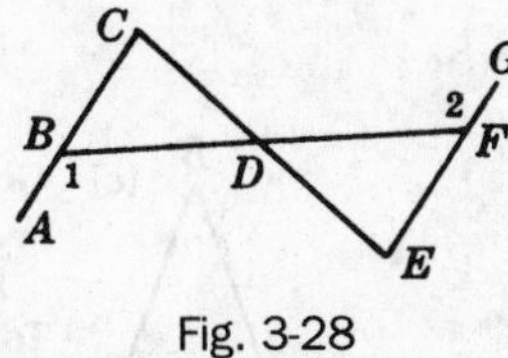

Fig. 3-28

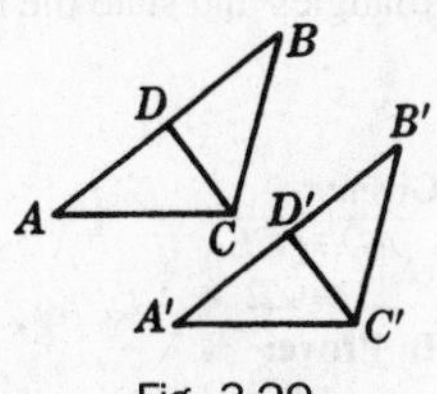

Fig. 3-29

3.7. Prove each of the following: (3.7)

(a) If a line bisects an angle of a triangle and is perpendicular to the opposite side, then it bisects that side.

(b) If the diagonals of a quadrilateral bisect each other, then its opposite sides are congruent.

(c) If the base and a leg of one isosceles triangle are congruent to the base and a leg of another isosceles triangle, then their vertex angles are congruent.

(d) Lines drawn from a point on the perpendicular bisector of a given line to the ends of the given line are congruent.

(e) If the legs of one right triangle are congruent respectively to the legs of another, their hypotenuses are congruent.

3.8. In each part of Fig. 3-30, name the congruent angles that are opposite sides of a triangle. (3.8)

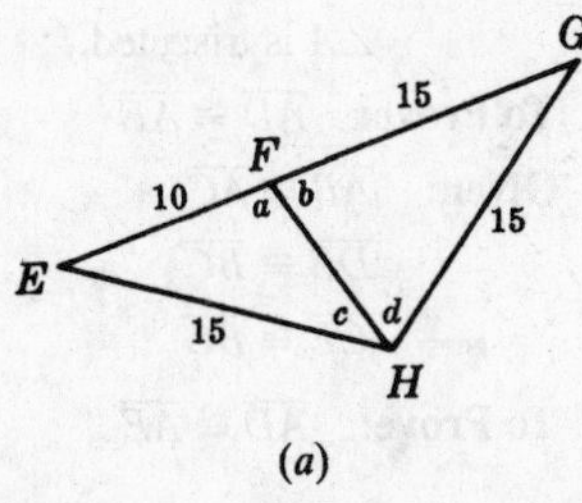

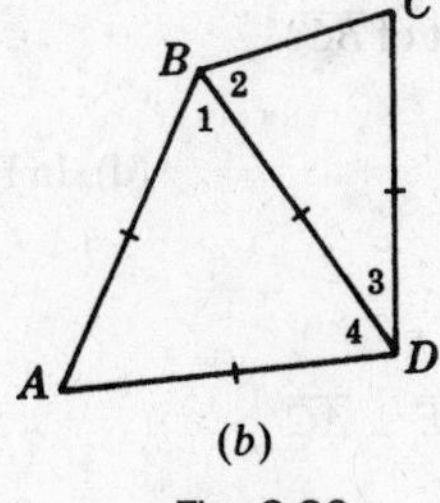

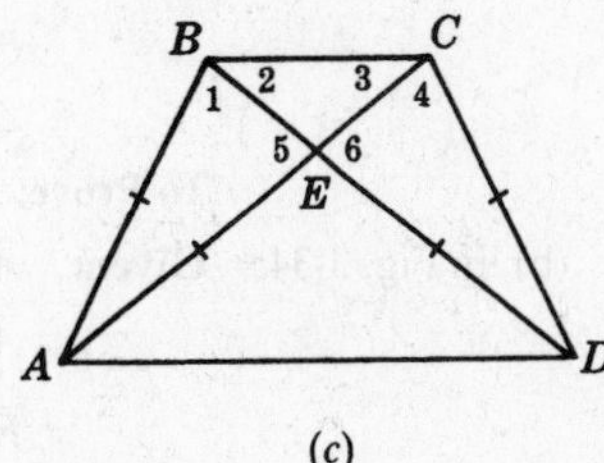

Fig. 3-30

3.9. In each part of Fig. 3-31, name the congruent sides that are opposite congruent angles of a triangle. (3.9)

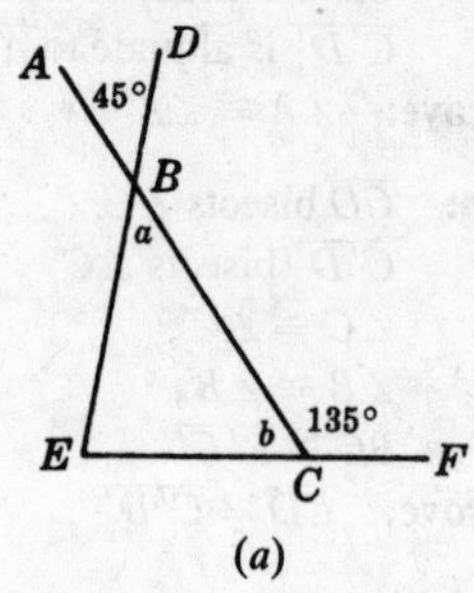

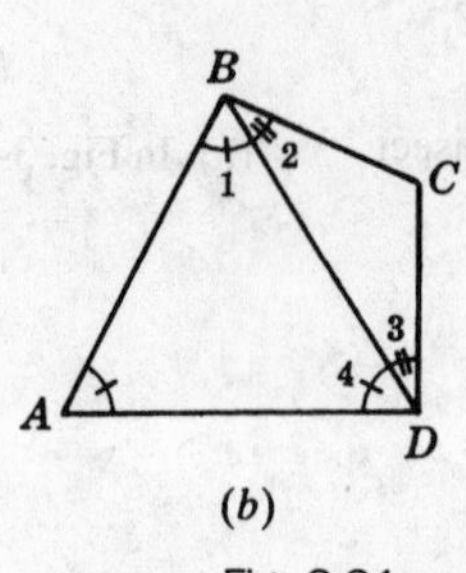

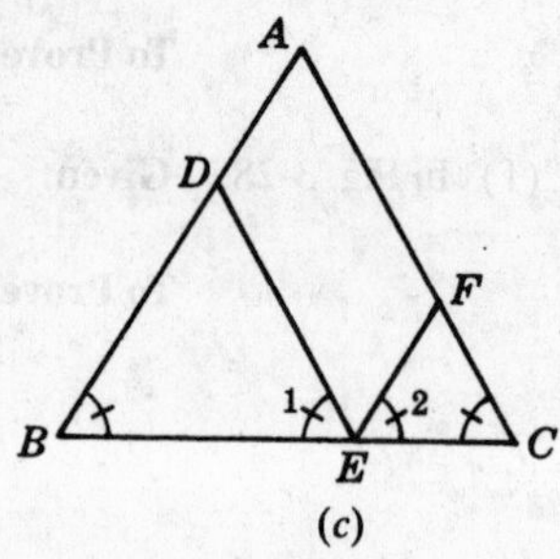

Fig. 3-31

3.10. In each part of Fig. 3-32, two triangles are to be proved congruent. Make a diagram showing the congruent parts of both triangles and state the reason for congruency. (3.10)

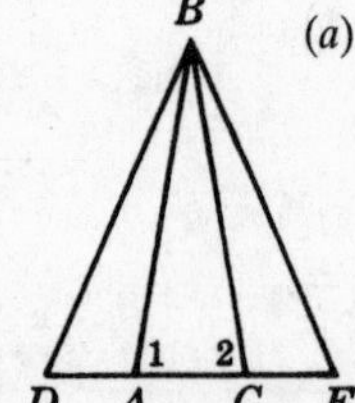

(*a*) **Given:** $\overline{AD} \cong \overline{CE}$, $\angle 1 \cong \angle 2$
To Prove: $\triangle ABD \cong \triangle CBE$

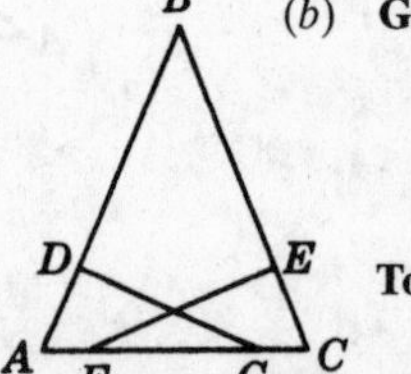

(*b*) **Given:** $\overline{AB} \cong \overline{BC}$, $\overline{DG} \perp \overline{AB}$, $\overline{EF} \perp \overline{BC}$, $\overline{BD} \cong \overline{BE}$
To Prove: $\triangle AGD \cong \triangle CFE$

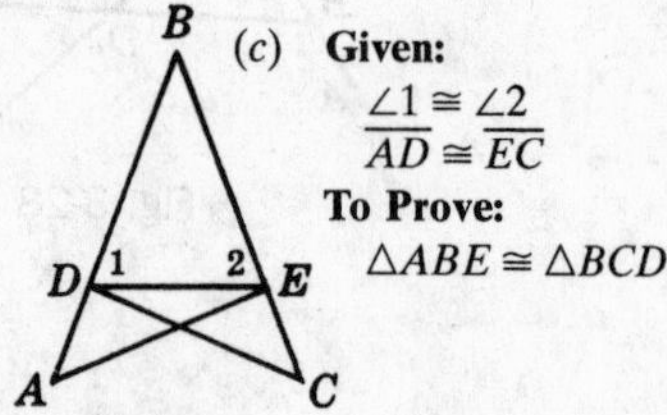

(*c*) **Given:** $\angle 1 \cong \angle 2$, $\overline{AD} \cong \overline{EC}$
To Prove: $\triangle ABE \cong \triangle BCD$

Fig. 3-32

3.11. In each part of Fig. 3-33, ΔI, ΔII, and ΔIII can be proved congruent. Make a diagram showing the congruent parts and state the reason for congruency. (3.10)

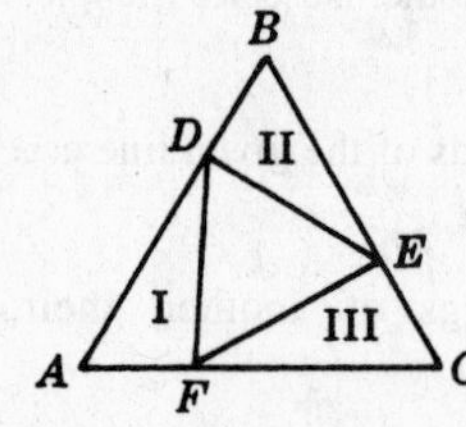

(*a*) **Given:** $\triangle ABC$ is equilateral. $\overline{AF} \cong \overline{BD} \cong \overline{CE}$
To Prove: ΔI ≅ ΔII ≅ ΔIII

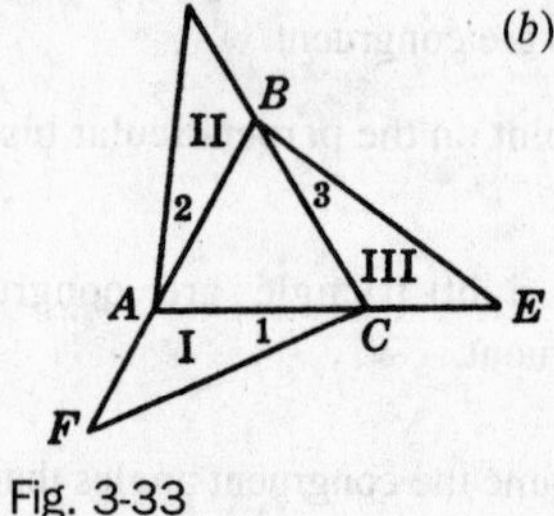

(*b*) **Given:** $\triangle ABC$ is equilateral. $\overline{AF}$, $\overline{BD}$, and $\overline{CE}$ are extensions of the sides of $\triangle ABC$. $\angle 1 \cong \angle 2 \cong \angle 3$
To Prove: ΔI ≅ ΔII ≅ ΔIII

Fig. 3-33

3.12. Prove each of the following: (3.11)

(a) In Fig. 3-34: **Given:** $\overline{AB} \cong \overline{AC}$, F is midpoint of $\overline{BC}$. $\angle 1 \cong \angle 2$
To Prove: $\overline{FD} \cong \overline{FE}$

(b) In Fig. 3-34: **Given:** $\overline{AB} \cong \overline{AC}$, $\overline{AD} \cong \overline{AE}$, $\overline{FD} \perp \overline{AB}, \overline{FE} \perp \overline{AC}$
To Prove: $\overline{BF} \cong \overline{FC}$

(c) In Fig. 3-35: **Given:** $\overline{AB} \cong \overline{AC}$, $\angle A$ is trisected.
To Prove: $\overline{AD} \cong \overline{AE}$

(d) In Fig. 3-35: **Given:** $\overline{AB} \cong \overline{AC}$, $\overline{DB} \cong \overline{BC}$, $\overline{CE} \cong \overline{BC}$
To Prove: $\overline{AD} \cong \overline{AE}$

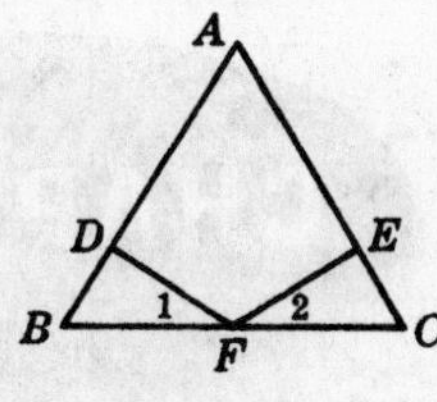

Fig. 3-34

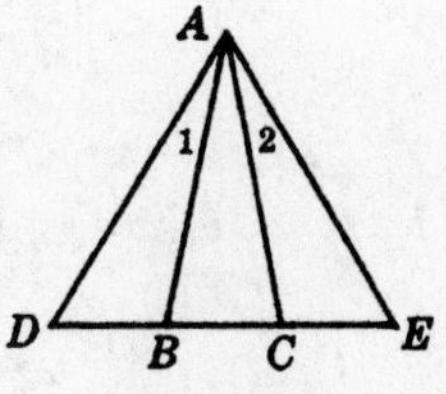

Fig. 3-35

3.13. Prove each of the following: (3.12)

(a) The median to the base of an isosceles triangle bisects the vertex angle.

(b) If the bisector of an angle of a triangle is also an altitude to the opposite side, then the other two sides of the triangle are congruent.

(c) If a median to a side of a triangle is also an altitude to that side, then the triangle is isosceles.

(d) In an isosceles triangle, the medians to the legs are congruent.

(e) In an isosceles triangle, the bisectors of the base angles are congruent.

Parallel Lines, Distances, and Angle Sums

4.1 Parallel Lines

Parallel lines are straight lines which lie in the same plane and do not intersect however far they are extended. The symbol for parallel is $\|$; thus, $\overleftrightarrow{AB} \| \overleftrightarrow{CD}$ is read "line $\overleftrightarrow{AB}$ is parallel to line $\overleftrightarrow{CD}$." In diagrams, arrows are used to indicate that lines are parallel (see Fig. 4-1).

Fig. 4-1 Fig. 4-2

A *transversal* of two or more lines is a line that cuts across these lines. Thus, $\overleftrightarrow{EF}$ is a transversal of $\overleftrightarrow{AB}$ and $\overleftrightarrow{CD}$, in Fig. 4-2.

The *interior angles* formed by two lines cut by a transversal are the angles between the two lines, while the *exterior angles* are those outside the lines. Thus, of the eight angles formed by $\overleftrightarrow{AB}$ and $\overleftrightarrow{CD}$ cut by $\overleftrightarrow{EF}$ in Fig. 4-2, the interior angles are $\angle 1$, $\angle 2$, $\angle 3$, and $\angle 4$; the exterior angles are $\angle 5$, $\angle 6$, $\angle 7$, and $\angle 8$.

4.1A Pairs of Angles Formed by Two Lines Cut by a Transversal

Corresponding angles of two lines cut by a transversal are angles on the same side of the transversal and on the same side of the lines. Thus, $\angle 1$ and $\angle 2$ in Fig. 4-3 are corresponding angles of $\overleftrightarrow{AB}$ and $\overleftrightarrow{CD}$ cut by transversal $\overleftrightarrow{EF}$. Note that in this case the two angles are both to the right of the transversal and both below the lines.

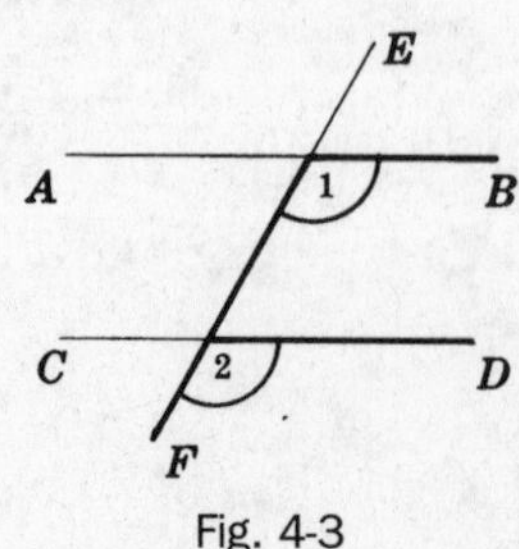

Fig. 4-3

When two parallel lines are cut by a transversal, the sides of two corresponding angles form a capital F in varying positions, as shown in Fig. 4-4.

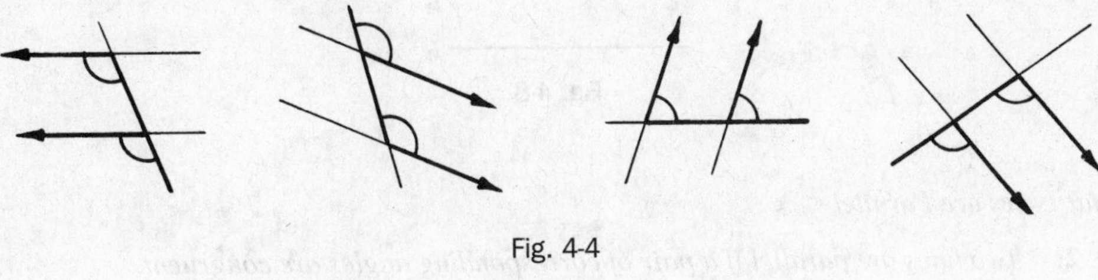

Fig. 4-4

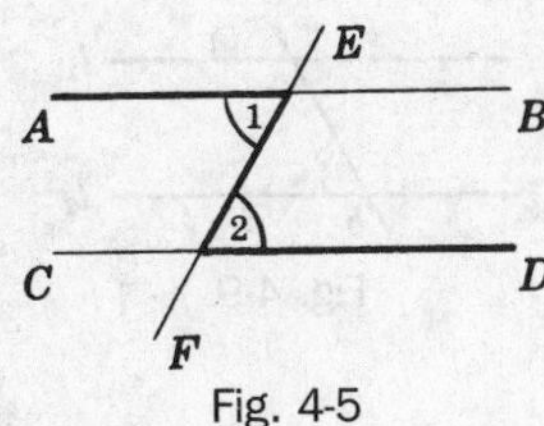

Fig. 4-5

Alternate interior angles of two lines cut by a transversal are nonadjacent angles between the two lines and on opposite sides of the transversal. Thus, ∠1 and ∠2 in Fig. 4-5 are alternate interior angles of $\overleftrightarrow{AB}$ and $\overleftrightarrow{CD}$ cut by $\overleftrightarrow{EF}$. When parallel lines are cut by a transversal, the sides of two alternate interior angles form a capital Z or N in varying positions, as shown in Fig. 4-6.

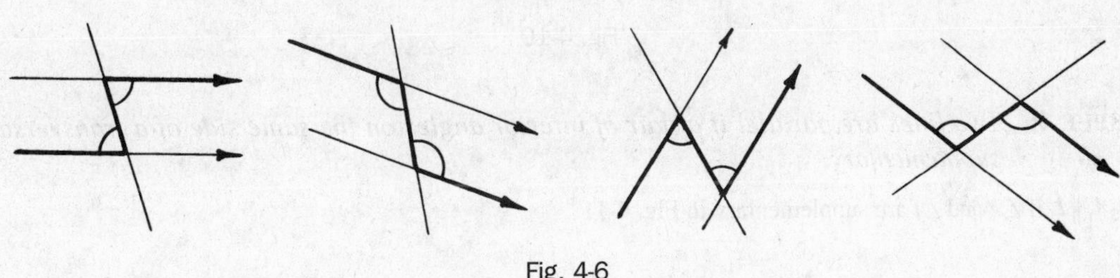

Fig. 4-6

When parallel lines are cut by a transversal, *interior angles on the same side of the transversal* can be readily located by noting the capital U formed by their sides (Fig. 4-7).

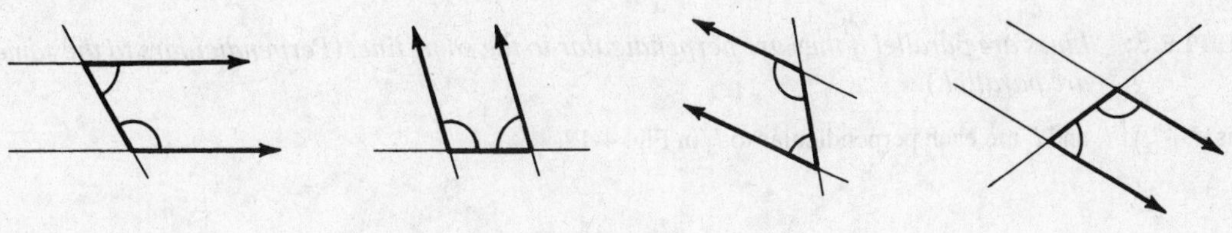

Fig. 4-7

4.1B Principles of Parallel Lines

PRINCIPLE 1: *Through a given point not on a given line, one and only one line can be drawn parallel to a given line.* (Parallel-Line Postulate)

Thus, either l_1 or l_2 but not both may be parallel to l_3 in Fig. 4-8.

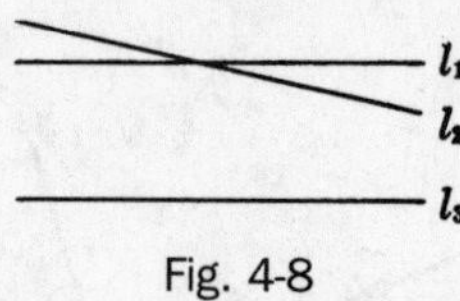

Fig. 4-8

Proving that Lines are Parallel

PRINCIPLE 2: *Two lines are parallel if a pair of corresponding angles are congruent.*

Thus, $l_1 \parallel l_2$ if $\angle a \cong \angle b$ in Fig. 4-9.

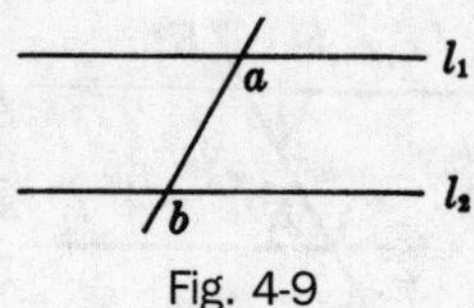

Fig. 4-9

PRINCIPLE 3: *Two lines are parallel if a pair of alternate interior angles are congruent.*

Thus, $l_1 \parallel l_2$ if $\angle c \cong \angle d$ in Fig. 4-10.

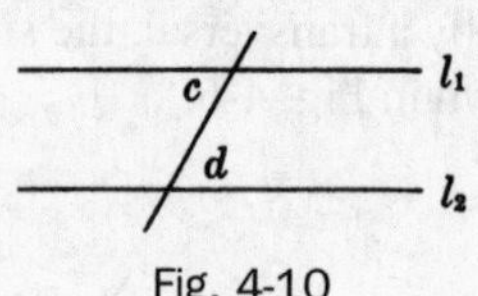

Fig. 4-10

PRINCIPLE 4: *Two lines are parallel if a pair of interior angles on the same side of a transversal are supplementary.*

Thus, $l_1 \parallel l_2$ if $\angle e$ and $\angle f$ are supplementary in Fig. 4-11.

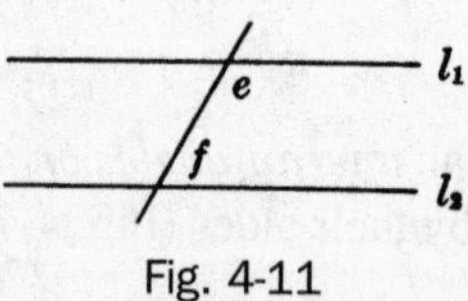

Fig. 4-11

PRINCIPLE 5: *Lines are parallel if they are perpendicular to the same line.* (*Perpendiculars to the same line are parallel.*)

Thus, $l_1 \parallel l_2$ if l_1 and l_2 are each perpendicular to l_3 in Fig. 4-12.

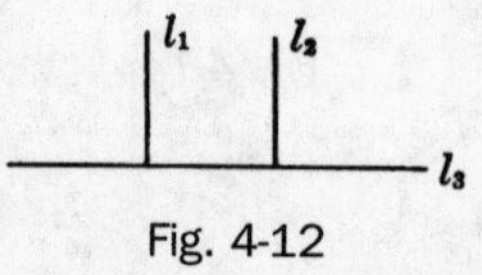

l_1 l_2 l_3

Fig. 4-12

PRINCIPLE 6: *Lines are parallel if they are parallel to the same line.* (*Parallels to the same line are parallel.*)

Thus, $l_1 \parallel l_2$ if l_1 and l_2 are each parallel to l_3 in Fig. 4-13.

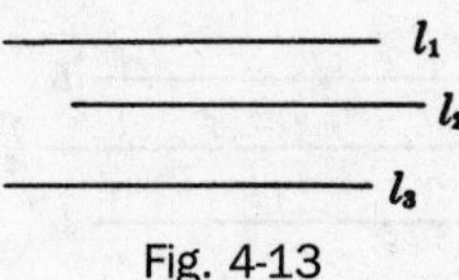

Fig. 4-13

Properties of Parallel Lines

PRINCIPLE 7: *If two lines are parallel, each pair of corresponding angles are congruent. (Corresponding angles of parallel lines are congruent.)*

Thus, if $l_1 \parallel l_2$, then $\angle a \cong \angle b$ in Fig. 4-14.

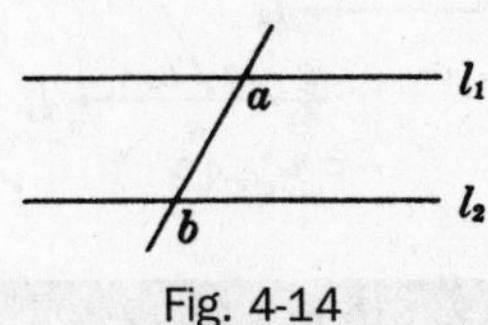

Fig. 4-14

PRINCIPLE 8: *If two lines are parallel, each pair of alternate interior angles are congruent. (Alternate interior angles of parallel lines are congruent.)*

Thus, if $l_1 \parallel l_2$, then $\angle c \cong \angle d$ in Fig. 4-15.

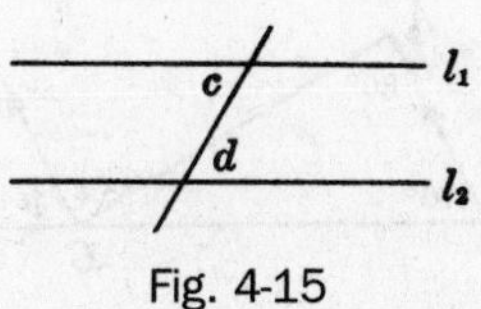

Fig. 4-15

PRINCIPLE 9: *If two lines are parallel, each pair of interior angles on the same side of the transversal are supplementary.*

Thus, if $l_1 \parallel l_2$, $\angle e$ and $\angle f$ are supplementary in Fig. 4-16.

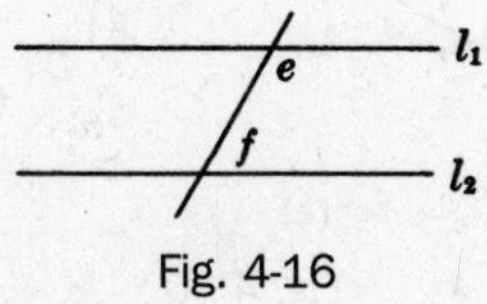

Fig. 4-16

PRINCIPLE 10: *If lines are parallel, a line perpendicular to one of them is perpendicular to the others also.*

Thus, if $l_1 \parallel l_2$ and $l_3 \perp l_1$, then $l_3 \perp l_2$ in Fig. 4-17.

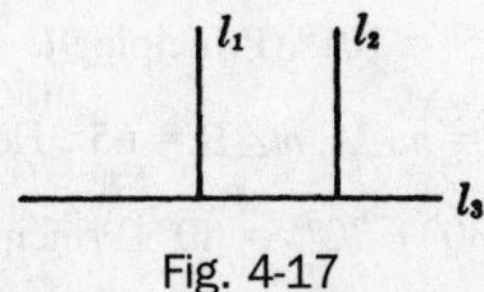

Fig. 4-17

PRINCIPLE 11: *If lines are parallel, a line parallel to one of them is parallel to the others also.*

Thus, if $l_1 \parallel l_2$ and $l_3 \parallel l_1$, then $l_3 \parallel l_2$ in Fig. 4-18.

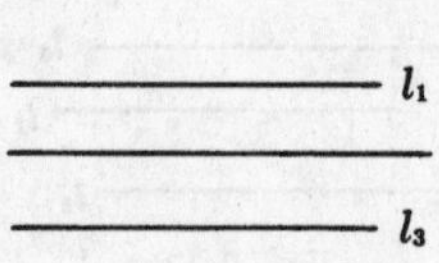

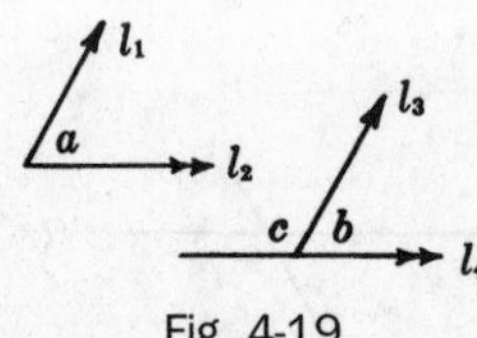

Fig. 4-18

PRINCIPLE 12: *If the sides of two angles are respectively parallel to each other, the angles are either congruent or supplementary.*

Thus, if $l_1 \parallel l_3$ and $l_2 \parallel l_4$ in Fig. 4-19, then $\angle a \cong \angle b$ and $\angle a$ and $\angle c$ are supplementary.

Fig. 4-19

SOLVED PROBLEMS

4.1 Numerical applications of parallel lines

In each part of Fig. 4-20, find the measure x and the measure y of the indicated angles.

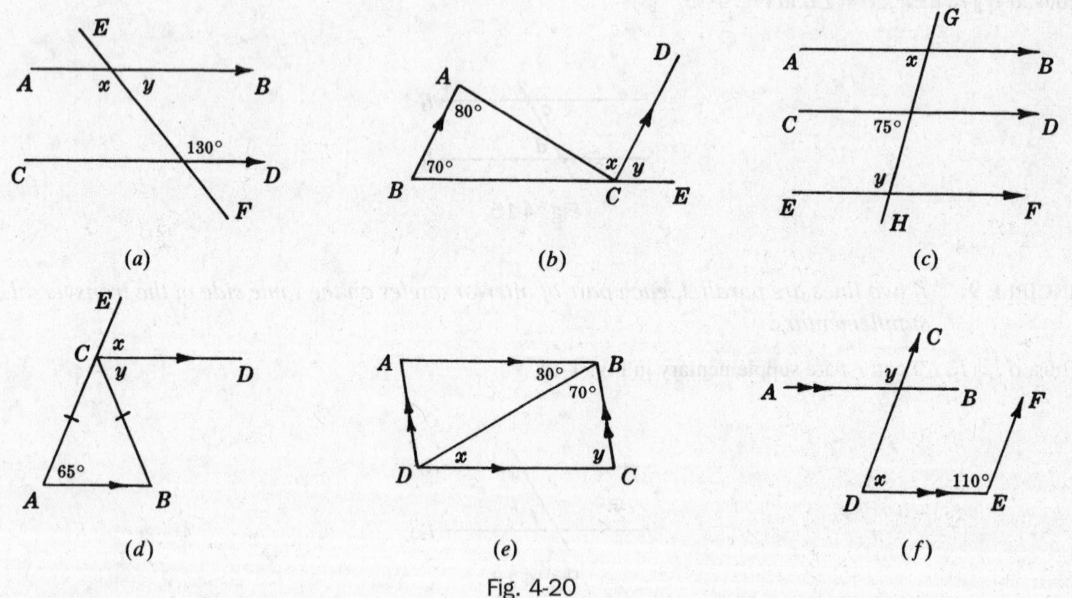

Fig. 4-20

Solutions

(a) $x = 130°$ (Principle 8). $y = 180° - 130° = 50°$ (Principle 9).

(b) $x = 80°$ (Principle 8). $y = 70°$ (Principle 7).

(c) $x = 75°$ (Principle 7). $y = 180° - 75° = 105°$ (Principle 9).

(d) $x = 65°$ (Principle 7). Since $m\angle B = m\angle A$, $m\angle B = 65°$. Hence, $y = 65°$ (Principle 8).

(e) $x = 30°$ (Principle 8). $y = 180° - (30° + 70°) = 80°$ (Principle 9).

(f) $x = 180° - 110° = 70°$ (Principle 9). $y = 110°$ (Principle 12).

4.2 Applying parallel line principles and their converses

The following short proofs refer to Fig. 4-21. In each, the first statement is given. State the parallel-line principle needed as the reason for each of the remaining statements.

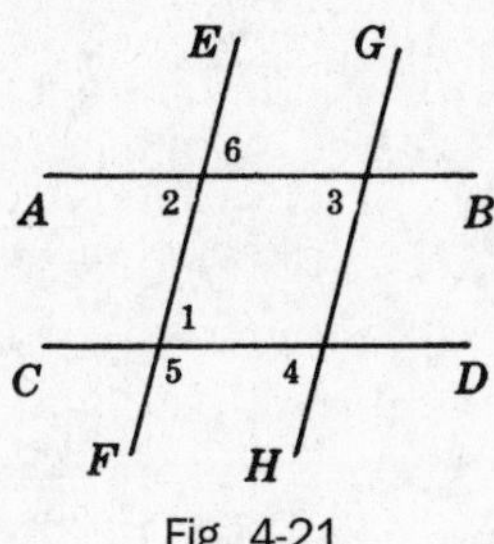

Fig. 4-21

(a) 1. $\angle 1 \cong \angle 2$ 1. Given
2. $\overline{AB} \parallel \overline{CD}$ 2. ?
3. $\angle 3 \cong \angle 4$ 3. ?

(b) 1. $\angle 2 \cong \angle 3$ 1. Given
2. $\overline{EF} \parallel \overline{GH}$ 2. ?
3. $\angle 4$ sup. $\angle 5$ 3. ?

(c) 1. $\angle 5$ sup. $\angle 4$ 1. Given
2. $\overline{EF} \parallel \overline{GH}$ 2. ?
3. $\angle 3 \cong \angle 6$ 3. ?

(d) 1. $\overline{EF} \perp \overline{AB}$, $\overline{GH} \perp \overline{AB}$, $\overline{EF} \perp \overline{CD}$ 1. Given
2. $\overline{EF} \parallel \overline{GH}$ 2. ?
3. $\overline{CD} \perp \overline{GH}$ 3. ?

Solutions

(a) 2: Principle 3; 3: Principle 7.

(b) 2: Principle 2; 3: Principle 9.

(c) 2: Principle 4; 3: Principle 8.

(d) 2: Principle 5; 3: Principle 10.

4.3 Algebraic applications of parallel lines

In each part of Fig. 4-22, find x and y. Provide the reason for each equation obtained from the diagram.

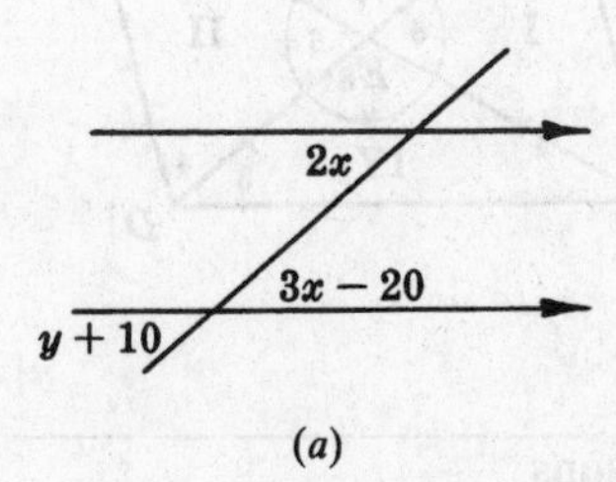

(a)

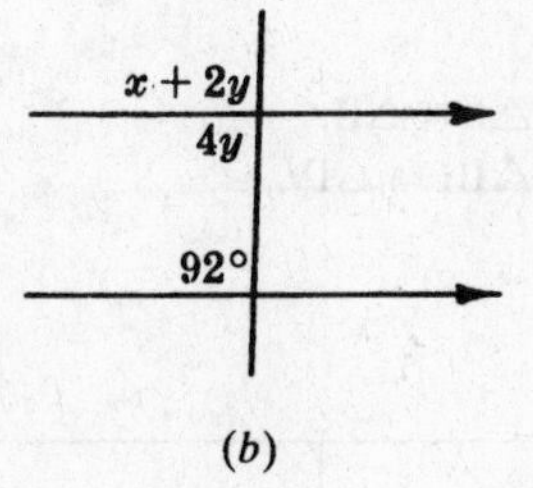

(b)

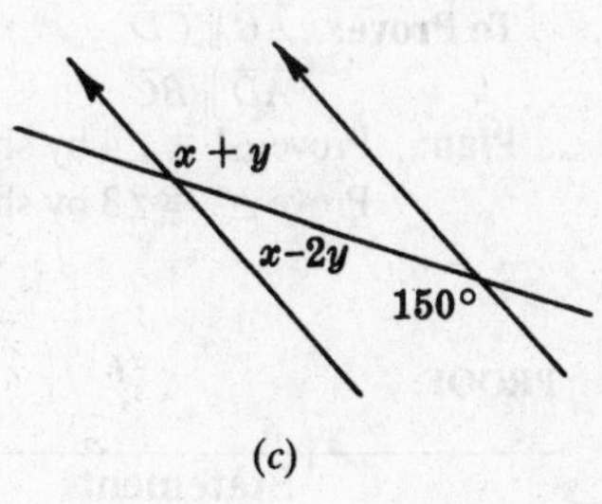

(c)

Fig. 4-22

Solutions

(a) $3x - 20 = 2x$ (Principle 8)
$x = 20°$
$y + 10 = 2x$ (Principle 7)
$y + 10 = 40$
$y = 30°$

(b) $4y = 180 - 92 = 88$ (Principle 9)
$y = 22°$
$x + 2y = 92$ (Principle 7)
$x + 44 = 92$
$x = 48°$

(c) (1) $x + y = 150$ (Principle 8)
(2) $x + 2y = 30$ (Principle 9)
$3y = 120$ (Subt. Postulate)
$y = 40°$
$x + 40 = 150$
$x = 110°$

4.4 Proving a parallel-line problem

Given: $\overline{AB} \cong \overline{AC}$
$\overrightarrow{AE} \parallel \overline{BC}$

To Prove: $\overrightarrow{AE}$ bisects $\angle DAC$

Plan: Show that $\angle 1$ and $\angle 2$ are congruent to the congruent angles B and C.

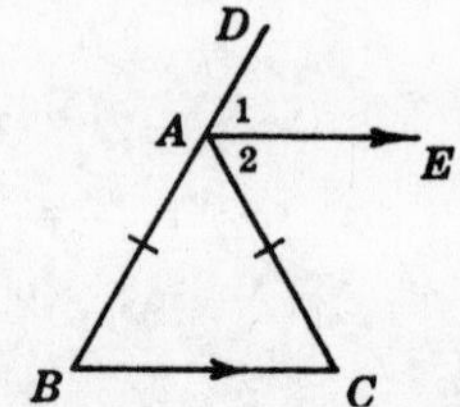

PROOF:

Statements	Reasons
1. $\overrightarrow{AE} \parallel \overline{BC}$	1. Given
2. $\angle 1 \cong \angle B$	2. Corresponding ∠s of ∥ lines are ≅.
3. $\angle 2 \cong \angle C$	3. Alternate interior ∠s of ∥ lines are ≅.
4. $\overline{AB} \cong \overline{AC}$	4. Given
5. $\angle B \cong \angle C$	5. In a △, ∠s opposite ≅ sides are ≅.
6. $\angle 1 \cong \angle 2$	6. Things ≅ to ≅ things are ≅ to each other.
7. $\overrightarrow{AE}$ bisects $\angle DAC$.	7. To divide into two congruent parts is to bisect.

4.5 Proving a parallel-line problem stated in words

Prove that if the diagonals of a quadrilateral bisect each other, the opposite sides are parallel.

Given: Quadrilateral $ABCD$
$\overline{AC}$ and $\overline{BD}$ bisect each other.

To Prove: $\overline{AB} \parallel \overline{CD}$
$\overline{AD} \parallel \overline{BC}$

Plan: Prove $\angle 1 \cong \angle 4$ by showing △I ≅ △II.
Prove $\angle 2 \cong \angle 3$ by showing △III ≅ △IV.

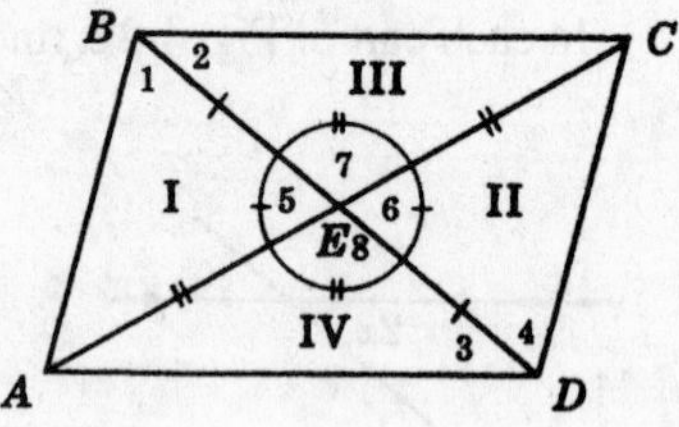

PROOF:

Statements	Reasons
1. $\overline{AC}$ and $\overline{BD}$ bisect each other.	1. Given
2. $\overline{BE} \cong \overline{ED}$ $\overline{AE} \cong \overline{EC}$	2. To bisect is to divide into two congruent parts.
3. $\angle 5 \cong \angle 6$, $\angle 7 \cong \angle 8$	3. Vertical ∠s are ≅.
4. △I ≅ △II, △III ≅ △IV	4. SAS
5. $\angle 1 \cong \angle 4$, $\angle 2 \cong \angle 3$	5. Corresponding parts of congruent △s are ≅.
6. $\overline{AB} \parallel \overline{CD}$, $\overline{BC} \parallel \overline{AD}$	6. Lines cut by a transversal are ∥ if alternate interior ∠s are ≅.

4.2 Distances

4.2A Distances Between Two Geometric Figures

The distance between two geometric figures is the straight line segment which is the *shortest segment between the figures*.

1. The distance *between two points*, such as P and Q in Fig. 4-23(a), is the line segment $\overline{PQ}$ between them.

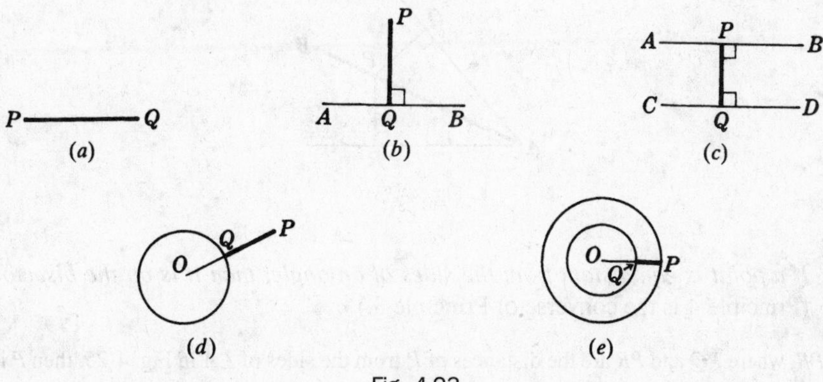

Fig. 4-23

2. The distance *between a point and a line*, such as P and $\overleftrightarrow{AB}$ in (b), is the line segment PQ, the perpendicular from the point to the line.
3. The distance *between two parallels*, such as $\overleftrightarrow{AB}$ and $\overleftrightarrow{CD}$ in (c), is the segment $\overline{PQ}$, a perpendicular between the two parallels.
4. The distance *between a point and a circle*, such as P and circle O in (d), is $\overline{PQ}$, the segment of $\overline{OP}$ between the point and the center of the circle.
5. The distance *between two concentric circles*, such as two circles whose center is O, is $\overline{PQ}$, the segment of the larger radius that lies between the two circles, as shown in (e).

4.2B Distance Principles

PRINCIPLE 1: *If a point is on the perpendicular bisector of a line segment, then it is equidistant from the ends of the line segment.*

Thus if P is on $\overleftrightarrow{CD}$, the $\perp$ bisector of $\overline{AB}$ in Fig. 4-24, then $\overline{PA} \cong \overline{PB}$.

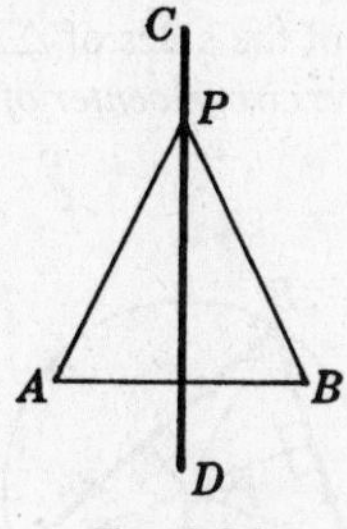

Fig. 4-24

PRINCIPLE 2: *If a point is equidistant from the ends of a line segment, then it is on the perpendicular bisector of the line segment.* (Principle 2 is the converse of Principle 1.)

Thus if $\overline{PA} \cong \overline{PB}$ in Fig. 4-24, then P is on $\overleftrightarrow{CD}$, the $\perp$ bisector of $\overline{AB}$.

PRINCIPLE 3: *If a point is on the bisector of an angle, then it is equidistant from the sides of the angle.*

Thus if P is on $\overleftrightarrow{AB}$, the bisector of $\angle A$ in Fig. 4-25, then $\overline{PQ} \cong \overline{PR}$, where PQ and PR are the distances of P from the sides of the angle.

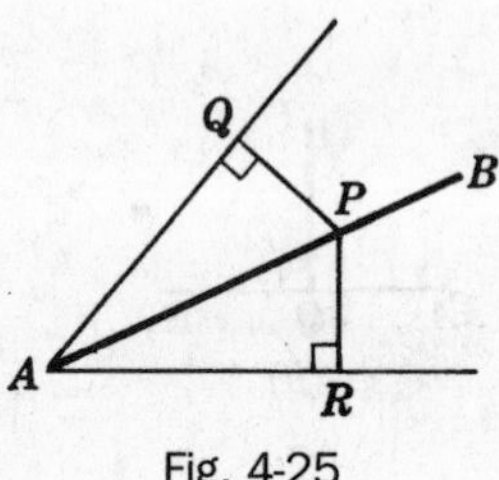

Fig. 4-25

PRINCIPLE 4: *If a point is equidistant from the sides of an angle, then it is on the bisector of the angle.* (Principle 4 is the converse of Principle 3.)

Thus if $PQ = PR$, where PQ and PR are the distances of P from the sides of $\angle A$ in Fig. 4-25, then P is on $\overleftrightarrow{AB}$, the bisector of $\angle A$.

PRINCIPLE 5: *Two points each equidistant from the ends of a line segment determine the perpendicular bisector of the line segment.* (The line joining the vertices of two isosceles triangles having a common base is the perpendicular bisector of the base.)

Thus if $\overline{PA} \cong \overline{PB}$ and $\overline{QA} \cong \overline{QB}$ in Fig. 4-26, then P and Q determine $\overleftrightarrow{CD}$, the $\perp$ bisector of $\overline{AB}$.

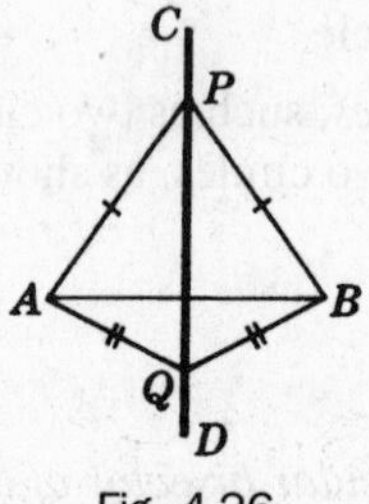

Fig. 4-26

PRINCIPLE 6: *The perpendicular bisectors of the sides of a triangle meet in a point which is equidistant from the vertices of the triangle.*

Thus if P is the intersection of the $\perp$ bisectors of the sides of $\triangle ABC$ in Fig. 4-27, then $\overline{PA} \cong \overline{PB} \cong \overline{PC}$. P is the center of the circumscribed circle and is called the *circumcenter of* $\triangle ABC$.

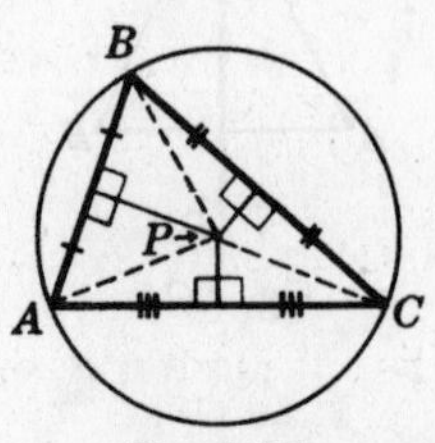

Fig. 4-27

PRINCIPLE 7: *The bisectors of the angles of a triangle meet in a point which is equidistant from the sides of the triangle.*

Thus if Q is the intersection of the bisectors of the angles of $\triangle ABC$ in Fig. 4-28, then $\overline{QR} \cong \overline{QS} \cong \overline{QT}$, where these are the distances from Q to the sides of $\triangle ABC$. Q is the center of the inscribed circle and is called the *incenter* of $\triangle ABC$.

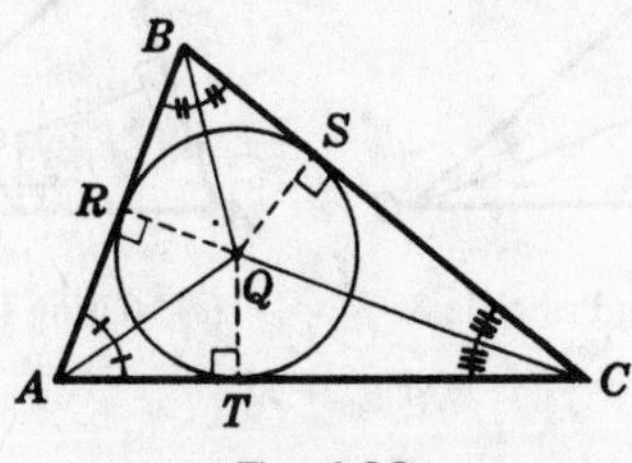

Fig. 4-28

SOLVED PROBLEMS

4.6 Finding distances

In each of the following, find the distance and indicate the kind of distance involved. In Fig. 4-29(a), find the distance (a) from P to A; (b) from P to $\overleftrightarrow{CD}$; (c) from A to $\overleftrightarrow{BC}$; (*d*) from $\overleftrightarrow{AB}$ to $\overleftrightarrow{CD}$. In Fig. 4-29(b), find the distance (e) from P to inner circle O; (f) from P to outer circle O; (g) between the concentric circles.

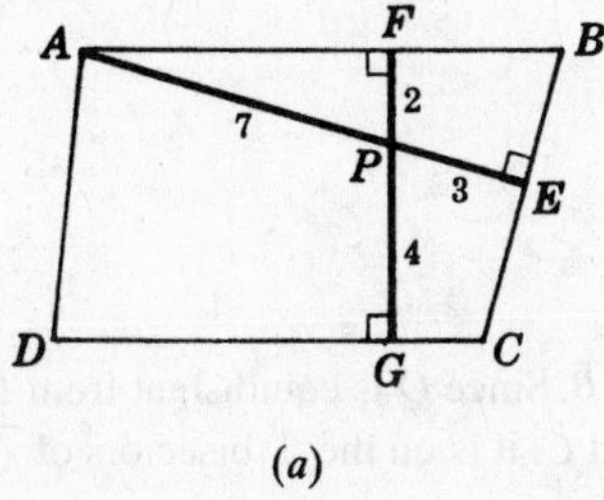

Fig. 4-29

Solutions

(a) $PA = 7$, distance between two points

(b) $PG = 4$, distance from a point to a line

(c) $AE = 10$, distance from a point to a line

(d) $FG = 6$, distance between two parallel lines

(e) $PQ = 12 - 3 = 9$, distance from a point to a circle

(f) $PR = 12 - 5 = 7$, distance from a point to a circle

(g) $QR = 5 - 3 = 2$, distance between two concentric circles

4.7 Locating a point satisfying given conditions

In Fig. 4-30.

(a) Locate P, a point on $\overleftrightarrow{BC}$ and equidistant from A and C.

(b) Locate Q, a point on $\overleftrightarrow{AB}$ and equidistant from $\overline{BC}$ and $\overline{AC}$.

(c) Locate R, the center of the circle circumscribed about $\triangle ABC$.

(d) Locate S, the center of the circle inscribed in $\triangle ABC$.

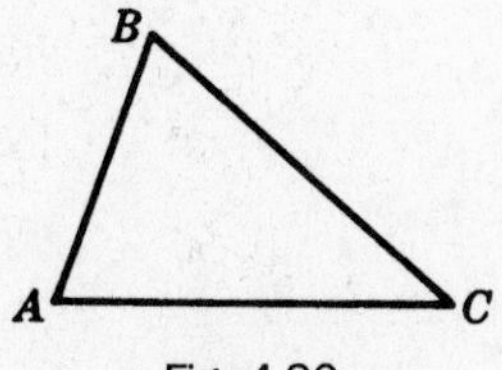

Fig. 4-30

Solutions

See Fig. 4-31.

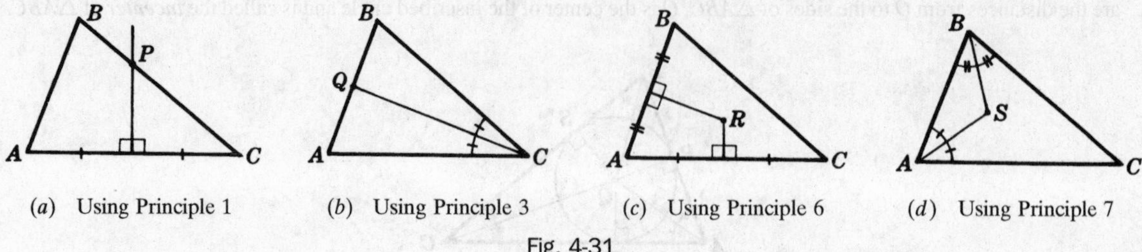

(*a*) Using Principle 1 (*b*) Using Principle 3 (*c*) Using Principle 6 (*d*) Using Principle 7

Fig. 4-31

4.8 Applying principles 2 and 4

For each $\triangle ABC$ in Fig. 4-32, describe P, Q, and R as equidistant points, and locate each on a bisector.

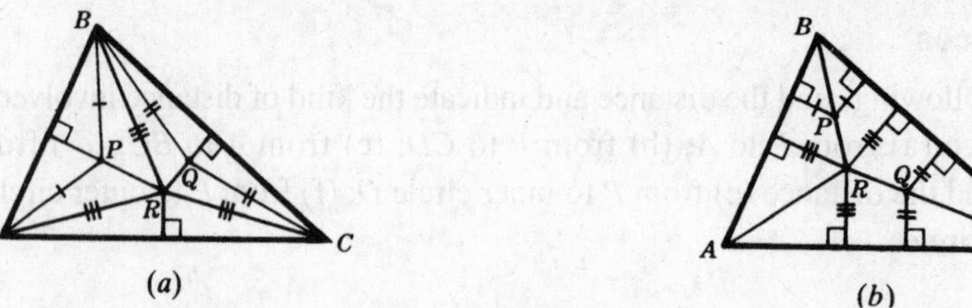

Fig. 4-32

Solutions

(a) Since P is equidistant from A and B, it is on the $\perp$ bisector of $\overline{AB}$. Since Q is equidistant from B and C, it is on the $\perp$ bisector of BC. Since R is equidistant from A, B, and C, it is on the $\perp$ bisectors of $\overline{AB}$, $\overline{BC}$, and $\overline{AC}$.

(b) Since P is equidistant from $\overleftrightarrow{AB}$ and $\overleftrightarrow{BC}$, it is on the bisector of $\angle B$. Since Q is equidistant from $\overleftrightarrow{AC}$ and $\overleftrightarrow{BC}$, it is on the bisector of $\angle C$. Since R is equidistant from $\overleftrightarrow{AB}$, $\overleftrightarrow{BC}$, and $\overleftrightarrow{AC}$, it is on the bisectors of $\angle A$, $\angle B$, and $\angle C$.

4.9 Applying principles 1, 3, 6, and 7

For each $\triangle ABC$ in Fig. 4-33, describe P, Q, and R as equidistant points. Also, describe R as the center of a circle.

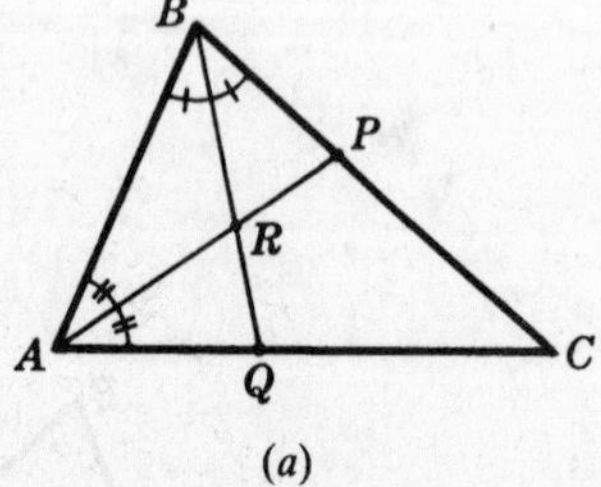

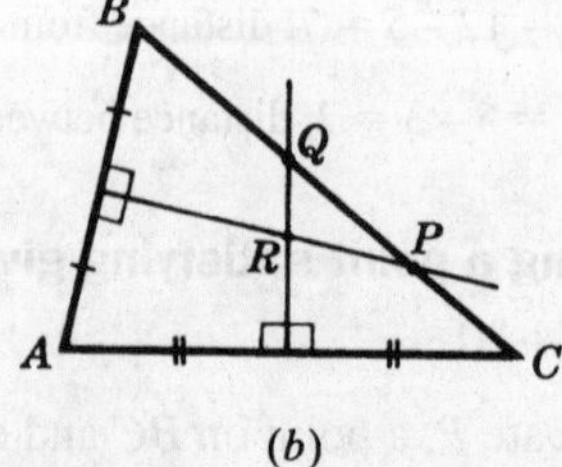

Fig. 4-33

Solutions

(a) Since P is on the bisector of $\angle A$, it is equidistant from $\overleftrightarrow{AB}$ and $\overleftrightarrow{AC}$. Since Q is on the bisectors of $\angle B$, it is equidistant from $\overleftrightarrow{AB}$ and $\overleftrightarrow{BC}$. Since R is on the bisectors of $\angle A$ and $\angle B$, it is equidistant from $\overleftrightarrow{AB}$, $\overleftrightarrow{BC}$, and $\overleftrightarrow{AC}$. R is the incenter of $\triangle ABC$, that is, the center of its inscribed circle.

(b) Since P is on the $\perp$ bisector of $\overline{AB}$, it is equidistant from A and B. Since Q is on the $\perp$ bisector of $\overline{AC}$, it is equidistant from A and C. Since R is on the $\perp$ bisectors of $\overline{AB}$ and $\overline{AC}$, it is equidistant from A, B, and C. R is the circumcenter of $\triangle ABC$, that is, the center of its circumscribed circle.

4.10 Applying principles 1, 3, 6, and 7

In each part of Fig. 4-34, find two points equidistant from the ends of a line segment, and find the perpendicular bisector determined by the two points.

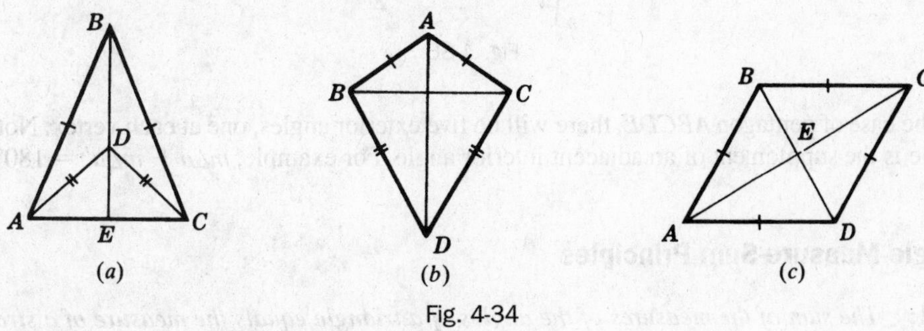

Fig. 4-34

Solutions

(a) B and D are equidistant from A and C; hence, $\overline{BE}$ is the $\perp$ bisector of $\overline{AC}$.

(b) A and D are equidistant from B and C; hence, $\overline{AD}$ is the $\perp$ bisector of $\overline{BC}$.

(c) B and D are equidistant from A and C; hence, $\overline{BD}$ is the $\perp$ bisector of $\overline{AC}$. A and C are equidistant from B and D; hence, $\overline{AC}$ is the $\perp$ bisector of $\overline{BD}$.

4.3 Sum of the Measures of the Angles of a Triangle

The angles of any triangle may be torn off, as in Fig. 4-35(a), and then fitted together as shown in (b). The three angles will form a straight angle.

We can prove that the sum of the measures of the angles of a triangle equals 180° by drawing a line through one vertex of the triangle parallel to the side opposite the vertex. In Fig. 4-35(c), $\overleftrightarrow{MN}$ is drawn through B parallel to AC. Note that the measure of the straight angle at B equals the sum of the measures of the angles of $\triangle ABC$; that is, $a° + b° + c° = 180°$. Each pair of congruent angles is a pair of alternate interior angles of parallel lines.

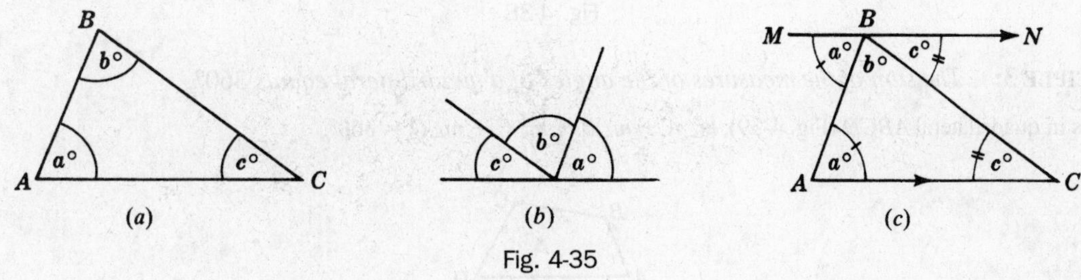

Fig. 4-35

4.3A Interior and Exterior Angles of a Polygon

An exterior angle of a polygon is formed whenever one of its sides is extended through a vertex. If each of the sides of a polygon is extended, as shown in Fig. 4-36, an exterior angle will be formed at each vertex. Each of these exterior angles is the supplement of its adjacent interior angle.

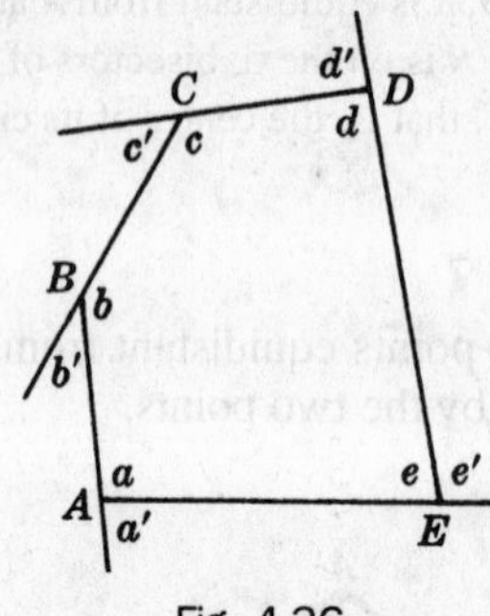

Fig. 4-36

Thus, in the case of pentagon *ABCDE*, there will be five exterior angles, one at each vertex. Note that each exterior angle is the supplement of an adjacent interior angle. For example, $m\angle a + m\angle a' = 180°$.

4.3B Angle-Measure-Sum Principles

PRINCIPLE 1: *The sum of the measures of the angles of a triangle equals the measure of a straight angle or* 180°.

Thus in $\triangle ABC$ of Fig. 4-37, $m\angle A + m\angle B + m\angle C = 180°$.

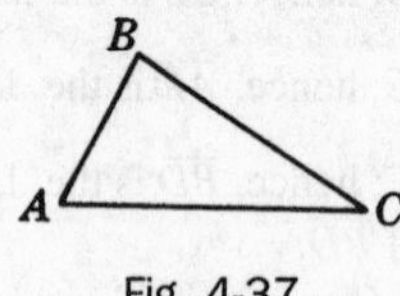

Fig. 4-37

PRINCIPLE 2: *If two angles of one triangle are congruent respectively to two angles of another triangle, the remaining angles are congruent.*

Thus in $\triangle ABC$ and $\triangle A'B'C'$ in Fig. 4-38, if $\angle A \cong \angle A'$ and $\angle B \cong \angle B'$, then $\angle C \cong \angle C'$.

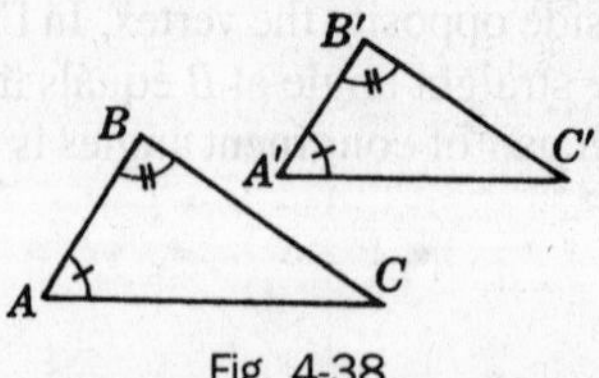

Fig. 4-38

PRINCIPLE 3: *The sum of the measures of the angles of a quadrilateral equals* 360°.

Thus in quadrilateral *ABCD* (Fig. 4-39), $m\angle A + m\angle B + m\angle C + m\angle D = 360°$.

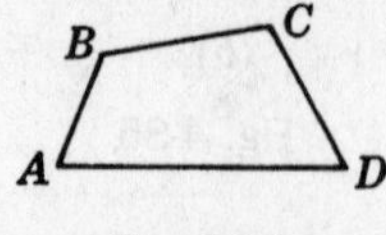

Fig. 4-39

PRINCIPLE 4: *The measure of each exterior angle of a triangle equals the sum of the measures of its two nonadjacent interior angles.*

Thus in $\triangle ABC$ in Fig. 4-40, $m\angle ECB = m\angle A + m\angle B$.

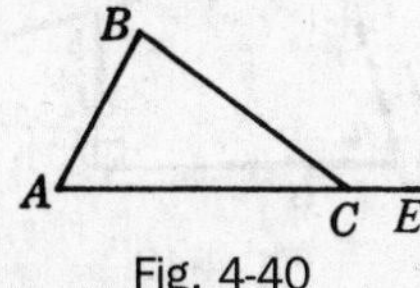

Fig. 4-40

PRINCIPLE 5: *The sum of the measures of the exterior angles of a triangle equals* 360°.

Thus in $\triangle ABC$ in Fig. 4-41, $m\angle a' + m\angle b' + m\angle c' = 360°$.

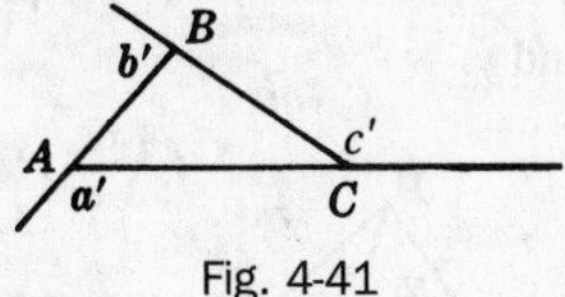

Fig. 4-41

PRINCIPLE 6: *The measure of each angle of an equilateral triangle equals* 60°.

Thus if $\triangle ABC$ in Fig. 4-42 is equilateral, then $m\angle A = 60°$, $m\angle B = 60°$, and $m\angle C = 60°$.

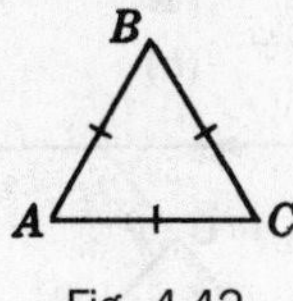

Fig. 4-42

PRINCIPLE 7: *The acute angles of a right triangle are complementary.*

Thus in rt. $\triangle ABC$ in Fig. 4-43, if $m\angle C = 90°$, then $m\angle A + m\angle B = 90°$.

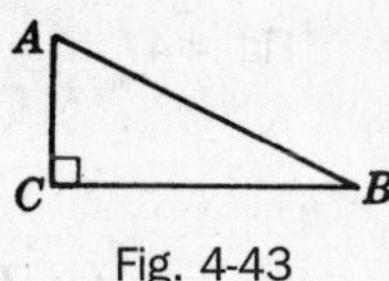

Fig. 4-43

PRINCIPLE 8: *The measure of each acute angle of an isosceles right triangle equals* 45°.

Thus in isos. rt. $\triangle ABC$ in Fig. 4-44, if $m\angle C = 90°$, then $m\angle A = 45°$ and $m\angle B = 45°$.

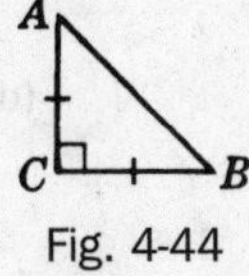

Fig. 4-44

PRINCIPLE 9: *A triangle can have no more than one right angle.*

Thus in rt. $\triangle ABC$ in Fig. 4-43, if $m\angle C = 90°$, then $\angle A$ and $\angle B$ cannot be rt. ∠s

PRINCIPLE 10: *A triangle can have no more than one obtuse angle.*

Thus in obtuse $\triangle ABC$ in Fig. 4-45, if $\angle C$ is obtuse, then $\angle A$ and $\angle B$ cannot be obtuse angles.

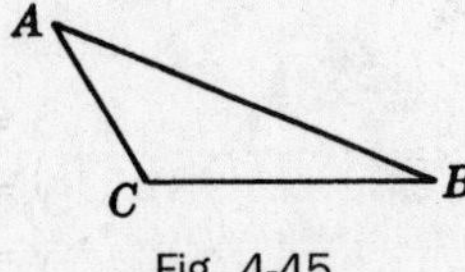

Fig. 4-45

PRINCIPLE 11: *Two angles are supplementary if their sides are respectively perpendicular to each other.*

Thus if $l_1 \perp l_3$ and $l_2 \perp l_4$ in Fig. 4-46, then $\angle a \cong \angle b$, and a and $\angle c$ are supplementary.

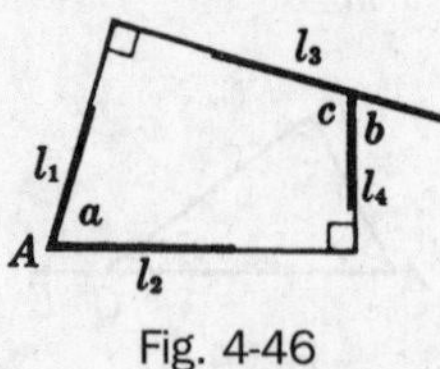

Fig. 4-46

SOLVED PROBLEMS

4.11 Numerical applications of angle-measure-sum principles

In each part of Fig. 4-47, find x and y.

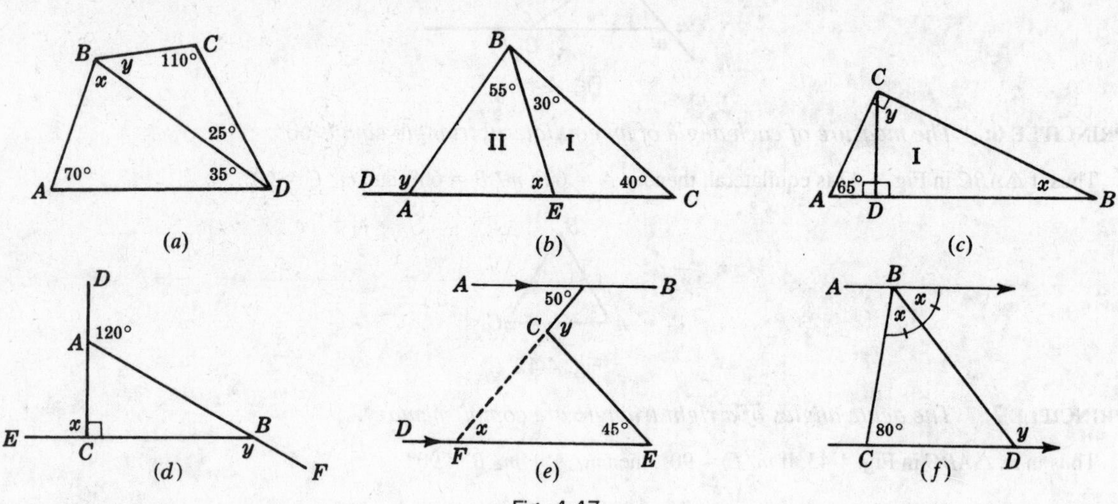

Fig. 4-47

Solutions

(a) $x + 35 + 70 = 180$ (Pr. 1)

$x = 75°$

$y + 110 + 25 = 180$ (Pr. 1)

$y = 45°$

Check: The sum of the measures of the angles of quad. $ABCD$ should equal 360°.

$70 + 120 + 110 + 60 \stackrel{?}{=} 360$

$360 = 360$

(b) x is ext. ∠ of △I.

$x = 30 + 40$ (Pr. 4)

$x = 70°$

y is an ext. ∠ of △ABC.

$y = m\angle B + 40$ (Pr. 4)

$y = 85 + 40 = 125°$

(c) In △ABC, $x + 65 = 90$ (Pr. 7)

$x = 25°$

In △I, $x + y = 90$ (Pr. 7)

$25 + y = 90$

$y = 65°$

(d) Since $\overleftrightarrow{DC} \perp \overleftrightarrow{EB}$, $x = 90°$

$x + y + 120 = 360$ (Pr. 7)

$90 + y + 120 = 360$

$y = 150°$

(e) Since $\overleftrightarrow{AB} \parallel \overleftrightarrow{DE}$, $x = 50°$

$y = x + 45$ (Pr. 4)

$y = 50 + 45 = 95°$

(f) Since $\overleftrightarrow{AB} \parallel \overleftrightarrow{CD}$, $2x + 80 = 180$

$2x = 100$

$x = 50°$

$y = x + 80°$ (Pr. 4)

$y = 50 + 80 = 130°$

4.12 Applying angle-measure-sum principles to isosceles and equilateral triangles

Find x and y in each part of Fig. 4-48.

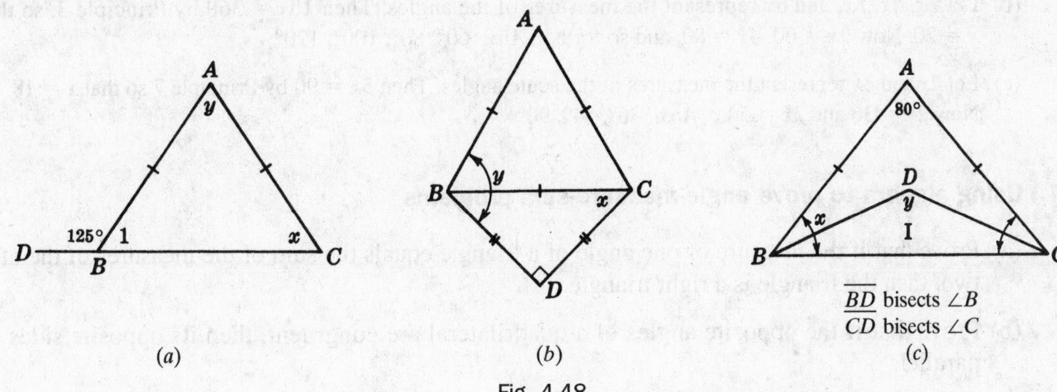

Fig. 4-48

Solutions

(a) Since $\overline{AB} \cong \overline{AC}$, we have $\angle 1 \cong \angle x$

$x = 180 - 125 = 55°$

$2x + y = 180$ (Pr. 1)

$110 + y = 180$

$y = 70°$

(b) By Pr. 8, $x = 45°$

Since $m\angle ABC = 60°$ (Pr. 6)

and $m\angle CBD = 45°$ (Pr. 8)

$y = 60 + 45 = 105°$

(c) Since $\overline{AB} \cong \overline{AC}$, $\angle ABC \cong \angle ACB$

$2x + 80 = 180$ (Pr. 1)

$x = 50°$

In $\triangle$I, $\frac{1}{2}x + \frac{1}{2}x + y = 180$ (Pr. 1)

$x + y = 180$

$50 + y = 180$

$y = 130°$

4.13 Applying ratios to angle-measure sums

Find the measure of each angle

(a) Of a triangle if its angle measures are in the ratio of 3:4:5 [Fig. 3-49(a)]

(b) Of a quadrilateral if its angle measures are in the ratio of 3:4:5:6 [(b)]

(c) Of a right triangle if the ratio of the measures of its acute angles is 2:3 [(c)]

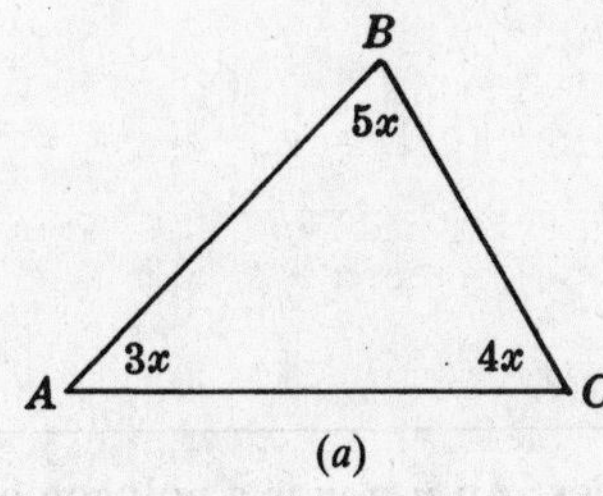

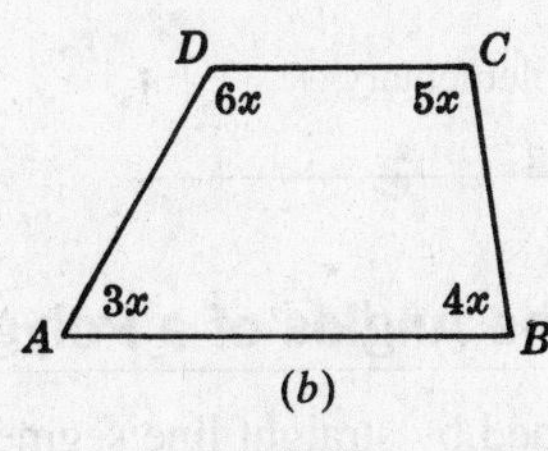

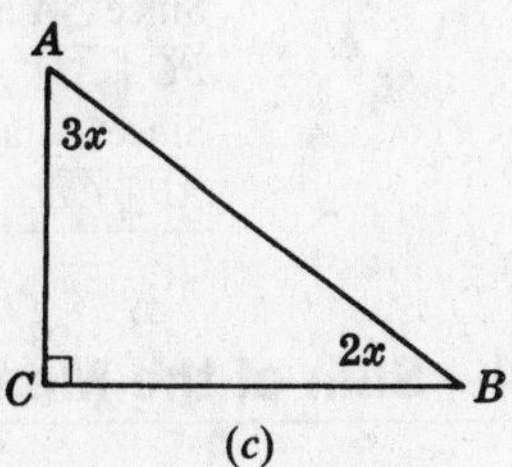

Fig. 4-49

Solutions

(a) Let $3x$, $4x$, and $5x$ represent the measures of the angles. Then $12x = 180$ by Principle 1, so that $x = 15$. Now $3x = 45$, $4x = 60$, and $5x = 75$. *Ans.* 45°, 60°, 75°

(b) Let $3x$, $4x$, $5x$, and $6x$ represent the measures of the angles. Then $18x = 360$ by Principle 3, so that $x = 20$. Now $3x = 60$, $4x = 80$, and so forth. *Ans.* 60°, 80°, 100°, 120°

(c) Let $2x$ and $3x$ represent the measures of the acute angles. Then $5x = 90$ by Principle 7 so that $x = 18$. Now $2x = 36$ and $3x = 54$. *Ans.* 36°, 54°, 90°

4.14 Using algebra to prove angle-measure-sum problems

(a) Prove that if the measure of one angle of a triangle equals the sum of the measures of the other two, then the triangle is a right triangle.

(b) Prove that if the opposite angles of a quadrilateral are congruent, then its opposite sides are parallel.

Solutions

(a) **Given:** $\triangle ABC$, $m\angle C = m\angle A + m\angle B$
To Prove: $\triangle ABC$ is a right triangle.
Plan: Prove $m\angle C = 90°$

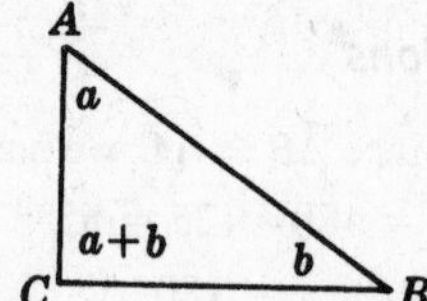

ALGEBRAIC PROOF:

Let a = number of degrees in $\angle A$
b = number of degrees in $\angle B$
Then $a + b$ = number of degrees in $\angle C$

$a + b + (a + b) = 180$ (Pr. 1)
$2a + 2b = 180$
$a + b = 90$

Since $m\angle C = 90°$, $\triangle ABC$ is a rt. $\triangle$.

(b) **Given:** Quadrilateral $ABCD$
$\angle A \cong \angle C$, $\angle B \cong \angle D$
To Prove: $\overline{AB} \parallel \overline{CD}$, $\overline{BC} \parallel \overline{AD}$
Plan: Prove int. ∠s on same side of transversal are supplementary.

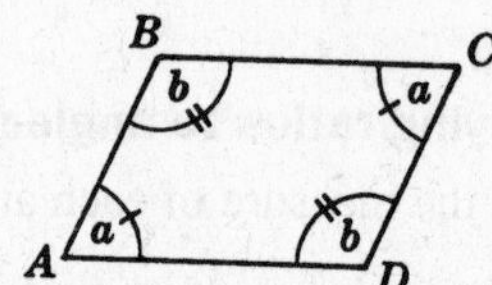

ALGEBRAIC PROOF:

Let a = number of degrees in $\angle A$ and $\angle C$,
b = number of degrees in $\angle B$ and $\angle D$.

$2a + 2b = 360$ (Pr. 3)
$a + b = 180$

Since $\angle A$ and $\angle B$ are supplementary, $\overline{BC} \parallel \overline{AD}$.
Since $\angle A$ and $\angle D$ are supplementary, $\overline{AB} \parallel \overline{CD}$.

4.4 Sum of the Measures of the Angles of a Polygon

A *polygon* is a closed plane figure bounded by straight line segments as sides. An *n-gon* is a polygon of n sides. Thus, a polygon of 20 sides is a 20-gon.

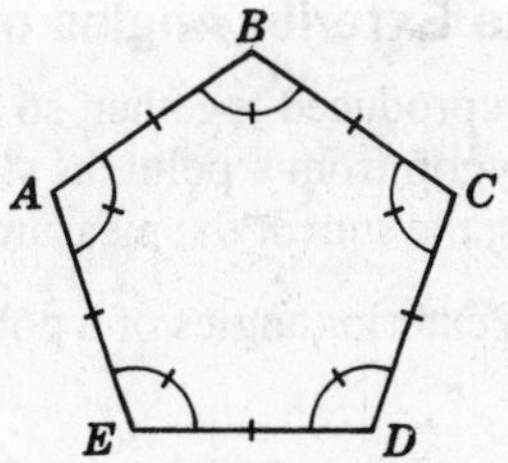

Regular Pentagon

Fig. 4-50

A *regular polygon* is an equilateral and equiangular polygon. Thus, a regular pentagon is a polygon having 5 congruent angles and 5 congruent sides (Fig. 4-50). A square is a regular polygon of 4 sides.

Names of Polygons According to the Number of Sides

Number of Sides	Polygon	Number of Sides	Polygon
3	Triangle	8	Octagon
4	Quadrilateral	9	Nonagon
5	Pentagon	10	Decagon
6	Hexagon	12	Dodecagon
7	Heptagon	n	n-gon

4.4A Sum of the Measures of the Interior Angles of a Polygon

By drawing diagonals from any vertex to each of the other vertices, as in Fig. 4-51, a polygon of 7 sides is divisible into 5 triangles. Note that each triangle has one side of the polygon, except the first and last triangles which have two such sides.

In general, this process will divide a polygon of n sides into $n - 2$ triangles; that is, the number of such triangles is always two less than the number of sides of the polygon.

The sum of the measures of the interior angles of the polygon equals the sum of the measures of the interior angles of the triangles. Hence:

$$\text{Sum of measures of interior angles of a polygon of } n \text{ sides} = (n - 2)180°$$

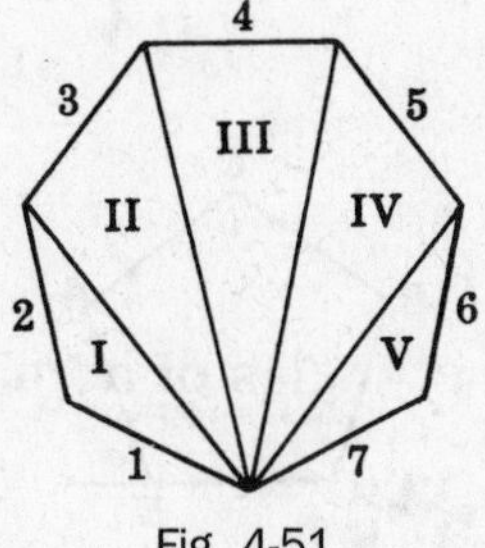

Fig. 4-51

4.4B Sum of the Measures of the Exterior Angles of a Polygon

The exterior angles of a polygon can be reproduced together, so that they have the same vertex. To do this, draw lines parallel to the sides of the polygon from a point, as shown in Fig. 4-52. If this is done, it can be seen that regardless of the number of sides, the sum of the measures of the exterior angle equals 360°. Hence:

Sum of measures of exterior angles of a polygon of n sides $= 360°$

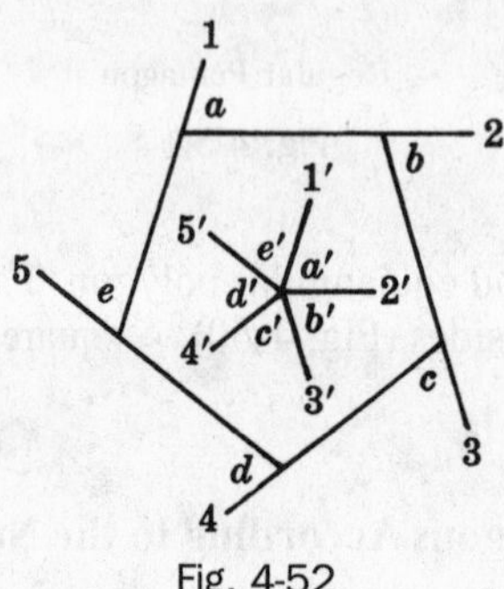

Fig. 4-52

4.4C Polygon-Angle Principles

For any polygon

PRINCIPLE 1: *If S is the sum of the measures of the interior angles of a polygon of n sides, then*

$$S = n - 2 \text{ straight angles} = (n - 2)180°$$

The sum of the measures of the interior angles of a polygon of 10 sides (decagon) equals 1440°, since $S = 8(180) = 1440$.

PRINCIPLE 2: *The sum of the measures of the exterior angles of any polygon equals* 360°.

Thus, the sum of the measures of the exterior angles of a polygon of 23 sides equals 360°.

For a regular polygon

PRINCIPLE 3: *If a regular polygon of n sides (Fig. 4-53) has an interior angle of measure i and an exterior angle of measure e (in degrees), then*

$$i = \frac{180(n-2)}{n}, \quad e = \frac{360}{n}, \text{ and } i + e = 180$$

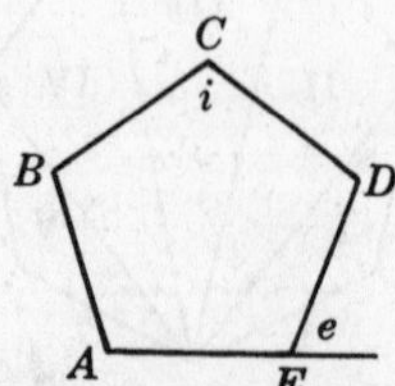

Regular Polygon

Fig. 4-53

Thus for a regular polygon of 20 sides,

$$i = \frac{180(20-2)}{20} = 162 \qquad e = \frac{360}{20} = 18 \qquad i + e = 162 + 18 = 180$$

SOLVED PROBLEMS

4.15 Applying angle-measure formulas to a polygon

(a) Find the sum of the measures of the interior angles of a polygon of 9 sides (express your answer in straight angles and in degrees).

(b) Find the number of sides a polygon has if the sum of the measures of the interior angles is 3600°.

(c) Is it possible to have a polygon the sum of whose angle measures is 1890°?

Solutions

(a) S (in straight angles) $= n - 2 = 9 - 2 = 7$ straight angles; $m\angle S = (n-2)180 = 7(180) = 1260°$.

(b) S (in degrees) $= (n-2)180$. Then $3600 = (n-2)180$, from which $n = 22$.

(c) Since $1890 = (n-2)180$, then $n = 12\frac{1}{2}$. A polygon cannot have $12\frac{1}{2}$ sides.

4.16 Applying angle-measure formulas to a regular polygon

(a) Find each exterior angle measure of a regular polygon having 9 sides.

(b) Find each interior angle measure of a regular polygon having 9 sides.

(c) Find the number of sides a regular polygon has if each exterior angle measure is 5°.

(d) Find the number of sides a regular polygon has if each interior angle measure is 165°.

Solutions

(a) Since $n = 9$, $m\angle e = \frac{360}{n} = \frac{360}{9} = 40$. *Ans.* 40°

(b) Since $n = 9$, $m\angle i = \frac{(n-2)180}{n} = \frac{(9-2)180}{9} = 140$. *Ans.* 140°

Another method: Since $i + e = 180$, $i = 180 - e = 180 - 40 = 140$.

(c) Substituting $e = 5$ in $e = \frac{360}{n}$, we have $5 = \frac{360}{n}$. Then $5n = 360$, so $n = 72$. *Ans.* 72 sides

(d) Substituting $i = 165$ in $i + e = 180$, we have $165 + e = 180$ or $e = 15$.

Then, using $e = \frac{360}{n}$ with $e = 15$, we have $15 = \frac{360}{n}$, or $n = 24$. *Ans.* 24 sides

4.17 Applying algebra to the angle-measure sums of a polygon

Find each interior angle measure of a quadrilateral (a) if its interior angles are represented by $x + 10$, $2x + 20$, $3x - 50$, and $2x - 20$; (b) if its exterior angles are in the ratio 2:3:4:6.

Solutions

(a) Since the sum of the measures of the interior $\angle$s is 360°, we add

$$(x + 10) + (2x + 20) + (3x - 50) + (2x - 20) = 360$$
$$8x - 40 = 360$$
$$x = 50$$

Then $x + 10 = 60$; $2x + 20 = 120$; $3x - 50 = 100$; $2x - 20 = 80$. *Ans.* 60°, 120°, 100°, 80°

(b) Let the exterior angles be represented respectively by $2x$, $3x$, $4x$, and $6x$. Then $2x + 3x + 4x + 6x = 360$. Solving gives us $15x = 360$ and $x = 24$. Hence, the exterior angles measure 48°, 72°, 96°, and 144°. The interior angles are their supplements. *Ans.* 132°, 108°, 84°, 36°

4.5 Two New Congruency Theorems

Three methods of proving triangles congruent have already been introduced here. These are

1. Side-Angle-Side (SAS)
2. Angle-Side-Angle (ASA)
3. Side-Side-Side (SSS)

Two additional methods of proving that triangles are congruent are

4. Side-Angle-Angle (SAA)
5. Hypotenuse-Leg (hy-leg)

4.5A Two New Congruency Principles

PRINCIPLE 1: (SAA) *If two angles and a side opposite one of them of one triangle are congruent to the corresponding parts of another, the triangles are congruent.*

Thus if $\angle A \cong \angle A'$, $\angle B \cong \angle B'$, and $\overline{BC} \cong \overline{B'C'}$ in Fig. 4-54, then $\triangle ABC \cong \triangle A'B'C'$.

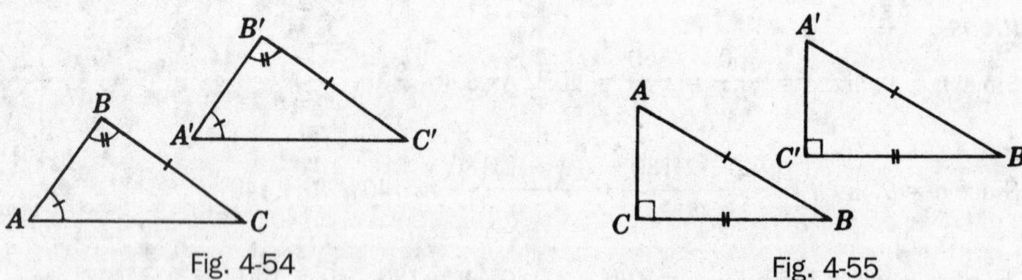

Fig. 4-54 Fig. 4-55

PRINCIPLE 2: (hy-leg) *If the hypotenuse and a leg of one right triangle are congruent to the corresponding parts of another right triangle, the triangles are congruent.*

Thus if hy$\overline{AB} \cong$ hy$\overline{A'B'}$ and leg $\overline{BC} \cong$ leg$\overline{B'C'}$ in Fig. 4-55, then rt. $\triangle ABC \cong$ *rt.* $\triangle A'B'C'$.

A proof of this principle is given in Chapter 16.

SOLVED PROBLEMS

4.18 Selecting congruent triangles using hy-leg or SAA

In (a) Fig. 4-56 and (b) Fig. 4-57, select congruent triangles and state the reason for the congruency.

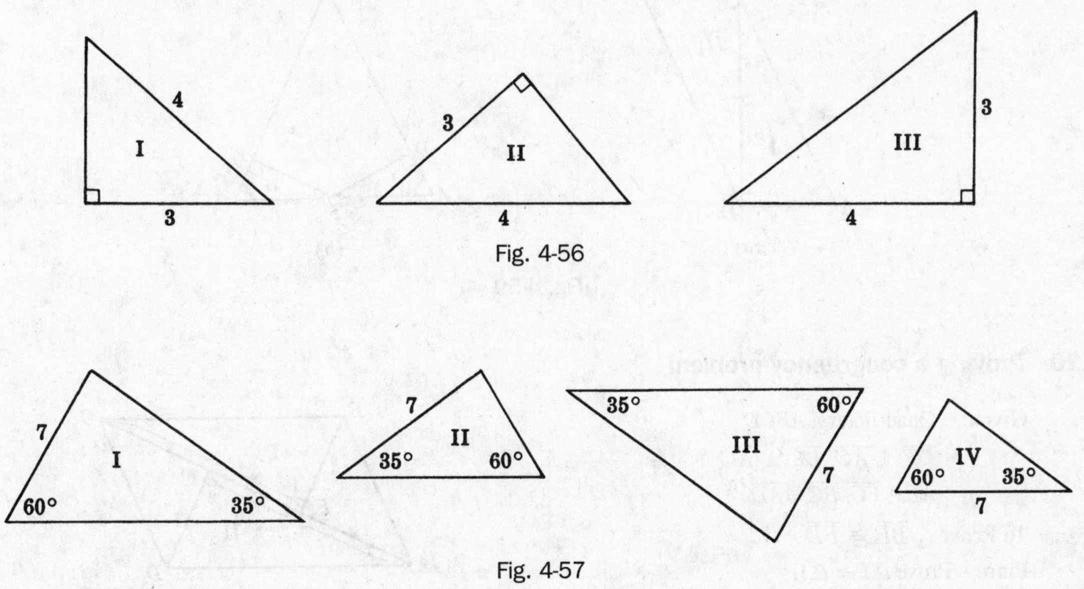

Fig. 4-56

Fig. 4-57

Solutions

(a) $\triangle I \cong \triangle II$ by hy-leg. In $\triangle III$, 4 is not a hypotenuse.

(b) $\triangle I \cong \triangle III$ by SAA. In $\triangle II$, 7 is opposite 60° instead of 35°. In $\triangle IV$, 7 is included between 60° and 35°.

4.19 Determining the reason for congruency of triangles

In each part of Fig. 4-58, $\triangle I$ can be proved congruent to $\triangle II$. Make a diagram showing the congruent parts of both triangles and state the reason for the congruency.

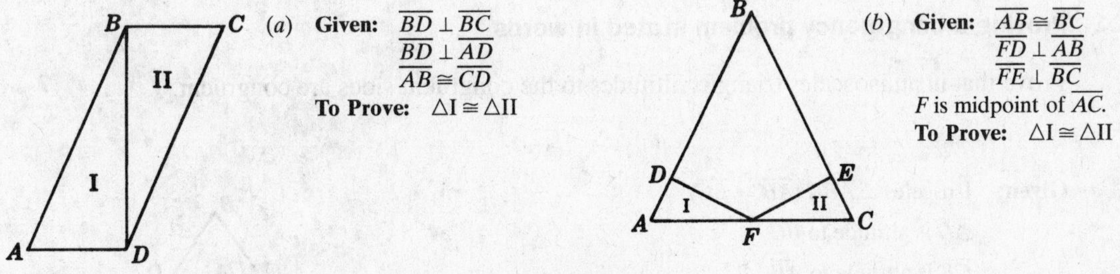

Fig. 4-58

Solutions

(a) See Fig. 4-59(a). $\triangle I \cong \triangle II$ by hy-leg.

(b) See Fig. 4-59(b). $\triangle I \cong \triangle II$ by SAA.

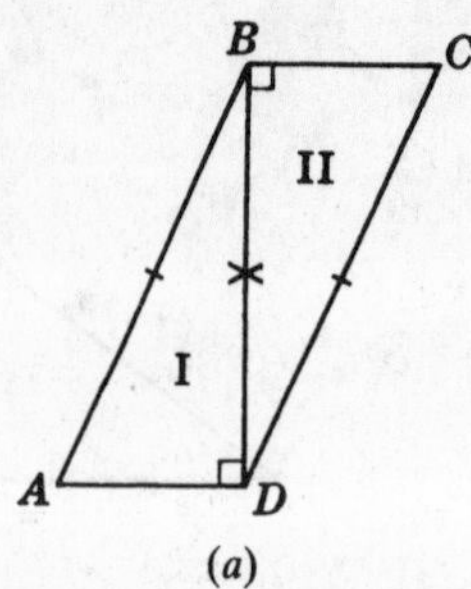

(*a*) (*b*)

Fig. 4-59

4.20 Proving a congruency problem

Given: Quadrilateral *ABCD*

$\overline{DF} \perp \overline{AC}$, $\overline{BE} \perp \overline{AC}$

$\overline{AE} \cong \overline{FC}$, $\overline{BC} \cong \overline{AD}$

To Prove: $\overline{BE} \cong \overline{FD}$

Plan: Prove $\triangle I \cong \triangle II$

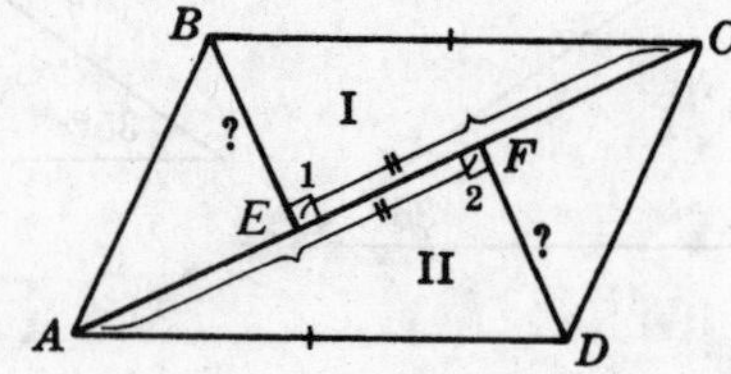

PROOF:

Statements	Reasons
1. $\overline{BC} \cong \overline{AD}$	1. Given
2. $\overline{DF} \perp \overline{AC}$, $\overline{BE} \perp \overline{AC}$	2. Given
3. $\angle 1 \cong \angle 2$	3. Perpendiculars form rt. ∠s, and all rt. ∠s are congruent.
4. $\overline{AE} \cong \overline{FC}$	4. Given
5. $\overline{EF} \cong \overline{EF}$	5. Identity
6. $\overline{AF} \cong \overline{EC}$	6. If equals are added to equals, the sums are equal. Definition of congruent segments.
7. $\triangle I \cong \triangle II$	7. Hy-leg
8. $\overline{BE} \cong \overline{FD}$	8. Corresponding parts of congruent △s are congruent.

4.21 Proving a congruency problem stated in words

Prove that in an isosceles triangle, altitudes to the congruent sides are congruent.

Given: Isosceles $\triangle ABC$ ($\overline{AB} \cong \overline{BC}$)

$\overline{AD}$ is altitude to $\overline{BC}$

$\overline{CE}$ is altitude to $\overline{AB}$

To Prove: $\overline{AD} \cong \overline{CE}$

Plan: Prove $\triangle ACE \cong \triangle CAD$

or $\triangle I \cong \triangle II$

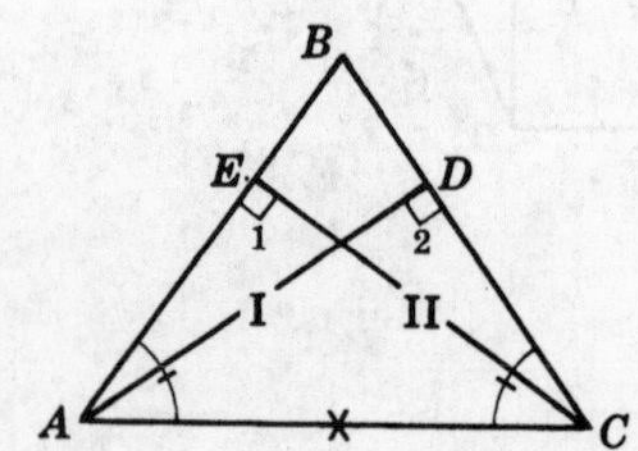

PROOF:

Statements	Reasons
1. $\overline{AB} \cong \overline{BC}$	1. Given
2. $\angle A \cong \angle C$	2. In a $\triangle$, angles opposite equal sides are equal.
3. $\overline{AD}$ is altitude to $\overline{BC}$, $\overline{CE}$ is altitude to $\overline{AB}$.	3. Given
4. $\angle 1 \cong \angle 2$	4. Altitudes form rt. $\triangle$s and rt. $\triangle$s are congruent.
5. $\overline{AC} \cong \overline{AC}$	5. Identity
6. $\triangle I \cong \triangle II$	6. SAA
7. $\overline{AD} \cong \overline{CE}$	7. Corresponding parts of congruent $\triangle$s are congruent.

SUPPLEMENTARY PROBLEMS

4.1. In each part of Fig. 4-60, find x and y. (4.1)

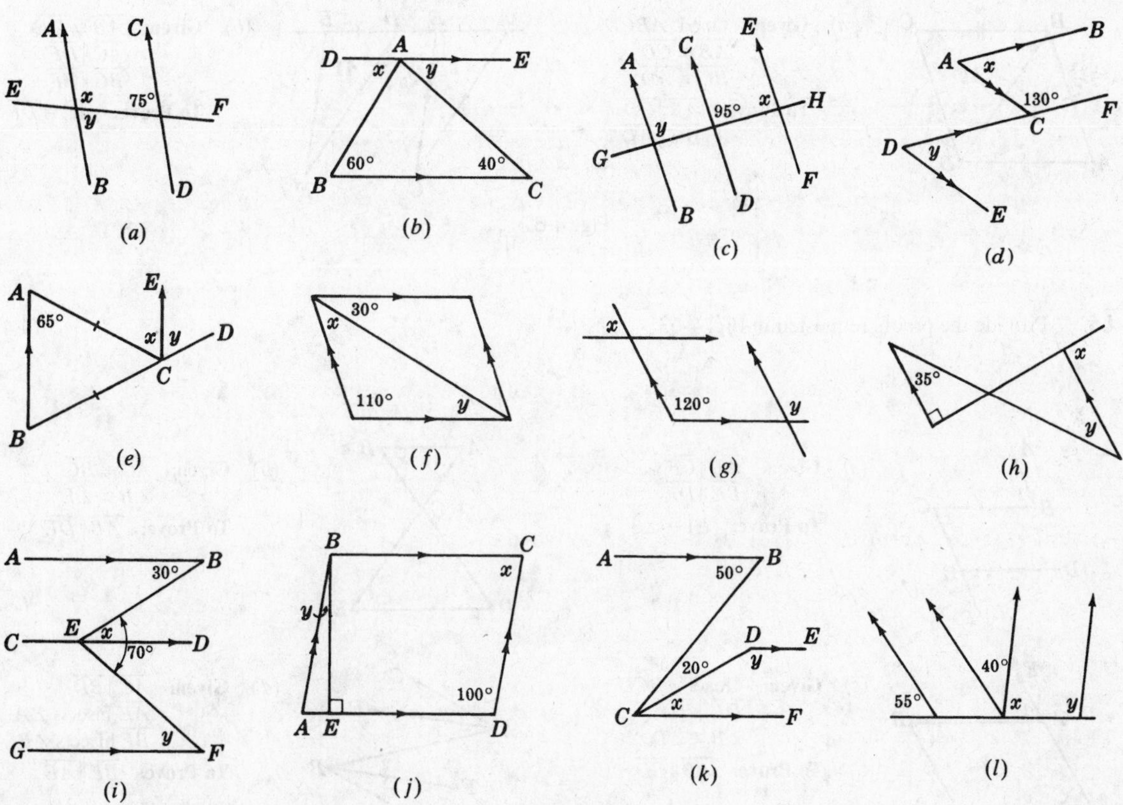

Fig. 4-60

4.2. In each part of Fig. 4-61, find x and y. (4.3)

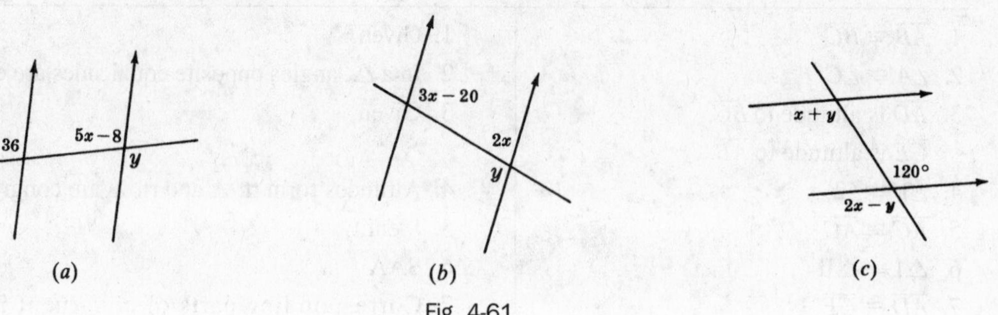

Fig. 4-61

4.3. If two parallel lines are cut by a transversal, find (4.3)

(a) Two alternate interior angles represented by $3x$ and $5x - 70$

(b) Two corresponding angles represented by $2x + 10$ and $4x - 50$

(c) Two interior angles on the same side of the transversal represented by $2x$ and $3x$

4.4. Provide the proofs requested in Fig. 4-62 (4.4)

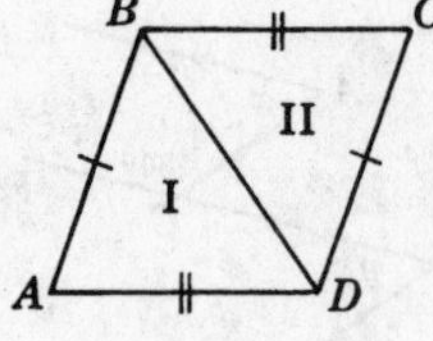

(*a*) **Given:** Quad. $ABCD$
$\overline{AB} \cong \overline{CD}$
$\overline{BC} \cong \overline{AD}$
To Prove: $\overline{AB} \parallel \overline{CD}$
$\overline{BC} \parallel \overline{AD}$

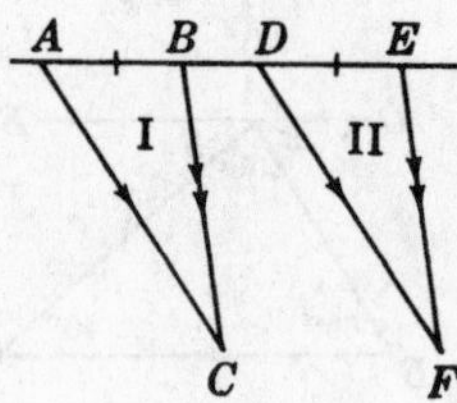

(*b*) **Given:** $\overline{AB} \cong \overline{DE}$
$\overline{AC} \parallel \overline{DF}$
$\overline{BC} \parallel \overline{EF}$
To Prove: $\overline{AC} \cong \overline{DF}$

Fig. 4-62

4.5. Provide the proofs requested in Fig. 4-63 (4.4)

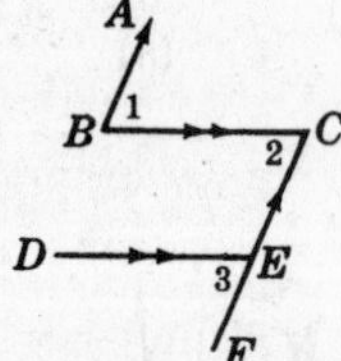

(*a*) **Given:** $\overline{AB} \parallel \overline{CF}$
$\overline{BC} \parallel \overline{DE}$
To Prove: $\angle 1 \cong \angle 3$

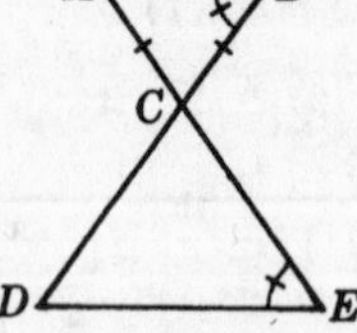

(*b*) **Given:** $\overline{AC} \cong \overline{BC}$
$\angle B \cong \angle E$
To Prove: $\overline{AB} \parallel \overline{DE}$

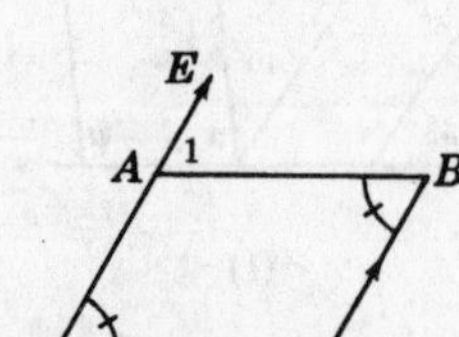

(*c*) **Given:** Quad. $ABCD$
$\overline{DE} \parallel \overline{BC}$
$\angle B \cong \angle D$
To Prove: $\overline{AB} \parallel \overline{CD}$

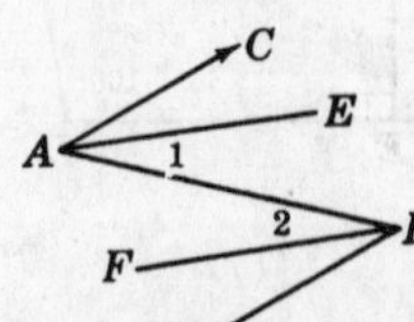

(*d*) **Given:** $\overline{AC} \parallel \overline{BD}$
$\overline{AE}$ bisects $\angle A$
$\overline{BF}$ bisects $\angle B$
To Prove: $\overline{BF} \parallel \overline{AE}$

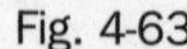
Fig. 4-63

4.6. Prove each of the following: (4.5)

(a) If the opposite sides of a quadrilateral are parallel, then they are also congruent.

(b) If $\overline{AB}$ and $\overline{CD}$ bisect each other at E, then $\overline{AC} \parallel \overline{BD}$.

(c) In quadrilateral $ABCD$, let $\overline{BC} \parallel \overline{AD}$. If the diagonals $\overline{AC}$ and $\overline{BD}$ intersect at E and $\overline{AE} \parallel \overline{DE}$, then $\overline{BE} \cong \overline{CE}$.

(d) $\overleftrightarrow{AB}$ and $\overleftrightarrow{CD}$ are parallel lines cut by a transversal at E and F. If $\overrightarrow{EG}$ and $\overrightarrow{FH}$ bisect a pair of corresponding angles, then $\overrightarrow{EG} \parallel \overrightarrow{FH}$.

(e) If a line through vertex B of $\triangle ABC$ *is parallel to* $\overline{AC}$ *and bisects the angle formed by extending* $\overline{AB}$ *through* B, *then* $\triangle ABC$ is isosceles.

4.7. In Fig. 4-64, find the distance from (a) A to B; (b) E to $\overline{AC}$; (c) A to $\overline{BC}$; (d) $\overline{ED}$ to $\overline{BC}$. (4.6)

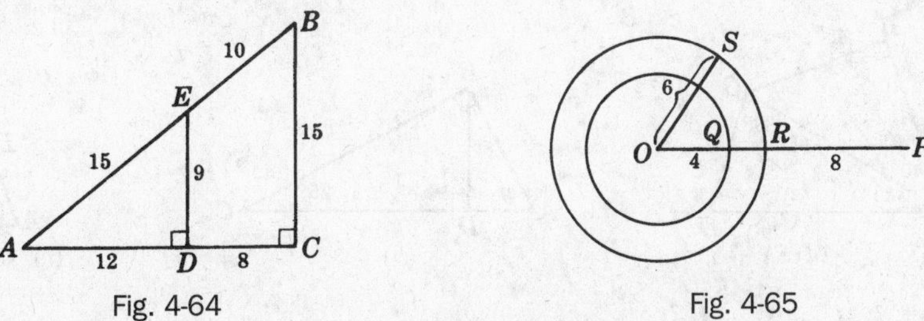

Fig. 4-64

Fig. 4-65

4.8. In Fig. 4-65, find the distance (a) from P to the outer circle; (b) from P to the inner circle; (c) between the concentric circles; (d) from P to O. (4.6)

4.9. In Fig. 4-66 (4.6)

(a) Locate P, a point on $\overline{AD}$, equidistant from B and C. Then locate Q, a point on $\overline{AD}$, equidistant from $\overline{AB}$ and $\overline{BC}$.

(b) Locate R, a point equidistant from A, B, and C. Then locate S, a point equidistant from B, C, and D.

(c) Locate T, a point equidistant from $\overline{BC}$, $\overline{CD}$, and $\overline{AD}$. Then locate U, a point equidistant from $\overline{AB}$, $\overline{BC}$, and $\overline{CD}$.

4.10. In each part of Fig. 4-67, describe P, Q, and R as equidistant points and locate them on a bisector.

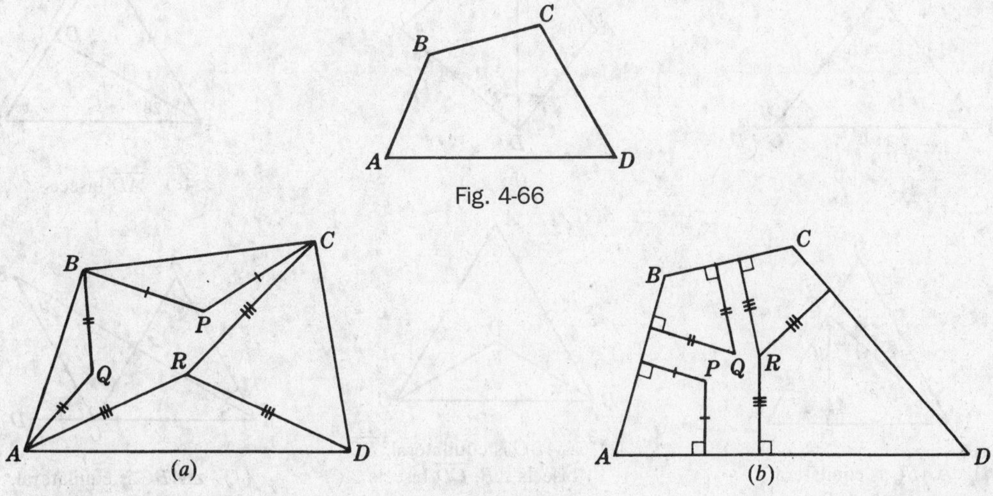

Fig. 4-66

Fig. 4-67

4.11. In each part of Fig. 4-68, describe P, Q, and R as equidistant points. (4.9)

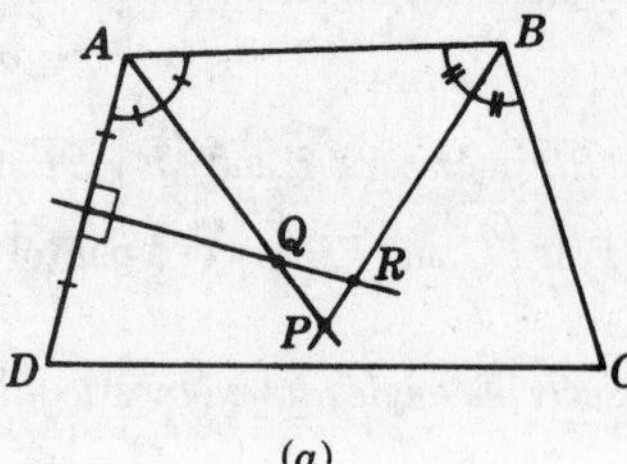

(*a*)

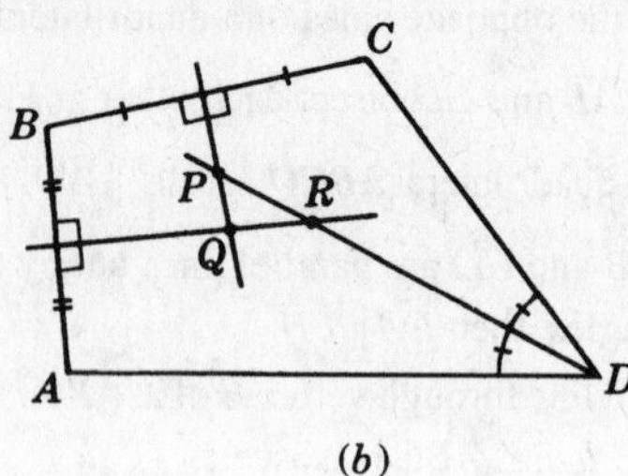

(*b*)

Fig. 4-68

4.12. Find x and y in each part of Fig. 4-69. (4.11)

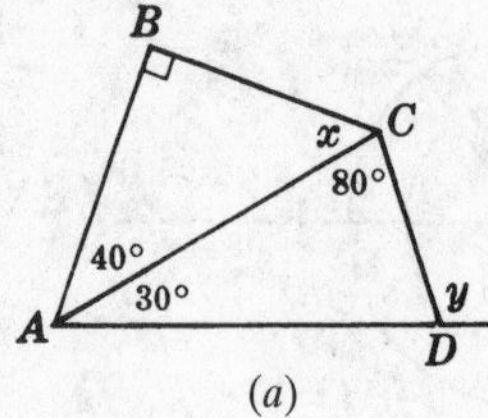

(*a*)

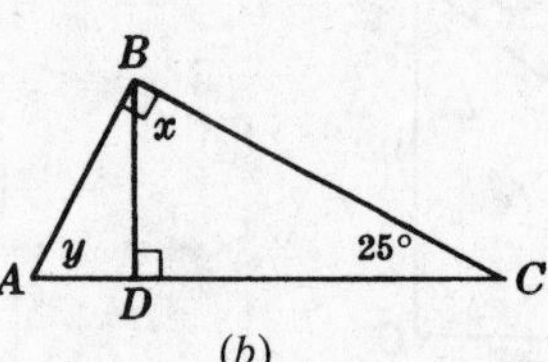

(*b*)

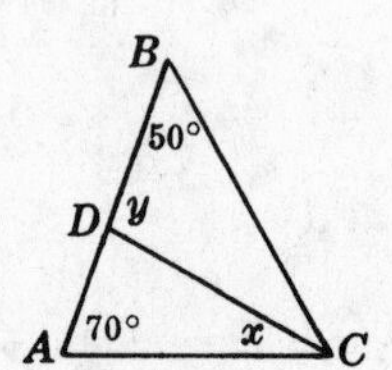

(*c*) $\overline{CD}$ bisects $\angle C$.

(*d*) $\overline{BD}$ bisects $\angle B$; $\overline{CD}$ bisects $\angle C$.

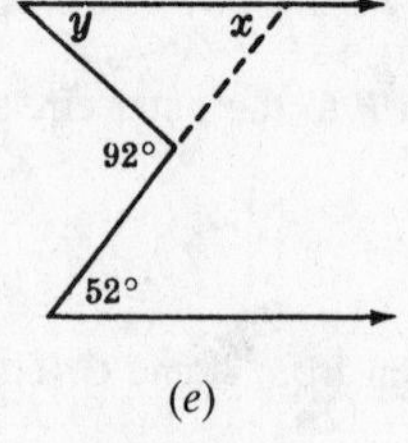

(*e*)

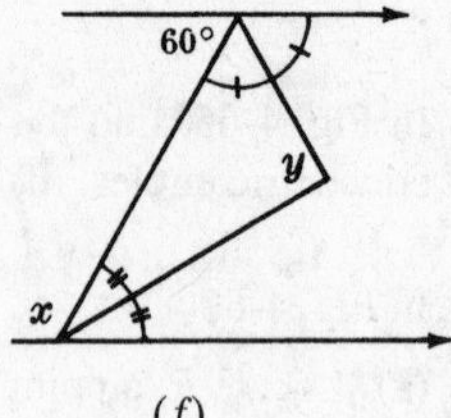

(*f*)

Fig. 4-69

4.13. Find x and y in each part of Fig. 4-70. (4.12)

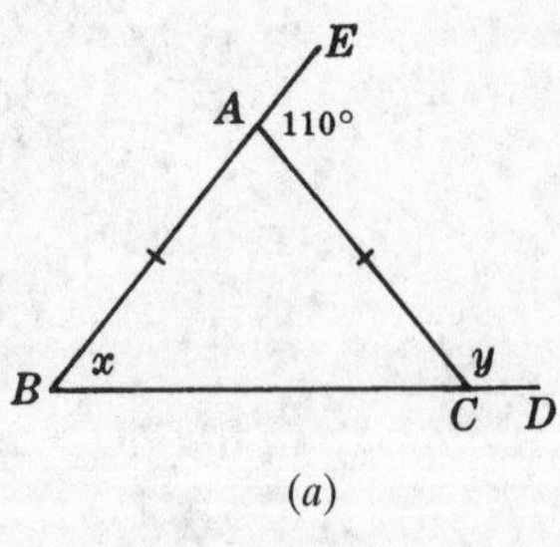

(*a*)

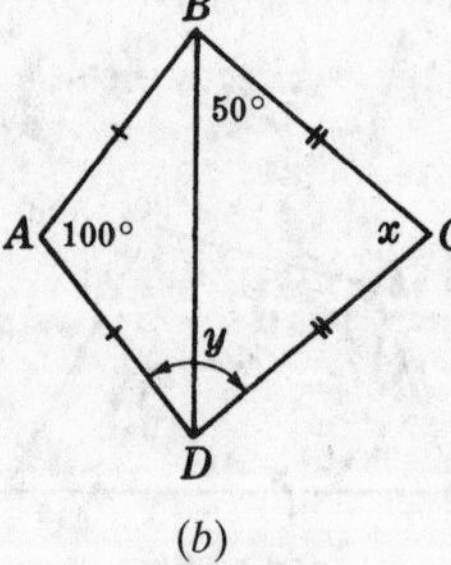

(*b*)

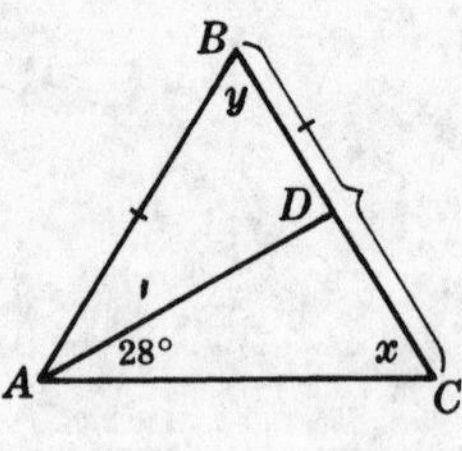

(*c*) $\overline{AD}$ bisects $\angle A$.

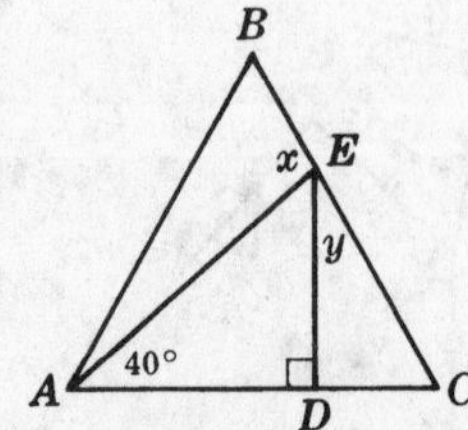

(*d*) $\triangle ABC$ is equilateral.

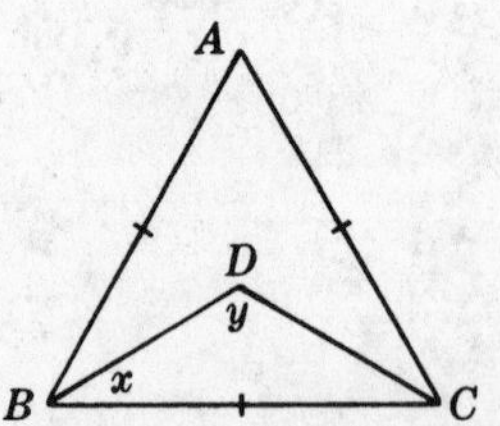

(*e*) $\triangle ABC$ is equilateral. $\overline{BD}$ bisects $\angle B$, $\overline{CD}$ bisects $\angle C$.

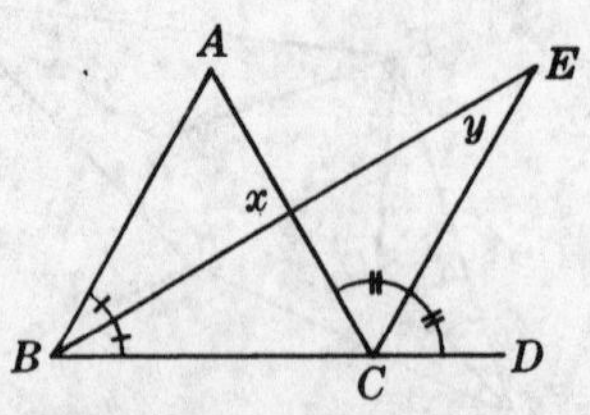

(*f*) $\triangle ABC$ is equilateral.

Fig. 4-70

4.14. Find the measure of each angle (4.13)

(a) Of a triangle if its angle measures are in the ratio 1:3:6

(b) Of a right triangle if its acute angle measures are in the ratio 4:5

(c) Of an isosceles triangle if the ratio of the measures of its base angle to a vertex angle is 1:3

(d) Of a quadrilateral if its angle measures are in the ratio 1:2:3:4

(e) Of a triangle, one of whose angles measures 55° and whose other two angle measures are in the ratio 2:3

(f) Of a triangle if the ratio of the measures of its exterior angles is 2:3:4

4.15. Prove each of the following: (4.14)

(a) In quadrilateral $ABCD$ if $\angle A \cong \angle D$ and $\angle B \cong \angle C$, then $\overline{BC} \parallel \overline{AD}$.

(b) Two parallel lines are cut by a transversal. Prove that the bisectors of two interior angles on the same side of the transversal are perpendicular to each other.

4.16. Show that a triangle is (4.14)

(a) Equilateral if its angles are represented by $x + 15$, $3x - 75$, and $2x - 30$

(b) Isosceles if its angles are represented by $x + 15$, $3x - 35$, and $4x$

(c) A right triangle if its angle measures are in the ratio 2:3:5

(d) An obtuse triangle if one angle measures 64° and the larger of the other two measures 10° less than five times the measure of the smaller

4.17. (a) Find the sum of the measures of the interior angles (in straight angles) of a polygon of 9 sides; of 32 sides. (4.15)

(b) Find the sum of the measures of the interior angles (in degrees) of a polygon of 11 sides; of 32 sides; of 1002 sides.

(c) Find the number of sides a polygon has if the sum of the measures of the interior angles is 28 straight angles; 20 right angles; 4500°; 36,000°.

4.18. (a) Find the measure of each exterior angle of a regular polygon having 18 sides; 20 sides; 40 sides. (4.16)

(b) Find the measure of each interior angle of a regular polygon having 18 sides; 20 sides; 40 sides.

(c) Find the number of sides a regular polygon has if each exterior angle measures 120°; 40°; 18°; 2°.

(d) Find the number of sides a regular polygon has if each interior angle measures 60°, 150°; 170°, 175°; 179°.

4.19. (a) Find each interior angle of a quadrilateral if its interior angles are represented by $x - 5$, $x + 20$, $2x - 45$, and $2x - 30$. (4.17)

(b) Find the measure of each interior angle of a quadrilateral if the measures of its exterior angles are in the ratio of 1:2:3:3.

4.20. In each part of Fig. 4-71, select congruent triangles and state the reason for the congruency. (4.18)

4.21. In each part of Fig. 4-72, two triangles can be proved congruent. Make a diagram showing the congruent parts of both triangles, and state the reason for the congruency. (4.19)

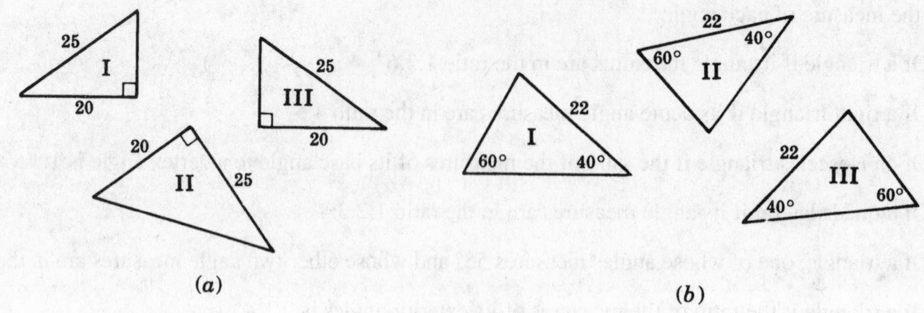

Fig. 4-71

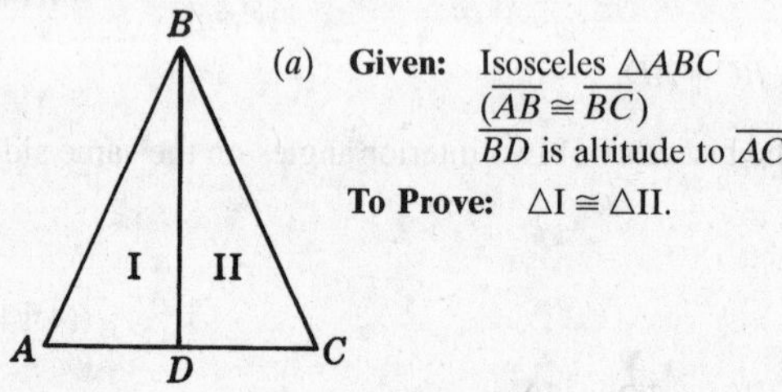

(*a*) **Given:** Isosceles $\triangle ABC$ ($\overline{AB} \cong \overline{BC}$)
$\overline{BD}$ is altitude to $\overline{AC}$

To Prove: $\triangle I \cong \triangle II$.

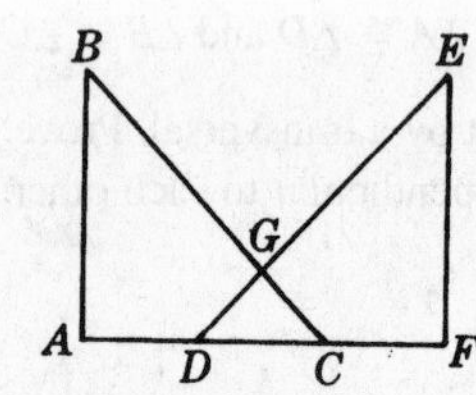

(*b*) **Given:** $\overline{AB} \cong \overline{EF}$
$\overline{AB} \perp \overline{AF}$, $\overline{EF} \perp \overline{AF}$
$\overline{DG} \cong \overline{GC}$

To Prove: $\triangle ABC \cong \triangle DEF$

Fig. 4-72

4.22. Provide the proofs requested in Fig. 4-73. (4.20)

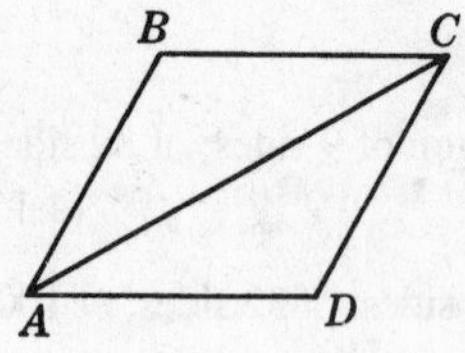

(*a*) **Given:** $\angle B \cong \angle D$
$\overline{BC} \parallel \overline{AD}$

To Prove: $\overline{BC} \cong \overline{AD}$

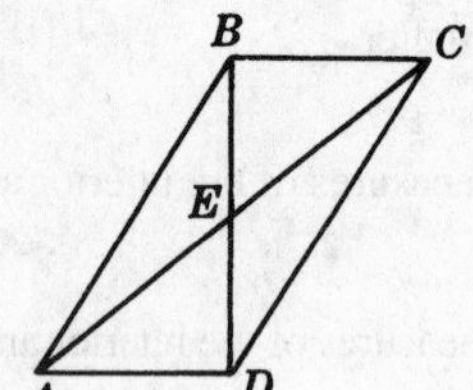

(*b*) **Given:** $\overline{AE} \cong \overline{EC}$
$\overline{BD} \perp \overline{BC}$
$\overline{BD} \perp \overline{AD}$

To Prove: $\overline{BE} \cong \overline{ED}$

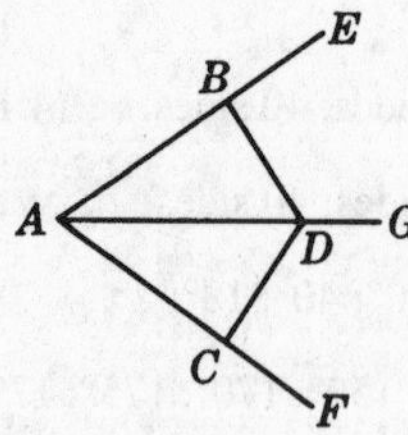

(*c*) **Given:** $\overline{BD} \cong \overline{DC}$
$\overline{BD} \perp \overline{AE}$
$\overline{DC} \perp \overline{AF}$

To Prove: $\overline{AG}$ bisects $\angle A$.

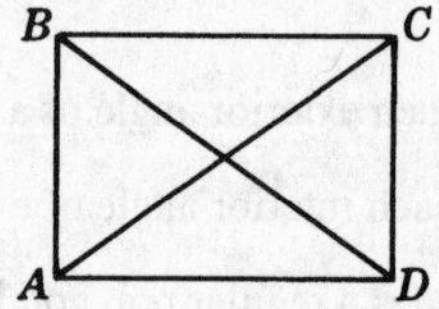

(*d*) **Given:** $\overline{AC} \cong \overline{BD}$
$\overline{AB} \perp \overline{AD}$
$\overline{CD} \perp \overline{AD}$

To Prove: $\overline{AB} \cong \overline{CD}$

Fig. 4-73

4.23. Prove each of the following: (4.21)

(a) If the perpendiculars to two sides of a triangle from the midpoint of the third side are congruent, then the triangle is isosceles.

(b) Perpendiculars from a point in the bisector of an angle to the sides of the angle are congruent.

(c) If the altitudes to two sides of a triangle are congruent, then the triangle is isosceles.

(d) Two right triangles are congruent if the hypotenuse and an acute angle of one are congruent to the corresponding parts of the other.

Parallelograms, Trapezoids, Medians, and Midpoints

5.1 Trapezoids

A *trapezoid* is a quadrilateral having two, and only two, parallel sides. The *bases* of the trapezoid are its parallel sides; the *legs* are its nonparallel sides. The *median* of the trapezoid is the segment joining the midpoints of its legs.

Thus in trapezoid *ABCD* in Fig. 5-1, the bases are $\overline{AD}$ and $\overline{BC}$, and the legs are $\overline{AB}$ and $\overline{CD}$. If *M* and *N* are midpoints, then $\overline{MN}$ is the median of the trapezoid.

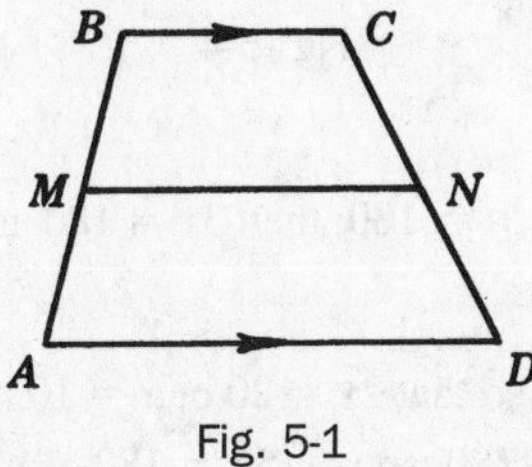

Fig. 5-1

An *isosceles trapezoid* is a trapezoid whose legs are congruent. Thus in isosceles trapezoid *ABCD* in Fig. 5-2 $\overline{AB} \cong \overline{CD}$.

The *base angles* of a trapezoid are the angles at the ends of its longer base: $\angle A$ *and* $\angle D$ are the base angles of isosceles trapezoid *ABCD*.

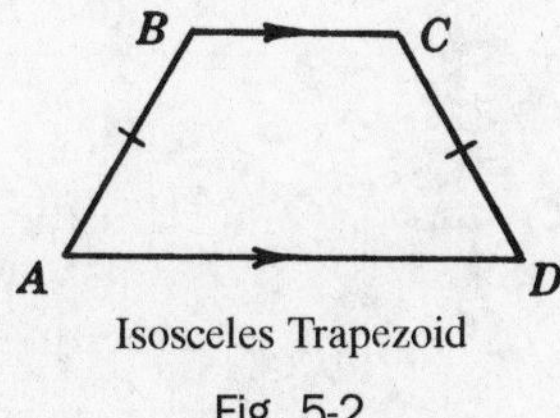

Isosceles Trapezoid

Fig. 5-2

5.1A Trapezoid Principles

PRINCIPLE 1: *The base angles of an isosceles trapezoid are congruent.*

Thus in trapezoid *ABCD* of Fig. 5-3 if $\overline{AB} \cong \overline{CD}$, then $\angle A \cong \angle D$.

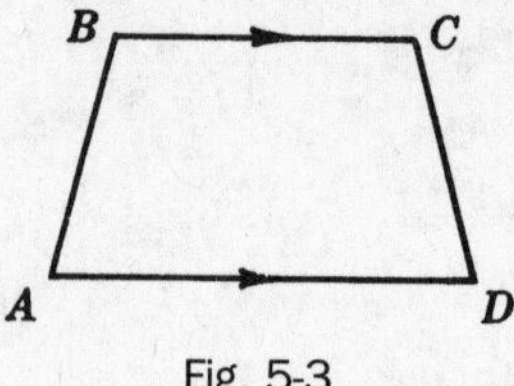

Fig. 5-3

PRINCIPLE 2: *If the base angles of a trapezoid are congruent, the trapezoid is isosceles.*

Thus in Fig. 5-3 if $\angle A \cong \angle D$, then $\overline{AB} \cong \overline{CD}$.

SOLVED PROBLEMS

5.1 Applying algebra to the trapezoid

In each trapezoid in Fig. 5-4, find x and y.

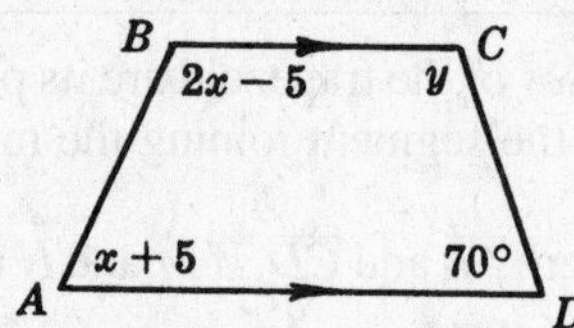

(*a*) *ABCD* is a trapezoid.

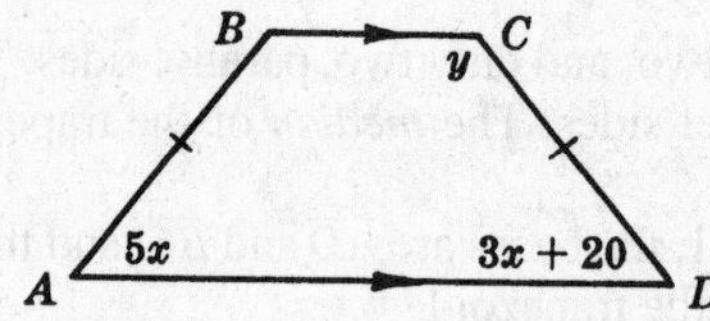

(*b*) *ABCD* is an isosceles trapezoid.

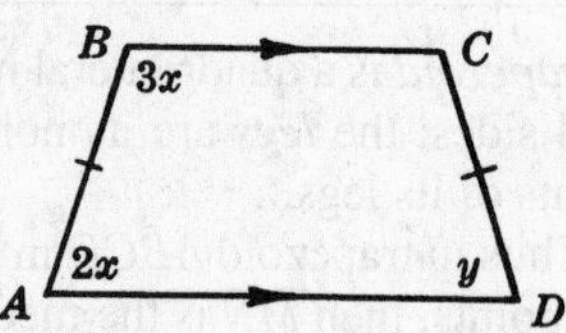

(*c*) *ABCD* is an isosceles trapezoid.

Fig. 5-4

Solutions

(a) Since $\overline{AD} \parallel \overline{BC}$, $(2x - 5) + (x + 5) = 180$; then $3x = 180$ and $x = 60$.
Also, $y + 70 = 180$ or $y = 110$.

(b) Since $\angle A \cong \angle D$, $5x = 3x + 20$, so that $2x = 20$ or $x = 10$.
Since $\overline{BC} \parallel \overline{AD}$, $y + (3x + 20) = 180$, so $y + 50 = 180$ or $y = 130$.

(c) Since $\overline{BC} \parallel \overline{AD}$, $3x + 2x = 180$ or $x = 36$.
Since $\angle D \cong \angle A$, $y = 2x$ or $y = 72$.

5.2 Proof of a trapezoid principle stated in words

Prove that the base angles of an isosceles trapezoid are congruent.

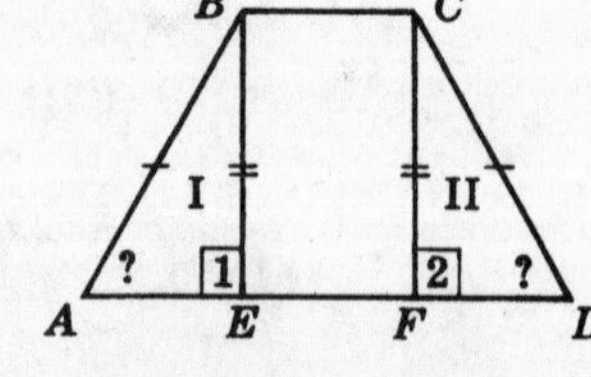

Given: Isosceles trapezoid *ABCD*
($\overline{BC} \parallel \overline{AD}$, $\overline{AB} \cong \overline{CD}$)

To Prove: $\angle A \cong \angle D$

Plan: Draw ⊥s to base from *B* and *C*.
Prove $\triangle$I $\cong$ II.

PROOF:

Statements	Reasons
1. Draw $\overline{BE} \perp \overline{AD}$ and $\overline{CF} \perp \overline{AD}$.	1. A ⊥ may be drawn to a line from an outside point.
2. $\overline{BC} \parallel \overline{AD}$, $\overline{AB} \cong \overline{CD}$	2. Given
3. $\overline{BE} \cong \overline{CF}$	3. Parallel lines are everywhere equidistant. Definition of congruent segments.
4. $\angle 1 \cong \angle 2$	4. Perpendiculars form rt. ∠s. All rt. ∠s are congruent.
5. $\triangle$I $\cong$ $\triangle$II	5. Hy-leg
6. $\angle A \cong \angle D$	6. Corresponding parts of congruent △s are congruent.

5.2 Parallelograms

A parallelogram is a quadrilateral whose opposite sides are parallel. The symbol for parallelogram is ▱. Thus in ▱*ABCD* in Fig. 5-5, $\overline{AB} \parallel \overline{CD}$ in $\overline{AD} \parallel \overline{BC}$.

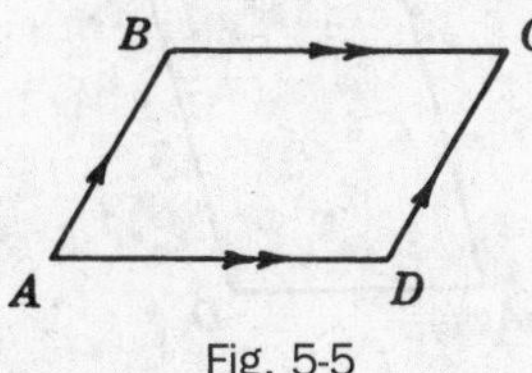

Fig. 5-5

If the opposite sides of a quadrilateral are parallel, then it is a parallelogram. (This is the converse of the above definition.) Thus if $\overline{AB} \parallel \overline{CD}$ and $\overline{AD} \parallel \overline{BC}$, then *ABCD* is a ▱.

5.2A Principles Involving Properties of Parallelograms

PRINCIPLE 1: *The opposite sides of a parallelogram are parallel.* (This is the definition.)

PRINCIPLE 2: *A diagonal of a parallelogram divides it into two congruent triangles.*

$\overline{BD}$ is a diagonal of ▱*ABCD* in Fig. 5-6, so $\triangle\text{I} \cong \triangle\text{II}$.

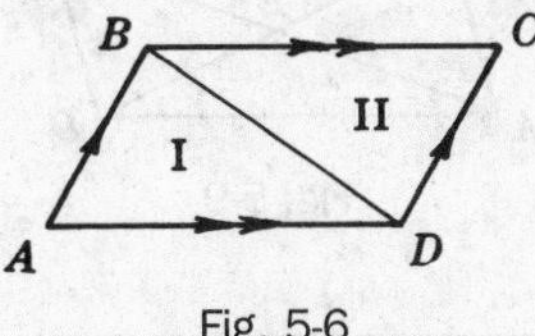

Fig. 5-6

PRINCIPLE 3: *The opposite sides of a parallelogram are congruent.*

Thus in ▱*ABCD* in Fig. 5-5, $\overline{AB} \cong \overline{CD}$ and $\overline{AD} \cong \overline{BC}$.

PRINCIPLE 4: *The opposite angles of a parallelogram are congruent.*

Thus in ▱*ABCD*, $\angle A \cong \angle C$ and $\angle B \cong \angle D$.

PRINCIPLE 5: *The consecutive angles of a parallelogram are supplementary.*

Thus in ▱*ABCD*, $\angle A$ *is the supplement of both* $\angle B$ and $\angle D$.

PRINCIPLE 6: *The diagonals of a parallelogram bisect each other.*

Thus in ▱*ABCD* in Fig. 5-7, $\overline{AE} \cong \overline{EC}$ and $\overline{BE} \cong \overline{ED}$.

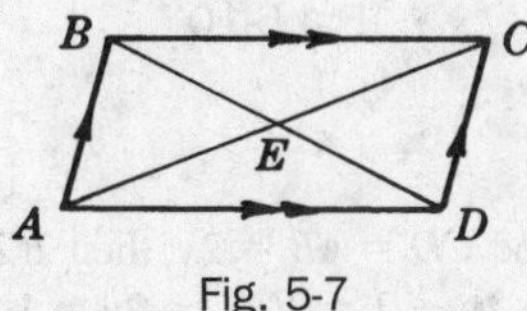

Fig. 5-7

5.2B Proving a Quadrilateral is a Parallelogram

PRINCIPLE 7: *A quadrilateral is a parallelogram if its opposite sides are parallel.*

Thus if $\overline{AB} \parallel \overline{CD}$ and $\overline{AD} \parallel \overline{BC}$, then *ABCD* is a ▱.

PRINCIPLE 8: *A quadrilateral is a parallelogram if its opposite sides are congruent.*

Thus if $\overline{AB} \cong \overline{CD}$ and $\overline{AD} \cong \overline{BC}$ in Fig. 5-8, then *ABCD* is a ▱.

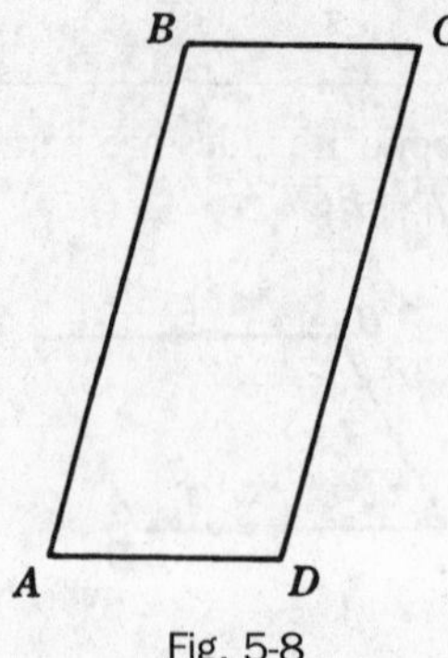

Fig. 5-8

PRINCIPLE 9: *A quadrilateral is a parallelogram if two sides are congruent and parallel.*

Thus if $\overline{BC} \cong \overline{AD}$ and $\overline{BC} \parallel \overline{AD}$ in Fig. 5-8, then $ABCD$ is a ▱.

PRINCIPLE 10: *A quadrilateral is a parallelogram if its opposite angles are congruent.*

Thus if $\angle A \cong \angle C$ and $\angle B \cong \angle D$ in Fig. 5-8, then $ABCD$ is a ▱.

PRINCIPLE 11: *A quadrilateral is a parallelogram if its diagonals bisect each other.*

Thus if $\overline{AE} \cong \overline{EC}$ and $\overline{BE} \cong \overline{ED}$ in Fig. 5-9, then $ABCD$ is a ▱.

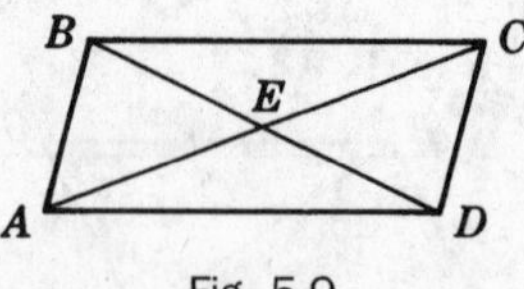

Fig. 5-9

SOLVED PROBLEMS

5.3 Applying properties of parallelograms

Assuming $ABCD$ is a parallelogram, find x and y in each part of Fig. 5-10.

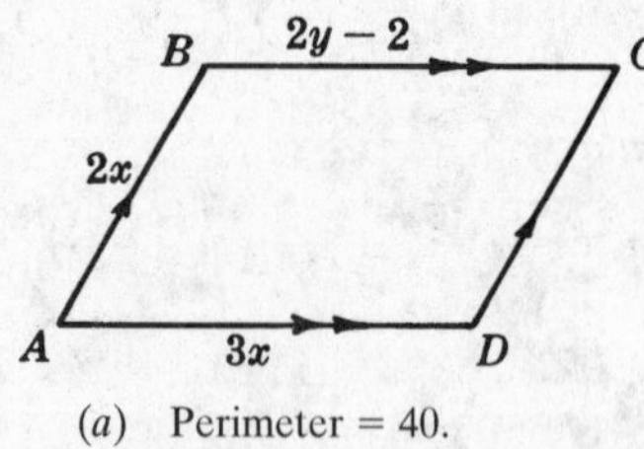

(*a*) Perimeter = 40.

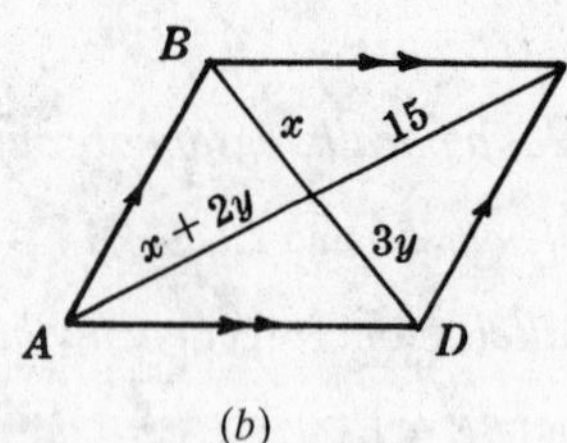

(*b*)

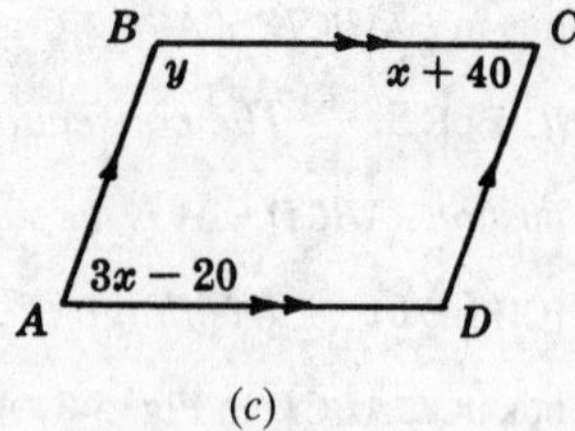

(*c*)

Fig. 5-10

Solutions

(a) By Principle 3, $BC = AD = 3x$ and $CD = AB = 2x$; then $2(2x + 3x) = 40$, so that $10x = 40$ or $x = 4$. By Principle 3, $2y - 2 = 3x$; then $2y - 2 = 3(4)$, so $2y = 14$ or $y = 7$.

(b) By Principle 6, $x + 2y = 15$ and $x = 3y$. Substituting $3y$ for x in the first equation yields $3y + 2y = 15$ or $y = 3$. Then $x = 3y = 9$.

(c) By Principle 4, $3x - 20 = x + 40$, so $2x = 60$ for $x = 30$. By Principle 5, $y + (x + 40) = 180$. Then $y + (30 + 40) = 180$ or $y = 110$.

5.4 Applying principle 7 to determine parallelograms

Name the parallelograms in each part of Fig. 5-11.

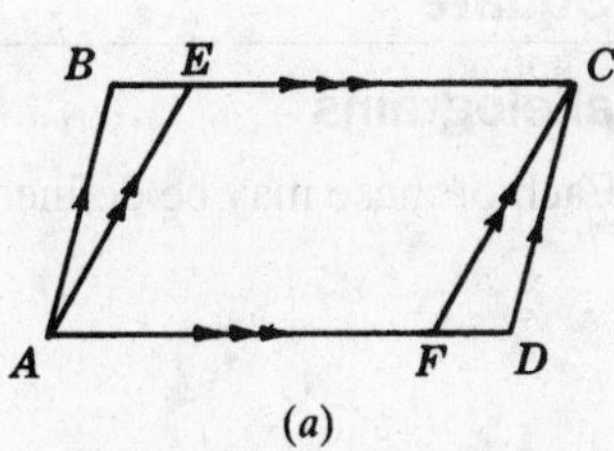

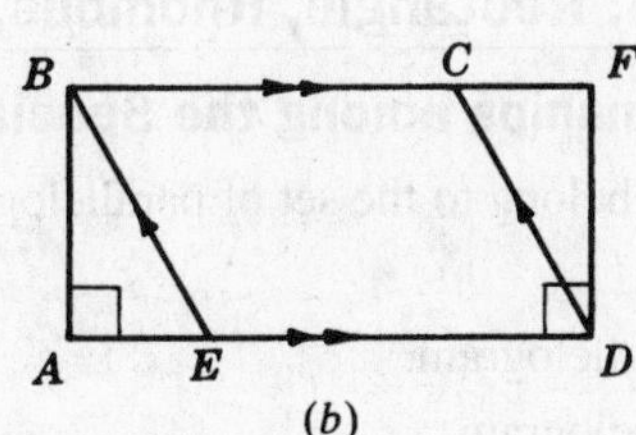

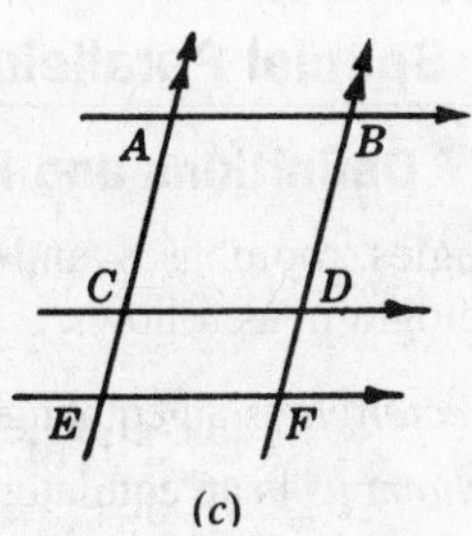

Fig. 5-11

Solutions

(a) *ABCD*, *AECF*; (b) *ABFD*, *BCDE*; (c) *ABDC*, *CDFE*, *ABFE*.

5.5 Applying principles 9, 10, and 11

State why *ABCD* is a parallelogram in each part of Fig. 5-12.

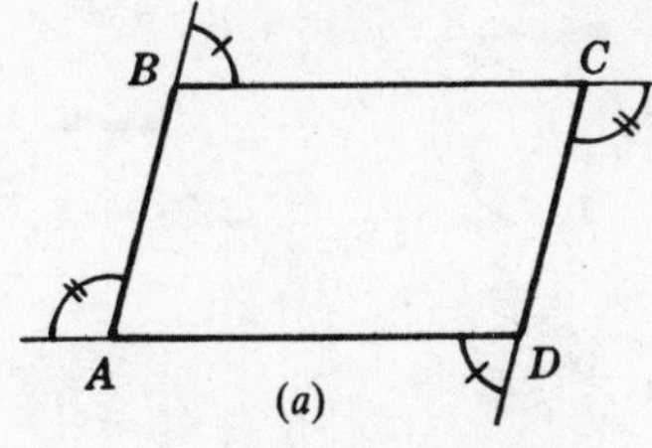

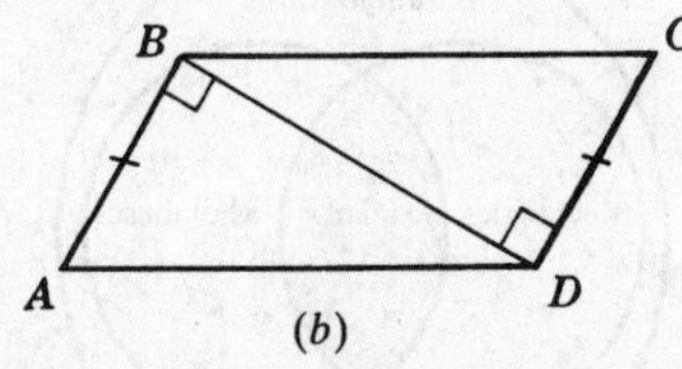

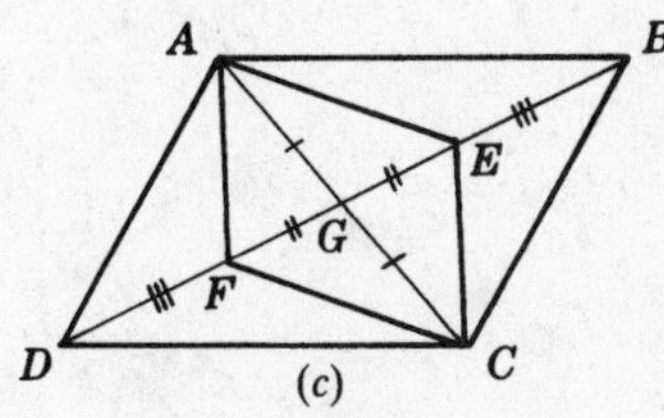

Fig. 5-12

Solutions

(a) Since supplements of congruent angles are congruent, opposite angles of *ABCD* are congruent. Thus by Principle 10, *ABCD* is a parallelogram.

(b) Since perpendiculars to the same line are parallel, $\overline{AB} \parallel \overline{CD}$. Hence by Principle 9, *ABCD* is a parallelogram.

(c) By the addition axiom, $\overline{DG} \cong \overline{GB}$. Hence by Principle 11, *ABCD* is a parallelogram.

5.6 Proving a parallelogram problem

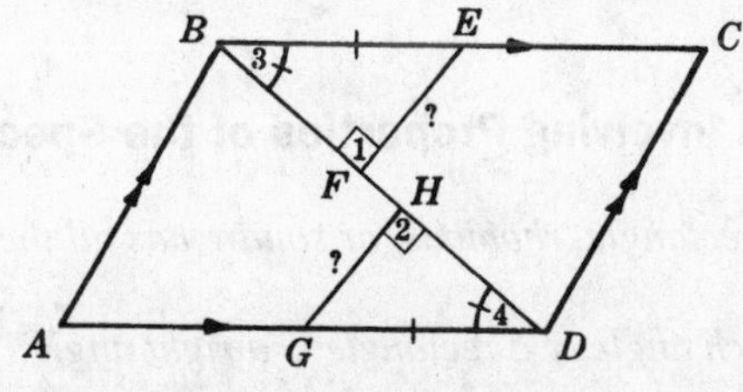

Given: $\square ABCD$
E is midpoint of $\overline{BC}$.
G is midpoint of $\overline{AD}$.
$\overline{EF} \perp \overline{BD}$, $\overline{GH} \perp \overline{BD}$

To Prove: $\overline{EF} \cong \overline{GH}$

Plan: Prove $\triangle BFE \cong \triangle GHD$

PROOF:

Statements	Reasons
1. *E* is midpoint of $\overline{BC}$. *G* is midpoint of $\overline{AD}$.	1. Given
2. $BE = \frac{1}{2}BC$, $GC = \frac{1}{2}AD$	2. A midpoint cuts a segment in half.
3. *ABCD* is a $\square$.	3. Given
4. $\overline{BC} \cong \overline{AD}$	4. Opposite sides of a $\square$ are congruent.
5. $\overline{BE} \cong \overline{GD}$	5. Halves of equals are equal.
6. $\overline{EF} \perp \overline{BD}$, $\overline{GH} \perp \overline{BD}$	6. Given
7. $\angle 1 \cong \angle 2$	7. Perpendiculars form rt. ∠s. Rt. ∠s are ≅.
8. $\overline{BC} \parallel \overline{AD}$	8. Opposite sides of a $\square$ are ∥.
9. $\angle 3 \cong \angle 4$	9. Alternate interior ∠s of ∥ lines are ≅.
10. $\triangle BFE \cong \triangle GHD$	10. SAA
11. $\overline{EF} \cong \overline{GH}$	11. Corresponding parts of congruent △s are ≅.

5.3 Special Parallelograms: Rectangle, Rhombus, and Square

5.3A Definitions and Relationships among the Special Parallelograms

Rectangles, rhombuses, and squares belong to the set of parallelograms. Each of these may be defined as a parallelogram, as follows:

1. A *rectangle* is an equiangular parallelogram.
2. A *rhombus* is an equilateral parallelogram.
3. A *square* is an equilateral and equiangular parallelogram. Thus, a square is both a rectangle and a rhombus.

The relations among the special parallelograms can be pictured by using a circle to represent each set. Note the following in Fig. 5-13:

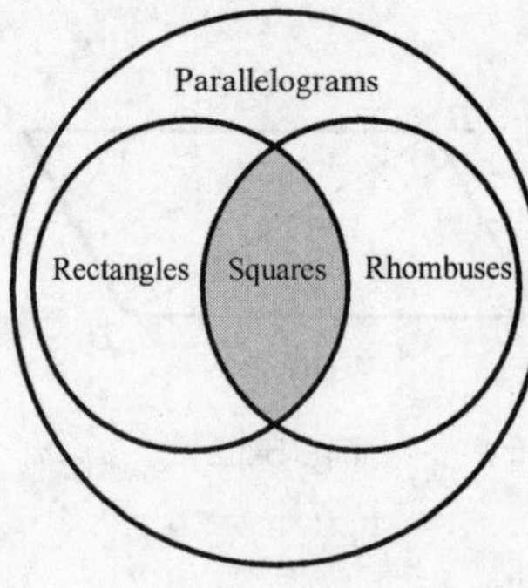

Fig. 5-13

1. Since every rectangle and every rhombus must be a parallelogram, the circle for the set of rectangles and the circle for the set of rhombuses must be inside the circle for the set of parallelograms.
2. Since every square is both a rectangle and a rhombus, the overlapping shaded section must represent the set of squares.

5.3B Principles Involving Properties of the Special Parallelograms

PRINCIPLE 1: *A rectangle, rhombus, or square has all the properties of a parallelogram.*

PRINCIPLE 2: *Each angle of a rectangle is a right angle.*

PRINCIPLE 3: *The diagonals of a rectangle are congruent.*

Thus in rectangle *ABCD* in Fig. 5-14, $\overline{AC} \cong \overline{BD}$.

PRINCIPLE 4: *All sides of a rhombus are congruent.*

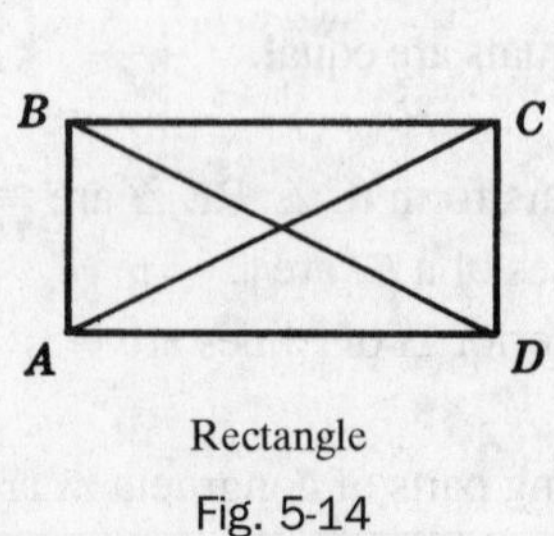

Rectangle

Fig. 5-14

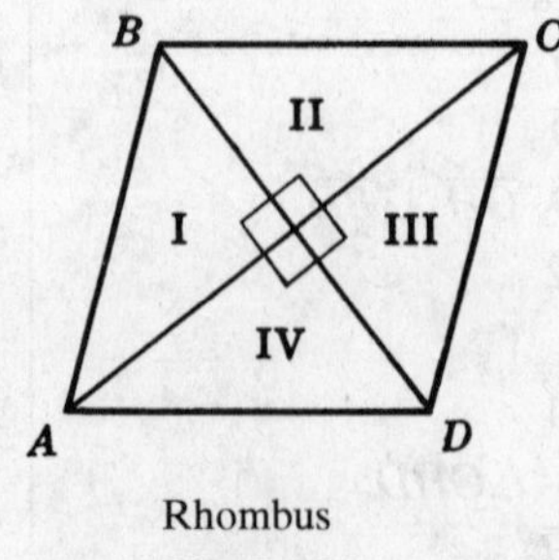

Rhombus

Fig. 5-15

PRINCIPLE 5: *The diagonals of a rhombus are perpendicular bisectors of each other.*

Thus in rhombus *ABCD* in Fig. 5-15, $\overline{AC}$ and $\overline{BD}$ are $\perp$ bisectors of each other.

PRINCIPLE 6: *The diagonals of a rhombus bisect the vertex angles.*

Thus in rhombus *ABCD*, $\overline{AC}$ bisects $\angle A$ and $\angle C$.

PRINCIPLE 7: *The diagonals of a rhombus form four congruent triangles.*

Thus in rhombus *ABCD*, $\triangle I \cong \triangle II \cong \triangle III \cong \triangle IV$.

PRINCIPLE 8: *A square has all the properties of both the rhombus and the rectangle.*

By definition, a square is both a rectangle and a rhombus.

5.3C Diagonal Properties of Parallelograms, Rectangles, Rhombuses, and Squares

Each check in the following table indicates a diagonal property of the figure.

Diagonal Properties	Parallelogram	Rectangle	Rhombus	Square
Diagonals bisect each other.	✓	✓	✓	✓
Diagonals are congruent.		✓		✓
Diagonals are perpendicular.			✓	✓
Diagonals bisect vertex angles.			✓	✓
Diagonals form 2 pairs of congruent triangles.	✓	✓	✓	✓
Diagonals form 4 congruent triangles.			✓	✓

5.3D Proving that a Parallelogram is a Rectangle, Rhombus, or Square

Proving that a Parallelogram is a Rectangle

The basic or minimum definition of a rectangle is this: *A rectangle is a parallelogram having one right angle.* Since the consecutive angles of a parallelogram are supplementary, if one angle is a right angle, the remaining angles must be right angles.

The converse of this basic definition provides a useful method of proving that a parallelogram is a rectangle, as follows:

PRINCIPLE 9: *If a parallelogram has one right angle, then it is a rectangle.*

Thus if *ABCD* in Fig. 5-16 is a ▱ and $m\angle A = 90°$, then *ABCD* is a rectangle.

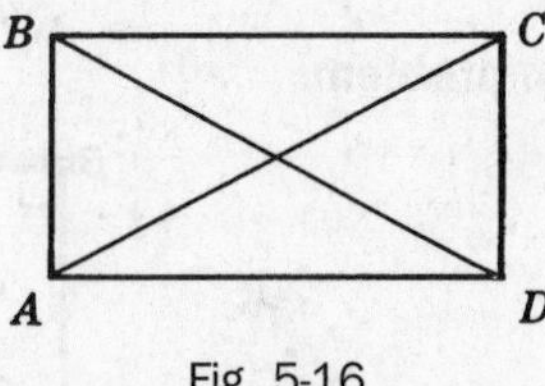

Fig. 5-16

PRINCIPLE 10: *If a parallelogram has congruent diagonals, then it is a rectangle.*

Thus if *ABCD* is a ▱ and $\overline{AC} \cong \overline{BD}$, then *ABCD* is a rectangle.

Proving that a Parallelogram is a Rhombus

The basic or minimum definition of a rhombus is this: *A rhombus is a parallelogram having two congruent adjacent sides.*

The converse of this basic definition provides a useful method of proving that a parallelogram is a rhombus, as follows:

PRINCIPLE 11: *If a parallelogram has congruent adjacent sides, then it is a rhombus.*

Thus if *ABCD* in Fig. 5-17 is a ▱ and $\overline{AB} \cong \overline{BC}$, then *ABCD* is a rhombus.

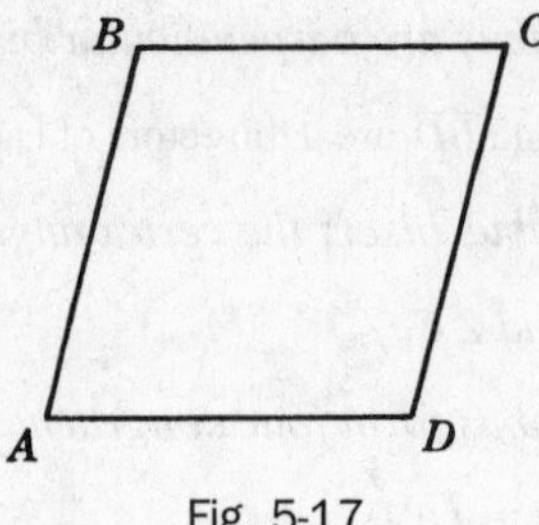

Fig. 5-17

Proving that a Parallelogram is a Square

PRINCIPLE 12: *If a parallelogram has a right angle and two congruent adjacent sides, then it is a square.*

This follows from the fact that a square is both a rectangle and a rhombus.

SOLVED PROBLEMS

5.7 Applying algebra to the rhombus

Assuming $ABCD$ is a rhombus, find x and y in each part of Fig. 5-18.

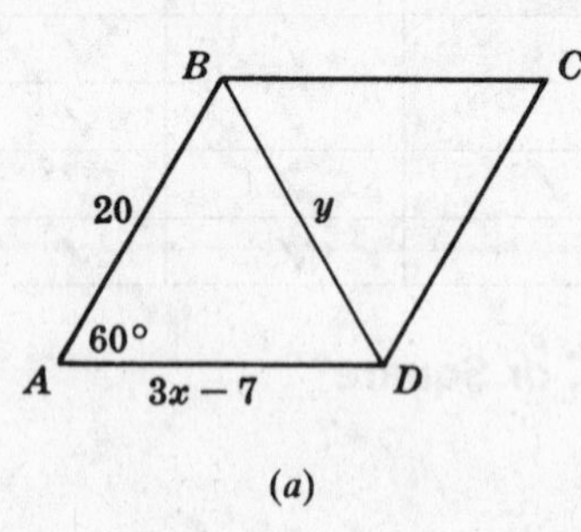

(a)

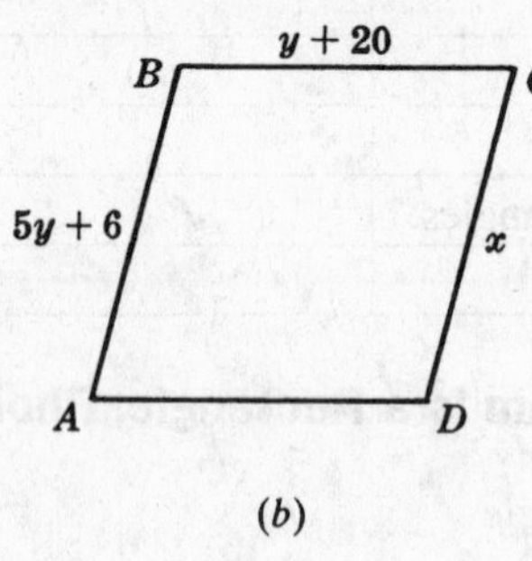

(b)

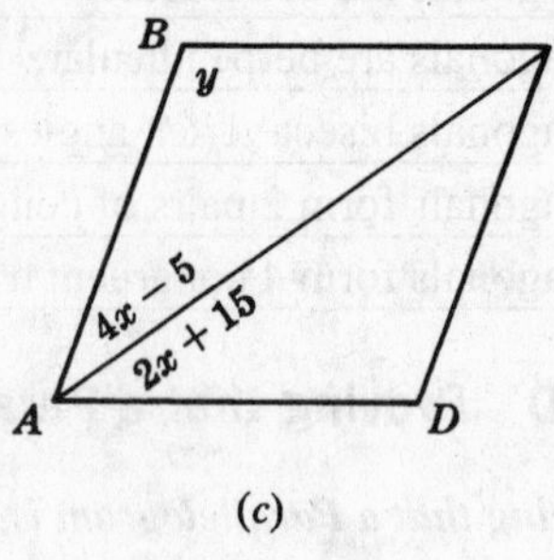

(c)

Fig. 5-18

Solutions

(a) Since $\overline{AB} \cong \overline{AD}$, $3x - 7 = 20$ or $x = 9$. Since $\triangle ABD$ is equiangular it is equilateral, and so $y = 20$.

(b) Since $\overline{BC} \cong \overline{AB}$, $5y + 6 = y + 20$ or $y = 3\frac{1}{2}$. Since $\overline{CD} \cong \overline{BC}$, $x = y + 20$ or $x = 23\frac{1}{2}$.

(c) Since $\overline{AC}$ bisects $\angle A$, $4x - 5 = 2x + 15$ or $x = 10$. Hence, $2x + 15 = 35$ and $m\angle A = 2(35°) = 70°$. Since $\angle B$ *and* $\angle A$ are supplementary, $y + 70 = 180$ or $y = 110$.

5.8 Proving a special parallelogram problem

Given: Rectangle $ABCD$
E is midpoint of $\overline{BC}$.
To Prove: $\overline{AE} \cong \overline{ED}$
Plan: Prove $\triangle AEB \cong \triangle DEC$

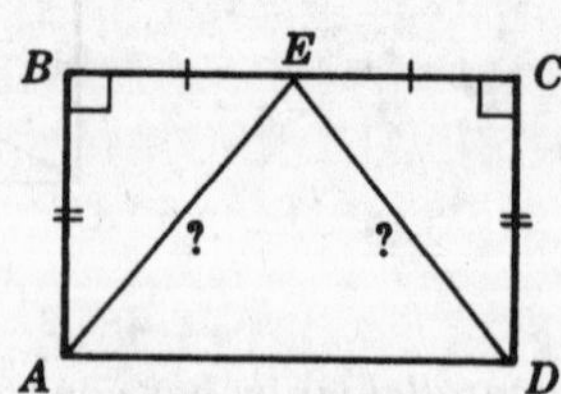

PROOF:

Statements	Reasons
1. $ABCD$ is a rectangle.	1. Given
2. E is midpoint of $\overline{BC}$.	2. Given
3. $\overline{BE} \cong \overline{EC}$	3. A midpoint divides a line into two congruent parts.
4. $\angle B \cong \angle C$	4. A rectangle is equiangular.
5. $\overline{AB} \cong \overline{CD}$	5. Opposite sides of a ▱ are congruent.
6. $\triangle AEB \cong \triangle DEC$	6. SAS
7. $\overline{AE} \cong \overline{ED}$	7. Corresponding parts of congruent ⚠ are congruent.

5.9 Proving a special parallelogram problem stated in words

Prove that a diagonal of a rhombus bisects each vertex angle through which it passes.

Solution

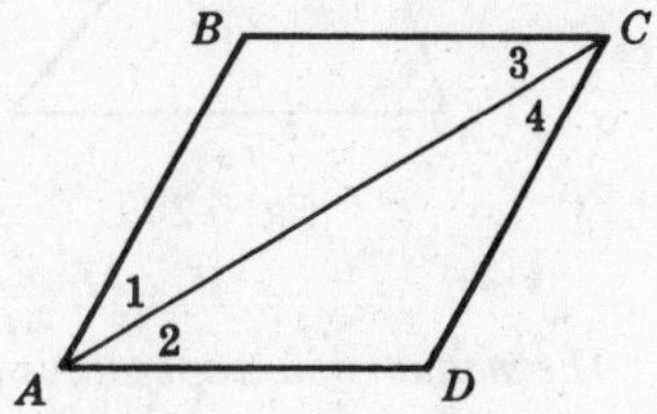

Given: Rhombus $ABCD$
$\overline{AC}$ is a diagonal.

To Prove: $\overline{AC}$ bisects $\angle A$ and $\angle C$.

Plan: Prove (1) $\angle 1$ and $\angle 2$ are congruent to $\angle 3$.
(2) $\angle 3$ and $\angle 4$ are congruent to $\angle 1$.

PROOF:

Statements	Reasons
1. $ABCD$ is a rhombus.	1. Given
2. $\overline{AB} \cong \overline{BC}$	2. A rhombus is equilateral.
3. $\angle 1 \cong \angle 3$	3. In a $\triangle$, angles opposite congruent sides are congruent.
4. $\overline{BC} \parallel \overline{AD}$, $\overline{AB} \parallel \overline{CD}$	4. Opposite sides of a $\square$ are $\parallel$.
5. $\angle 2 \cong \angle 3$, $\angle 1 \cong \angle 4$	5. Alternate interior $\angle$s of $\parallel$ lines are congruent.
6. $\angle 1 \cong \angle 2$, $\angle 3 \cong \angle 4$	6. Things congruent to the same thing are congruent to each other.
7. $\overline{AC}$ bisects $\angle A$ and $\angle C$.	7. To divide into two congruent parts is to bisect.

5.4 Three or More Parallels; Medians and Midpoints

5.4A Three or More Parallels

PRINCIPLE 1: *If three or more parallels cut off congruent segments on one transversal, then they cut off congruent segments on any other transversal.*

Thus if $l_1 \parallel l_2 \parallel l_3$ in Fig. 5-19 and segments a and b of transversal $\overleftrightarrow{AB}$ are congruent, then segments c and d of transversal $\overleftrightarrow{CD}$ are congruent.

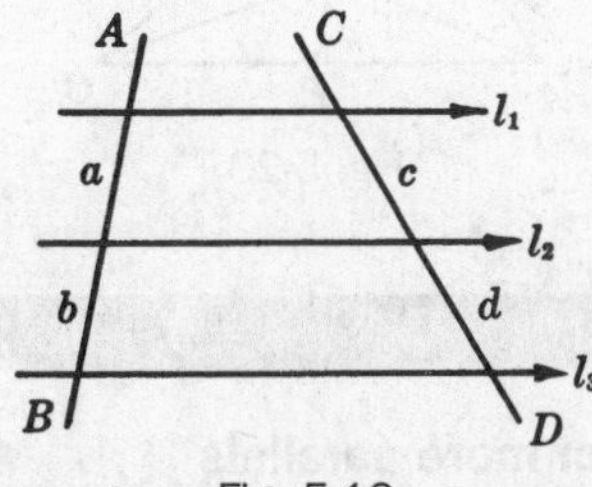

Fig. 5-19

5.4B Midpoint and Median Principles of Triangles and Trapezoids

PRINCIPLE 2: *If a line is drawn from the midpoint of one side of a triangle and parallel to a second side, then it passes through the midpoint of the third side.*

Thus in $\triangle ABC$ in Fig. 5-20 if M is the midpoint of $\overline{AB}$ and $\overline{MN} \parallel \overline{AC}$, then N is the midpoint of $\overline{BC}$.

PRINCIPLE 3: *If a line joins the midpoints of two sides of a triangle, then it is parallel to the third side and its length is one-half the length of the third side.*

Thus in $\triangle ABC$, if M and N are the midpoints of $\overline{AB}$ and $\overline{BC}$, *then* $\overline{MN} \parallel \overline{AC}$ and $MN = \frac{1}{2}AC$.

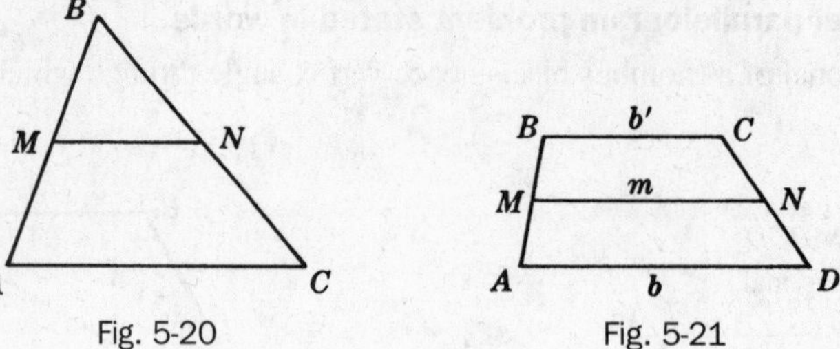

Fig. 5-20 Fig. 5-21

PRINCIPLE 4: *The median of a trapezoid is parallel to its bases, and its length is equal to one-half of the sum of their lengths.*

Thus if $\overline{MN}$ is the median of trapezoid $ABCD$ in Fig. 5-21 then $\overline{MN} \parallel \overline{AD}$, $\overline{MN} \parallel \overline{BC}$, and $m = \frac{1}{2}(b + b')$.

PRINCIPLE 5: *The length of the median to the hypotenuse of a right triangle equals one-half the length of the hypotenuse.*

Thus in rt. $\triangle ABC$ in Fig. 5-22, if $\overline{CM}$ is the median to hypotenuse $\overline{AB}$, then $CM = \frac{1}{2}AB$; that is, $\overline{CM} \cong \overline{AM} \cong \overline{MB}$.

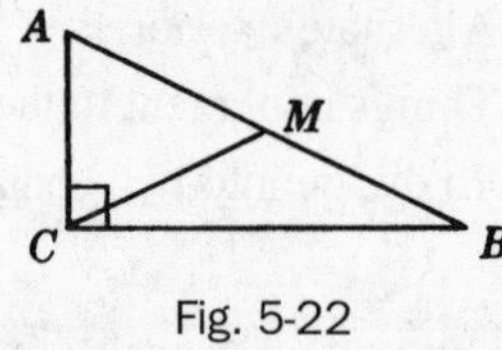

Fig. 5-22

PRINCIPLE 6: *The medians of a triangle meet in a point which is two-thirds of the distance from any vertex to the midpoint of the opposite side.*

Thus if $\overline{AN}$, $\overline{BP}$, and $\overline{CM}$ are medians of $\triangle ABC$ in Fig. 5-23, then they meet in a point G which is two-thirds of the distance from A to N, B to P, and C to M.

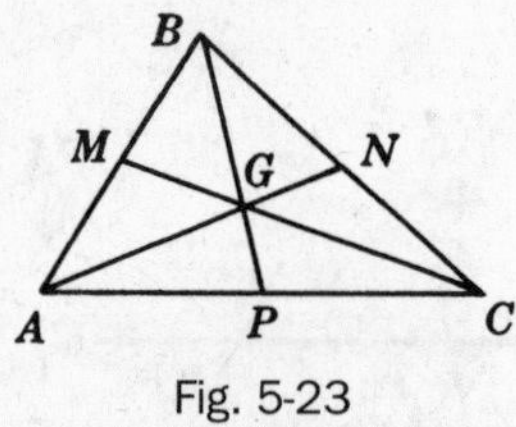

Fig. 5-23

SOLVED PROBLEMS

5.10 Applying principle 1 to three or more parallels

Find x and y in each part of Fig. 5-24.

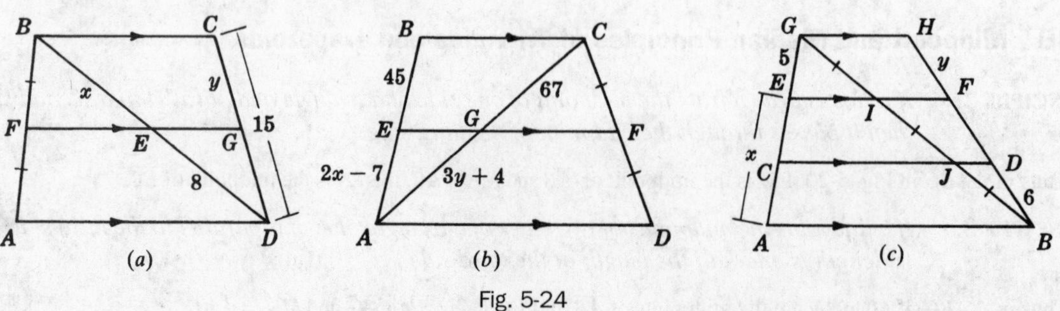

Fig. 5-24

Solutions

(a) Since $BE = ED$ and $GC = \frac{1}{2}CD$, $x = 8$ and $y = 7\frac{1}{2}$.

(b) Since $BE = EA$ and $CG = AG$, $2x - 7 = 45$ and $3y + 4 = 67$. Hence $x = 26$ and $y = 21$.

(c) Since $AC = CE = EG$ and $HF = FD = DB$, $x = 10$ and $y = 6$.

5.11 Applying principles 2 and 3

Find x and y in each part of Fig. 5-25.

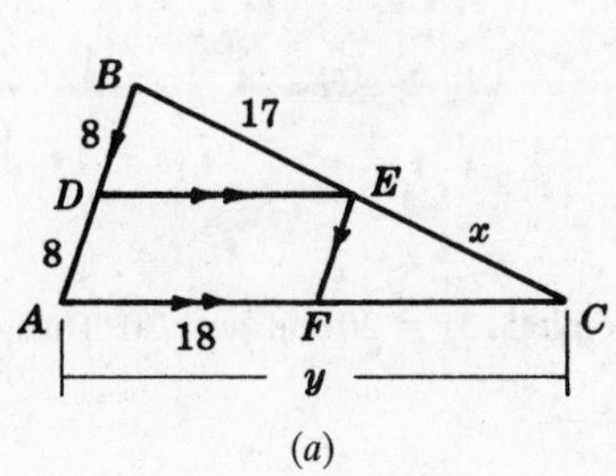

(a)

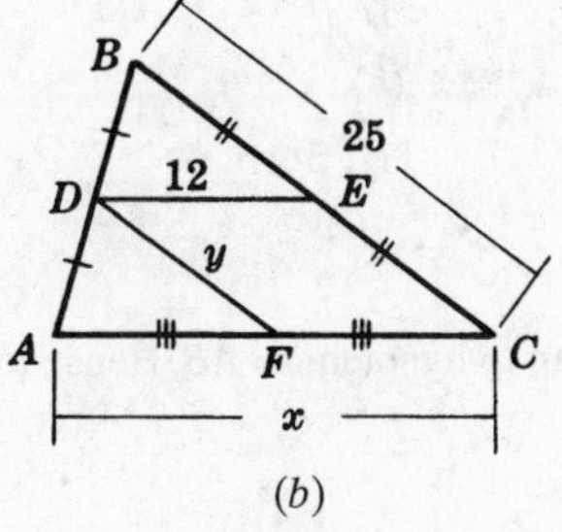

(b)

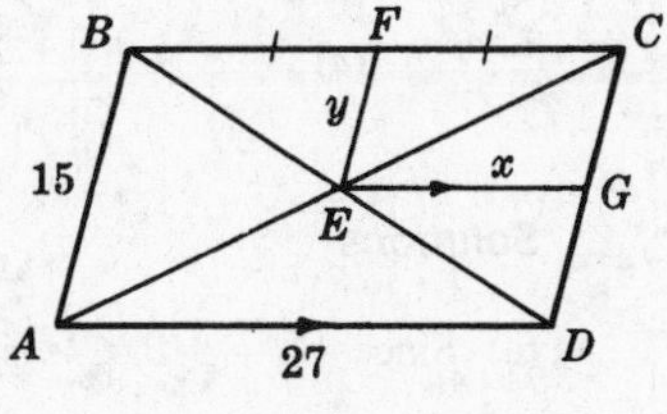

(c) *ABCD* is a parallelogram.

Fig. 5-25

Solutions

(a) By Principle 2, E is the midpoint of $\overline{BC}$ and F is the midpoint of $\overline{AC}$. Hence $x = 17$ and $y = 36$.

(b) By Principle 3, $DE = \frac{1}{2}AC$ and $DF = \frac{1}{2}BC$. Hence $x = 24$ and $y = 12\frac{1}{2}$.

(c) Since $ABCD$ is a parallelogram, E is the midpoint of $\overline{AC}$. Then by Principle 2, G is the midpoint of $\overline{CD}$. By Principle 3, $x = \frac{1}{2}(27) = 13\frac{1}{2}$ and $y = \frac{1}{2}(15) = 7\frac{1}{2}$.

5.12 Applying principle 4 to the median of a trapezoid

If $\overline{MP}$ is the median of trapezoid $ABCD$ in Fig. 5-26,

(a) Find m if $b = 20$ and $b' = 28$.

(b) Find b' if $b = 30$ and $m = 26$.

(c) Find b if $b' = 35$ and $m = 40$.

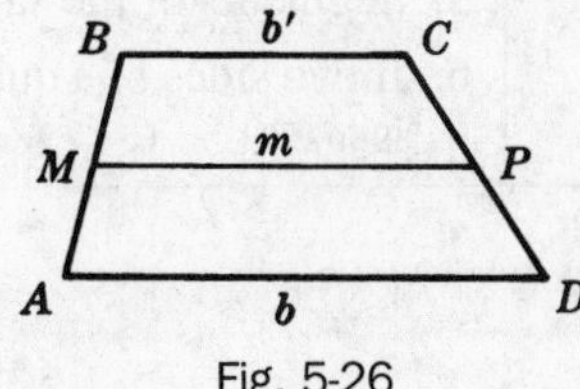

Fig. 5-26

Solutions

In each case, we apply the formula $m = \frac{1}{2}(b + b')$. The results are

(a) $m = \frac{1}{2}(20 + 28)$ or $m = 24$

(b) $26 = \frac{1}{2}(30 + b')$ or $b' = 22$

(c) $40 = \frac{1}{2}(b + 35)$ or $b = 45$

5.13 Applying principles 5 and 6 to the medians of a triangle

Find x and y in each part of Fig. 5-27.

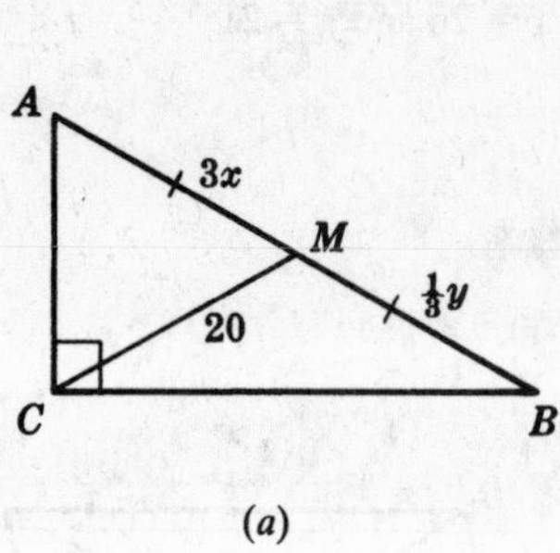

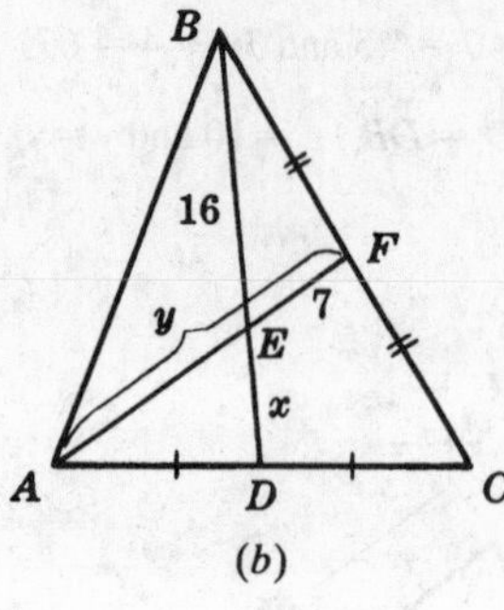

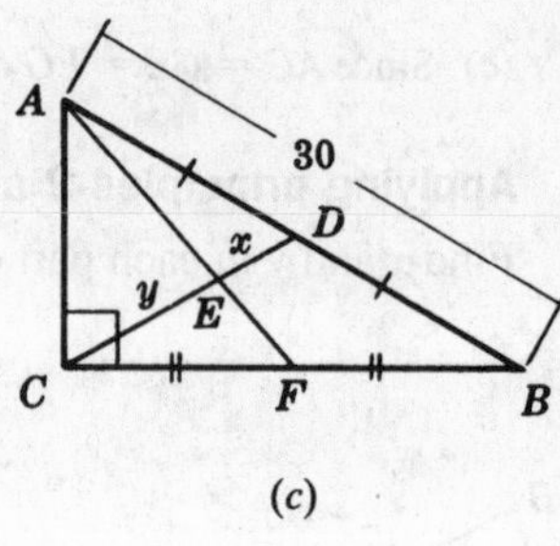

Fig. 5-27

Solutions

(a) Since $AM = MB$, $\overline{CM}$ is the median to hypotenuse $\overline{AB}$. Hence by Principle 5, $3x = 20$ and $\frac{1}{3}y = 20$. Thus, $x = 6\frac{2}{3}$ and $y = 60$.

(b) $\overline{BD}$ and $\overline{AF}$ are medians of $\triangle ABC$. Hence by Principle 6, $x = \frac{1}{2}(16) = 8$ and $y = 3(7) = 21$.

(c) $\overline{CD}$ is the median to hypotenuse $\overline{AB}$; hence by Principle 5, $CD = 15$.
$\overline{CD}$ and $\overline{AF}$ are medians of $\triangle ABC$; hence by Principle 6, $x = \frac{1}{3}(15) = 5$ and $y = \frac{2}{3}(15) = 10$.

5.14 Proving a midpoint problem

Given: Quadrilateral $ABCD$
E, F, G, and H are midpoints of $\overline{AD}$, $\overline{AB}$, $\overline{BC}$, and $\overline{CD}$, respectively.

To Prove: $EFGH$ is a ▱.

Plan: Prove $\overline{EF}$ and $\overline{GH}$ are congruent and parallel.

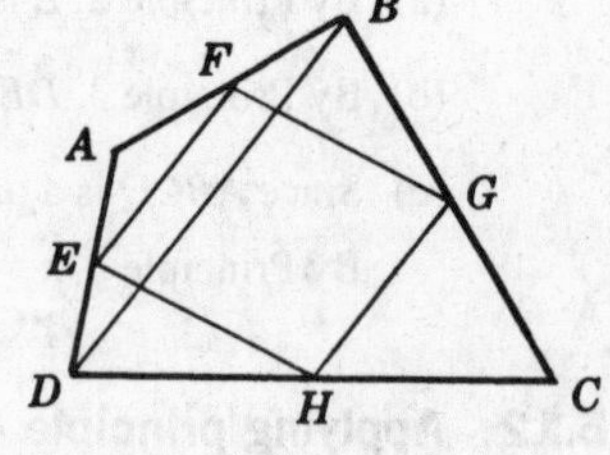

PROOF:

Statements	Reasons
1. Draw $\overline{BD}$.	1. A segment may be drawn between any two points.
2. E, F, G, and H are midpoints.	2. Given
3. $\overline{EF} \parallel \overline{BD}$ and $\overline{GH} \parallel \overline{BD}$ $EF = \frac{1}{2}BD$ and $GH = \frac{1}{2}BD$	3. A line segment joining the midpoints of two sides of a △ is parallel to the third side and equal in length to half the third side.
4. $\overline{EF} \parallel \overline{GH}$	4. Two lines parallel to a third line are parallel to each other.
5. $\overline{EF} \cong \overline{GH}$	5. Segments of the same length are congruent.
6. $EFGH$ is a ▱.	6. If two sides of a quadrilateral are ≅ and ∥, the quadrilateral is a ▱.

SUPPLEMENTARY PROBLEMS

5.1. Find x and y in each part of Fig. 5-28. (5.1)

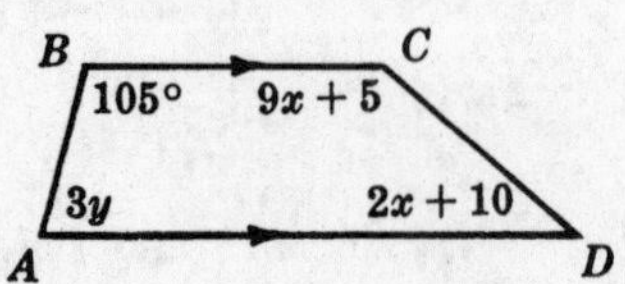

(*a*) *ABCD* is a trapezoid.

B
C
y
2x + 10
4x − 30
A
D

(*b*) *ABCD* is an isosceles trapezoid.

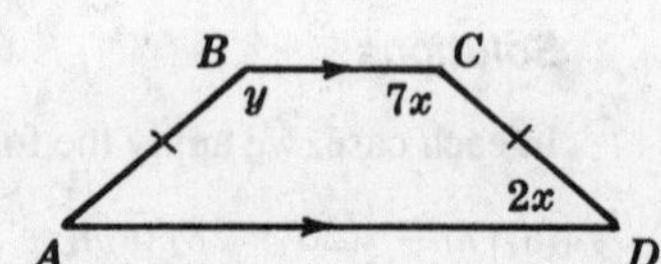

(*c*) *ABCD* is an isosceles trapezoid.

Fig. 5-28

5.2. Prove that if the base angles of a trapezoid are congruent, the trapezoid is isosceles. (5.2)

5.3. Prove that (a) the diagonals of an isosceles trapezoid are congruent; (b) if the nonparallel sides $\overline{AB}$ and $\overline{CD}$ of an isosceles trapezoid are extended until they meet at *E*, triangle *ADE* thus formed is isosceles. (5.2)

5.4. Name the parallelograms in each part of Fig. 5-29. (5.4)

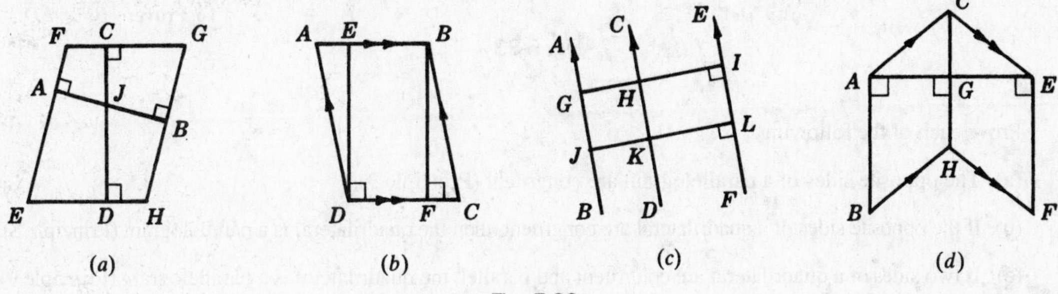

Fig. 5-29

5.5. State why *ABCD* in each part of Fig. 5-30 is a parallelogram. (5.5)

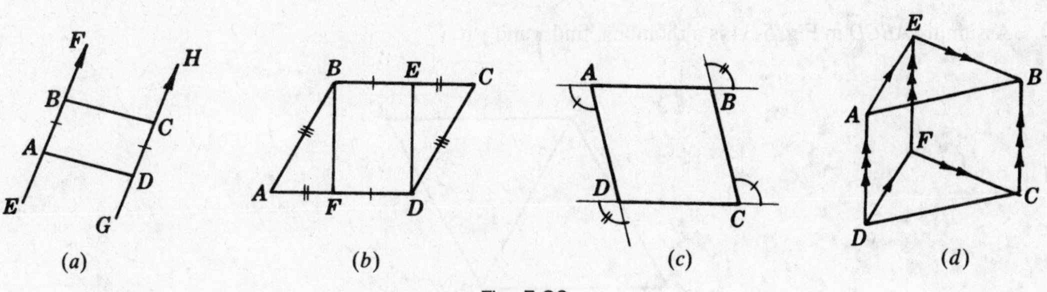

Fig. 5-30

5.6. Assuming *ABCD* in Fig. 5-31 is a parallelogram, find *x* and *y* if (5.3)

(a) $AD = 5x, AB = 2x, CD = y$, perimeter $= 84$

(b) $AB = 2x, BC = 3y + 8, CD = 7x - 25, AD = 5y - 10$

(c) $m\angle A = 4y - 60, m\angle C = 2y, m\angle D = x$

(d) $m\angle A = 3x, m\angle B = 10x - 15, m\angle C = y$

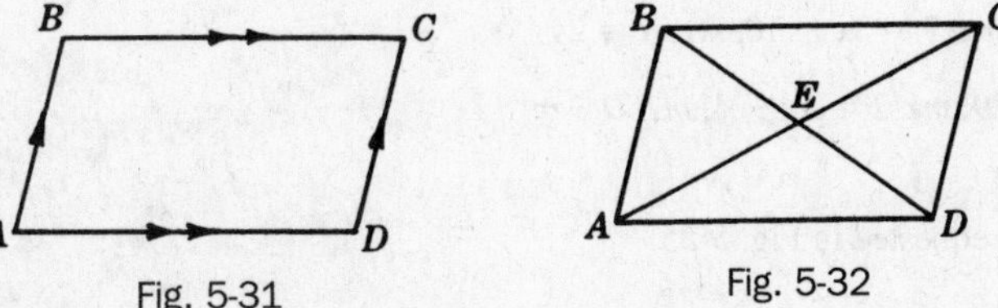

Fig. 5-31 Fig. 5-32

5.7. Assuming *ABCD* in Fig. 5-32 is a parallelogram, find *x* and *y* if (5.3)

(a) $AE = x + y, EC = 20, BE = x - y, ED = 8$

(b) $AE = x, EC = 4y, BE = x - 2y, ED = 9$

(c) $AE = 3x - 4, EC = x + 12, BE = 2y - 7, ED = x - y$

(d) $AE = 2x + y, AC = 30, BE = x + y, BD = 24$

5.8. Provide the proofs requested in Fig. 5-33. (5.6)

(*a*) **Given:** ▱$ABCD$
G is midpoint of $\overline{AB}$.
F is midpoint of $\overline{CD}$.
$\overline{HG} \perp \overline{AB}$, $\overline{EF} \perp \overline{CD}$
To Prove: $\overline{EF} \cong \overline{GH}$

(*b*) **Given:** ▱$ABCD$
$\overline{BE} \cong \overline{FD}$
To Prove: $\overline{BF} \cong \overline{ED}$

(*c*) **Given:** ▱$ABCD$
$\overline{BF}$ bisects $\angle B$,
$\overline{ED}$ bisects $\angle D$.
To Prove: $\overline{BF} \cong \overline{ED}$

Fig. 5-33

5.9. Prove each of the following:

(a) The opposite sides of a parallelogram are congruent (Principle 3).

(b) If the opposite sides of a quadrilateral are congruent, then the quadrilateral is a parallelogram (Principle 8).

(c) If two sides of a quadrilateral are congruent and parallel, the quadrilateral is a parallelogram (Principle 9).

(d) The diagonals of a parallelogram bisect each other (Principle 6).

(e) If the diagonals of a quadrilateral bisect each other, then the quadrilateral is a parallelogram (Principle 11).

5.10. Assuming $ABCD$ in Fig. 5-34 is a rhombus, find x and y if (5.7)

Fig. 5-34

(a) $BC = 35$, $CD = 8x - 5$, $BD = 5y$, $m\angle C = 60°$

(b) $AB = 43$, $AD = 4x + 3$, $BD = y + 8$, $m\angle B = 120°$

(c) $AB = 7x$, $AD = 3x + 10$, $BC = y$

(d) $AB = x + y$, $AD = 2x - y$, $BC = 12$

(e) $m\angle B = 130°$, $m\angle 1 = 3x - 10$, $m\angle A = 2y$

(f) $m\angle 1 = 8x - 29$, $m\angle 2 = 5x + 4$, $m\angle D - y$

5.11. Provide the proofs requested in Fig. 5-35. (5.8)

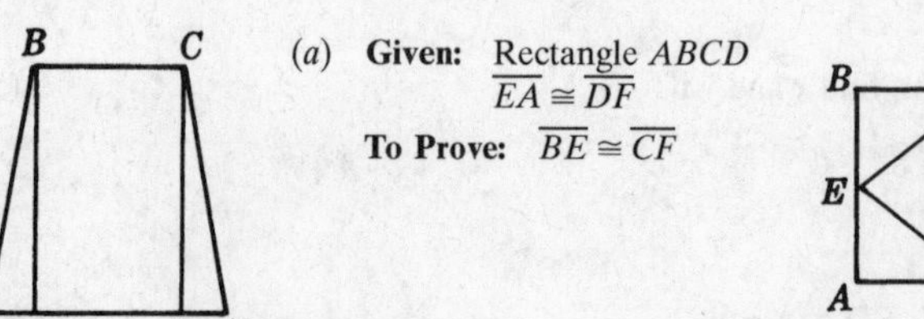

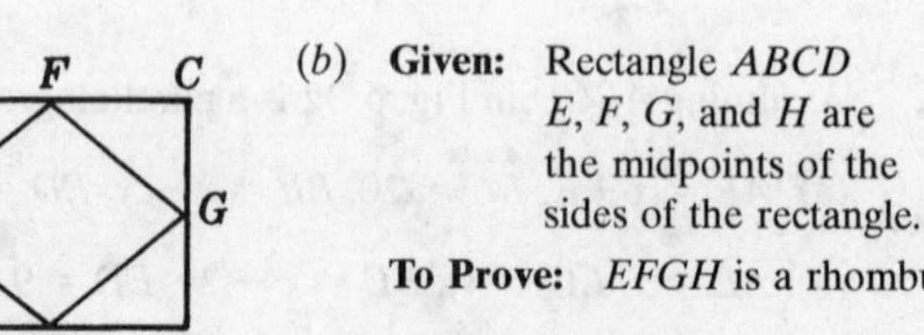

Fig. 5-35

5.12. Prove each of the following: (5.9)

(a) If the diagonals of a parallelogram are congruent, the parallelogram is a rectangle.

(b) If the diagonals of a parallelogram are perpendicular to each other, the parallelogram is a rhombus.

(c) If a diagonal of a parallelogram bisects a vertex angle, then the parallelogram is a rhombus.

(d) The diagonals of a rhombus divide it into four congruent triangles.

(e) The diagonals of a rectangle are congruent.

5.13. Find x and y in each part of Fig. 5-36. (5.10)

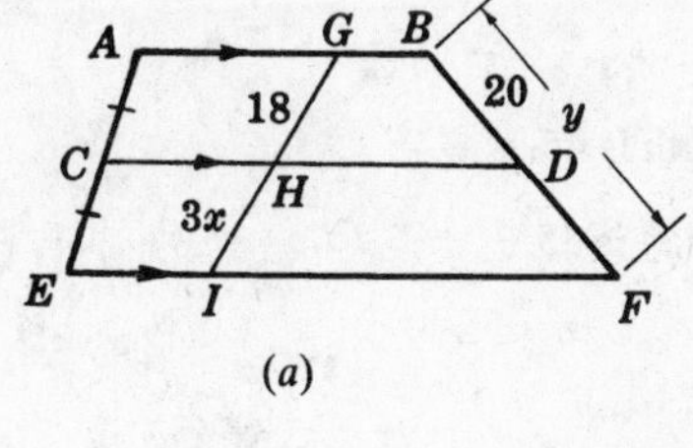

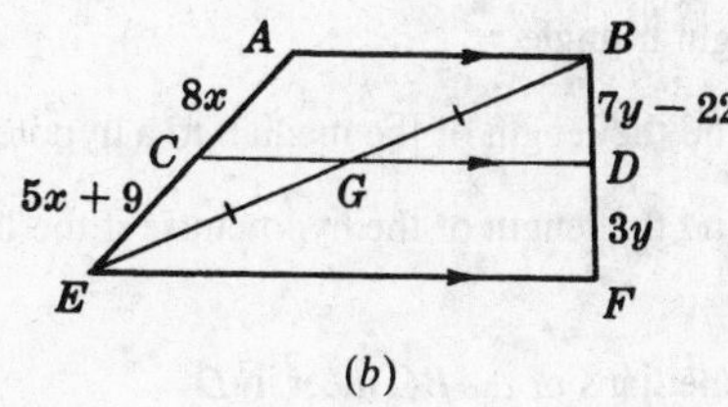

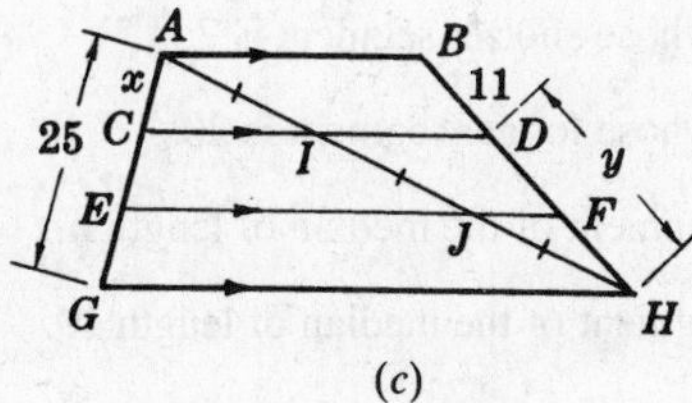

Fig. 5-36

5.14. Find x and y in each part of Fig. 5-37. (5.11)

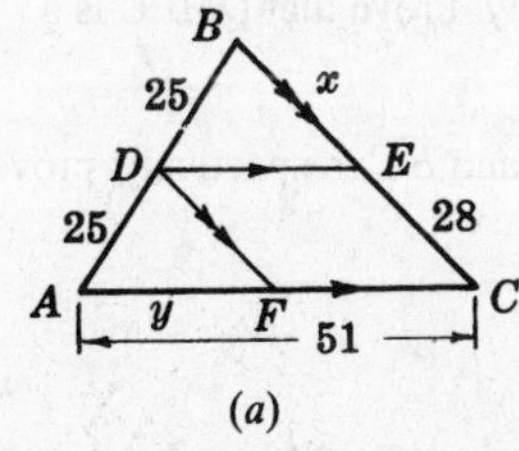

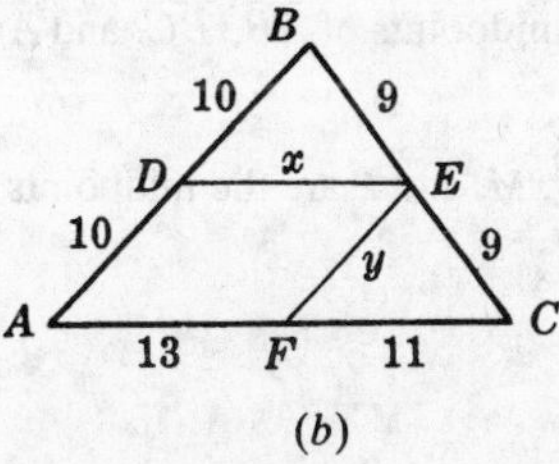

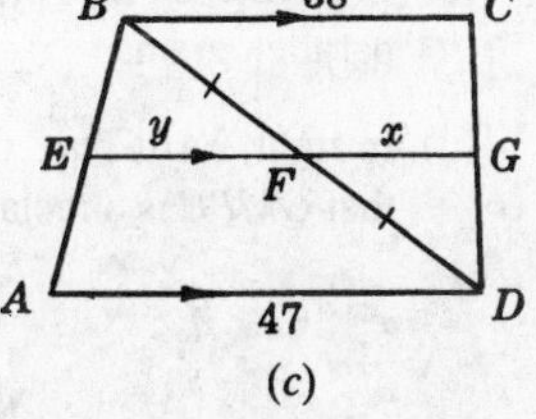

Fig. 5-37

5.15. If $\overline{MP}$ is the median of trapezoid $ABCD$ in Fig. 5-38 (5.12)

(a) Find m if $b = 23$ and $b' = 15$.

(b) Find b' if $b = 46$ and $m = 41$.

(c) Find b if $b' = 51$ and $m = 62$.

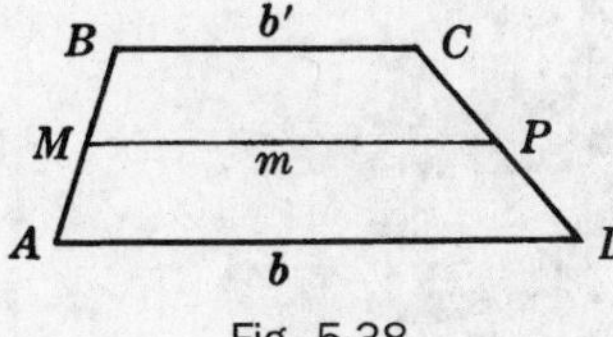

Fig. 5-38

5.16. Find x and y in each part of Fig. 5-39. (5.11 and 5.12)

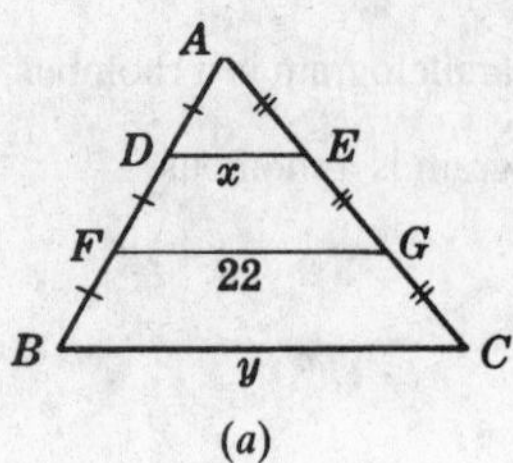

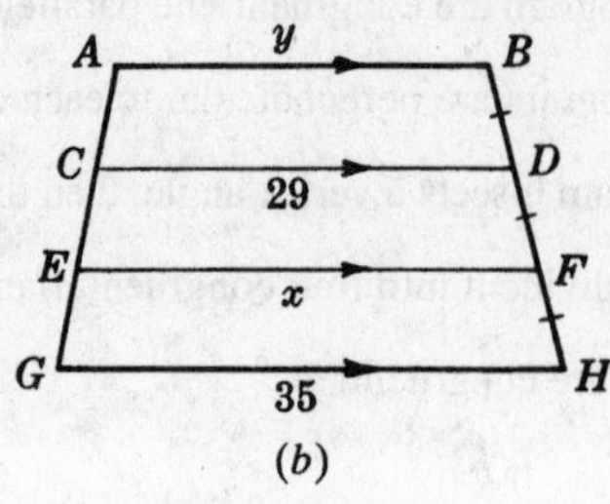

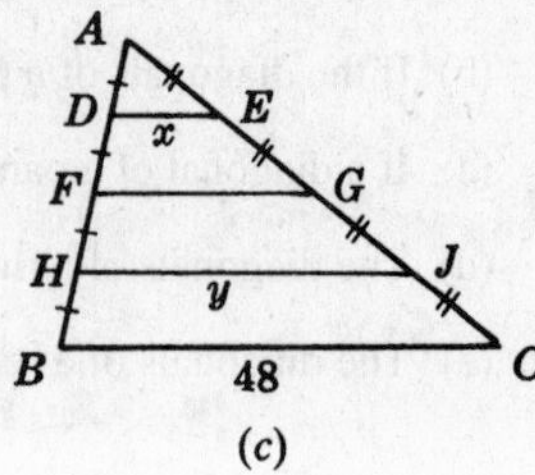

Fig. 5-39

5.17. In a right triangle (5.13)

(a) Find the length of the median to a hypotenuse whose length is 45.

(b) Find the length of the hypotenuse if the length of its median is 35.

5.18. If the medians of $\triangle ABC$ meet in D (5.13)

(a) Find the length of the median whose shorter segment is 7.

(b) Find the length of the median whose longer segment is 20.

(c) Find the length of the shorter segment of the median of length 42.

(d) Find the length of the longer segment of the median of length 39.

5.19. Prove each of the following: (5.14)

(a) If the midpoints of the sides of a rhombus are joined in order, the quadrilateral formed is a rectangle.

(b) If the midpoints of the sides of a square are joined in order, the quadrilateral formed is a square.

(c) In $\triangle ABC$, let M, P, and Q be the midpoints of $\overline{AB}$, $\overline{BC}$, and $\overline{AC}$, respectively. Prove that $QMPC$ is a parallelogram.

(d) In right $\triangle ABC$, $m\angle C = 90°$. If Q, M, and P are the midpoints of $\overline{AC}$, $\overline{AB}$, and $\overline{BC}$, respectively, prove that $QMPC$ is a rectangle.

CHAPTER 6

Circles

6.1 The Circle; Circle Relationships

The following terms are associated with the circle. Although some have been defined previously, they are repeated here for ready reference.

A *circle* is the set of all points in a plane that are at the same distance from a fixed point called the center. The symbol for circle is ⊙; for circles ⓢ.

The *circumference* of a circle is the distance around the circle. It contains 360°.

A *radius* of a circle is a line segment joining the center to a point on the circle.

Note: Since all radii of a given circle have the same length, we may at times use the word *radius* to mean the number that is "the length of the radius."

A *central angle* is an angle formed by two radii.

An *arc* is a continuous part of a circle. The symbol for arc is ⌒. A *semicircle* is an arc measuring one-half the circumference of a circle.

A *minor arc* is an arc that is less than a semicircle. A *major arc* is an arc that is greater than a semicircle.

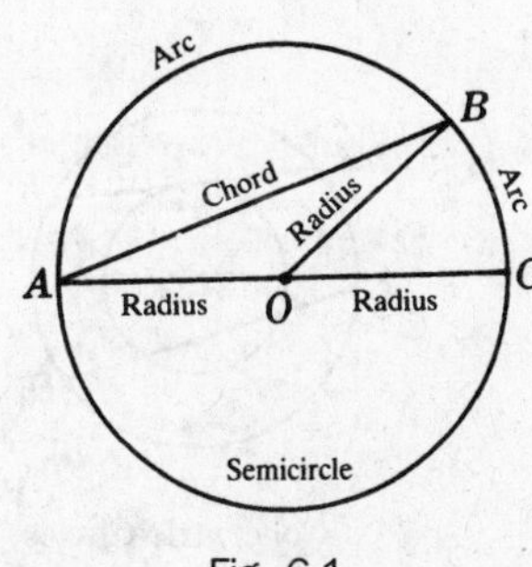

Fig. 6-1

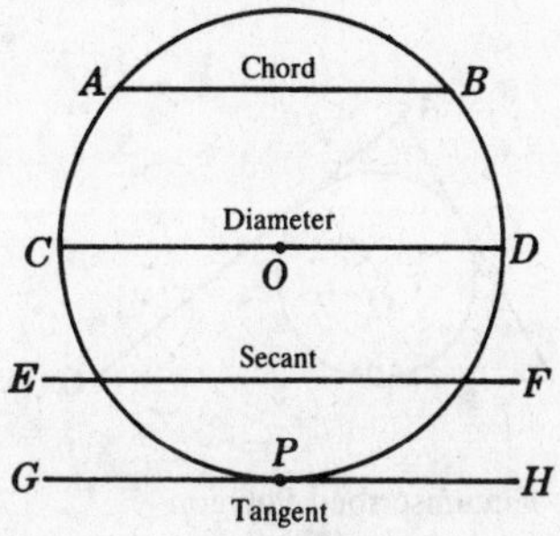

Fig. 6-2

Thus in Fig. 6-1, $\overset{\frown}{BC}$ is a minor arc and $\overset{\frown}{BAC}$ is a major arc. Three letters are needed to indicate a major arc.

To intercept an arc is to cut off the arc.

Thus in Fig. 6-1, $\angle BAC$ and $\angle BOC$ intercept $\overset{\frown}{BC}$.

A *chord* of a circle is a line segment joining two points of the circumference.

Thus in Fig. 6-2, $\overline{AB}$ is a chord.

A *diameter* of a circle is a chord through the center. A *secant* of a circle is a line that intersects the circle at two points. A *tangent* of a circle is a line that touches the circle at one and only one point no matter how far produced.

Thus, $\overline{CD}$ is a diameter of circle O in Fig. 6-2, $\overleftrightarrow{EF}$ is a secant, and $\overleftrightarrow{GH}$ is a tangent to the circle at P. P is the point of contact or the point of tangency.

An *inscribed polygon* is a polygon all of whose sides are chords of a circle. A *circumscribed circle* is a circle passing through each vertex of a polygon.

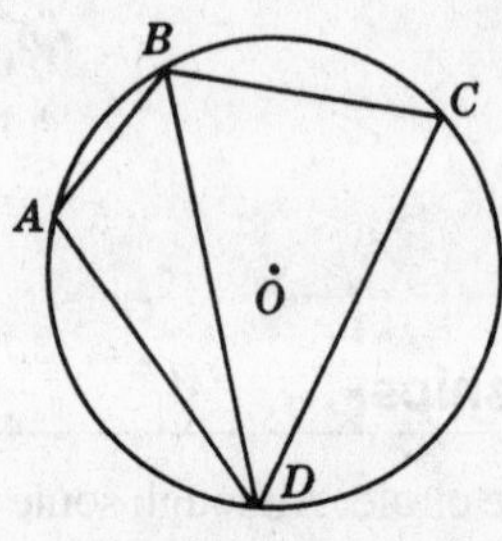

Inscribed Polygons
Circumscribed Circle

Fig. 6-3

Thus $\triangle ABD$, $\triangle BCD$, and quadrilateral $ABCD$ are inscribed polygons of circle O in Fig. 6-3. Circle O is a circumscribed circle of quadrilateral $ABCD$.

A *circumscribed polygon* is a polygon all of whose sides are tangents to a circle. An *inscribed circle* is a circle to which all the sides of a polygon are tangents.

Thus, $\triangle ABC$ is a circumscribed polygon of circle O in Fig. 6-4. Circle O is an inscribed circle of $\triangle ABC$.

Concentric circles are circles that have the same center.

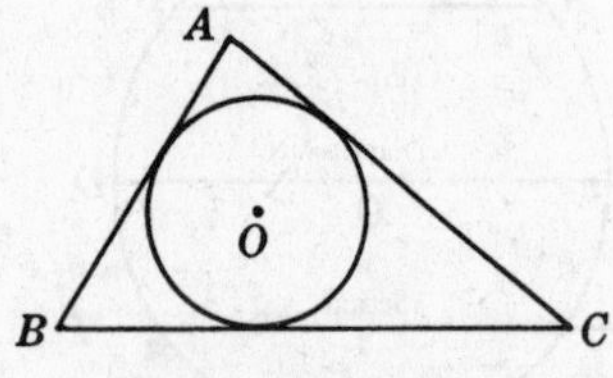

Circumscribed Polygon
Inscribed Circle

Fig. 6-4

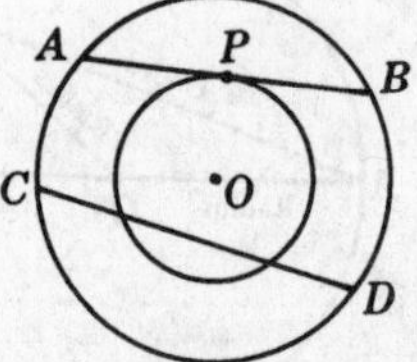

Concentric Circles

Fig. 6-5

Thus, the two circles shown in Fig. 6-5 are concentric circles. $\overline{AB}$ is a tangent of the inner circle and a chord of the outer one. $\overline{CD}$ is a secant of the inner circle and a chord of the outer one.

Two circles are *equal* if their radii are equal in length; two circles are *congruent* if their radii are congruent.

Two arcs are congruent if they have equal degree measure and length. We use the notation $m\overset{\frown}{AC}$ to denote "measure of arc AC."

6.1A Circle Principles

PRINCIPLE 1: *A diameter divides a circle into two equal parts.*

Thus, diameter $\overline{AB}$ divides circle O of Fig. 6-6 into two congruent semicircles, $\overset{\frown}{ACB}$ and $\overset{\frown}{ADB}$.

PRINCIPLE 2: *If a chord divides a circle into two equal parts, then it is a diameter.* (This is the converse of Principle 1.)

Thus if $\widehat{ACB} \cong \widehat{ADB}$ in Fig. 6-6, then $\overline{AB}$ is a diameter.

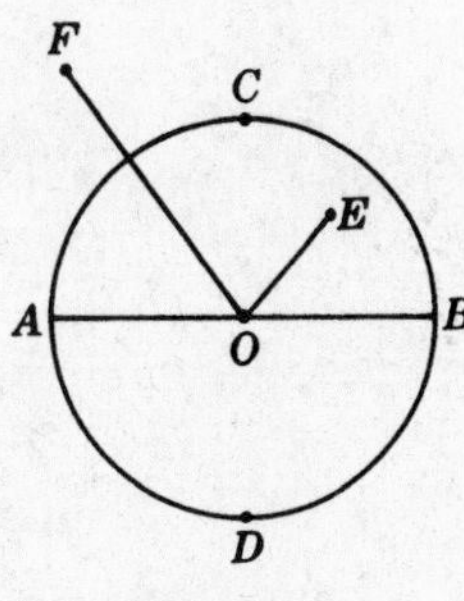

Fig. 6-6

PRINCIPLE 3: *A point is outside, on, or inside a circle according to whether its distance from the center is greater than, equal to, or smaller than the radius.*

F is outside circle O in Fig. 6-6, since $\overline{FO}$ is greater in length than a radius. E is inside circle O since $\overline{EO}$ is smaller in length than a radius. A is on circle O since $\overline{AO}$ is a radius.

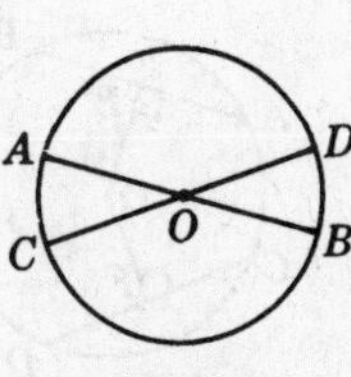

Fig. 6-7

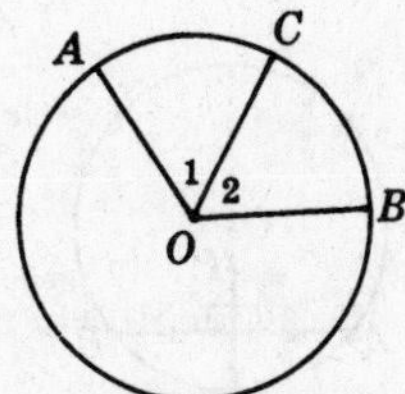

Fig. 6-8

PRINCIPLE 4: *Radii of the same or congruent circles are congruent.*

Thus in circle O of Fig. 6-7, $\overline{OA} \cong \overline{OC}$.

PRINCIPLE 5: *Diameters of the same or congruent circles are congruent.*

Thus in circle O of Fig. 6-7, $\overline{AB} \cong \overline{CD}$.

PRINCIPLE 6: *In the same or congruent circles, congruent central angles have congruent arcs.*

Thus in circle O of Fig. 6-8, if $\angle 1 \cong \angle 2$, then $\widehat{AC} \cong \widehat{CB}$.

PRINCIPLE 7: *In the same or congruent circles, congruent arcs have congruent central angles.*

Thus in circle O of Fig. 6-8, if $\widehat{AC} \cong \widehat{CB}$, then $\angle 1 \cong \angle 2$.

(Principles 6 and 7 are converses of each other.)

PRINCIPLE 8: *In the same or congruent circles, congruent chords have congruent arcs.*

Thus in circle O of Fig. 6-9, if $\overline{AB} \cong \overline{AC}$, then $\widehat{AB} \cong \widehat{AC}$.

PRINCIPLE 9: *In the same or congruent circles, congruent arcs have congruent chords.*

Thus in circle O of Fig. 6-9, if $\widehat{AB} \cong \widehat{AC}$, then $\overline{AB} \cong \overline{AC}$.

(Principles 8 and 9 are converses of each other.)

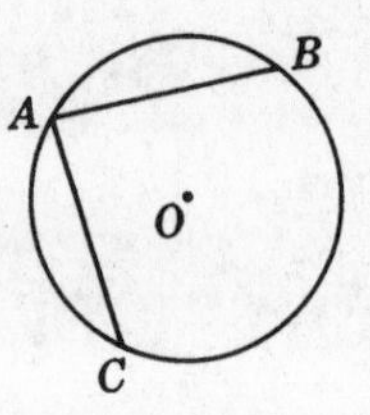

Fig. 6-9

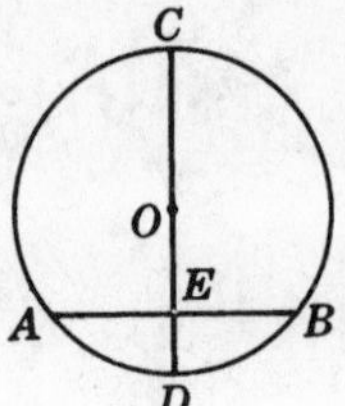

Fig. 6-10

PRINCIPLE 10: *A diameter perpendicular to a chord bisects the chord and its arcs.*

Thus in circle O of Fig. 6-10, if $\overline{CD} \perp \overline{AB}$, then $\overline{CD}$ bisects $\overline{AB}$, $\widehat{AB}$, and $\widehat{ACB}$.

A proof of this principle is given in Chapter 16.

PRINCIPLE 11: *A perpendicular bisector of a chord passes through the center of the circle.*

Thus in circle O of Fig. 6-11, if $\overline{PD}$ is the perpendicular bisector of $\overline{AB}$, then $\overline{PD}$ passes through center O.

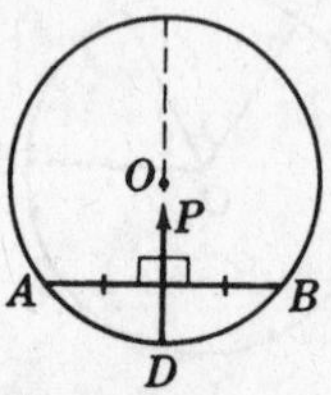

Fig. 6-11

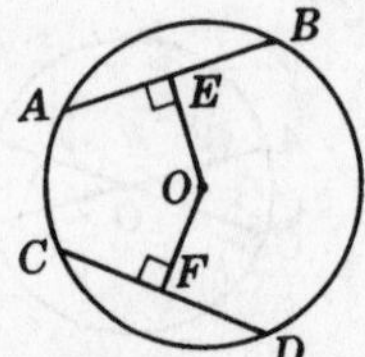

Fig. 6-12

PRINCIPLE 12: *In the same or congruent circles, congruent chords are equally distant from the center.*

Thus in circle O of Fig. 6-12, if $\overline{AB} \cong \overline{CD}$, if $\overline{OE} \perp \overline{AB}$, and if $\overline{OF} \perp \overline{CD}$, then $\overline{OE} \cong \overline{OF}$.

PRINCIPLE 13: *In the same or congruent circles, chords that are equally distant from the center are congruent.*

Thus in circle O of Fig. 6-12, if $\overline{OE} \cong \overline{OF}$, $\overline{OE} \perp \overline{AB}$, and $\overline{OF} \perp \overline{CD}$, then $\overline{AB} \cong \overline{CD}$.

(Principles 12 and 13 are converses of each other.)

SOLVED PROBLEMS

6.1 Matching test of circle vocabulary

Match each part of Fig. 6-13 on the left with one of the names on the right:

(a) $\overline{OE}$ — 1. Radius

(b) $\overline{FG}$ — 2. Central angle

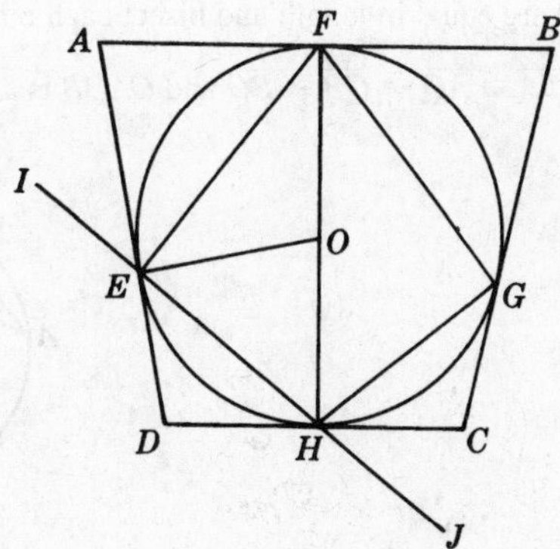

Fig. 6-13

(c) $\overline{FH}$	3. Semicircle
(d) $\overline{CD}$	4. Minor arc
(e) $\overline{IJ}$	5. Major arc
(f) $\widehat{EF}$	6. Chord
(g) $\widehat{FGH}$	7. Diameter
(h) $\widehat{FEG}$	8. Secant
(i) $\angle EOF$	9. Tangent
(j) Circle O about $EFGH$	10. Inscribed polygon
(k) Circle O in $ABCD$	11. Circumscribed polygon
(l) Quadrilateral $EFGH$	12. Inscribed circle
(m) Quadrilateral $ABCD$	13. Circumscribed circle

Solutions

(a) 1 (c) 7 (e) 8 (g) 3 (i) 2 (k) 12 (m) 11

(b) 6 (d) 9 (f) 4 (h) 5 (j) 13 (l) 10

6.2 Applying principles 4 and 5

In Fig. 6-14, (a) what kind of triangle is OCD; (b) what kind of quadrilateral is $ABCD$? (c) In Fig. 6-15 if circle O = circle Q, what kind of quadrilateral is $OAQB$?

Solutions

Radii or diameters of the same or equal circles have equal lengths.

(a) Since $\overline{OC} \cong \overline{OD}$, $\triangle OCD$ is isosceles.

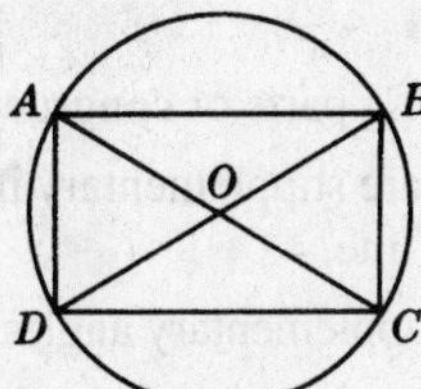

Fig. 6-14

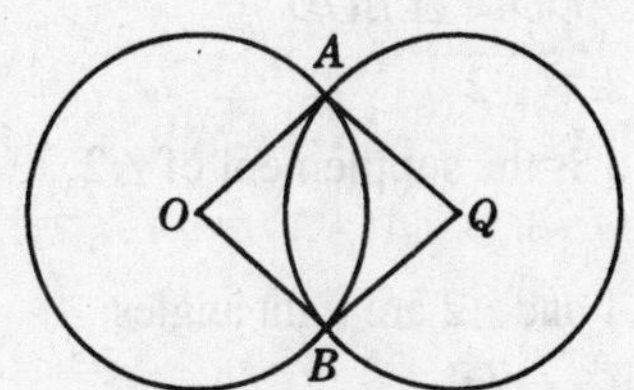

Fig. 6-15

(b) Since diagonals $\overline{AC}$ and $\overline{BD}$ are equal in length and bisect each other, $ABCD$ is a rectangle.

(c) Since the circles are equal, $\overline{OA} \cong \overline{AQ} \cong \overline{QB} \cong \overline{BO}$ and $OAQB$ is a rhombus.

6.3 Proving a circle problem

Given: $\overline{AB} \cong \overline{DE}$
$\overline{BC} \cong \overline{EF}$

To Prove: $\angle B \cong \angle E$

Plan: Prove $\triangle I \cong \triangle II$.

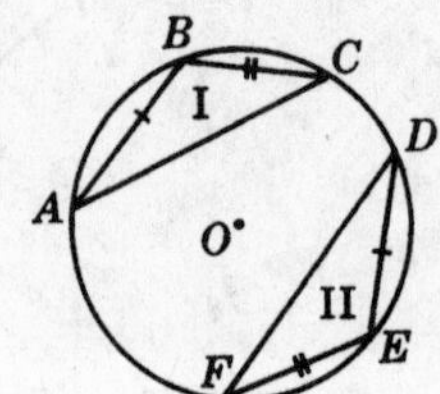

PROOF:

Statements	Reasons
1. $\overline{AB} \cong \overline{DE}, \overline{BC} \cong \overline{EF}$	1. Given
2. $\overset{\frown}{AB} \cong \overset{\frown}{DE}, \overset{\frown}{BC} \cong \overset{\frown}{EF}$	2. In a circle, ≅ chords have ≅ arcs.
3. $\overset{\frown}{ABC} \cong \overset{\frown}{DEF}$	3. If equals are added to equals, the sums are equal. Definition of ≅ arcs.
4. $\overline{AC} \cong \overline{DF}$	4. In a circle, ≅ arcs have ≅ chords.
5. $\triangle I \cong \triangle II$	5. SSS
6. $\angle B \cong \angle E$	6. Corresponding parts of congruent △s are ≅.

6.4 Proving a circle problem stated in words

Prove that if a radius bisects a chord, then it is perpendicular to the chord.

Solution

Given: Circle O
$\overline{OC}$ bisects $\overline{AB}$.

To Prove: $\overline{OC} \perp \overline{AB}$

Plan: Prove $\triangle AOD \cong \triangle BOD$ to show $\angle 1 \cong \angle 2$.
Also, $\angle 1$ and $\angle 2$ are supplementary.

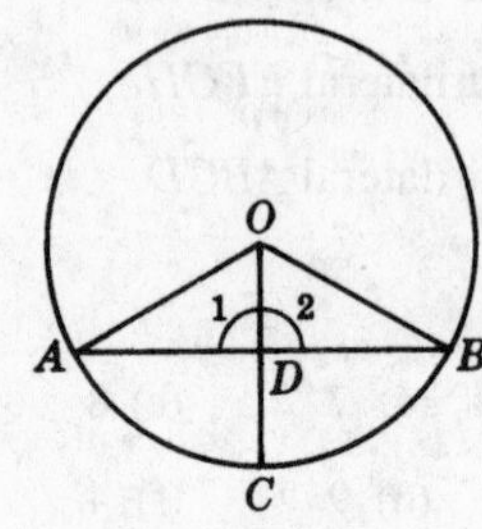

PROOF:

Statements	Reasons
1. Draw $\overline{OA}$ and $\overline{OB}$.	1. A straight line segment may be drawn between any two points.
2. $\overline{OA} \cong \overline{OB}$	2. Radii of a circle are congruent.
3. $\overline{OC}$ bisects $\overline{AB}$.	3. Given
4. $\overline{AD} \cong \overline{DB}$	4. To bisect is to divide into two ≅ parts.
5. $\overline{OD} \cong \overline{OD}$	5. Reflexive property
6. $\triangle AOD \cong \triangle BOD$	6. SSS
7. $\angle 1 \cong \angle 2$	7. Corresponding parts of congruent △s are ≅.
8. $\angle 1$ is the supplement of $\angle 2$.	8. Adjacent ∠s are supplementary if exterior sides lie in a straight line.
9. $\angle 1$ and $\angle 2$ are right angles.	9. Congruent supplementary angles are right angles.
10. $\overline{OC} \perp \overline{AB}$.	10. Rt. ∠s are formed by perpendiculars.

6.2 Tangents

The *length of a tangent* from a point to a circle is the length of the segment of the tangent from the given point to the point of tangency. Thus, *PA* is the length of the tangent from *P* to circle *O* in Fig. 6-16.

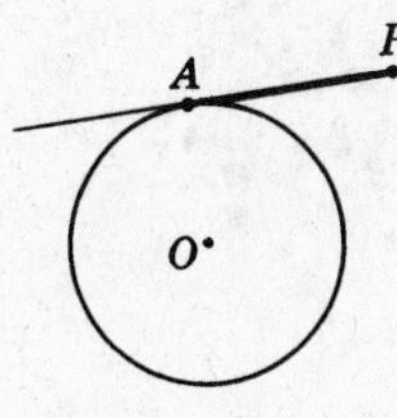

Fig. 6-16

6.2A Tangent Principles

PRINCIPLE 1: *A tangent is perpendicular to the radius drawn to the point of contact.*

Thus if $\overleftrightarrow{AB}$ is a tangent to circle *O* at *P* in Fig. 6-17, and $\overline{OP}$ is drawn, then $\overline{AB} \perp \overline{OP}$.

PRINCIPLE 2: *A line is tangent to a circle if it is perpendicular to a radius at its outer end.*

Thus if $\overleftrightarrow{AB} \perp$ radius $\overline{OP}$ at *P* of Fig. 6-17, then $\overleftrightarrow{AB}$ is tangent to circle *O*.

Fig. 6-17 Fig. 6-18

PRINCIPLE 3: *A line passes through the center of a circle if it is perpendicular to a tangent at its point of contact.*

Thus if $\overleftrightarrow{AB}$ is tangent to circle *O* at *P* in Fig. 6-18, and $\overline{CP} \perp \overleftrightarrow{AB}$ at *P*, then $\overline{CP}$ extended will pass through the center *O*.

PRINCIPLE 4: *Tangents to a circle from an outside point are congruent.*

Thus if $\overline{AP}$ and $\overline{AQ}$ are tangent to circle *O* at *P* and *Q* (Fig. 6-19), then $\overline{AP} \cong \overline{AQ}$.

PRINCIPLE 5: *The segment from the center of a circle to an outside point bisects the angle between the tangents from the point to the circle.*

Thus $\overline{OA}$ bisects $\angle PAQ$ in Fig. 6-19 if $\overline{AP}$ and $\overline{AQ}$ are tangents to circle *O*.

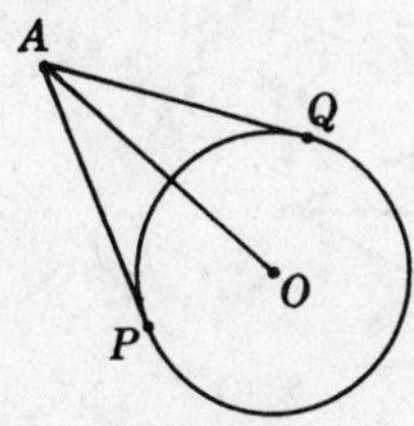

Fig. 6-19

6.2B Two Circles in Varying Relative Positions

The *line of centers of two circles* is the line joining their centers. Thus, $\overline{OO'}$ is the line of centers of circles O and O' in Fig. 6-20.

Fig. 6-20 Fig. 6-21

Circles Tangent Externally

Circles O and O' in Fig. 6-21 are tangent externally at P. $\overleftrightarrow{AB}$ is the common internal tangent of both circles. The line of centers $\overline{OO'}$ passes through P, is perpendicular to $\overleftrightarrow{AB}$, and is equal in length to the sum of the radii, $R + r$. Also $\overleftrightarrow{AB}$ bisects each of the common external tangents, $\overline{CD}$ and $\overline{EF}$.

Circles Tangent Internally

Circles O and O' in Fig. 6-22 are tangent internally at P. $\overleftrightarrow{AB}$ is the common external tangent of both circles. The line of centers $\overline{OO'}$ if extended passes through P, is perpendicular to $\overleftrightarrow{AB}$, and is equal in length to the difference of the radii, $R - r$.

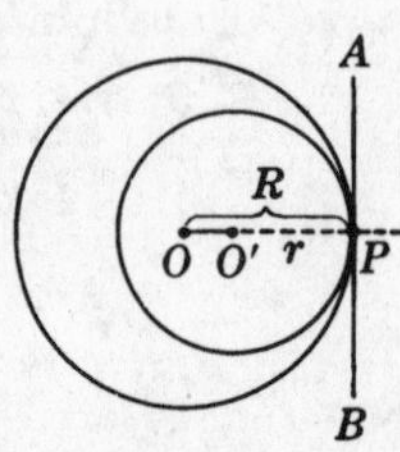

Fig. 6-22

Overlapping Circles

Circles O and O' in Fig. 6-23 overlap. Their common chord is $\overline{AB}$. If the circles are unequal, their (equal) common external tangents $\overrightarrow{CD}$ and $\overrightarrow{EF}$ meet at P. The line of centers $\overline{OO'}$ is the perpendicular bisector of $\overline{AB}$ and, if extended, passes through P.

Circles Outside Each Other

Circles O and O' in Fig. 6-24 are entirely outside each other. The common internal tangents, $\overline{AB}$ and $\overline{CD}$ meet at P. If the circles are unequal, their common external tangents, $\overline{EF}$ and $\overline{GH}$ if extended, meet at P'. The line of centers $\overline{OO'}$ passes through P and P'. Also, $AB = CD$ and $EF = GH$.

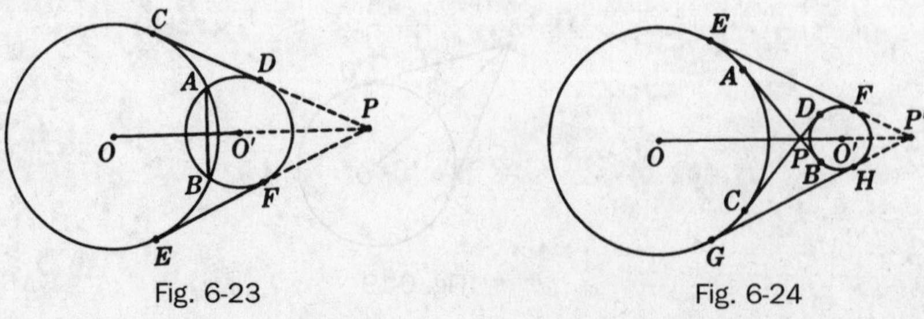

Fig. 6-23 Fig. 6-24

SOLVED PROBLEMS

6.5 Triangles and quadrilaterals having tangent sides

Points P, Q, and R in Fig. 6-25 are points of tangency.

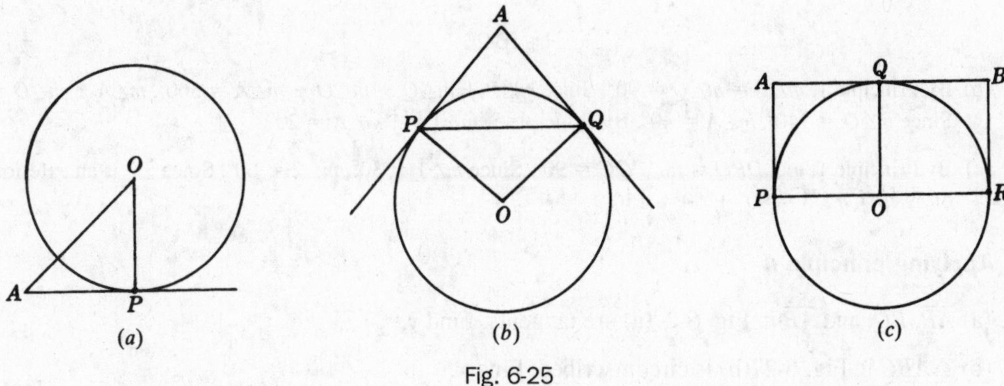

Fig. 6-25

(a) In Fig. 6-25(a), if $AP = OP$, what kind of triangle is OPA?

(b) In Fig. 6-25(b), if $AP = PQ$, what kind of triangle is APQ?

(c) In Fig. 6-25(b), if $AP = OP$, what kind of quadrilateral is $OPAQ$?

(d) In Fig. 6-25(c), if $\overline{OQ} \perp \overline{PR}$, what kind of quadrilateral is $PABR$?

Solutions

(a) $\overline{AP}$ is tangent to the circle at P; then by Principle 1, $\angle OPA$ is a right angle. Also, $AP = OP$. Hence, $\triangle OAP$ is an isosceles right triangle.

(b) $\overline{AP}$ and $\overline{AQ}$ are tangents from a point to the circle; hence by Principle 4, $AP = AQ$. Also, $AP = PQ$. Then $\triangle APQ$ is an equilateral triangle.

(c) By Principle 4, $AP = AQ$. Also, $\overline{OP}$ and $\overline{OQ}$ are $\cong$ radii. And $AP = OP$. By Principle 1, $\angle APO$ is a rt. $\angle$. Then $AP = AQ = OP = OQ$; hence, $OPAQ$ is a rhombus with a right angle, or a square.

(d) By Principle 1, $\overline{AP} \perp \overline{PR}$ and $\overline{BR} \perp \overline{PR}$. Then $\overline{AP} \parallel \overline{BR}$, since both are $\perp$ to $\overline{PR}$. By Principle 1, $\overline{AB} \perp \overline{OQ}$; also, $\overline{PR} \perp \overline{OQ}$ (Given). Then $\overline{AB} \parallel \overline{PR}$, since both are $\perp$ to $\overline{OQ}$. Hence, $PABR$ is a parallelogram with a right angle, or a rectangle.

6.6 Applying principle 1

(a) In Fig. 6-26(a), $\overline{AP}$ is a tangent. Find $\angle A$ if $m\angle A$: $m\angle O = 2{:}3$.

(b) In Fig. 6-26(b), $\overline{AP}$ and $\overline{AQ}$ are tangents. Find $m\angle 1$ if $m\angle O = 140°$.

(c) In Fig. 6-26(c), $\overline{DP}$ and $\overline{CQ}$ are tangents. Find $m\angle 2$ and $m\angle 3$ if $\angle OPD$ is trisected and $\overline{PQ}$ is a diameter.

Solutions

(a) By Principle 1, $m\angle P = 90°$. Then $m\angle A + m\angle O = 90°$. If $m\angle A = 2x$ and $m\angle O = 3x$, then $5x = 90$ and $x = 18$. Hence, $m\angle A = 36°$.

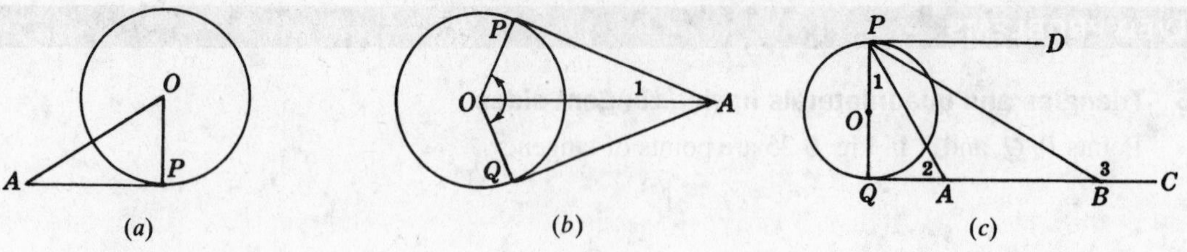

Fig. 6-26

(b) By Principle 1, $m\angle P = m\angle Q = 90°$. Since $m\angle P + m\angle Q + m\angle O + m\angle A = 360°$, $m\angle A + m\angle O = 180°$. Since $m\angle O = 140°$, $m\angle A = 40°$. By Principle 5, $m\angle 1 = \frac{1}{2}\, m\angle A = 20°$.

(c) By Principle 1, $m\angle DPQ = m\angle PQC = 90°$. Since $m\angle 1 = 30°$, $m\angle 2 = 60°$. Since $\angle 3$ is an exterior angle of $\triangle PBQ$, $m\angle 3 = 90° + 60° = 150°$.

6.7 Applying principle 4

(a) $\overline{AP}$, $\overline{BQ}$, and $\overline{AB}$ in Fig. 6-27(a) are tangents. Find y.

(b) $\triangle ABC$ in Fig. 6-27(b) is circumscribed. Find x.

(c) Quadrilateral $ABCD$ in Fig. 6-27(c) is circumscribed. Find x.

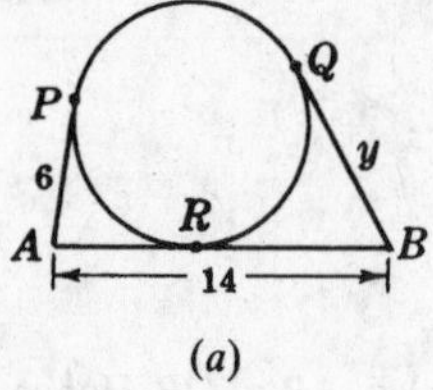

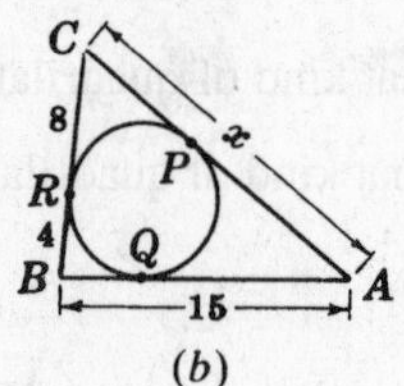

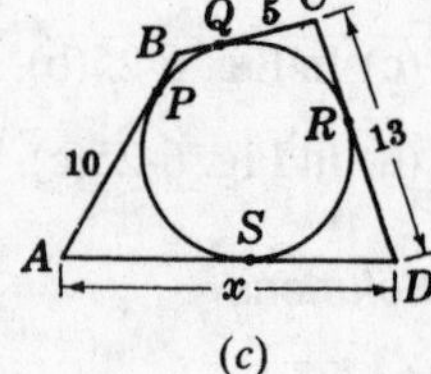

Fig. 6-27

Solutions

(a) By Principle 4, $AR = 6$, and $RB = y$. Then $RB = AB - AR = 14 - 6 = 8$. Hence, $y = RB = 8$.

(b) By Principle 4, $PC = 8$, $QB = 4$, and $AP = AQ$. Then $AQ = AB - QB = 11$. Hence, $x = AP + PC = 11 + 8 = 19$.

(c) By Principle 4, $AS = 10$, $CR = 5$, and $RD = SD$. Then $RD = CD - CR = 8$. Hence, $x = AS + SD = 10 + 8 = 18$.

6.8 Finding the line of centers

Two circles have radii of 9 and 4, respectively. Find the length of their line of centers (a) if the circles are tangent externally, (b) if the circles are tangent internally, (c) if the circles are concentric, (d) if the circles are 5 units apart. (See Fig. 6-28.)

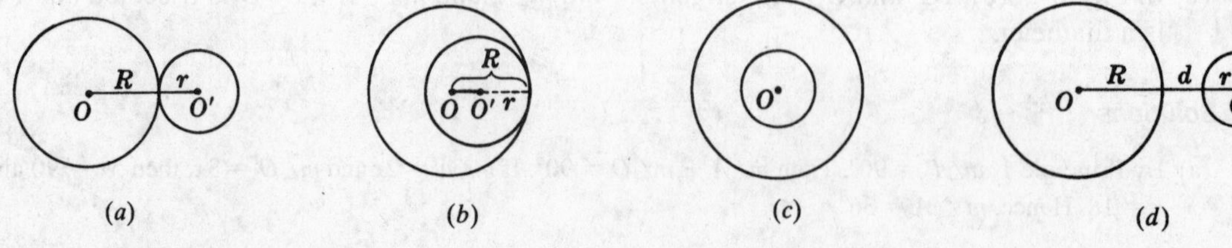

Fig. 6-28

Solutions

Let R = radius of larger circle, r = radius of smaller circle.

(a) Since $R = 9$ and $r = 4$, $OO' = R + r = 9 + 4 = 13$.

(b) Since $R = 9$ and $r = 4$, $OO' = R - r = 9 - 4 = 5$.

(c) Since the circles have the same center, their line of centers has zero length.

(d) Since $R = 9$, $r = 4$, and $d = 5$, $OO' = R + d + r = 9 + 5 + 4 = 18$.

6.9 Proving a tangent problem stated in words

Prove: Tangents to a circle from an outside point are congruent (Principle 4).

Given: Circle O
$\overline{AP}$ is tangent at P.
$\overline{AQ}$ is tangent at Q.

To Prove: $\overline{AP} \cong \overline{AQ}$

Plan: Draw $\overline{OP}$, $\overline{OQ}$, and $\overline{OA}$ and prove $\triangle AOP \cong \triangle AOQ$.

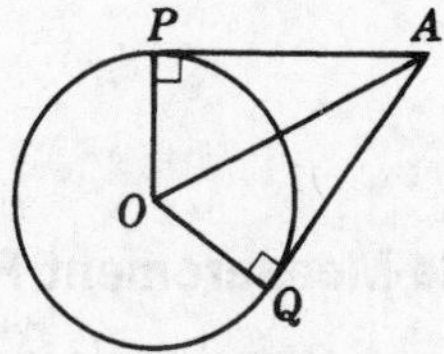

PROOF:

Statements	Reasons
1. Draw $\overline{OP}$, $\overline{OQ}$, and $\overline{OA}$.	1. A straight line may be drawn between any two points.
2. $\overline{OP} \cong \overline{OQ}$	2. Radii of a circle are congruent.
3. $\angle P$ and $\angle Q$ are right angles.	3. A tangent is $\perp$ to radius drawn to point of contact.
4. $\overline{OA} \cong \overline{OA}$	4. Reflexive property
5. $\triangle AOP \cong \triangle AOQ$	5. Hy-leg
6. $\overline{AP} \cong \overline{AQ}$	6. Corresponding parts of congruent $\triangle$s are congruent.

6.3 Measurement of Angles and Arcs in a Circle

A *central angle* has the same number of degrees as the arc it intercepts. Thus, as shown in Fig. 6-29, a central angle which is a right angle intercepts a 90° arc; a 40° central angle intercepts a 40° arc, and a central angle which is a straight angle intercepts a semicircle of 180°.

Since the numerical measures in degrees of both the central angle and its intercepted arc are the same, we may restate the above principle as follows: A central angle is measured by its intercepted arc. The symbol $\stackrel{\circ}{=}$ may be used to mean "is measured by." (Do not say that the central angle equals its intercepted arc. An angle cannot *equal* an arc.)

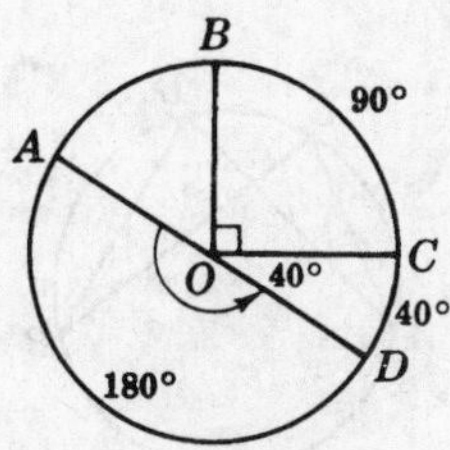

Fig. 6-29

An *inscribed angle* is an angle whose vertex is on the circle and whose sides are chords. *An angle inscribed in an arc* has its vertex on the arc and its sides passing through the ends of the arc. Thus, $\angle A$ in Fig. 6-30 is an inscribed angle whose sides are the chords $\overline{AB}$ and $\overline{AC}$. Note that $\angle A$ intercepts $\overparen{BC}$ and is inscribed in $\overparen{BAC}$.

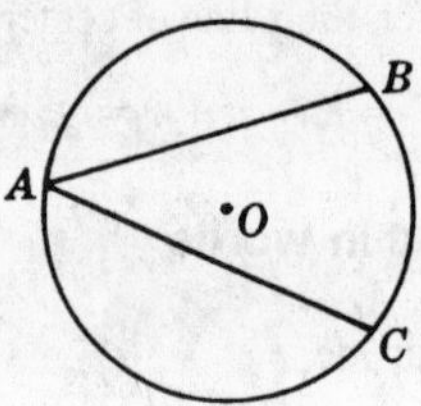

Fig. 6-30

6.3A Angle-Measurement Principles

PRINCIPLE 1: *A central angle is measured by its intercepted arc.*

PRINCIPLE 2: *An inscribed angle is measured by one-half its intercepted arc.*

A proof of this principle is given in Chapter 16.

PRINCIPLE 3: *In the same or congruent circles, congruent inscribed angles have congruent intercepted arcs.*

Thus in Fig. 6-31, if $\angle 1 \cong \angle 2$, then $\overparen{BC} \cong \overparen{DE}$.

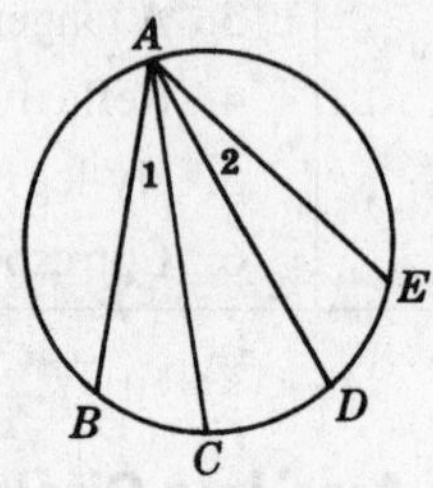

Fig. 6-31

PRINCIPLE 4: *In the same or congruent circles, inscribed angles having congruent intercepted arcs are congruent.* (This is the converse of Principle 3.)

Thus in Fig. 6-31, if $\overparen{BC} \cong \overparen{DE}$, then $\angle 1 \cong \angle 2$.

PRINCIPLE 5: *Angles inscribed in the same or congruent arcs are congruent.*

Thus in Fig. 6-32, if $\angle C$ and $\angle D$ are inscribed in $\overparen{ACB}$, then $\angle C \cong \angle D$.

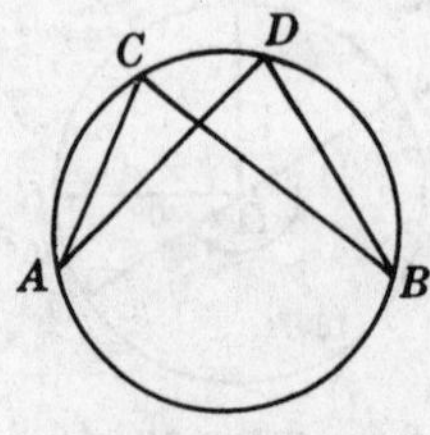

Fig. 6-32

PRINCIPLE 6: *An angle inscribed in a semicircle is a right angle.*

Thus in Fig. 6-33, since $\angle C$ is inscribed in semicircle $\widehat{ACD}$, $m\angle C = 90°$.

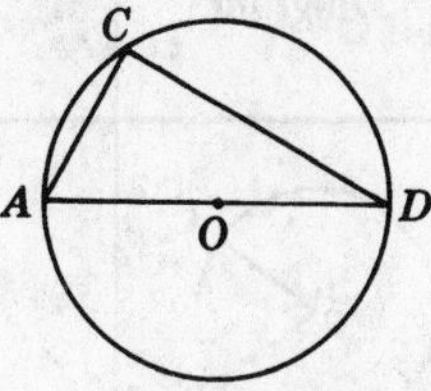

Fig. 6-33

PRINCIPLE 7: *Opposite angles of an inscribed quadrilateral are supplementary.*

Thus in Fig. 6-34, if *ABCD* is an inscribed quadrilateral, $\angle A$ is the supplement of $\angle C$.

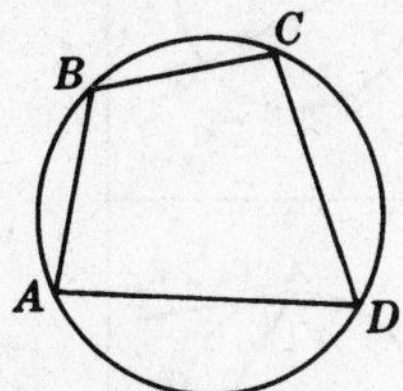

Fig. 6-34

PRINCIPLE 8: *Parallel lines intercept congruent arcs on a circle.*

Thus in Fig. 6-35, if $\overline{AB} \parallel \overline{CD}$, then $\widehat{AC} \cong \widehat{BD}$. If tangent $\overleftrightarrow{FG}$ is parallel to $\overline{CD}$, then $\widehat{PC} \cong \widehat{PD}$.

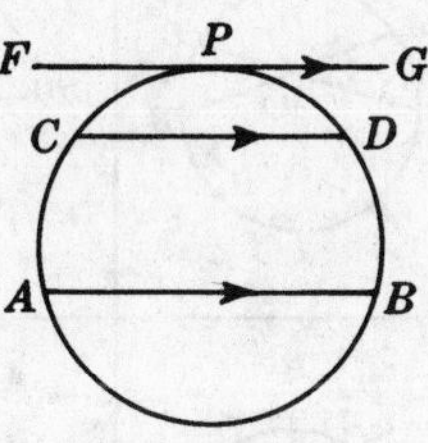

Fig. 6-35

PRINCIPLE 9: *An angle formed by a tangent and a chord is measured by one-half its intercepted arc.*

PRINCIPLE 10: *An angle formed by two intersecting chords is measured by one-half the sum of the intercepted arcs.*

PRINCIPLE 11: *An angle formed by two secants intersecting outside a circle is measured by one-half the difference of the intercepted arcs.*

PRINCIPLE 12: *An angle formed by a tangent and a secant intersecting outside a circle is measured by one-half the difference of the intercepted arcs.*

PRINCIPLE 13: *An angle formed by two tangents intersecting outside a circle is measured by one-half the difference of the intercepted arcs.*

Proofs of Principles 10 to 13 are given in Chapter 16.

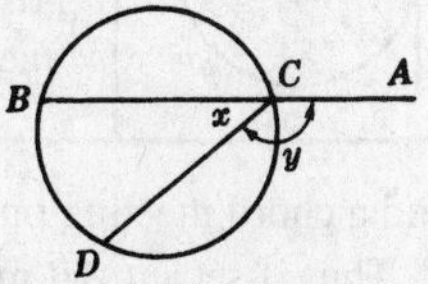

Fig. 6-36

6.3B Table of Angle-Measurement Principles

Position of Vertex	Kind of Angle	Diagram	Measurement Formula	Method of Measurement
Center of circle	Central angle (apply Principle 1)	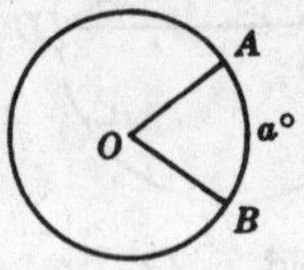	$\angle O \doteq \overset{\frown}{AB}$ $m\angle O = a°$	By intercepted arc
On the circle	Inscribed angle (apply Principle 2)	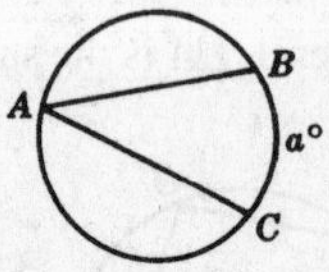	$\angle A \doteq \frac{1}{2}\overset{\frown}{BC}$ $m\angle A = \frac{1}{2}a°$	By one-half intercepted arc
	Angle formed by a tangent and a chord (apply Principle 9)	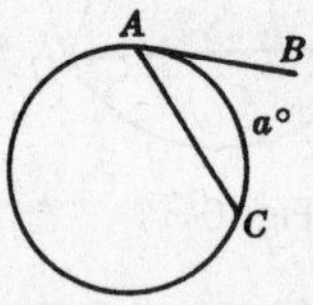		
Inside the circle	Angle formed by two intersecting chords (apply Principle 10)	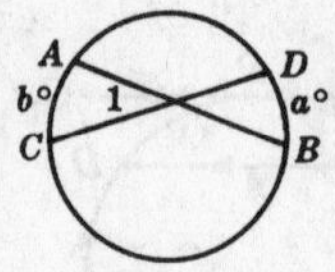	$\angle 1 \doteq \frac{1}{2}(\overset{\frown}{AC}+\overset{\frown}{BD})$ $m\angle 1 = \frac{1}{2}(a° + b°)$	By one-half sum of intercepted arcs
Outside the circle	Angle formed by two secants (apply Principle 10)	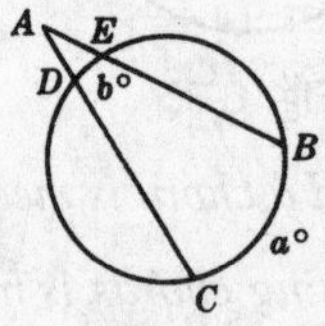	$\angle A \doteq \frac{1}{2}(\overset{\frown}{BC}-\overset{\frown}{DE})$ $m\angle A = \frac{1}{2}(a° - b°)$	By one-half difference of intercepted arcs
	Angle formed by a secant and a tangent (apply Principle 12)	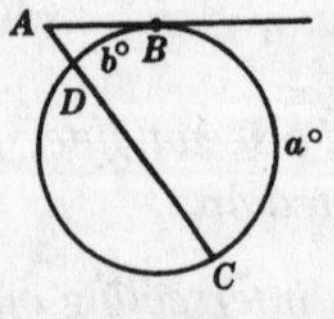	$\angle A \doteq \frac{1}{2}(\overset{\frown}{BC}-\overset{\frown}{BD})$ $m\angle A = \frac{1}{2}(a° - b°)$	
	Angle formed by two tangents (apply Principle 13)	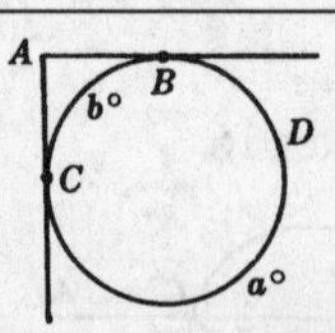	$\angle A \doteq \frac{1}{2}(\overset{\frown}{BDC}-\overset{\frown}{BC})$ $m\angle A = \frac{1}{2}(a° - b°)$ Also, $m\angle A = (180 - b)°$	

Note: To find the angle formed by a secant and a chord meeting on the circle, first find the measure of the inscribed angle adjacent to it and then subtract from 180°. Thus if secant $\overline{AB}$ meets chord $\overline{CD}$ at C on the circle in Fig. 6-36, to find $m\angle y$, first find the measure of inscribed $\angle x$. Obtain $m\angle y$ by subtracting $m\angle x$ from 180°.

SOLVED PROBLEMS

6.10 Applying principles 1 and 2

(a) In Fig. 6-37(a), if $m\angle y = 46°$, find $m\angle x$.

(b) In Fig. 6-37(b), if $m\angle y = 112°$, find $m\angle x$.

(c) In Fig. 6-37(c), if $m\angle x = 75°$, find $m\widehat{y}$.

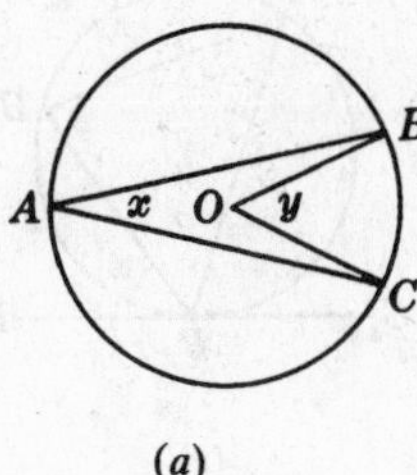

(a)

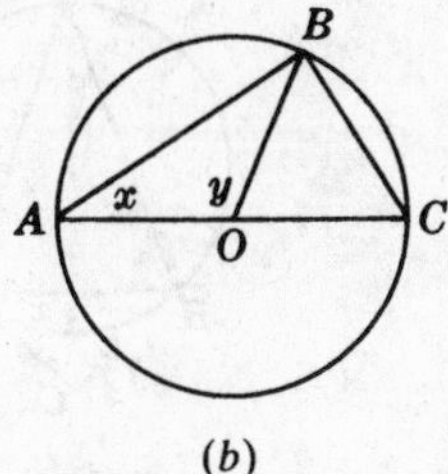

(b)

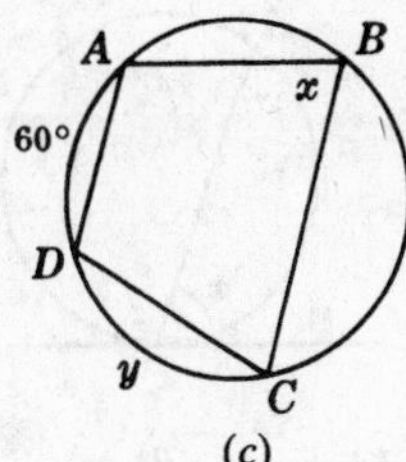

(c)

Fig. 6-37

Solutions

(a) $\angle y \doteq \widehat{BC}$, so $m\widehat{BC} = 46°$. Then $\angle x \doteq \frac{1}{2}\widehat{BC} = \frac{1}{2}(46°) = 23°$, so $m\angle x = 23°$.

(b) $\angle y \doteq \widehat{AB}$ so $m\widehat{AB} = 112°$.
$m\widehat{BC} = m(\widehat{ABC} - \widehat{AB}) = 180° - 112° = 68°$. Then $\angle x \doteq \frac{1}{2}\widehat{BC} = \frac{1}{2}(68°) = 34°$, so $m\angle x = 34°$.

(c) $\angle x \doteq \frac{1}{2}\widehat{ADC}$, so $m\widehat{ADC} = 150°$. Then $m\widehat{y} = m(\widehat{ADC} - \widehat{AD}) = 150° - 60° = 90°$.

6.11 Applying principles 3 to 8

Find x and y in each part of Fig. 6-38.

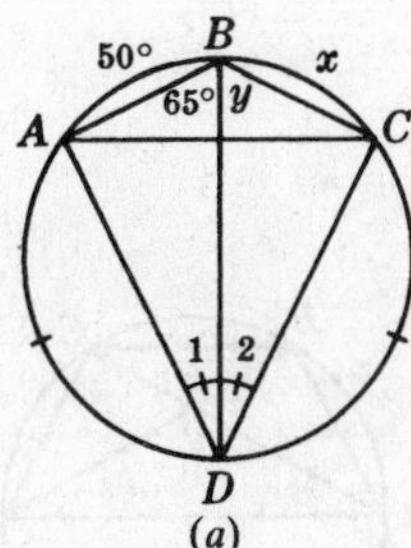

(a)

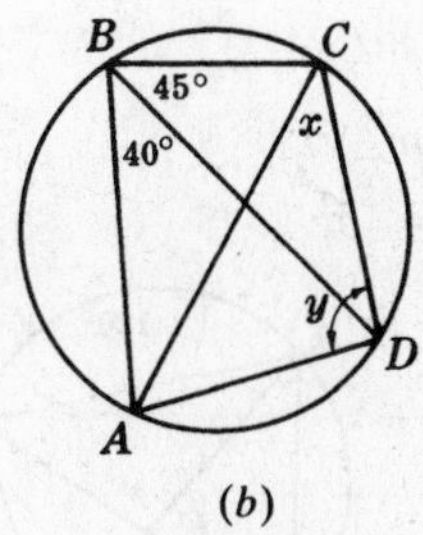

(b)

(c)

Fig. 6-38

Solutions

(a) Since $m\angle 1 = m\angle 2$, $m\widehat{x} = m\widehat{AB} = 50°$. Since $\widehat{AD} \cong \widehat{CD}$, $m\angle y = m\angle ABD = 65°$.

(b) $\angle ABD$ and $\angle x$ are inscribed in $\widehat{ABD}$; hence, $m\angle x = m\angle ABD = 40°$.
$ABCD$ is an inscribed quadrilateral; hence, $m\angle y = 180° - m\angle B = 95°$.

(c) Since $\angle x$ is inscribed in a semicircle, $m\angle x = 90°$. Since $\overline{AC} \parallel \overline{DE}$, $m\widehat{y} = m\widehat{CE} = 70°$.

6.12 Applying principle 9

In each part of Fig. 6-39, CD is a tangent at P.

(a) If $m\widehat{y} = 220°$ in part (a), find $m\angle x$.

(b) If $m\widehat{y} = 140°$ in part (b), find $m\angle x$.

(c) If $m\angle y = 75°$ in part (c), find $m\angle x$.

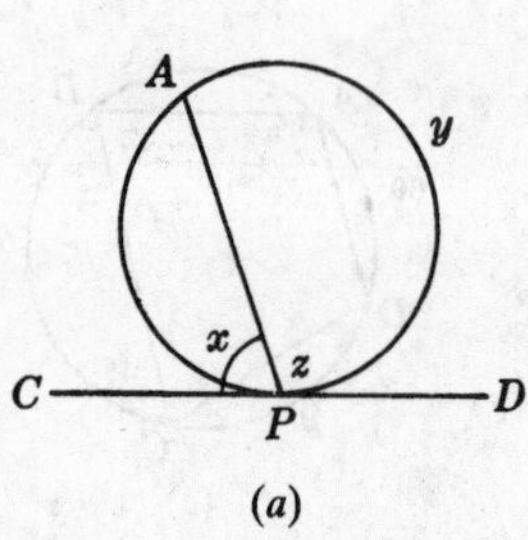

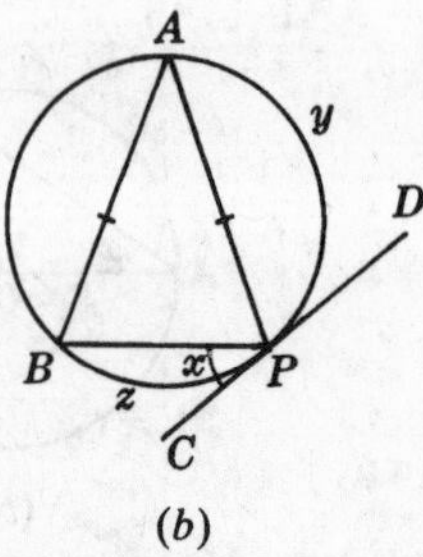

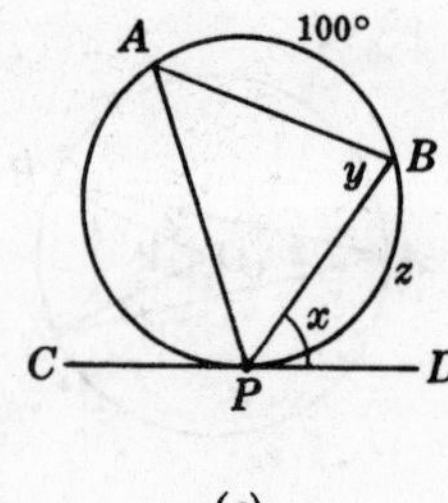

Fig. 6-39

Solutions

(a) $\angle z \overset{\circ}{=} \frac{1}{2}\widehat{y} = \frac{1}{2}(220°) = 110°$. So $m\angle x = 180° - 110° = 70°$.

(b) Since $AB = AP$, $m\widehat{AB} = m\widehat{y} = 140°$. Then $m\widehat{z} = 360° - 140° - 140° = 80°$. Since $\angle x \overset{\circ}{=} \frac{1}{2}\widehat{z} = 40°$, $m\angle x = 40°$.

(c) $\angle y \overset{\circ}{=} \frac{1}{2}\widehat{AP}$, so $m\widehat{AP} = 150°$. Then $m\widehat{z} = 360° - 100° - 150° = 110°$. Since $\angle x \overset{\circ}{=} \frac{1}{2}\widehat{z} = 55°$, $m\angle x = 55°$.

6.13 Applying Principle 10

(a) If $m\angle x = 95°$ in Fig. 6-40(a), find $m\widehat{y}$.

(b) If $m\widehat{y} = 80°$ in Fig. 6-40(b), find $m\angle x$.

(c) If $m\widehat{x} = 78°$ in Fig. 6-40(c), find $m\angle y$.

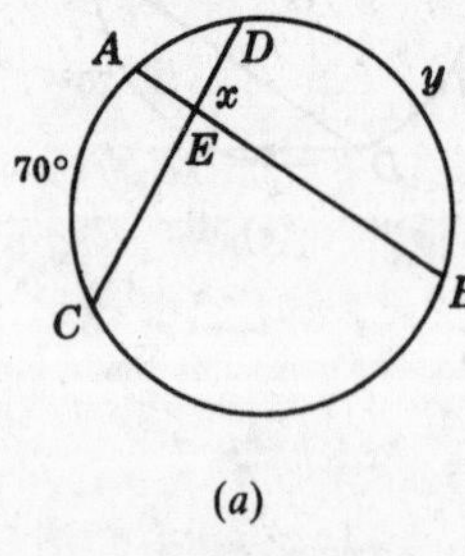

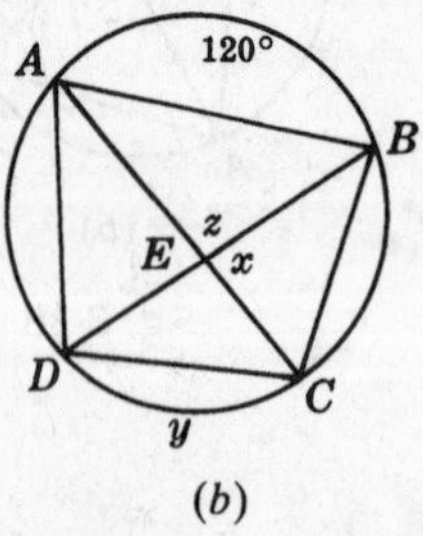

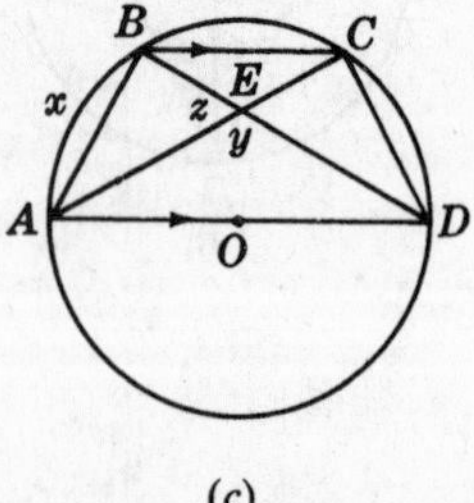

Fig. 6-40

Solutions

(a) $\angle x \overset{\circ}{=} \frac{1}{2}(\widehat{AC} + \widehat{y})$; thus $95° = \frac{1}{2}(70° + m\widehat{y})$, so $m\widehat{y} = 120°$.

(b) $\angle z \overset{\circ}{=} \frac{1}{2}(\widehat{y} + \widehat{AB}) = \frac{1}{2}(80° + 120°) = 100°$. Then $m\angle x = 180° - m\angle z = 80°$.

(c) Because $\overline{BC} \parallel \overline{AD}$, $m\widehat{CD} = m\widehat{x} = 78°$. Also, $\angle z \overset{\circ}{=} \frac{1}{2}(\widehat{x} + \widehat{CD}) = 78°$. Then $m\angle y = 180° - m\angle z = 102°$.

6.14 Applying principles 11 to 13

(a) If $m\angle x = 40°$ in Fig. 6-41(a), find $m\overparen{y}$.

(b) If $m\angle x = 67°$ in Fig. 6-41(b), find $m\overparen{y}$.

(c) If $m\angle x = 61°$ in Fig. 6-41(c), find $m\overparen{y}$.

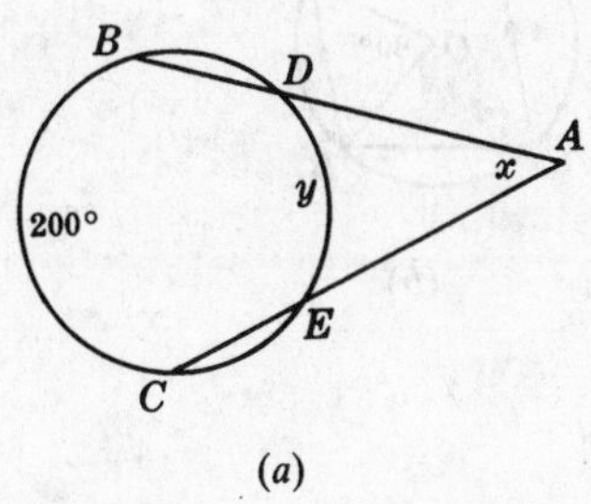

(a)

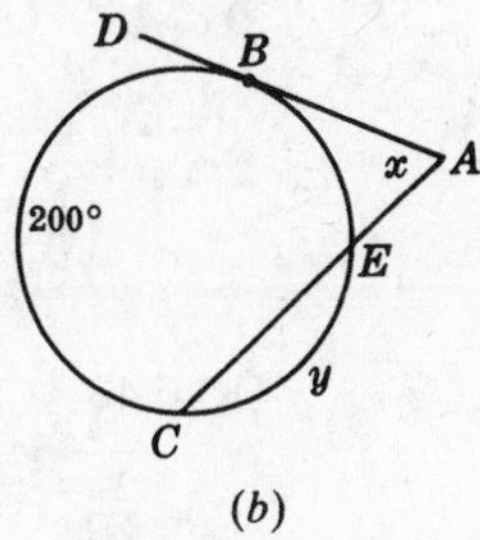

(b)

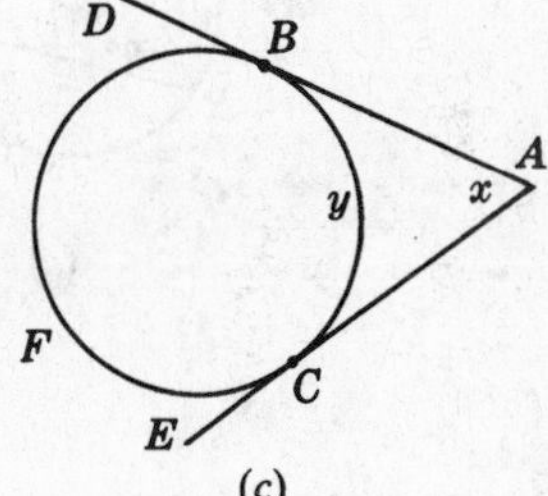

(c)

Fig. 6-41

Solutions

(a) $\angle x \overset{\circ}{=} \frac{1}{2}(\overparen{BC} - \overparen{y})$, so $40° = \frac{1}{2}(200° - m\overparen{y})$ or $m\overparen{y} = 120°$.

(b) $\angle x \overset{\circ}{=} \frac{1}{2}(\overparen{BC} - \overparen{BE})$, so $67° = \frac{1}{2}(200° - m\overparen{BE})$ or $m\overparen{BE} = 66°$.
Then $m\overparen{y} = 360° - 200° - 66° = 94°$.

(c) $\angle x \overset{\circ}{=} \frac{1}{2}(\overparen{BFC} - \overparen{y}$ and $m\overparen{BFC} = 360° - m\overparen{y}$. Then $61° = \frac{1}{2}[(360° - m\overparen{y}) - m\overparen{y}] = 180° - m\overparen{y}$. Thus $m\overparen{y} = 119°$.

6.15 Using equations in two unknowns to find arcs

In each part of Fig. 6-42, find x and y using equations in two unknowns.

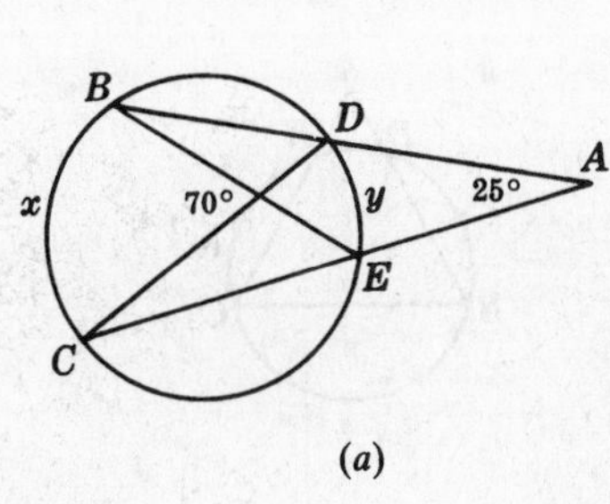

(a)

(b)

Fig. 6-42

Solutions

(a) By Principle 10, $70° = \frac{1}{2}(m\overparen{x} + m\overparen{y})$
By Principle 11, $25° = \frac{1}{2}(m\overparen{x} - m\overparen{y})$
If we add these two equations, we get $m\overparen{x} = 95°$. If we subtract one from the other, we get $m\overparen{y} = 45°$.

(b) Since $m\overparen{x} + m\overparen{y} = 360°$, $\frac{1}{2}(m\overparen{x} + m\overparen{y}) = 180°$
By Principle 13, $\frac{1}{2}(m\overparen{x} - m\overparen{y}) = 62°$
If we add these two equations, we find that $m\overparen{x} = 242°$. If we subtract one from the other, we get $m\overparen{y} = 118°$.

6.16 Measuring angles and arcs in general

Find x and y in each part of Fig. 6-43.

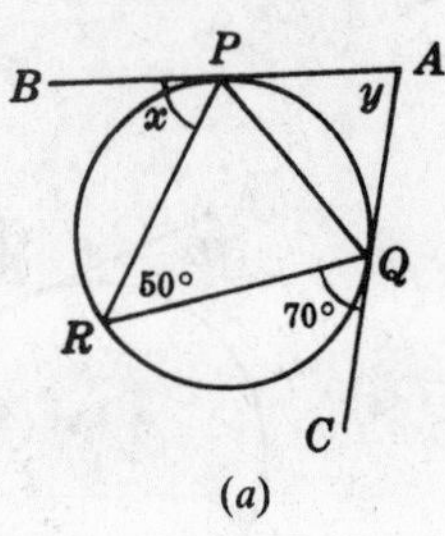

(a)

(b)

Fig. 6-43

Solutions

(a) By Principle 2, $50° = \frac{1}{2}m\overset{\frown}{PQ}$ or $m\overset{\frown}{PQ} = 100°$. Also, by Principle 9, $70° = \frac{1}{2}m\overset{\frown}{QR}$ or $m\overset{\frown}{QR} = 140°$.
Then $m\overset{\frown}{PR} = 360° - m\overset{\frown}{PQ} - m\overset{\frown}{QR} = 120°$.
By Principle 9, $x = \frac{1}{2}m\overset{\frown}{PR} = 60°$.
By Principle 13, $y = \frac{1}{2}(m\overset{\frown}{PRQ} - m\overset{\frown}{PQ}) = \frac{1}{2}(260° - 100°) = 80°$.

(b) By Principle 1, $m\overset{\frown}{AB} = 80°$. Also, by Principle 8, $m\overset{\frown}{BC} = m\overset{\frown}{PA} = 85°$. Then $m\overset{\frown}{PC} = 360° - m\overset{\frown}{PA} - m\overset{\frown}{AB} - m\overset{\frown}{BC} = 110°$.
By Principle 9, $x = \frac{1}{2}m\overset{\frown}{PC} = 55°$.
By Principle 12, $y = \frac{1}{2}(m\overset{\frown}{PCB} - m\overset{\frown}{PA}) = \frac{1}{2}(195° - 85°) = 55°$.

6.17 Proving an angle measurement problem

Given: $\overset{\frown}{BD} = \overset{\frown}{CE}$

To Prove: $AB = AC$

Plan: First prove $\overset{\frown}{CD} = \overset{\frown}{BE}$.
Use this to show that $\angle B \cong \angle C$.

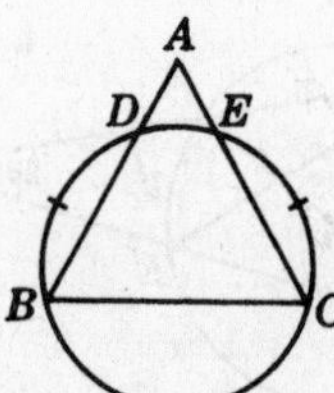

PROOF:

Statements	Reasons
1. $\overset{\frown}{BD} = \overset{\frown}{CE}$	1. Given
2. $\overset{\frown}{DE} = \overset{\frown}{DE}$	2. Reflexive property
3. $\overset{\frown}{BE} = \overset{\frown}{CD}$	3. If =s are added to =s, the sums are =.
4. $\angle B \cong \angle C$	4. In a circle, inscribed angles having equal intercepted arcs are ≅.
5. $AB = AC$	5. In a triangle, sides opposite ≅ angles are equal in length.

6.18 Proving an angle measurement problem stated in words

Prove that parallel chords drawn at the ends of a diameter are equal in length.

Solution

Given: Circle O

$\overline{AB}$ is a diameter.

$\overline{AC} \parallel \overline{BD}$

To Prove: $AC = BD$

Plan: Prove $\overset{\frown}{AC} \cong \overset{\frown}{BD}$

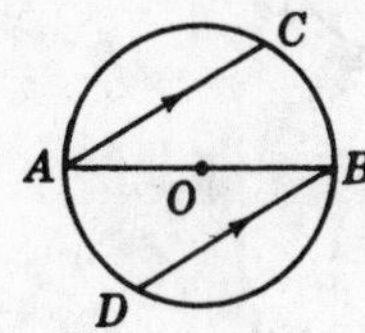

PROOF:

Statements	Reasons
1. $\overline{AB}$ is a diameter	1. Given
2. $\overset{\frown}{ACB} \cong \overset{\frown}{ADB}$	2. A diameter cuts a circle into two equal semicircles.
3. $\overline{AC} \parallel \overline{BD}$	3. Given
4. $\overset{\frown}{AD} \cong \overset{\frown}{BC}$	4. Parallel lines intercept $\cong$ arcs on a circle.
5. $\overset{\frown}{AC} \cong \overset{\frown}{BD}$	5. If equals are subtracted from equals, the differences are equal. Definition of $\cong$ arcs.
6. $AC = BD$	6. In a circle, equal arcs have chords which are equal in length.

SUPPLEMENTARY PROBLEMS

6.1. Provide the proofs requested in Fig. 6-44. (6.3)

(a) **Given:** $AB = DE$
$AC = DF$
To Prove: $\angle B \cong \angle E$

(b) **Given:** Circle O, $AB = BC$
Diameter $\overline{BD}$
To Prove: $\overline{BD}$ bisects $\angle AOC$.

(c) **Given:** Circle O
$\overset{\frown}{AB} = \overset{\frown}{CD}$
To Prove: $\angle AOC \cong \angle BOD$

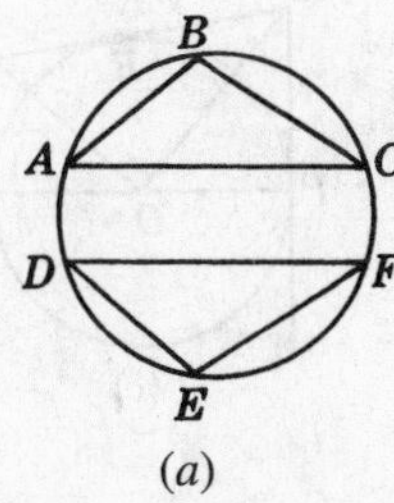

(a)

(b)

(c)

Fig. 6-44

6.2. Provide the proofs requested in Fig. 6-45. Please refer to figure 6-45(a) for problems 6.2(a) and (b); to figure 6-45(b) for problems 6.2(c) and (d); and figure 6-45(c) for problems 6.2(e) and (f). (6.3)

(a) **Given:** $AB = AC$
To Prove: $\overset{\frown}{ABC} \cong \overset{\frown}{ACB}$

(b) **Given:** $\overset{\frown}{ABC} \cong \overset{\frown}{ACB}$
To Prove: $AB = AC$

(c) **Given:** Circle O, $AB = AD$
Diameter $\overline{AC}$
To Prove: $BC = CD$

(d) **Given:** Circle O
$AB = AD$, $BC = CD$
To Prove: $\overline{AC}$ is a diameter.

(e) **Given:** $AD = BC$
To Prove: $AC = BD$

(f) **Given:** $AC = BD$
To Prove: $AD = BC$

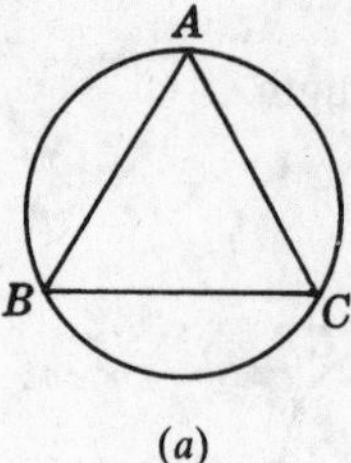

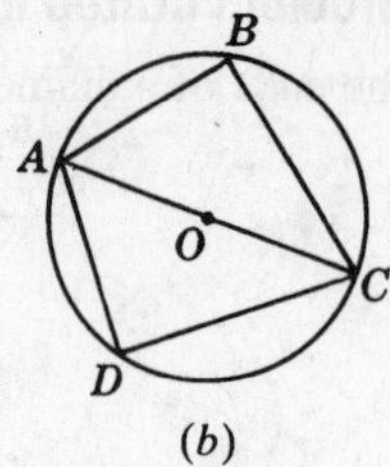

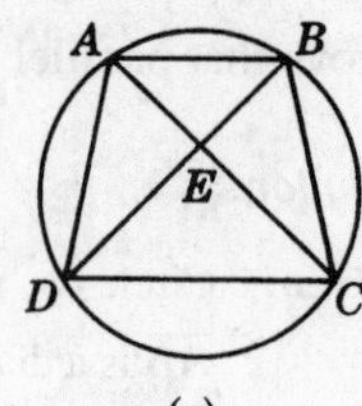

Fig. 6-45

6.3. Prove each of the following: (6.4)

(a) If a radius bisects a chord, then it bisects its arcs.

(b) If a diameter bisects the major arc of a chord, then it is perpendicular to the chord.

(c) If a diameter is perpendicular to a chord, it bisects the chord and its arcs.

6.4. Prove each of the following: (6.4)

(a) A radius through the point of intersection of two congruent chords bisects an angle formed by them.

(b) If chords drawn from the ends of a diameter make congruent angles with the diameter, the chords are congruent.

(c) In a circle, congruent chords are equally distant from the center of the circle.

(d) In a circle, chords that are equally distant from the center are congruent.

6.5. Determine each of the following, assuming t, t', and t'' in Fig. 6-46 are tangents. (6.5)

(a) If $m\angle A = 90°$ in Fig. 6-46(a), what kind of quadrilateral is $PAQO$?

(b) If $BR = RC$ in Fig. 6-46(b), what kind of triangle is ABC?

(c) What kind of quadrilateral is $PABQ$ in Fig. 6-46(c) if $\overline{PQ}$ is a diameter?

(d) What kind of triangle is AOB in Fig. 6-46(c)?

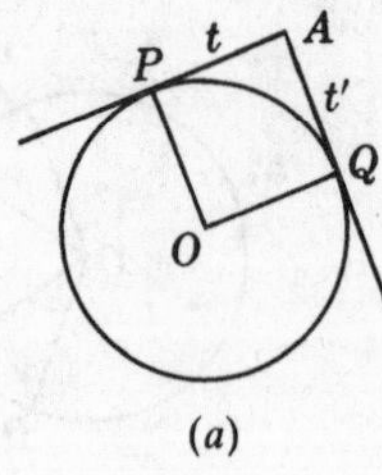

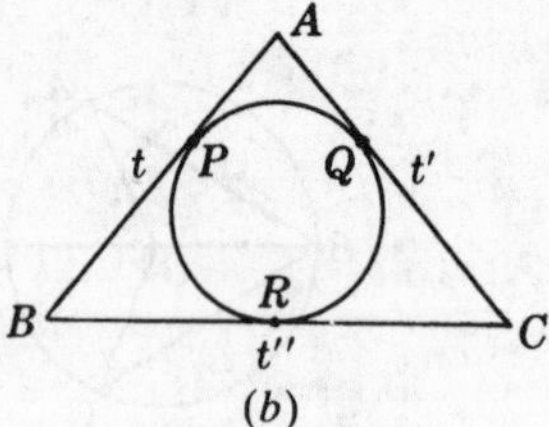

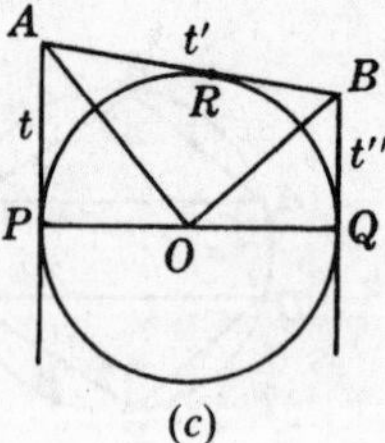

Fig. 6-46

6.6. In circle O, radii $\overline{OA}$ and $\overline{OB}$ are drawn to the points of tangency of $\overline{PA}$ and $\overline{PB}$. Find $m\angle AOB$ if $m\angle APB$ equals (a) 40°; (b) 120°; (c) 90°; (d) $x°$; (e) $(180 - x)°$; (f) $(90 - x)°$. (6.6)

6.7. Find each of the following (t and t' in Fig. 6-47 are tangents). (6.6)

In Fig. 6-47(a)

(a) If $m\angle POQ = 80°$, find $m\angle PAQ$.

(b) If $m\angle PBO = 25°$, find $m\angle 1$ and $m\angle PAQ$.

(c) If $m\angle PAQ = 72°$, find $m\angle 1$ and $m\angle PBO$.

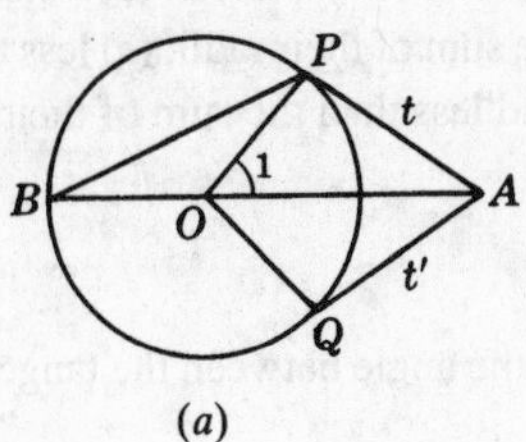

(b)

Fig. 6-47

In Fig. 6-47(b)

(d) If $\overline{PB}$ bisects $\angle APQ$, find $m\angle 2$.

(e) If $m\angle 1 = 35°$, find $m\angle 2$.

(f) If $PQ = QB$, find $m\angle 1$.

6.8. In Fig. 6-48(a), $\triangle ABC$ is circumscribed. (a) If $y = 9$, find x. (b) If $x = 25$, find y. In Fig. 6-48(b), quadrilateral $ABCD$ is circumscribed. (c) Find $AB + CD$. (d) Find perimeter of $ABCD$. In Fig. 6-48(c), quadrilateral $ABCD$ is circumscribed. (e) If $r = 10$, find x. (f) If $x = 25$, find r. (6.7)

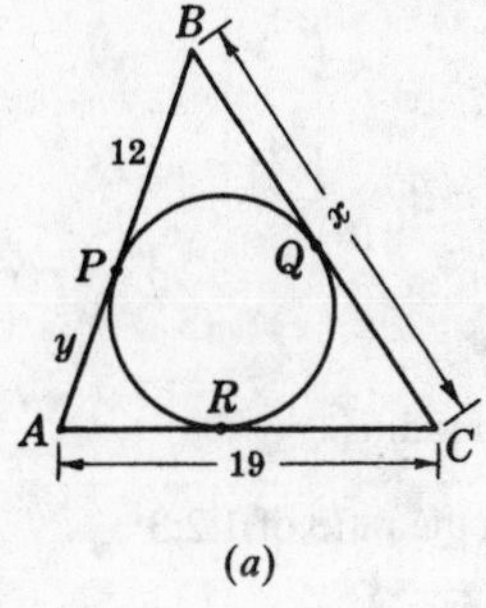

(a)

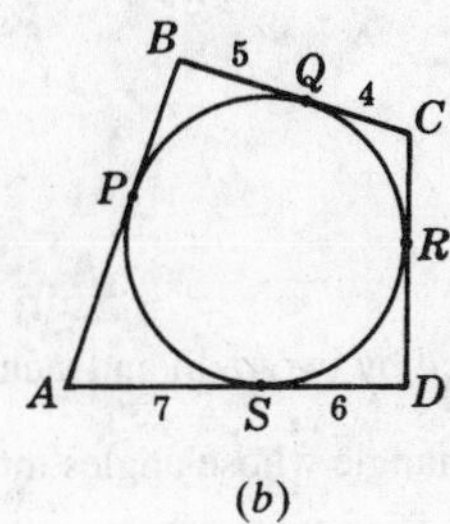

(b)

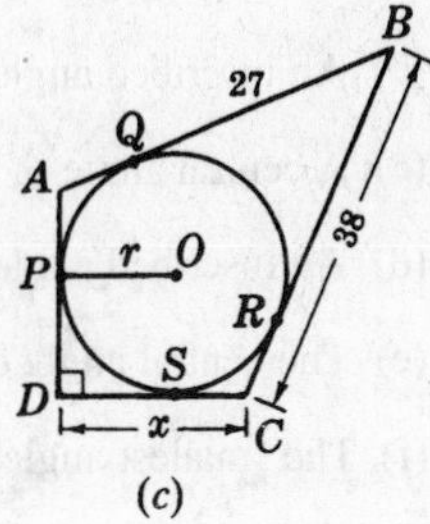

(c)

Fig. 6-48

6.9. If two circles have radii of 20 and 13, respectively, find their line of centers: (6.8)

(a) If the circles are concentric

(b) If the circles are 7 units apart

(c) If the circles are tangent externally

(d) If the circles are tangent internally

6.10. If the line of centers of two circles measures 30, what is the relation between the two circles: (6.8)

(a) If their radii are 25 and 5?

(b) If their radii are 35 and 5?

(c) If their radii are 20 and 5?

(d) If their radii are 25 and 10?

6.11. What is the relation between two circles if the length of their line of centers is (a) 0; (b) equal to the difference of their radii; (c) equal to the sum of their radii; (d) greater than the sum of their radii, (e) less than the difference of their radii and greater than 0; (f) greater than the difference and less than the sum of their radii? (6.8)

6.12. Prove each of the following: (6.9)

(a) The line from the center of a circle to an outside point bisects the angle between the tangents from the point to the circle.

(b) If two circles are tangent externally, their common internal tangent bisects a common external tangent.

(c) If two circles are outside each other, their common internal tangents are congruent.

(d) In a circumscribed quadrilateral, the sum of the lengths of the two opposite sides equals the sum of the lengths of the other two.

6.13. Find the number of degrees in a central angle which intercepts an arc of (a) 40°; (b) 90°; (c) 170°; (d) 180°; (e) $2x°$; (f) $(180 - x)°$; (g) $(2x - 2y)°$. (6.10)

6.14. Find the number of degrees in an inscribed angle which intercepts an arc of (a) 40°; (b) 90°; (c) 170°; (d) 180°; (e) 260°; (f) 348°; (g) $2x°$; (h) $(180 - x)°$; (i) $(2x - 2y)°$. (6.10)

6.15. Find the number of degrees in the arc intercepted by (6.10)

(a) A central angle of 85°

(b) An inscribed angle of 85°

(c) A central angle of $c°$

(d) An inscribed angle of $i°$

(e) The central angle of a triangle formed by two radii and a chord equal to a radius

(f) The smallest angle of an inscribed triangle whose angles intercept arcs in the ratio of 1:2:3

6.16. Find the number of degrees in each of the arcs intercepted by the angles of an inscribed triangle if the measures of these angles are in the ratio of (a) 1:2:3; (b) 2:3:4; (c) 5:6:7; (d) 1:4:5. (6.10)

6.17. (a) If $m\widehat{y} = 40°$ in Fig. 6-49(a), find $m\angle x$.

(b) If $m\angle x = 165°$ in Fig. 6-49(a), find $m\widehat{y}$.

(c) If $m\angle y = 115°$ in Fig. 6-49(b), find $m\angle x$.

(d) If $m\angle x = 108°$ in Fig. 6-49(b), find $m\angle y$.

(e) If $m\widehat{y} = 105°$ in Fig. 6-49(c), find $m\angle x$.

(f) If $m\angle x = 96°$ in Fig. 6-49(c), find $m\widehat{y}$.

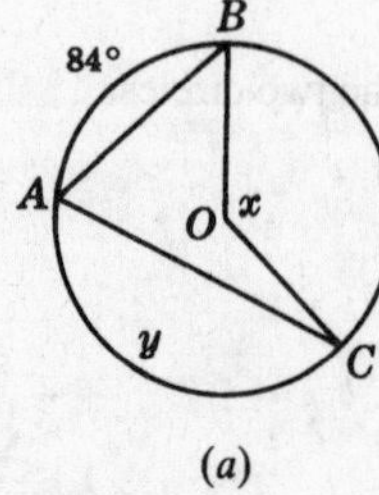

(a)

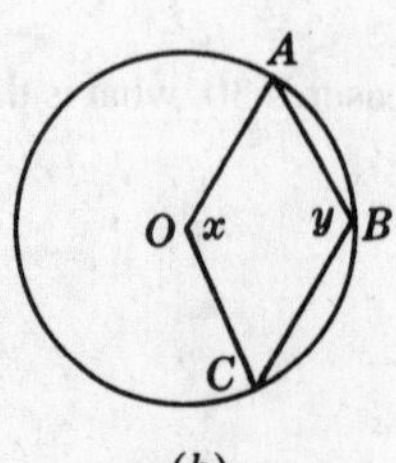

(b)

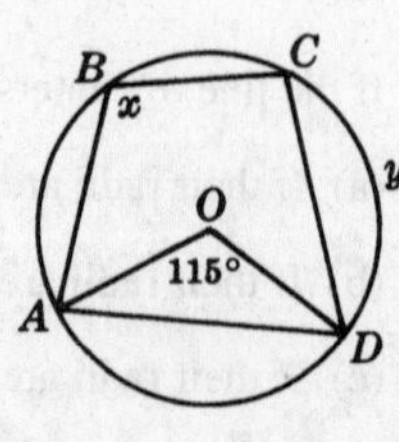

(c)

Fig. 6-49

6.18. If quadrilateral $ABCD$ is inscribed in a circle in Fig. 6-50, find (6.11)

(a) $m\angle A$ if $m\angle C = 45°$
(b) $m\angle B$ if $m\angle D = 90°$
(c) $m\angle C$ if $m\angle A = x°$
(d) $m\angle D$ if $m\angle B = (90 - x)°$
(e) $m\angle A$ if $m\widehat{BAD} = 160°$
(f) $m\angle B$ if $m\widehat{ABC} = 200°$
(g) $m\angle C$ if $m\widehat{BC} = 140°$ and $m\widehat{CD} = 110°$
(h) $m\angle D$ if $m\angle D{:}m\angle B = 2{:}3$

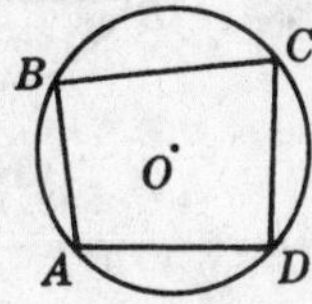

Fig. 6-50

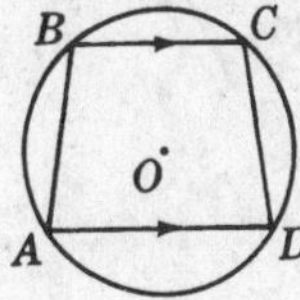

Fig. 6-51

6.19. If BC and AD are the parallel sides of inscribed trapezoid $ABCD$ in Fig. 6-51, find (6.11)

(a) $m\widehat{AB}$ if $m\widehat{CD} = 85°$
(b) $m\widehat{CD}$ if $m\widehat{AB} = y°$
(c) $m\widehat{AB}$ if $m\widehat{BC} = 60°$ and $m\widehat{AD} = 80°$
(d) $m\widehat{CD}$ if $m\widehat{AD} + m\widehat{BC} = 170°$
(e) $m\angle A$ if $m\angle D = 72°$
(f) $m\angle A$ if $m\angle C = 130°$
(g) $m\angle B$ if $m\angle C = 145°$
(h) $m\angle B$ if $m\widehat{AD} = 90°$ and $m\widehat{AB} = 84°$

6.20. A diameter is parallel to a chord. Find the number of degrees in an arc between the diameter and chord if the chord intercepts (a) a minor arc of 80°; (b) a major arc of 300°. (6.11)

6.21. Find x and y in each part of Fig. 6-52. (6.11)

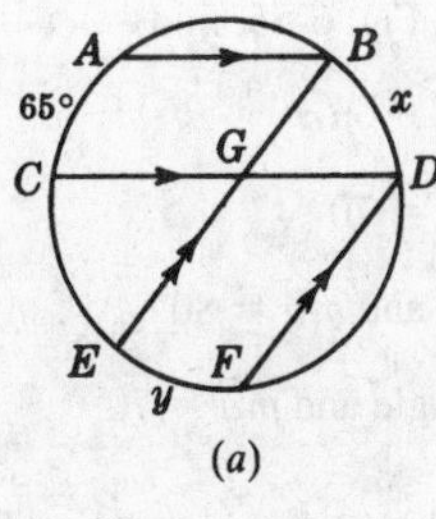

(a)

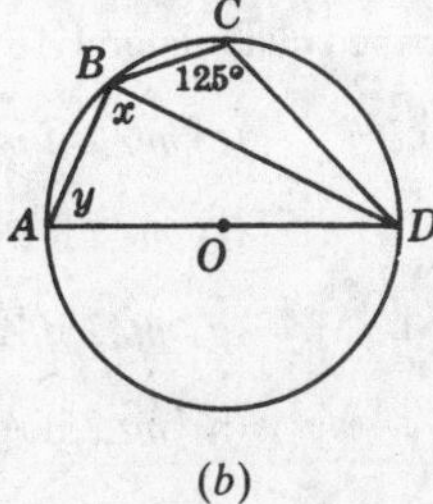

(b)

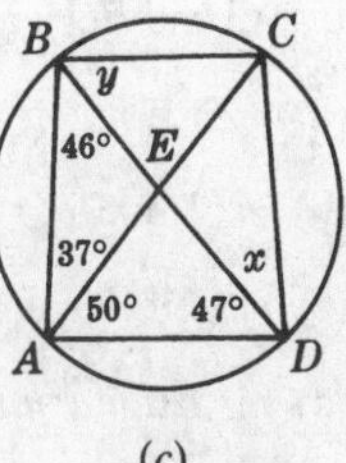

(c)

Fig. 6-52

6.22. Find the number of degrees in the angle formed by a tangent and a chord drawn to the point of tangency if the intercepted arc has measure (a) 38°; (b) 90°; (c) 138°; (d) 180°; (e) 250°; (f) 334°; (g) $x°$; (h) $(360 - x)°$; (i) $(2x + 2y)°$. (6.12)

6.23. Find the number of degrees in the arc intercepted by an angle formed by a tangent and a chord drawn to the point of tangency if the angle measures (a) 55°; (b) $67\frac{1}{2}°$; (c) 90°; (d) 135°; (e) $(90 - x)°$; (f) $(180 - x)°$; (g) $(x - y)°$; (h) $3\frac{1}{2}x°$. (6.12)

6.24. Find the number of degrees in the acute angle formed by a tangent through one vertex and an adjacent side of an inscribed (a) square; (b) equilateral triangle; (c) regular hexagon; (d) regular decagon. (6.12)

6.25. Find x and y in each part of Fig. 6-53 (t and t' are tangents). (6.12)

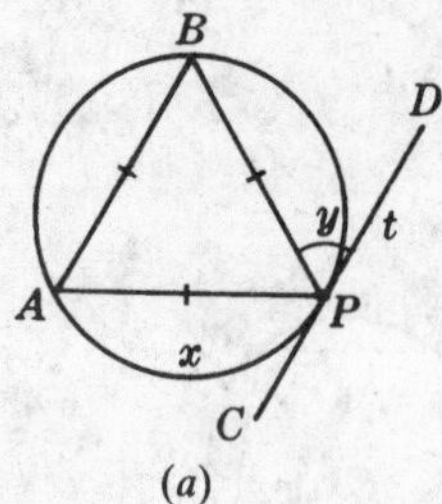

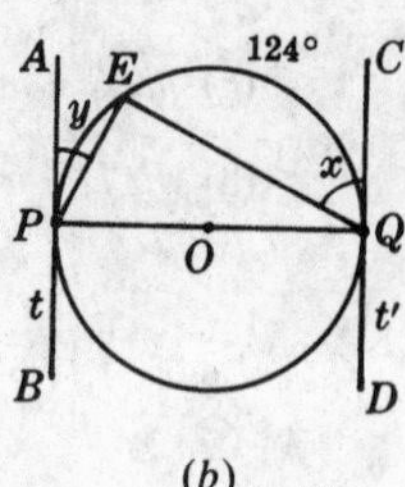

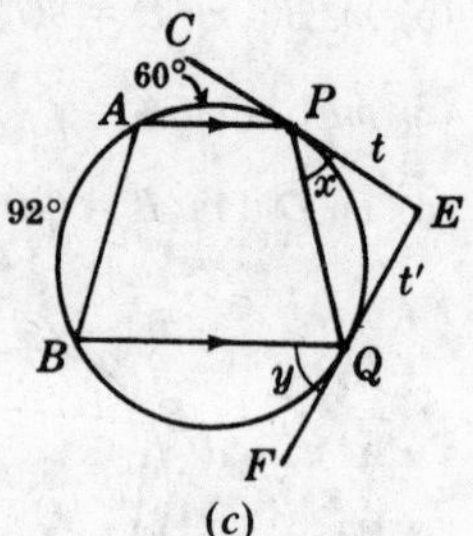

Fig. 6-53

6.26. If $\overline{AC}$ and $\overline{BD}$ are chords intersecting in a circle as shown in Fig. 6-54, find (6.13)

(a) $m\angle x$ if $m\overset{\frown}{AB} = 90°$ and $m\overset{\frown}{CD} = 60°$

(b) $m\angle x$ if $m\overset{\frown}{AB}$ and $m\overset{\frown}{CD}$ each equals 75°

(c) $m\angle x$ if $m\overset{\frown}{AB} + m\overset{\frown}{CD} = 230°$

(d) $m\angle x$ if $m\overset{\frown}{BC} + m\overset{\frown}{AD} = 160°$

(e) $m\overset{\frown}{AB} + m\overset{\frown}{CD}$ if $m\angle x = 70°$

(f) $m\overset{\frown}{BC} + m\overset{\frown}{AD}$ if $m\angle x = 65°$

(g) $m\overset{\frown}{BC}$ if $m\angle x = 60°$ and $m\overset{\frown}{AD} = 160°$

(h) $m\overset{\frown}{BC}$ if $m\angle y = 72°$ and $m\overset{\frown}{AD} = 2m\overset{\frown}{BC}$

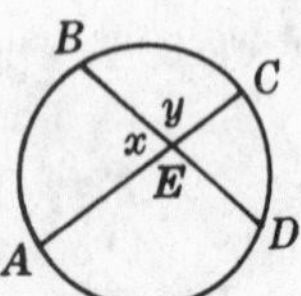

Fig. 6-54

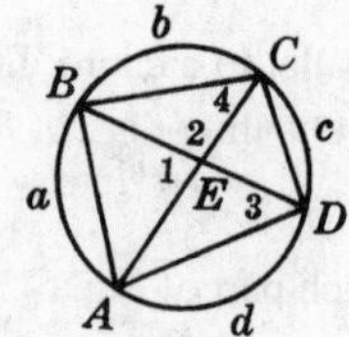

Fig. 6-55

6.27. If $\overline{AC}$ and $\overline{BD}$ are diagonals of an inscribed quadrilateral $ABCD$ as shown in Fig. 6-55, find (6.13)

(a) $m\angle 1$ if $m\overset{\frown}{a} = 95°$ and $m\overset{\frown}{c} = 75°$

(b) $m\angle 1$ if $m\overset{\frown}{b} = 88°$ and $m\overset{\frown}{d} = 66°$

(c) $m\angle 1$ if $m\overset{\frown}{b}$ and $m\overset{\frown}{d}$ each equals 100°

(d) $m\angle 1$ if $m\overset{\frown}{a}:m\overset{\frown}{b}:m\overset{\frown}{c}:m\overset{\frown}{d} = 1:2:3:4$

(e) $m\angle 2$ if $m\overset{\frown}{b} + m\overset{\frown}{d} = m\overset{\frown}{a} + m\overset{\frown}{c}$

(f) $m\angle 2$ if $\overset{\frown}{BC} \parallel \overset{\frown}{AD}$ and $m\overset{\frown}{a} = 70°$

(g) $m\angle 2$ if $\overline{AD}$ is a diameter and $m\overset{\frown}{b} = 80°$

(h) $m\angle 2$ if $ABCD$ is a rectangle and $m\overset{\frown}{a} = 70°$

6.28. Find x and y in each part of Fig. 6-56. (6.13)

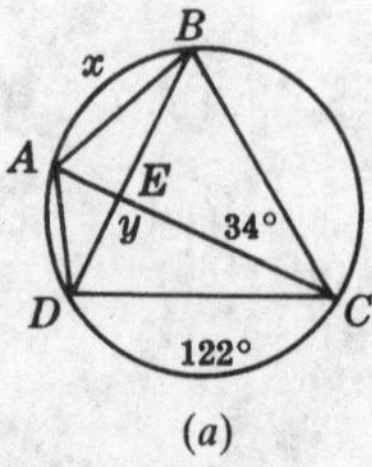

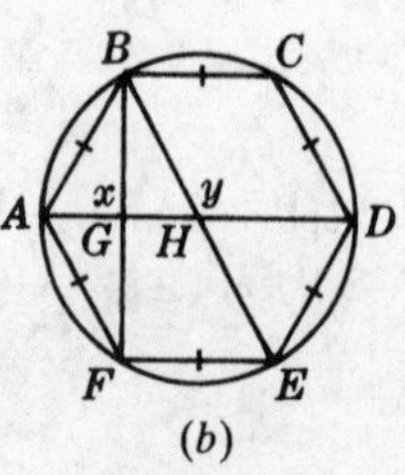

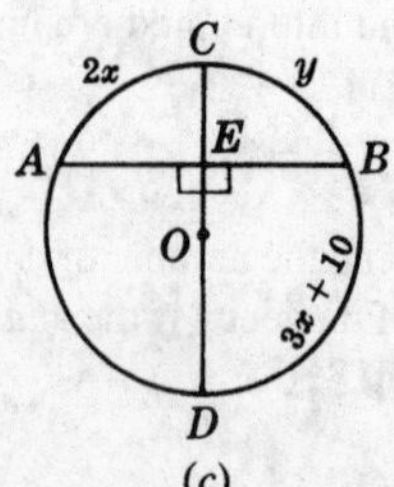

Fig. 6-56

6.29. If $\overline{AB}$ and $\overline{AC}$ are intersecting secants as shown in Fig. 6-57, find (6.14)

(a) $m\angle A$ if $m\widehat{c} = 100°$ and $m\widehat{a} = 40°$

(b) $m\angle A$ if $m\widehat{c} - m\widehat{a} = 74°$

(c) $m\angle A$ if $m\widehat{c} = m\widehat{a} + 40°$

(d) $m\angle A$ if $m\widehat{a}:m\widehat{b}:m\widehat{c}:m\widehat{d} = 1:4:3:2$

(e) $m\widehat{a}$ if $m\widehat{c} = 160°$ and $m\angle A = 20°$

(f) $m\widehat{c}$ if $m\widehat{a} = 60°$ and $m\angle A = 35°$

(g) $m\widehat{c} - m\widehat{a}$ if $m\angle A = 47°$

(h) $m\widehat{a}$ if $m\widehat{c} = 3\widehat{a}$ and $m\angle A = 25°$

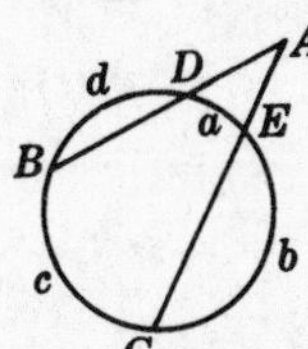

Fig. 6-57

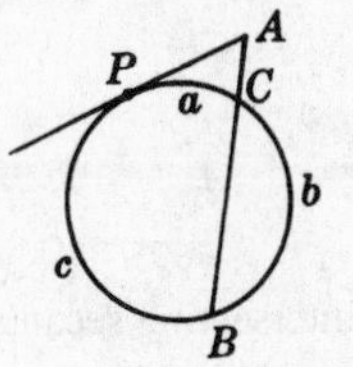

Fig. 6-58

6.30. If tangent $\overline{AP}$ and secant $\overline{AB}$ intersect as shown in Fig. 6-58, find (6.14)

(a) $m\angle A$ if $m\widehat{c} = 150°$ and $m\widehat{a} = 60°$

(b) $m\angle A$ if $m\widehat{c} = 200°$ and $m\widehat{b} = 110°$

(c) $m\angle A$ if $m\widehat{b} = 120°$ and $m\widehat{a} = 70°$

(d) $m\angle A$ if $m\widehat{c} - m\widehat{a} = 73°$

(e) $m\angle A$ if $m\widehat{a}:m\widehat{b}:m\widehat{c} = 1:4:7$

(f) $m\widehat{a}$ if $m\widehat{c} = 220°$ and $m\angle A = 40°$

(g) $m\widehat{c}$ if $m\widehat{a} = 55°$ and $m\angle A = 30°$

(h) $m\widehat{a}$ if $m\widehat{c} = 3\, m\widehat{a}$ and $m\angle A = 45°$

(i) $m\widehat{a}$ if $m\widehat{b} = 100°$ and $m\angle A = 50°$

6.31. If $\overrightarrow{AP}$ and $\overrightarrow{AQ}$ are intersecting tangents as shown in Fig. 6-59, find (6.14)

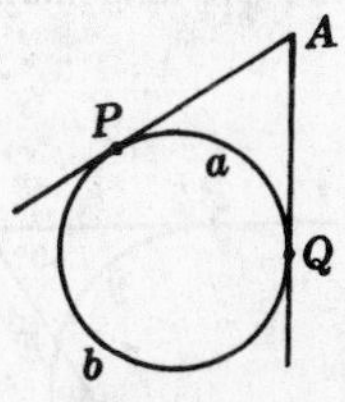

Fig. 6-59

(a) $m\angle A$ if $m\widehat{b} = 200°$

(b) $m\angle A$ if $m\widehat{a} = 95°$

(c) $m\angle A$ if $m\widehat{a} = x°$

(d) $m\angle A$ if $m\widehat{a} = (90 - x)°$

(e) $m\angle A$ if $m\widehat{b} = 3\, m\widehat{a}$

(f) $m\angle A$ if $m\widehat{b} = m\widehat{a} + 50°$

(g) $m\angle A$ if $m\widehat{b} - m\widehat{a} = 84°$

(h) $m\angle A$ if $m\widehat{b}:m\widehat{a} = 5:1$

(i) $m\angle A$ if $m\widehat{b}:m\widehat{a} = 7:3$

(j) $m\angle A$ if $m\widehat{b} = 5m\widehat{a} - 60°$

(k) $m\widehat{a}$ if $m\angle A = 35°$

(l) $m\widehat{a}$ if $m\angle A = y°$

(m) $m\widehat{b}$ if $m\angle A = 60°$

(n) $m\widehat{b}$ if $m\angle A = x°$

(o) $m\widehat{b}$ if $\overrightarrow{AP} \perp \overrightarrow{AQ}$

6.32. Find x and y in each part of Fig. 6-60 (t and t' are tangents). (6.14)

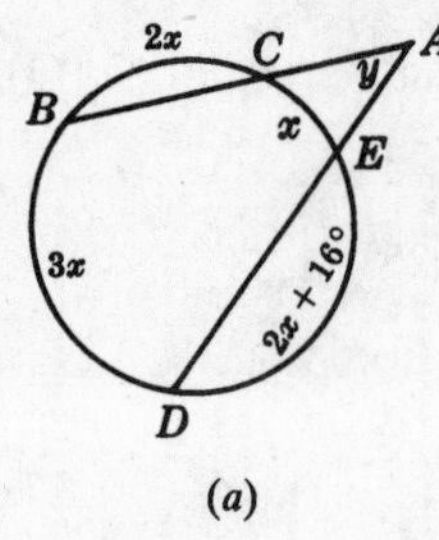

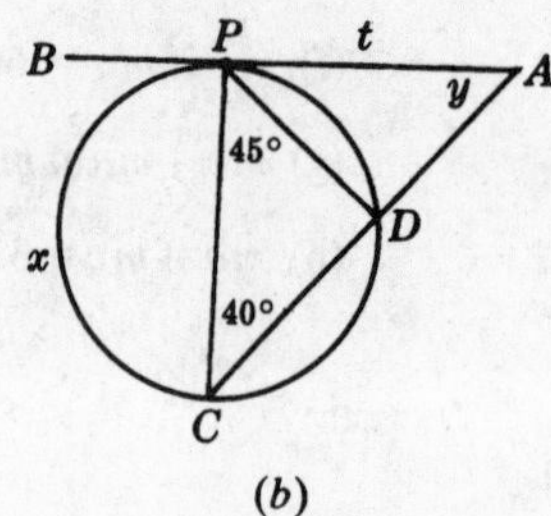

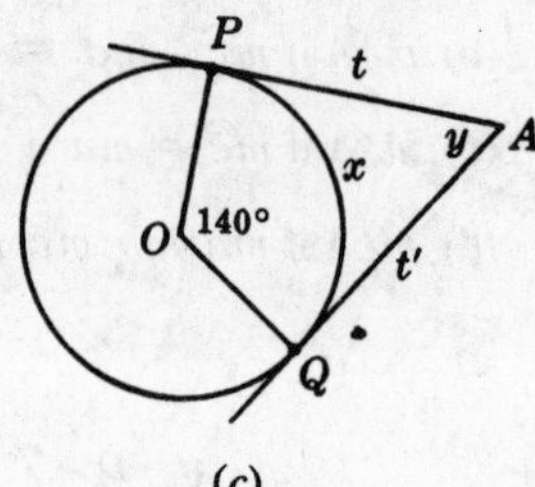

Fig. 6-60

6.33. If $\overline{AB}$ and $\overline{AC}$ are intersecting secants as shown in Fig. 6-61, find (6.15)

(a) $m\widehat{x}$ if $m\angle 1 = 80°$ and $m\angle A = 40°$

(b) $m\widehat{x}$ if $m\angle 1 + m\angle A = 150°$

(c) $m\widehat{x}$ if $\angle 1$ and $\angle A$ are supplementary

(d) $m\widehat{y}$ if $m\angle 1 = 95°$ and $m\angle A = 45°$

(e) $m\widehat{y}$ if $m\angle 1 - m\angle A = 22\frac{1}{2}°$

(f) $m\widehat{y}$ if $m\widehat{x} + m\widehat{y} = 190°$ and $m\angle A = 50°$

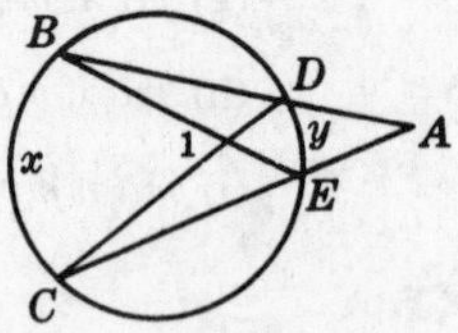

Fig. 6-61

6.34. Find x and y in each part of Fig. 6-62 (t and t' are tangents). (6.15)

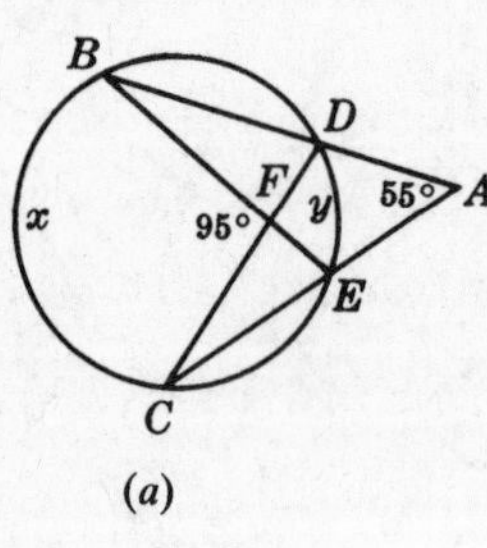

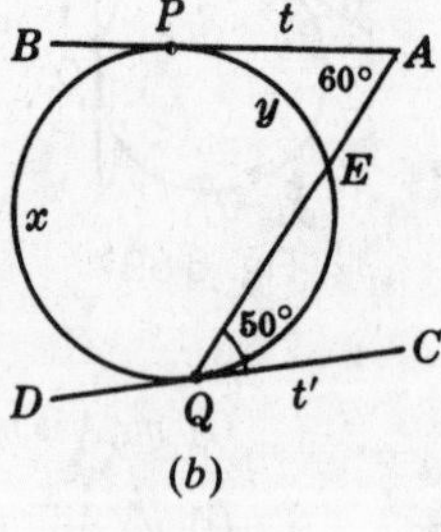

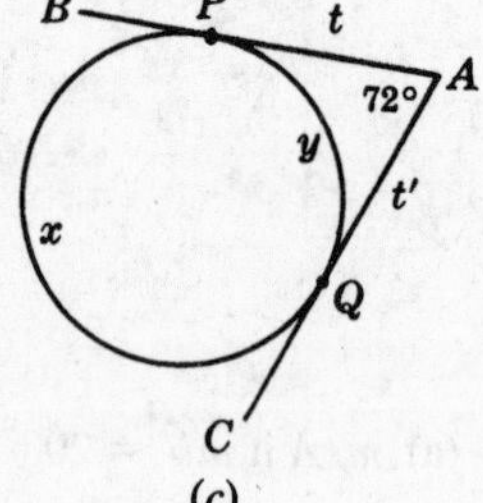

Fig. 6-62

6.35. If ABC is an inscribed triangle as shown in Fig. 6-63, find (6.16)

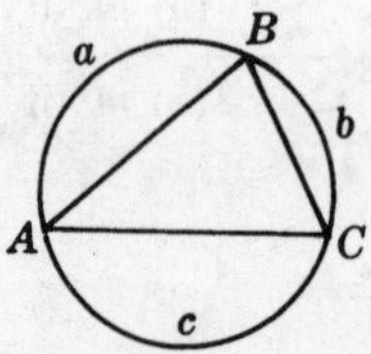

Fig. 6-63

(a) $m\angle A$ if $m\widehat{a} = 110°$ and $m\widehat{c} = 200°$

(b) $m\angle A$ if $\overline{AC} \perp \overline{BC}$ and $m\widehat{a} = 102°$

(c) $m\angle A$ if $\overline{AC}$ is a diameter and $m\widehat{a} = 80°$

(d) $m\angle A$ if $m\widehat{a}:m\widehat{b}:m\widehat{c} = 3:1:2$

(e) $m\angle A$ in $\overline{AC}$ is a diameter and $m\widehat{a}:m\widehat{b} = 5:4$

(f) $m\angle B$ if $m\widehat{ABC} = 208°$

(g) $m\angle B$ if $m\widehat{a} + m\widehat{b} = 3m\widehat{c}$

(h) $m\angle B$ if $m\widehat{a} = 75°$ and $m\widehat{c} = 2m\widehat{b}$

(i) $m\angle C$ if $\overline{AB} \perp \overline{BC}$ and $m\widehat{a} = 5m\widehat{b}$

(j) $m\widehat{c}$ if $m\angle A:m\angle B:m\angle C = 5:4:3$

6.36. If $ABCP$ is an inscribed quadrilateral, $\overleftrightarrow{PD}$ a tangent, and $\overrightarrow{AF}$ a secant in Fig. 6-64, find (6.16)

(a) $m\angle 1$ if $m\widehat{a} = 94°$ and $m\widehat{c} = 54°$

(b) $m\angle 2$ if $\overline{AP}$ is a diameter

(c) $m\angle 3$ if $m\widehat{CPA} = 250°$

(d) $m\angle 3$ if $m\angle ABC = 120°$

(e) $m\angle 4$ if $m\widehat{BCP} = 130°$ and $m\widehat{b} = 50°$

(f) $m\angle 4$ if $\overline{BC} \parallel \overline{AP}$ and $m\widehat{a} = 74°$

(g) $m\widehat{a}$ if $\overline{BC} \parallel \overline{AP}$ and $m\angle 6 = 42°$

(h) $m\widehat{a}$ if $\overline{AC}$ is a diameter and $m\angle 5 = 35°$

(i) $m\widehat{b}$ if $\overline{AC} \perp \overline{BP}$ and $m\angle 2 = 57°$

(j) $m\widehat{c}$ if $\overline{AC}$ and $\overline{BP}$ are diameters and $m\angle 5 = 41°$

(k) $m\widehat{d}$ if $m\angle 1 = 95°$ and $m\widehat{b} = 95°$

(l) $m\angle CPA$ if $m\angle 3 = 79°$

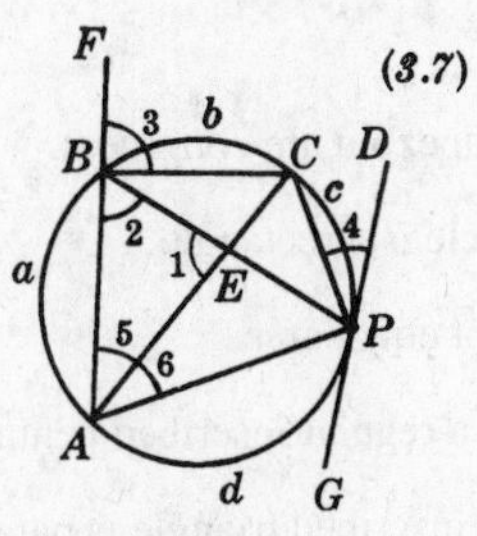

Fig. 6-64

6.37. Find x and y in each part of Fig. 6-65 (t and t' are tangents). 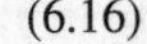

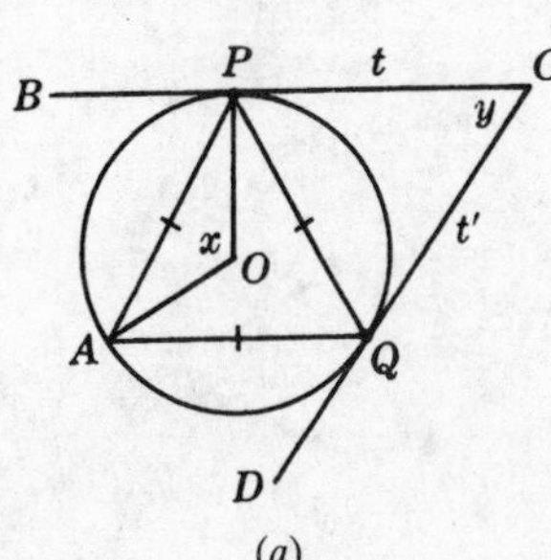

(a)

(b) PBCA is an inscribed square.

(c)

Fig. 6-65

6.38. Find x and y in each part of Fig. 6-66. (6.16)

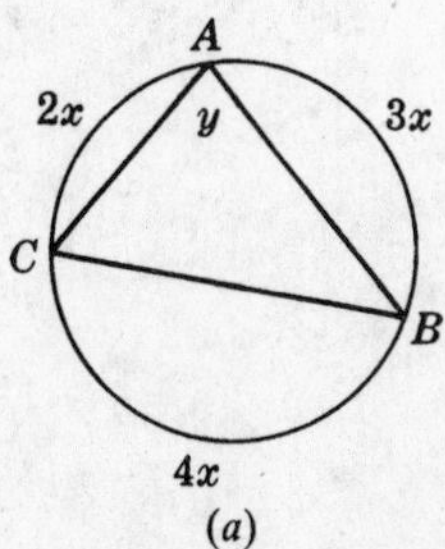

(a)

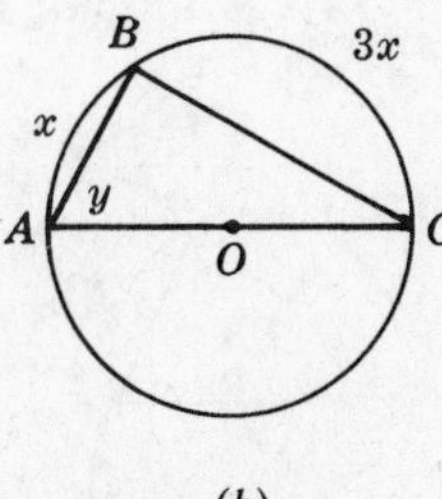

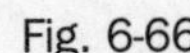

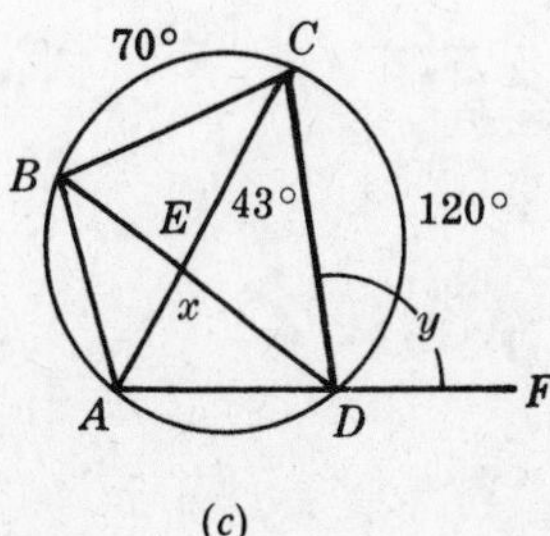

(c)

Fig. 6-66

6.39. Provide the proofs requested in Fig. 6-67. (6.17)

(a) **Given:** $\overline{AC}$ bisects $\angle A$
To Prove: $\overline{BC} \cong \overline{CD}$

(b) **Given:** $\overline{BC} \cong \overline{CD}$
To Prove: $\overline{AC}$ bisects $\angle A$

(c) **Given:** $\overline{AB} \parallel \overline{CD}$
$\overline{AB}$ is a tangent.
To Prove: $PC = PD$

(d) **Given:** $PC = PD$
$\overleftrightarrow{AB}$ is a tangent.
To Prove: $\overline{AB} \parallel \overline{CD}$

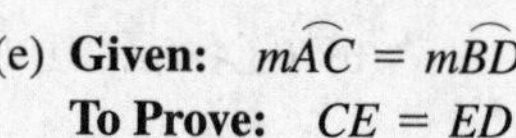

(e) **Given:** $m\widehat{AC} = m\widehat{BD}$
To Prove: $CE = ED$

(f) **Given:** $CE = EB$
To Prove: $\widehat{AC} \cong \widehat{BD}$

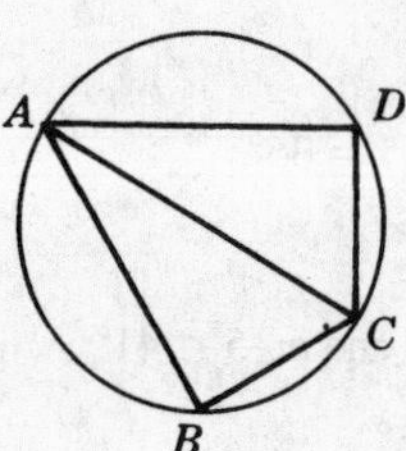

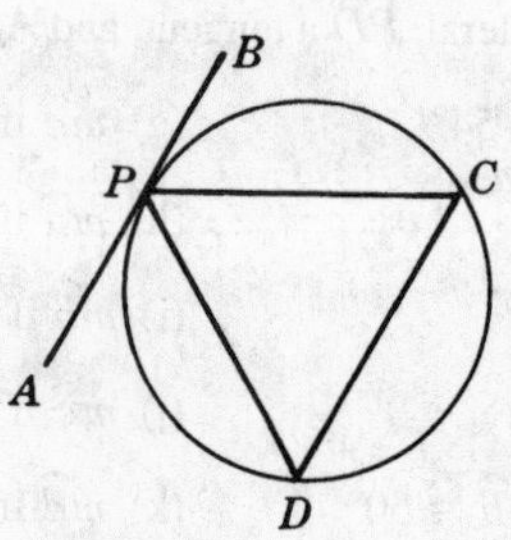

Fig. 6-67

6.40. Prove each of the following: (6.18)

(a) The base angles of an inscribed trapezoid are congruent.

(b) A parallelogram inscribed in a circle is a rectangle.

(c) In a circle, parallel chords intercept equal arcs.

(d) Diagonals drawn from a vertex of a regular inscribed pentagon trisect the vertex angle.

(e) If a tangent through a vertex of an inscribed triangle is parallel to its opposite side, the triangle is isosceles.

CHAPTER 7

Similarity

7.1 Ratios

Ratios are used to compare quantities by division. The ratio of two quantities is the first divided by the second. A ratio is an abstract number, that is, a number without a unit of measure. Thus, the ratio of 10 ft to 5 ft is 10 ft ÷ 5 ft, which equals 2.

A ratio can be expressed in the following ways: (1) using a colon, as in 3:4; (2) using "to" as in 3 to 4; (3) as a common fraction, as in $\frac{3}{4}$; (4) as a decimal, 0.75; and (5) as a percent, 75%.

The quantities involved in a ratio must have the same unit. A ratio should be simplified by reducing to lowest terms and eliminating fractions. Thus to find the ratio of 1 ft to 4 in, we first change the foot to 12 in, and then take the ratio of 12 in to 4 in; the result is a ratio of 3 to 1, or 3. Also, the ratio of $2\frac{1}{2}:\frac{1}{2}$ would be restated as 5:1 or 5.

The ratio of three or more quantities may be expressed as a *continued ratio*. Thus, the ratio of \$2 to \$3 to \$5 is the continued ratio 2:3:5. This enlarged ratio is a combination of three separate ratios; these are 2:3, 3:5, and 2:5.

Throughout this chapter, readers should use a calculator whenever they choose.

SOLVED PROBLEMS

7.1 Ratio of two quantities with the same unit

Express each of the following ratios in lowest terms: (a) 15° to 3°; (b) \$1.25 to \$5; (c) $2\frac{1}{2}$ years to 2 years.

Solutions

(a) $\frac{15}{3} = 5$

(b) $\frac{1.25}{5} = \frac{1}{4}$

(c) $\frac{2\frac{1}{2}}{2} = \frac{5}{4}$

7.2 Ratio of two quantities with different units

Express each of the following ratios in lowest terms: (a) 2 years to 3 months; (b) 80 cents to \$3.20.

Solutions

(a) 2 years to 3 months = 24 months to 3 months $= \frac{24}{3} = 8$

(b) 80 cents to \$3.20 = 80 cents to 320 cents $= \frac{80}{320} = \frac{1}{4}$

7.3 Continued ratio of three quantities

Express each of the following ratios in lowest terms: (a) 1 gal to 2 qt to 2 pt; (b) 1 ton to 1lb to 8 oz.

Solutions

(a) 1 gal to 2 qt to 2 pt = 4 qt to 2 qt to 1 qt = $4:2:1$

(b) 1 ton to 1 lb to 8 oz = 2000 lb to 1 lb to $\frac{1}{2}$lb = $2000:1:\frac{1}{2} = 4000:2:1$

7.4 Numerical and algebraic ratios

Express each of the following ratios in lowest terms: (a) 50 to 60; (b) 6.3 to 0.9; (c) 12 to $\frac{3}{8}$; (d) $2x$ to $5x$; (e) $5s^2$ to s^3; (f) x to $5x$ to $7x$.

Solutions

(a) $\frac{50}{60} = \frac{5}{6}$

(b) $\frac{6.3}{0.9} = 7$

(c) $12 \div \frac{3}{8} = 32$

(d) $\frac{2x}{5x} = \frac{2}{5}$

(e) $\frac{5s^2}{s^3} = \frac{5}{s}$

(f) $x:5x:7x = 1:5:7$

7.5 Using ratios in angle problems

If two angles are in the ratio of 3:2, find the angles if (a) they are adjacent and form an angle measuring 40°; (b) they are acute angles of a right triangle; (c) they are two angles of a triangle whose third angle measures 70°.

Solutions

Let the measures of the angles be $3x$ and $2x$. Then:

(a) $3x + 2x = 40$, so that $5x = 40$ or $x = 8$; hence, the angles measure 24° and 16°.

(b) $3x + 2x = 90$, so $5x = 90$ or $x = 18$; hence, the angles measure 54° and 36°.

(c) $3x + 2x + 70 = 180$, so $5x = 110$ or $x = 22$; hence, the angles measure 66° and 44°.

7.6 Three angles having a fixed ratio

Three angles are in the ratio of 4:3:2. Find the angles if (a) the first and the third are supplementary; (b) the angles are the three angles of a triangle.

Solutions

Let the measures of the angles be $4x$, $3x$, and $2x$. Then:

(a) $4x + 2x = 180$, so that $6x = 180$ for $x = 30$; hence, the angles measure 120°, 90°, and 60°.

(b) $4x + 3x + 2x = 180$, so $9x = 180$ or $x = 20$; hence, the angles measure 80°, 60°, and 40°.

7.2 Proportions

A *proportion* is an equality of two ratios. Thus, $2:5 = 4:10$ (or $\frac{2}{5} = \frac{4}{10}$) is a proportion.

The fourth term of a proportion is the *fourth proportional* to the other three taken in order. Thus in $2:3 = 4:x$, x is the fourth proportional to 2, 3, and 4.

The *means* of a proportion are its middle terms, that is, its second and third terms. The *extremes* of a proportion are its outside terms, that is, its first and fourth terms. Thus in $a:b = c:d$, the means are b and c, and the extremes are a and d.

If the two means of a proportion are the same, either mean is the *mean proportional* between the first and fourth terms. Thus in $9:3 = 3:1$, 3 is the mean proportional between 9 and 1.

7.2A Proportion Principles

PRINCIPLE 1: *In any proportion, the product of the means equals the product of the extremes.*

Thus if $a:b = c:d$, then $ad = bc$.

PRINCIPLE 2: *If the product of two numbers equals the product of two other numbers, either pair may be made the means of a proportion and the other pair may be made the extremes.*

Thus if $3x = 5y$, then $x:y = 5:3$ or $y:x = 3:5$ or $3:y = 5:x$ or $5:x = 3:y$.

7.2B Methods of Changing a Proportion into an Equivalent Proportion

PRINCIPLE 3: *(Inversion method) A proportion may be changed into an equivalent proportion by inverting each ratio.*

Thus if $\frac{1}{x} = \frac{4}{5}$, then $\frac{x}{1} = \frac{5}{4}$.

PRINCIPLE 4: *(Alternation method) A proportion may be changed into an equivalent proportion by interchanging the means or by interchanging the extremes.*

Thus if $\frac{x}{3} = \frac{y}{2}$, then $\frac{x}{y} = \frac{3}{2}$ or $\frac{2}{3} = \frac{y}{x}$.

PRINCIPLE 5: *(Addition method) A proportion may be changed into an equivalent proportion by adding terms in each ratio to obtain new first and third terms.*

Thus if $\frac{a}{b} = \frac{c}{d}$, then $\frac{a+b}{b} = \frac{c+d}{d}$. If $\frac{x-2}{2} = \frac{9}{1}$, then $\frac{x}{2} = \frac{10}{1}$.

PRINCIPLE 6: *(Subtraction method) A proportion may be changed into an equivalent proportion by subtracting terms in each ratio to obtain new first and third terms.*

Thus if $\frac{a}{b} = \frac{c}{d}$, then $\frac{a-b}{b} = \frac{c-d}{d}$. If $\frac{x+3}{3} = \frac{9}{1}$, then $\frac{x}{3} = \frac{8}{1}$.

7.2C Other Proportion Principles

PRINCIPLE 7: *If any three terms of one proportion equal the corresponding three terms of another proportion, the remaining terms are equal.*

Thus if $\frac{x}{y} = \frac{3}{5}$ and $\frac{x}{4} = \frac{3}{5}$, then $y = 4$.

PRINCIPLE 8: *In a series of equal ratios, the sum of any of the numerators is to the sum of the corresponding denominators as any numerator is to its denominator.*

Thus if $\frac{a}{b} = \frac{c}{d} = \frac{e}{f}$, then $\frac{a+c+e}{b+d+f} = \frac{a}{b}$. If $\frac{x-y}{4} = \frac{y-3}{5} = \frac{3}{1}$, then $\frac{x-y+y-3+3}{4+5+1} = \frac{3}{1}$ or $\frac{x}{10} = \frac{3}{1}$.

SOLVED PROBLEMS

7.7 Finding unknowns in proportions

Solve the following proportions for x:

(a) $x:4 = 6:8$ (c) $x:5 = 2x:(x+3)$ (e) $\frac{x}{2x-3} = \frac{3}{5}$

(b) $3:x = x:27$ (d) $\frac{3}{x} = \frac{2}{5}$ (f) $\frac{x-2}{4} = \frac{7}{x+2}$

Solutions

(a) Since $4(6) = 8x$, $8x = 24$ or $x = 3$.

(b) Since $x^2 = 3(27)$, $x^2 = 81$ or $x = \pm 9$.

(c) Since $5(2x) = x(x + 3)$, we have $10x = x^2 + 3x$. Then $x^2 - 7x = 0$, so $x = 0$ or 7.

(d) Since $2x = 3(5)$, $2x = 15$ or $x = 7\frac{1}{2}$.

(e) Since $3(2x - 3) = 5x$, we have $6x - 9 = 5x$, so $x = 9$.

(f) Since $4(7) = (x - 2)(x + 2)$, we have $28 = x^2 - 4$. Then $x^2 = 32$, so $x = \pm 4\sqrt{2}$.

7.8 Finding fourth proportionals to three given numbers

Find the fourth proportional to (a) 2, 4, 6; (b) 4, 2, 6; (c) $\frac{1}{2}$, 3, 4; (d) b, d, c.

Solutions

(a) We have $2:4 = 6:x$, so $2x = 24$ or $x = 12$.

(b) We have $4:2 = 6:x$, so $4x = 12$ or $x = 3$.

(c) We have $\frac{1}{2}:3 = 4:x$, so $\frac{1}{2}x = 12$ or $x = 24$.

(d) We have $b:d = c:x$, so $bx = cd$ or $x = cd/b$.

7.9 Finding the mean proportional to two given numbers

Find the positive mean proportional x between (a) 5 and 20; (b) $\frac{1}{2}$ and $\frac{8}{9}$.

Solutions

(a) We have $5:x = x:20$, so $x^2 = 100$ or $x = 10$.

(b) We have $\frac{1}{2}:x = x:\frac{8}{9}$, so $x^2 = \frac{4}{9}$ or $x = \frac{2}{3}$.

7.10 Changing equal products into proportions

(a) Form a proportion whose fourth term is x and such that $2bx = 3s^2$.

(b) Find the ratio x to y if $ay = bx$.

Solutions

(a) $2b:3s = s:x$ or $2b:3 = s^2:x$ or $2b:s^2 = 3:x$ (b) $x:y = a:b$

7.11 Changing proportions into new proportions

Use each of the following to form a new proportion whose first term is x:

(a) $\dfrac{15}{x} = \dfrac{3}{4}$ (b) $\dfrac{x-6}{6} = \dfrac{5}{3}$ (c) $\dfrac{x+8}{8} = \dfrac{4}{3}$ (d) $\dfrac{5}{2} = \dfrac{15}{x}$

Solutions

(a) By Principle 3, $\dfrac{x}{15} = \dfrac{4}{3}$ (c) By Principle 6, $\dfrac{x}{8} = \dfrac{1}{3}$

(b) By Principle 5, $\dfrac{x}{6} = \dfrac{8}{3}$ (d) By Principle 4, $\dfrac{x}{2} = \dfrac{15}{5}$; thus, $\dfrac{x}{2} = \dfrac{3}{1}$

7.12 Combining numerators and denominators of proportions

Use Principle 8 to find x in each of the following proportions:

(a) $\dfrac{x-2}{9} = \dfrac{2}{3}$ (b) $\dfrac{x+y}{8} = \dfrac{x-y}{4} = \dfrac{2}{3}$ (c) $\dfrac{3x-y}{15} = \dfrac{y-3}{10} = \dfrac{3}{5}$

Solutions

(a) Adding numerators and denominators yields $\frac{x - 2 + 2}{9 + 3} = \frac{2}{3}$ or $\frac{x}{12} = \frac{2}{3}$, so $x = 8$.

(b) Here we have $\frac{(x + y) + (x - y)}{8 + 4} = \frac{2}{3}$, which gives $\frac{2x}{12} = \frac{2}{3}$, so $x = 4$.

(c) We use all three ratios to get $\frac{(3x - y) + (y - 3) + 3}{15 + 10 + 5} = \frac{3}{5}$ or $\frac{3x}{30} = \frac{3}{5}$, so $x = 6$.

7.3 Proportional Segments

If two segments are divided proportionately, (1) the corresponding new segments are in proportion, and (2) the two original segments and either pair of corresponding new segments are in proportion.

Thus if $\overline{AB}$ and $\overline{AC}$ in Fig. 7-1 are divided proportionately by $\overline{DE}$, we may write a proportion such as $\frac{a}{b} = \frac{c}{d}$ using the four segments; or we may write a proportion such as $\frac{a}{AB} = \frac{c}{AC}$ using the two original segments and two of their new segments.

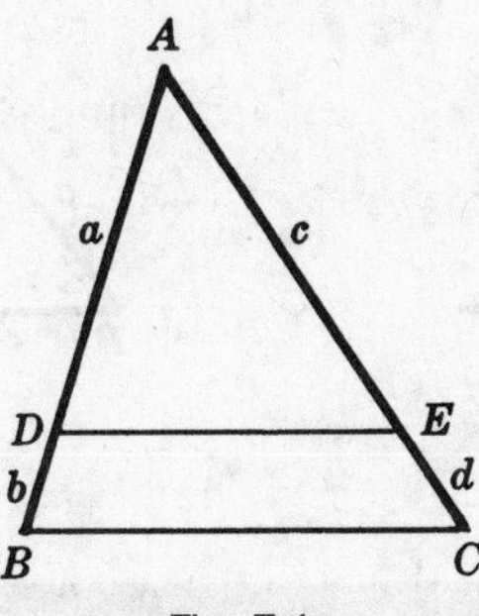

Fig. 7-1

7.3A Obtaining the Eight Arrangements of Any Proportion

A proportion such as $\frac{a}{b} = \frac{c}{d}$ can be arranged in eight ways. To obtain the eight variations, we let each term of the proportion represent one of the new segments of Fig. 7-1. Two of the possible proportions are then obtained from each direction, as follows:

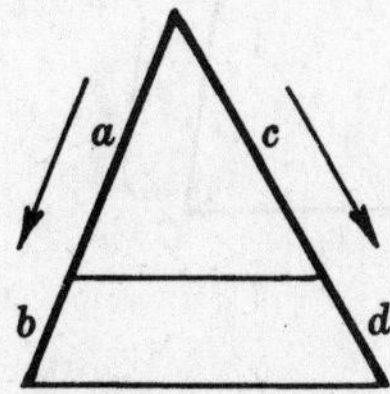

Direction: Down

$\frac{a}{b} = \frac{c}{d}$ or $\frac{c}{d} = \frac{a}{b}$

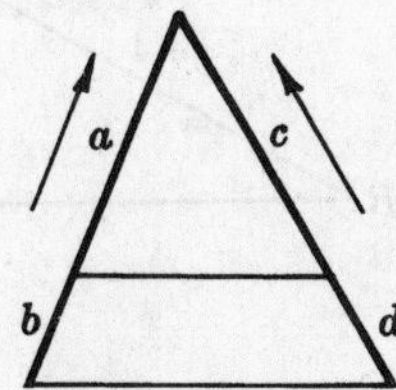

Direction: Up

$\frac{b}{a} = \frac{d}{c}$ or $\frac{d}{c} = \frac{b}{a}$

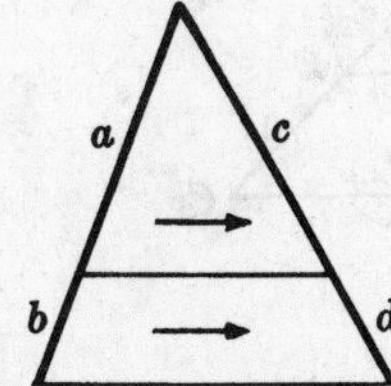

Direction: Right

$\frac{a}{c} = \frac{b}{d}$ or $\frac{b}{d} = \frac{a}{c}$

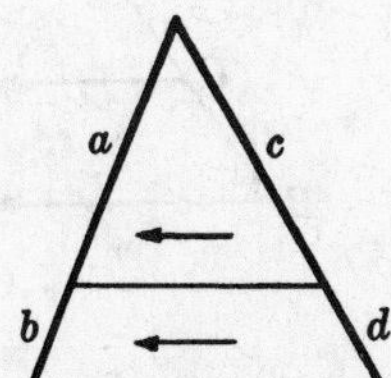

Direction: Left

$\frac{c}{a} = \frac{d}{b}$ or $\frac{d}{b} = \frac{c}{a}$

7.3B Principles of Proportional Segments

PRINCIPLE 1: *If a line is parallel to one side of a triangle, then it divides the other two sides proportionately.*

Thus in $\triangle ABC$ of Fig. 7-2, if $\overline{DE} \parallel \overline{BC}$, then $\frac{a}{b} = \frac{c}{d}$.

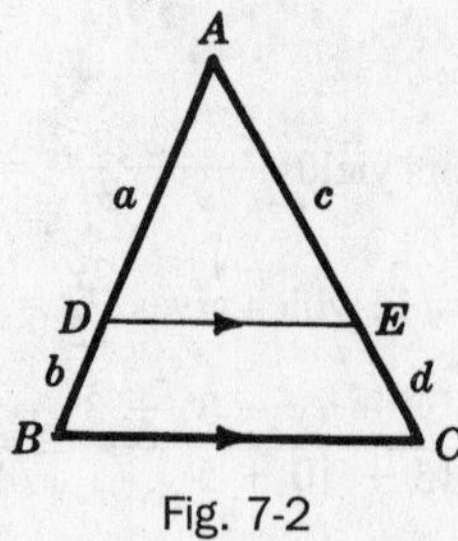

Fig. 7-2

PRINCIPLE 2: *If a line divides two sides of a triangle proportionately, it is parallel to the third side. (Principles 1 and 2 are converses.)*

Thus in △ABC (Fig. 7-2), if $\frac{a}{b} = \frac{c}{d}$, then $\overline{DE} \parallel \overline{BC}$.

PRINCIPLE 3: *Three or more parallel lines divide any two transversals proportionately.*

Thus if $\overleftrightarrow{AB} \parallel \overleftrightarrow{EF} \parallel \overleftrightarrow{CD}$ in Fig. 7-3, then $\frac{a}{b} = \frac{c}{d}$.

PRINCIPLE 4: *A bisector of an angle of a triangle divides the opposite side into segments which are proportional to the adjacent sides.*

Thus in $\triangle ABC$ of Fig. 7-4, if $\overline{CD}$ bisects $\angle C$, then $\frac{a}{b} = \frac{c}{d}$.

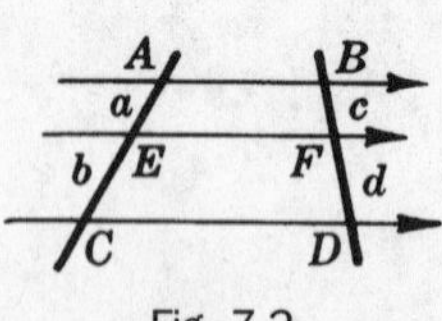

Fig. 7-3

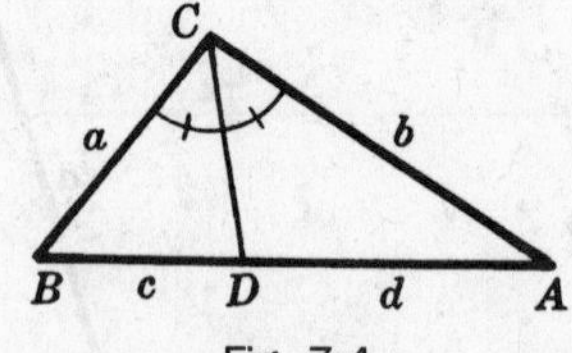

Fig. 7-4

SOLVED PROBLEMS

7.13 Applying principle 1

Find x in each part of Fig. 7-5.

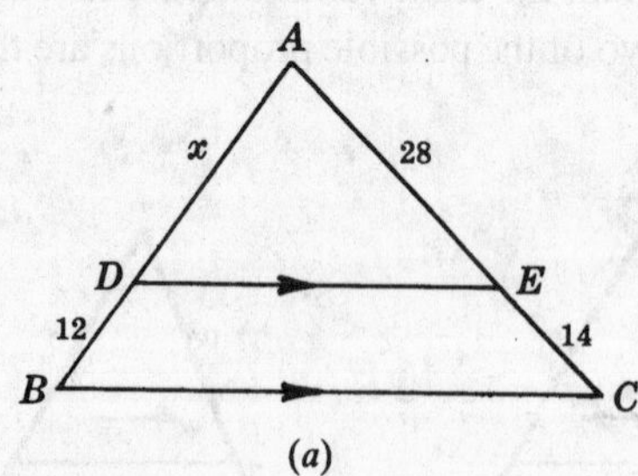

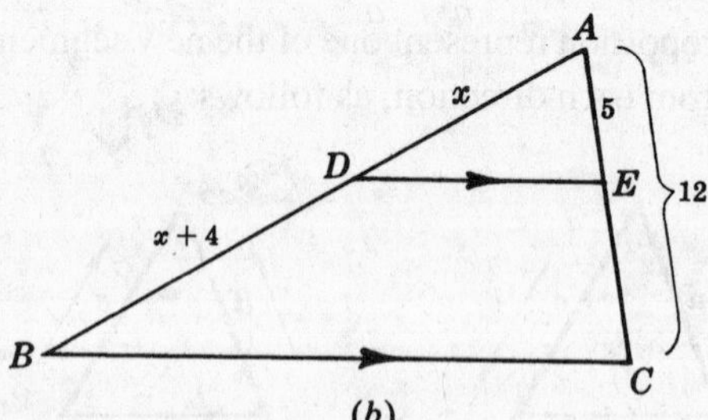

Fig. 7-5

Solutions

(a) $\overline{DE} \parallel \overline{BC}$; hence $\frac{x}{12} = \frac{28}{14}$, so that $x = 24$.

(b) We have $EC = 7$ and $\overline{DE} \parallel \overline{BC}$; hence, $\frac{x}{x + 4} = \frac{5}{7}$. Then $7x = 5x + 20$ and $x = 10$.

7.14 Applying principle 3

Find x in each part of Fig. 7-6.

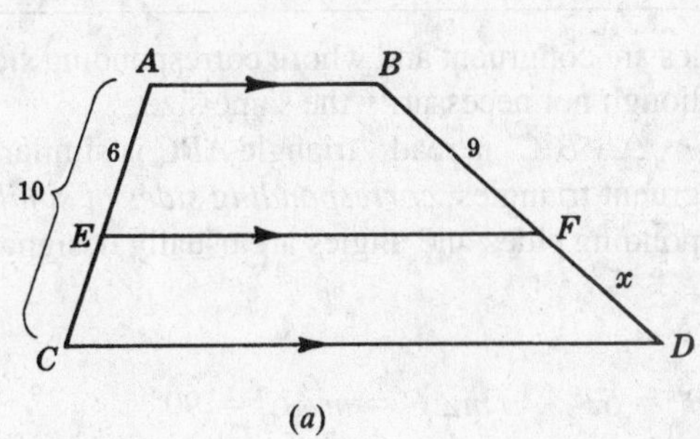

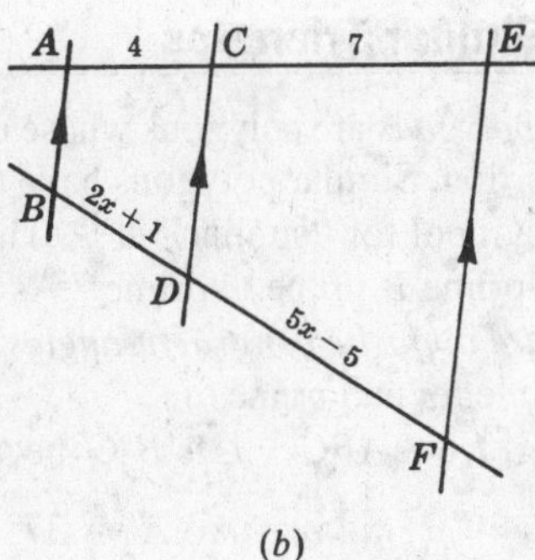

Fig. 7-6

Solutions

(a) We have $EC = 4$ and $\overline{AB} \| \overline{EF} \| \overline{CD}$; hence, $\frac{x}{9} = \frac{4}{6}$ and $x = 6$.

(b) $\overleftrightarrow{AB} \| \overleftrightarrow{CD} \| \overleftrightarrow{EF}$; hence, $\frac{5x - 5}{2x + 1} = \frac{7}{4}$, from which $20x - 20 = 14x + 7$. Then $6x = 27$ and $x = 4\frac{1}{2}$.

7.15 Applying principle 4

Find x in each part of Fig. 7-7.

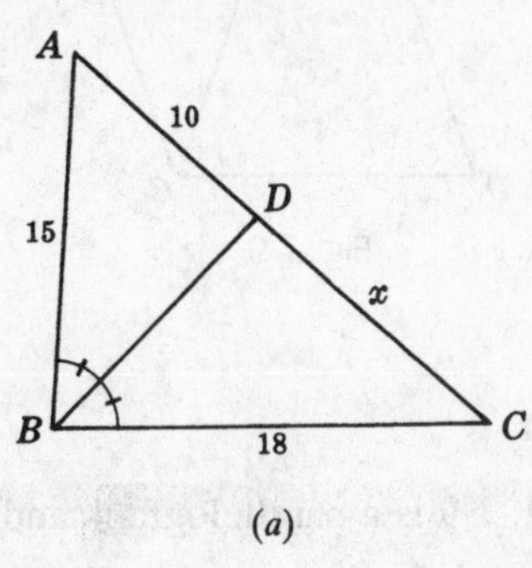

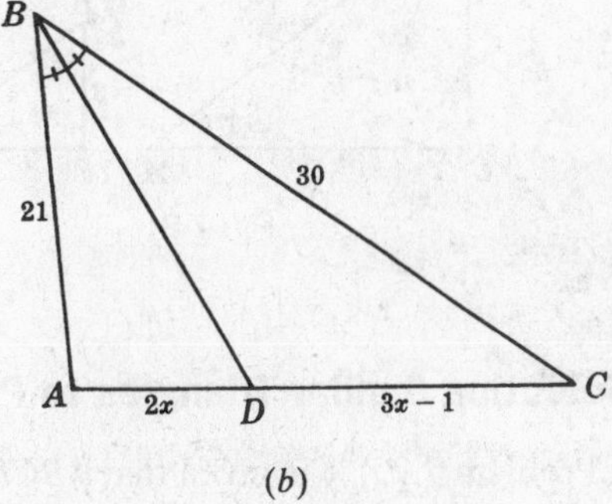

Fig. 7-7

Solutions

(a) $\overline{BD}$ bisects $\angle B$; hence, $\frac{x}{10} = \frac{18}{15}$ and $x = 12$.

(b) $\overline{BD}$ bisects $\angle B$; hence, $\frac{3x - 1}{2x} = \frac{30}{21} = \frac{10}{7}$. Thus, $21x - 7 = 20x$ and $x = 7$.

7.16 Proving a proportional-segments problem

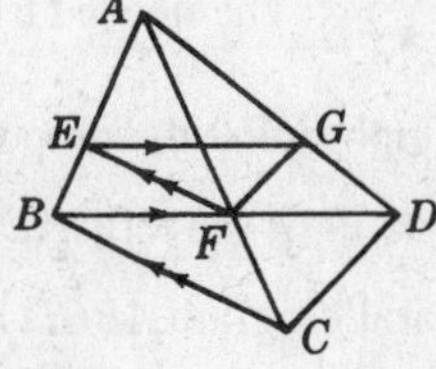

Given: $\overline{EG} \| \overline{BD}, \overline{EF} \| \overline{BC}$
To Prove: $\overline{FG} \| \overline{CD}$
Plan: Prove that FG divides $\overline{AC}$ and $\overline{AD}$ proportionately.

PROOF:

Statements	Reasons
1. $\overline{EG} \| \overline{BD}, \overline{EF} \| \overline{BC}$	1. Given
2. $\frac{AE}{EB} = \frac{AG}{GD}, \frac{AE}{EB} = \frac{AF}{FC}$	2. A line (segment) parallel to one side of a triangle divides the other two sides proportionately.
3. $\frac{AF}{FC} = \frac{AG}{GD}$	3. Substitution postulate
4. $\overline{FG} \| \overline{CD}$	4. If a line divides two sides of a triangle proportionately, it is parallel to the third side.

7.4 Similar Triangles

Similar polygons are polygons whose corresponding angles are congruent and whose corresponding sides are in proportion. Similar polygons have the same shape although not necessarily the same size.

The symbol for "similar" is ~. The notation $\triangle ABC \sim \triangle A'B'C'$ is read "triangle *ABC* is similar to triangle *A*-prime *B*-prime *C*-prime." As in the case of congruent triangles, *corresponding sides of similar triangles are opposite congruent angles*. (Note that corresponding sides and angles are usually designated by the same letter and primes.)

In Fig. 7-8 $\triangle ABC \sim \triangle A'B'C'$ because

$$m\angle A = m\angle A' = 37^\circ \qquad m\angle B = m\angle B' = 53^\circ \qquad m\angle C = m\angle C' = 90^\circ$$

and

$$\frac{a}{a'} = \frac{b}{b'} = \frac{c}{c'} \quad \text{or} \quad \frac{6}{3} = \frac{8}{4} = \frac{10}{5}$$

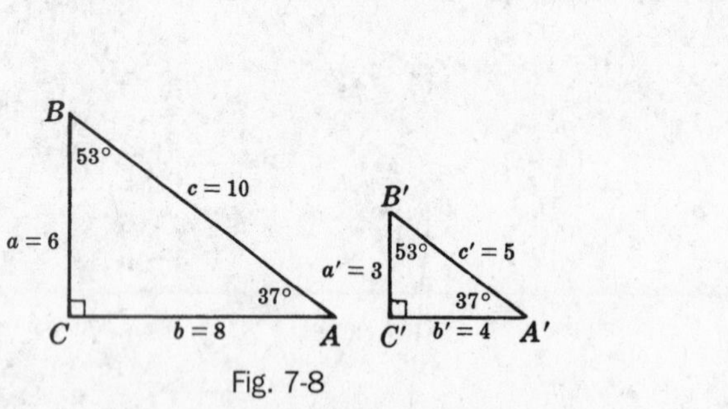

Fig. 7-8

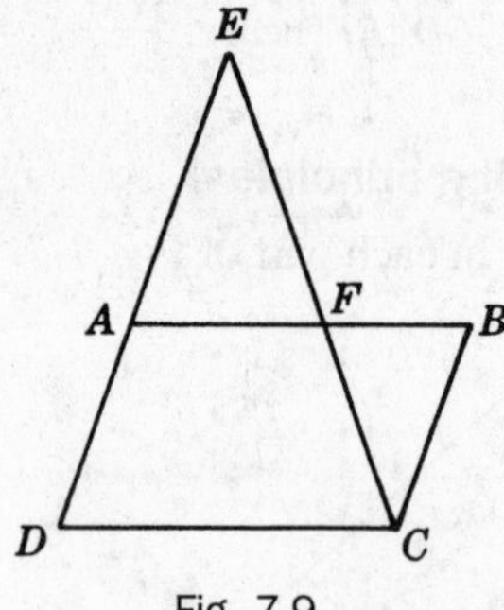

Fig. 7-9

7.4A Selecting Similar Triangles to Prove a Proportion

In Solved Problem 7.25, it is given that *ABCD* in a figure like Fig. 7-9 is a parallelogram, and we must prove that $\frac{AE}{BC} = \frac{AF}{FB}$. To prove this proportion, it is necessary to find similar triangles whose sides are in the proportion. This can be done simply by selecting the triangle whose letters *A*, *E*, and *F* are in the numerators and the triangle whose letters *B*, *C*, and *F* are in the denominators. Hence, we would prove $\triangle AEF \sim \triangle BCF$.

Suppose that the proportion to be proved is $\frac{AE}{AF} = \frac{BC}{FB}$. In such a case, interchanging the means leads to $\frac{AE}{BC} = \frac{AF}{FB}$. The needed triangles can then be selected based on the numerators and the denominators.

Suppose that the proportion to be proved is $\frac{AE}{AD} = \frac{AF}{FB}$. Then our method of selecting triangles could not be used until the term *AD* were replaced by *BC*. This is possible, because $\overline{AD}$ and $\overline{BC}$ are opposite sides of the parallelogram *ABCD* and therefore are congruent.

7.4B Principles of Similar Triangles

PRINCIPLE 1: *Corresponding angles of similar triangles are congruent (by the definition).*

PRINCIPLE 2: *Corresponding sides of similar triangles are in proportion (by the definition).*

PRINCIPLE 3: *Two triangles are similar if two angles of one triangle are congruent respectively to two angles of the other.*

Thus in Fig. 7-10, if $\angle A \cong \angle A'$ and $\angle B \cong \angle B'$, then $\triangle ABC \sim \triangle A'B'C'$.

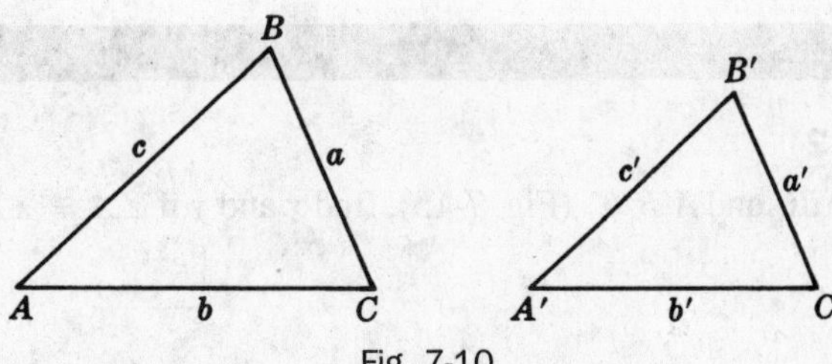

Fig. 7-10

PRINCIPLE 4: *Two triangles are similar if an angle of one triangle is congruent to an angle of the other and the sides including these angles are in proportion.*

Thus in Fig. 7-10, if $\angle C \cong \angle C'$ and $\frac{a}{a'} = \frac{b}{b'}$, then $\triangle ABC \sim \triangle A'B'C'$.

PRINCIPLE 5: *Two triangles are similar if their corresponding sides are in proportion.*

Thus in Fig. 7-10, if $\frac{a}{a'} = \frac{b}{b'} = \frac{c}{c'}$, then $\triangle ABC \sim A'B'C'$.

PRINCIPLE 6: *Two right triangles are similar if an acute angle of one is congruent to an acute angle of the other (corollary of Principle 3).*

PRINCIPLE 7: *A line parallel to a side of a triangle cuts off a triangle similar to the given triangle.*

Thus in Fig. 7-11, if $\overline{DE} \parallel \overline{BC}$, then $\triangle ADE \sim \triangle ABC$.

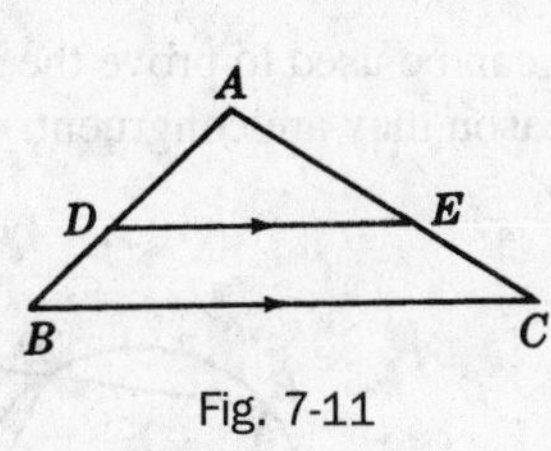

Fig. 7-11

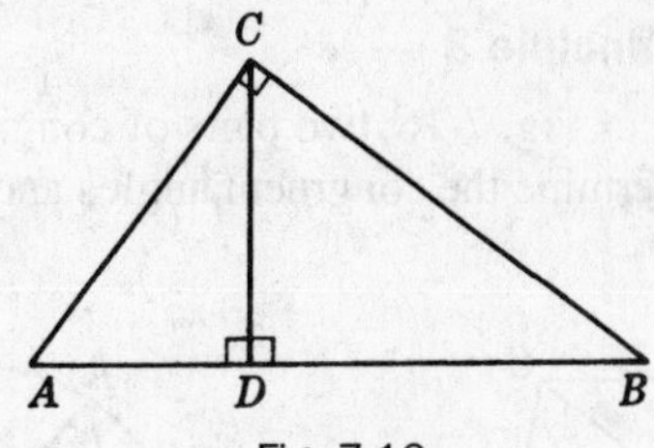

Fig. 7-12

PRINCIPLE 8: *Triangles similar to the same triangle are similar to each other.*

PRINCIPLE 9: *The altitude to the hypotenuse of a right triangle divides it into two triangles which are similar to the given triangle and to each other.*

Thus in Fig. 7-12, $\triangle ABC \sim \triangle ACD \sim \triangle CBD$.

PRINCIPLE 10: *Triangles are similar if their sides are respectively parallel to each other.*

Thus in Fig. 7-13, $\triangle ABC \sim \triangle A'B'C'$.

PRINCIPLE 11: *Triangles are similar if their sides are respectively perpendicular to each other.*

Thus in Fig. 7-14, $\triangle ABC \sim \triangle A'B'C'$.

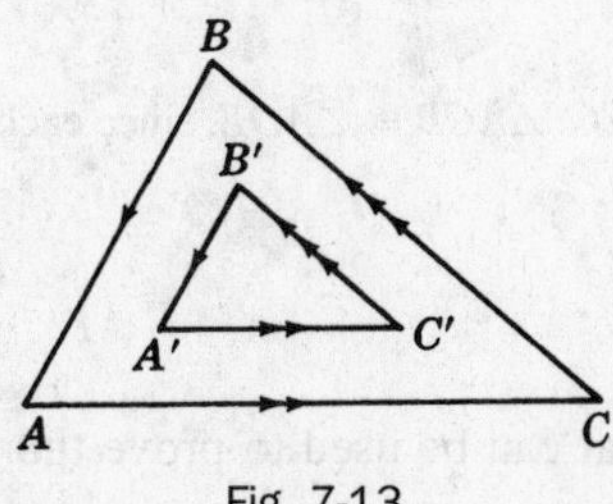

Fig. 7-13

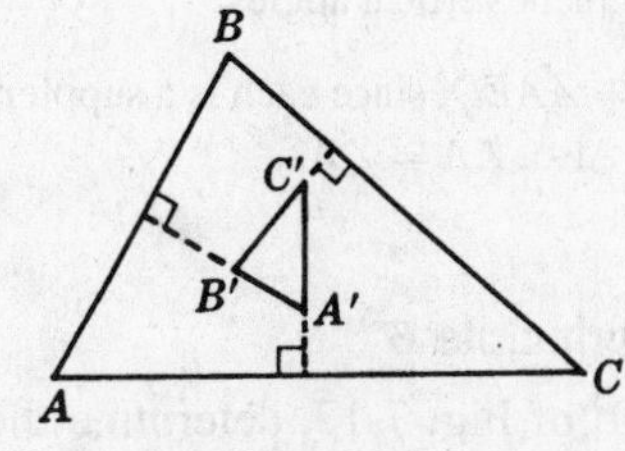

Fig. 7-14

SOLVED PROBLEMS

7.17 Applying principle 2

In similar triangles ABC and $A'B'C'$ (Fig. 7-15), find x and y if $\angle A \cong \angle A'$ and $\angle B \cong \angle B'$.

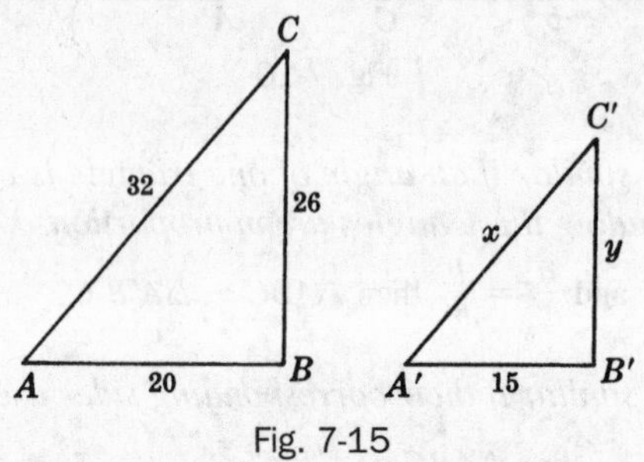

Fig. 7-15

Solution

Since $\angle A \cong \angle A'$ and $\angle B \cong \angle B'$, x and y correspond to 32 and 26, respectively. Hence, $\frac{x}{32} = \frac{15}{20}$, from which $x = 24$; also $\frac{y}{26} = \frac{15}{20}$ so $y = 19\frac{1}{2}$.

7.18 Applying principle 3

In each part of Fig. 7-16, two pairs of congruent angles can be used to prove the indicated triangles similar. Determine the congruent angles and state the reason they are congruent.

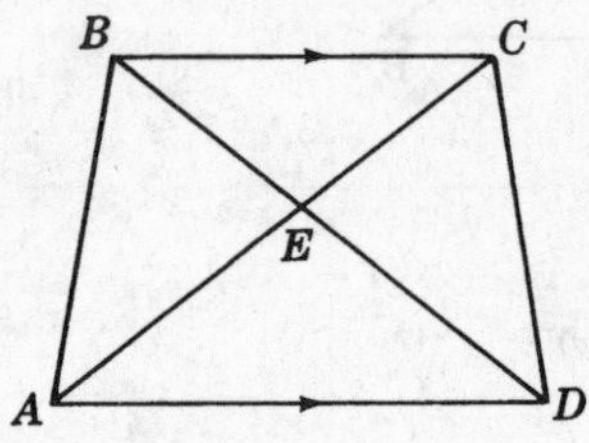

(*a*) $\triangle BEC \sim \triangle AED$
$ABCD$ is a trapezoid.

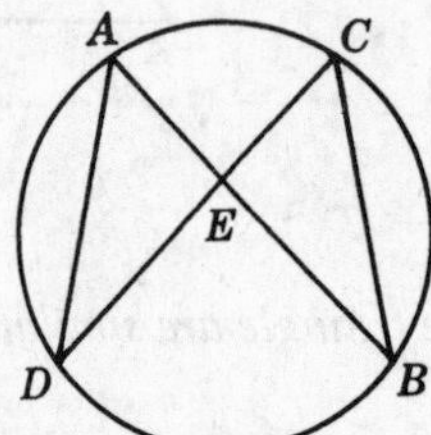

(*b*) $\triangle AED \sim \triangle CEB$

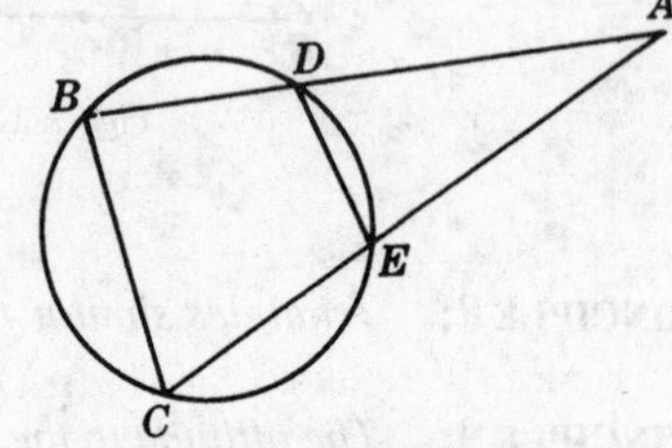

(*c*) $\triangle ABC \sim \triangle AED$

Fig. 7-16

Solutions

(a) $\angle CBD \cong \angle BDA$ and $\angle BCA \cong \angle CAD$, since alternate interior angles of parallel lines are congruent ($\overline{BC} \parallel \overline{AD}$). Also, $\angle BEC$ and $\angle AED$ are congruent vertical angles.

(b) $\angle A \cong \angle C$ and $\angle B \cong \angle D$, since angles inscribed in the same arc are congruent. Also, $\angle AED$ and $\angle CEB$ are congruent vertical angles.

(c) $\angle ABC \cong \angle AED$, since each is a supplement of $\angle DEC$. $\angle ACB \cong \angle ADE$, since each is a supplement of $\angle BDE$. Also, $\angle A \cong \angle A$.

7.19 Applying principle 6

In each part of Fig. 7-17, determine the angles that can be used to prove the indicated triangles similar.

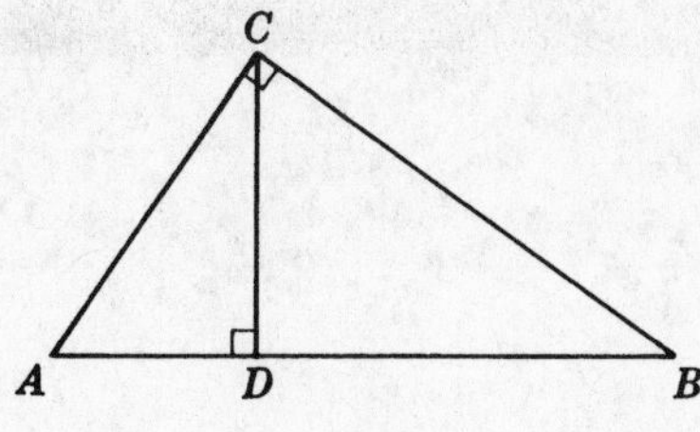

(a) $\triangle ACB \sim \triangle ADC$

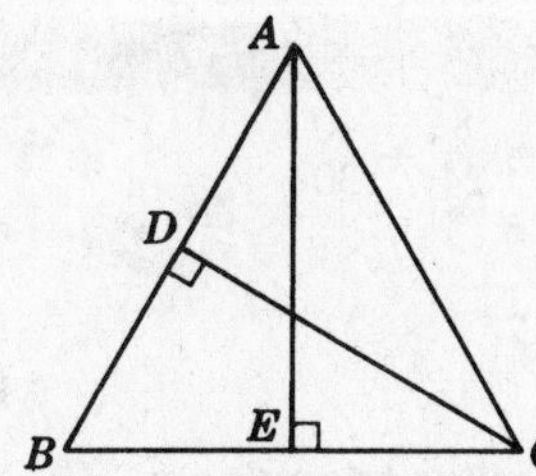

(b) $\triangle AEC \sim \triangle CDB$
$AB = AC$

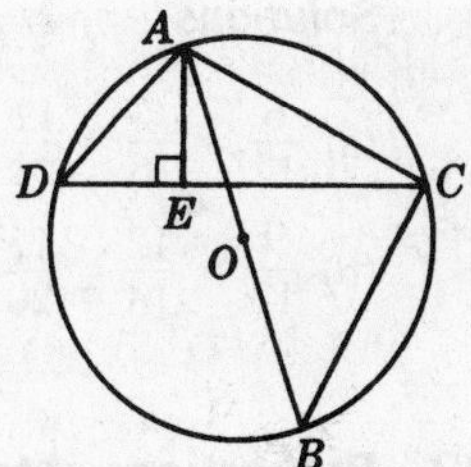

(c) $\triangle ADE \sim \triangle ABC$
$\overline{AB}$ is a diameter.

Fig. 7-17

Solutions

(a) $\angle ACB$ and $\angle ADC$ are right angles. $\angle A \cong \angle A$.

(b) $\angle AEC$ and $\angle BDC$ are right angles. $\angle B \cong \angle ACE$, since angles in a triangle opposite congruent sides are congruent.

(c) $\angle ACB$ is a right angle, since it is inscribed in a semicircle. Hence, $\angle AED \cong \angle ACB$. $\angle D \cong \angle B$, since angles inscribed in the same arc are congruent.

7.20 Applying principle 4

In each part of Fig. 7-18, determine the pair of congruent angles and the proportion needed to prove the indicated triangles similar.

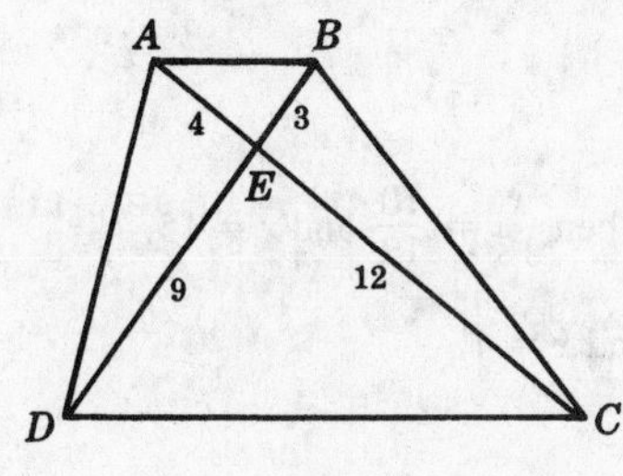

(a) $\triangle AEB \sim \triangle DEC$

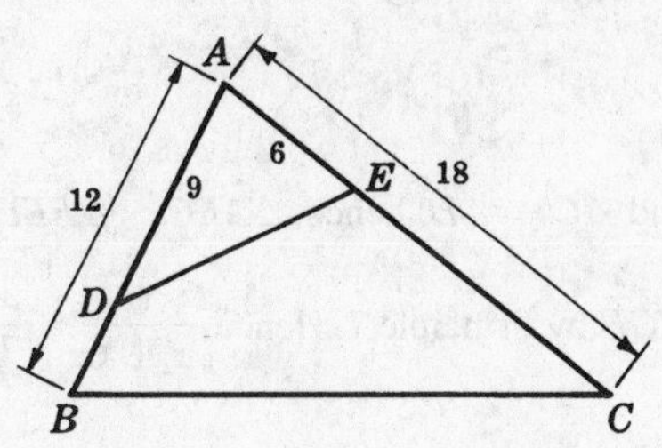

(b) $\triangle AED \sim \triangle ABC$

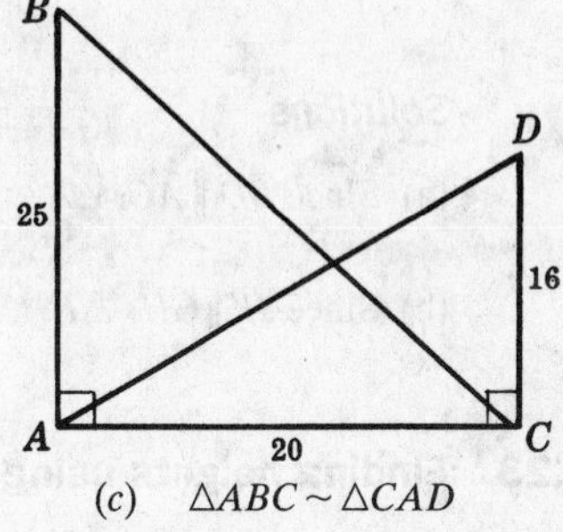

(c) $\triangle ABC \sim \triangle CAD$

Fig. 7-18

Solutions

(a) $\angle AEB \cong \angle DEC; \dfrac{3}{9} = \dfrac{4}{12}$ (b) $\angle A \cong \angle A; \dfrac{6}{12} = \dfrac{9}{18}$ (c) $\angle BAC \cong \angle ACD; \dfrac{20}{16} = \dfrac{25}{20}$

7.21 Applying principle 5

In each part of Fig. 7-19, determine the proportion needed to prove the indicated triangles similar.

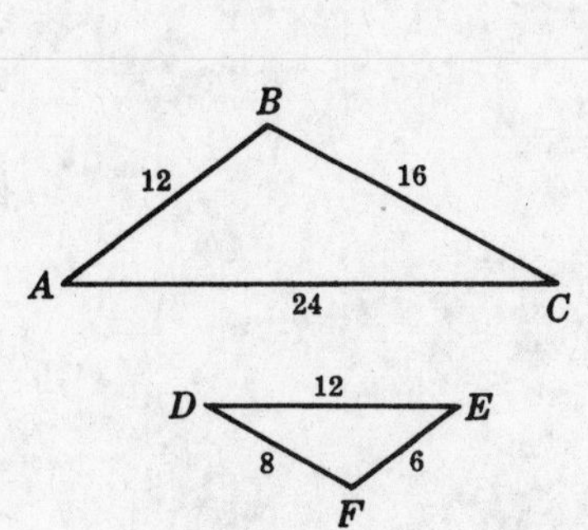

(a) $\triangle ABC \sim \triangle EFD$

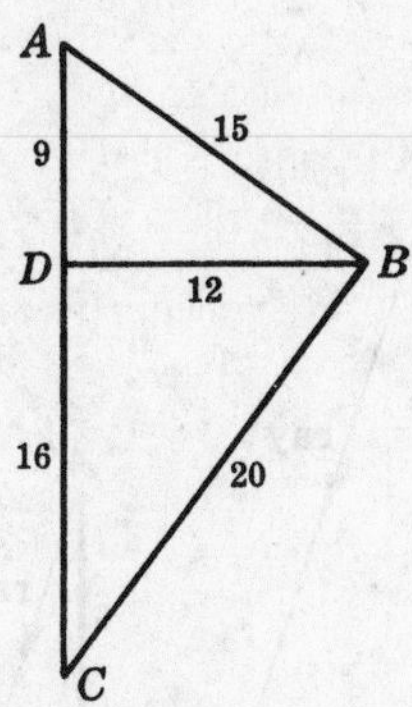

(b) $\triangle ABD \sim \triangle BCD$

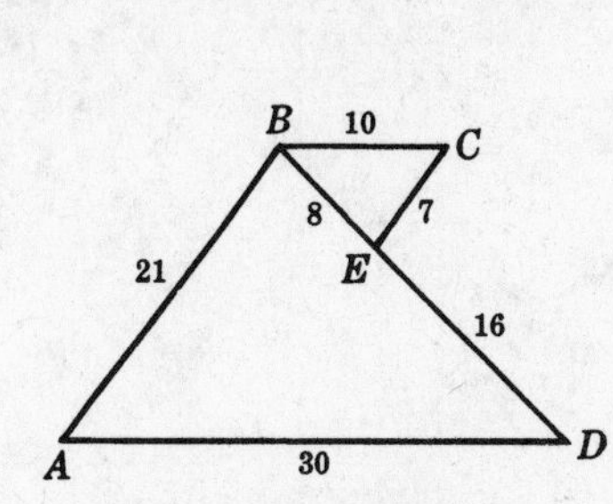

(c) $\triangle ABD \sim \triangle CEB$

Fig. 7-19

Solutions

(a) $\frac{6}{12} = \frac{8}{16} = \frac{12}{24}$ (c) $\frac{7}{21} = \frac{8}{24} = \frac{10}{30}$

(b) $\frac{9}{12} = \frac{12}{16} = \frac{15}{20}$

7.22 Proportions obtained from similar triangles

Find x in each part of Fig. 7-20.

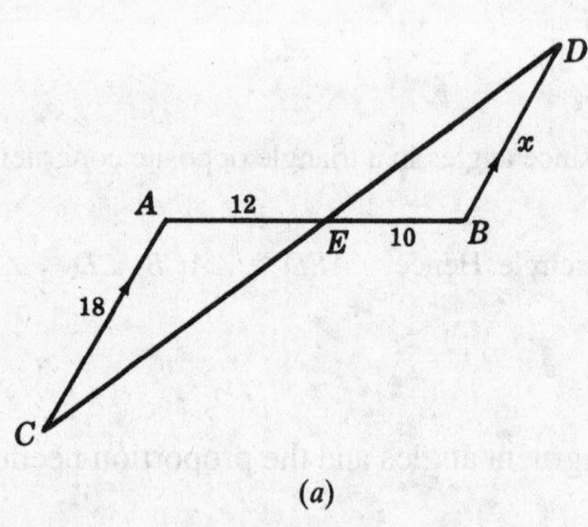

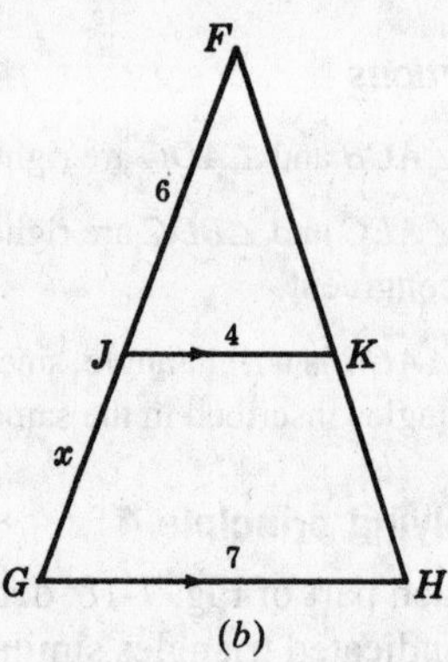

Fig. 7-20

Solutions

(a) Since $\overline{BD} \parallel \overline{AC}$, $\angle A \cong \angle B$ and $\angle C \cong \angle D$; hence, $\triangle AEC \sim \triangle BED$. Then $\frac{x}{18} = \frac{10}{12}$ and $x = 15$.

(b) Since $\overline{JK} \parallel \overline{GH}$, $\triangle FJK \sim \triangle FGH$ by Principle 7. Hence, $\frac{6}{x+6} = \frac{4}{7}$ and $x = 4\frac{1}{2}$.

7.23 Finding heights using ground shadows

A tree casts a 15-ft shadow at a time when a nearby upright pole 6 ft high casts a shadow of 2 ft. Find the height of the tree if both tree and pole make right angles with the ground.

Solution

At the same time in localities near each other, the rays of the sun strike the ground at equal angles; hence, $\angle B \cong \angle B'$ in Fig. 7-21. Since the tree and the pole make right angles with the ground, $\angle C \cong \angle C'$. Hence, $\triangle ABC \sim \triangle A'B'C'$, so $\frac{h}{6} = \frac{15}{2}$ and $h = 45$ ft.

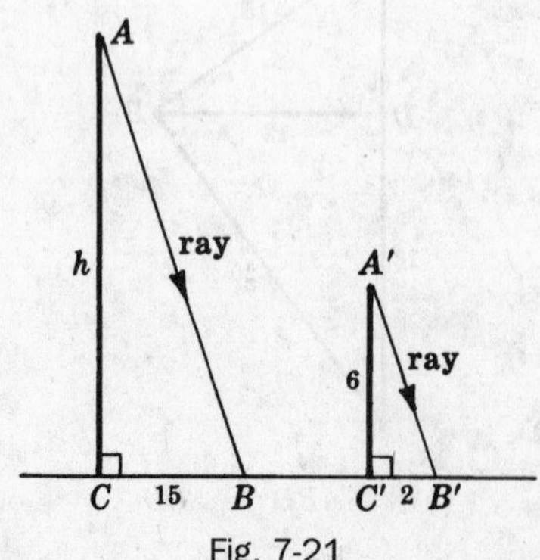

Fig. 7-21

7.24 Proving a similar-triangle problem stated in words

Prove that two isosceles triangles are similar if a base angle of one is congruent to a base angle of the other.

Solution

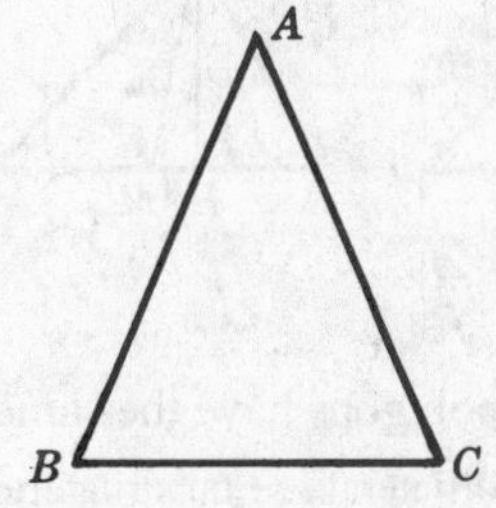

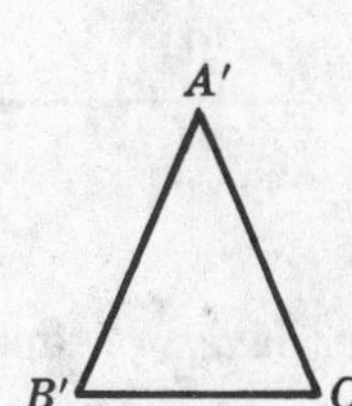

Given: Isosceles $\triangle ABC$ ($AB = AC$)
Isosceles $\triangle A'B'C'$ ($A'B' = A'C'$)
$\angle B \cong \angle B'$

To Prove: $\triangle ABC \sim \triangle A'B'C'$

Plan: Prove $\angle C \cong \angle C'$ and use Principle 3.

PROOF:

Statements	Reasons
1. $\angle B \cong \angle B'$	1. Given
2. $\angle B \cong \angle C$, $\angle B' \cong \angle C'$	2. Base angles of an isosceles triangle are congruent.
3. $\angle C \cong \angle C'$	3. Things $\cong$ to $\cong$ things are $\cong$ to each other.
4. $\triangle ABC \sim \triangle A'B'C'$	4. Two triangles are similar if two angles of one triangle are congruent to two angles of the other.

7.25 Proving a proportion problem involving similar triangles

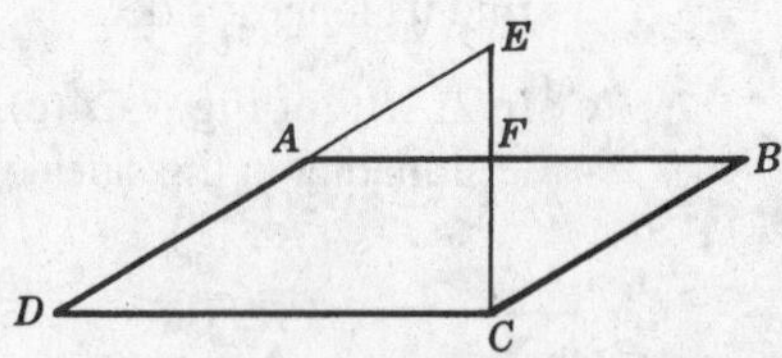

Given: Parallelogram $ABCD$

To Prove: $\dfrac{AE}{BC} = \dfrac{AF}{BF}$

Plan: Prove ($AEF \sim \triangle BCF$)

PROOF:

Statements	Reasons
1. $ABCD$ is a parallelogram.	1. Given
2. $\overline{ED} \parallel \overline{BC}$	2. Opposite sides of a parallelogram are parallel.
3. $\angle DEC \cong \angle ECB$	3. Alternate interior angles of parallel lines are congruent.
4. $\angle EFA \cong \angle BFC$	4. Vertical angles are congruent.
5. $\triangle AEF \sim \triangle BCF$	5. Two triangles are similar if two angles of one triangle are congruent to two angles of the other.
6. $\dfrac{AE}{BC} = \dfrac{AF}{BF}$	6. Corresponding sides of similar triangles are in proportion.

7.5 Extending A Basic Proportion Principle

PRINCIPLE 1: *Corresponding sides of similar triangles are in proportion.*

PRINCIPLE 2: *Corresponding segments of similar triangles are in proportion.*

PRINCIPLE 3: *Corresponding segments of similar polygons are in proportion.*

When *segments* replaces *sides*, Principle 1 becomes the more general Principle 2. When *polygons* replaces *triangles*, Principle 2 becomes the even more general Principle 3.

By *segments* we mean straight or curved segments such as altitudes, medians, angle bisectors, radii of inscribed or circumscribed circles, and circumferences of inscribed or circumscribed circles.

The *ratio of similitude* of two similar polygons is the ratio of any pair of corresponding lines.

Corollaries of Principles 2 and 3, such as the following, can be devised for any combination of corresponding lines:

1. Corresponding *altitudes* of similar triangles have the same ratio as any two correponding *medians*.

   Thus if $\triangle ABC \sim \triangle A'B'C'$ in Fig. 7-22, then $\dfrac{h}{h'} = \dfrac{m}{m'}$.

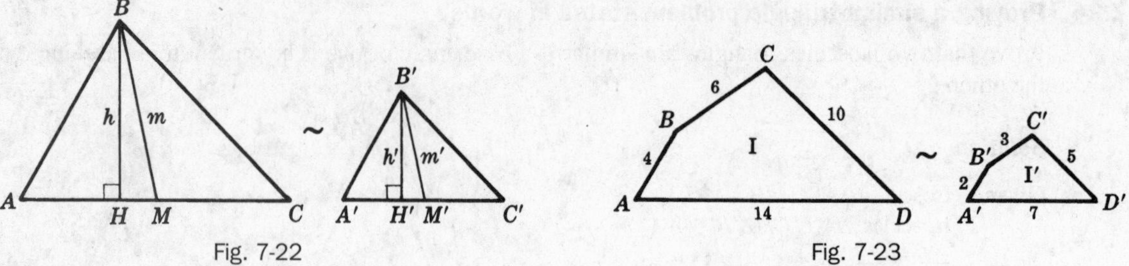

Fig. 7-22 Fig. 7-23

2. *Perimeters* of similar polygons have the same ratio as any two corresponding *sides*.
Thus in Fig. 7-23, if quadrilateral I ~ quadrilateral I′, then $\frac{34}{17} = \frac{4}{2} = \frac{6}{3} = \frac{10}{5} = \frac{14}{7}$.

SOLVED PROBLEMS

7.26 Line ratios from similar triangles

(a) In two similar triangles, corresponding sides are in the ratio 3:2. Find the ratio of corresponding medians [see Fig. 7-24(a)].

(b) The sides of a triangle are 4, 6, and 7 [Fig. 7-24(b)]. If the perimeter of a similar triangle is 51, find its longest side.

(c) In $\triangle ABC$ of Fig. 7-24(c), $BC = 25$ and the measure of the altitude to $\overline{BC}$ is 10. A line segment terminating in the sides of the triangle is parallel to $\overline{BC}$ and 3 units from A. Find its length.

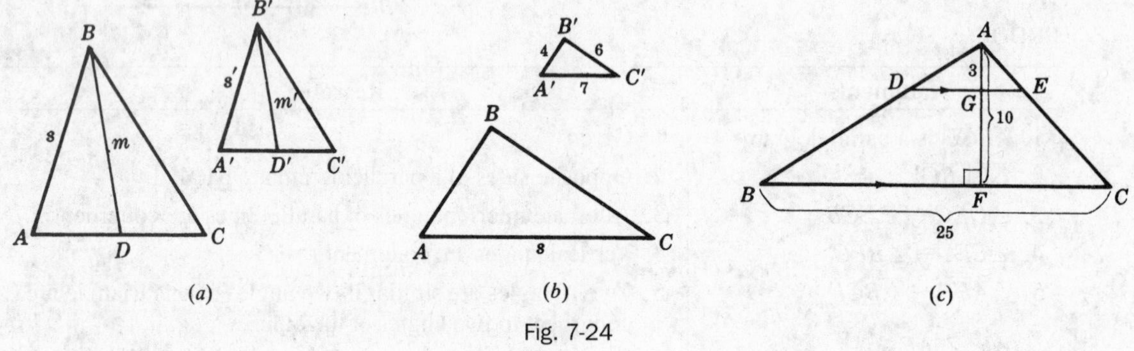

Fig. 7-24

Solutions

(a) If $\triangle ABC \sim \triangle A'B'C'$ and $\frac{s}{s'} = \frac{3}{2}$, then $\frac{m}{m'} = \frac{3}{2}$.

(b) The perimeter of $\triangle A'B'C'$ is $4 + 6 + 7 = 17$. Since $\triangle ABC \sim \triangle A'B'C'$, $\frac{s}{7} = \frac{51}{17}$ and $s = 21$.

(c) Since $\triangle ADE \sim ABC$, $\frac{DE}{25} = \frac{3}{10}$ and $DE = 7\frac{1}{2}$.

7.27 Line ratios from similar polygons

Complete each of the following statements:

(a) If corresponding sides of two similar polygons are in the ratio of 4:3, then the ratio of their perimeters is $\underline{?}$.

(b) The perimeters of two similar quadrilaterals are 30 and 24. If a side of the smaller quadrilateral is 8, the corresponding side of the larger is $\underline{?}$.

(c) If each side of a pentagon is tripled and the angles remain the same, then each diagonal is $\underline{?}$.

Solutions

(a) Since the polygons are similar, $\frac{p}{p'} = \frac{s}{s'} = \frac{4}{3}$.

(b) Since the quadrilaterals are similar, $\frac{s}{s'} = \frac{p}{p'}$. Then $\frac{s}{8} = \frac{30}{24}$ and $s = 10$.

(c) Tripled, since polygons are similar if their corresponding angles are congruent and their corresponding sides are in proportion.

7.6 Proving Equal Products of Lengths of Segments

In a problem, to prove that the product of the lengths of two segments equals the product of the lengths of another pair of segments, it is necessary to set up the proportion which will lead to the two equal products.

SOLVED PROBLEM

7.28 Proving an equal-products problem

Prove that if two secants intersect outside a circle, the product of the lengths of one of the secants and its external segment equals the product of the lengths of the other secant and its external segment.

Solution

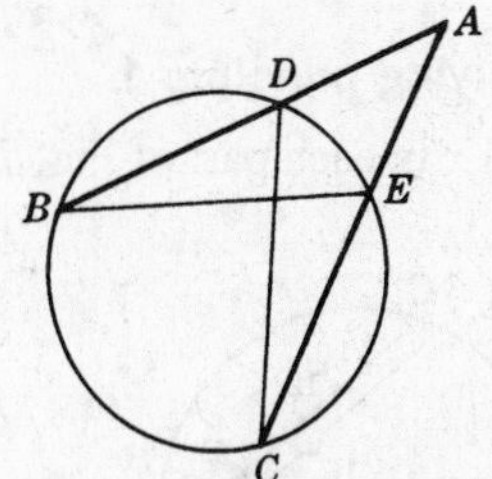

Given: Secants $\overline{AB}$ and $\overline{AC}$.
To Prove: $AB \times AD = AC \times AE$
Plan: Prove $\triangle ABE \sim \triangle ACD$ to obtain $\frac{AB}{AC} = \frac{AE}{AD}$.

PROOF:

Statements	Reasons
1. Draw $\overline{BE}$ and $\overline{CD}$.	1. A segment may be drawn between any two points.
2. $\angle A \cong \angle A$	2. Reflexive property.
3. $\angle B \cong \angle C$	3. Angles inscribed in the same arc are congruent.
4. $\triangle AEB \sim \triangle ADC$	4. Two triangles are similar if two angles of one triangle are congruent respectively to two angles of the other.
5. $\frac{AB}{AC} = \frac{AE}{AD}$	5. Corresponding sides of similar triangles are in proportion.
6. $AB \times AD = AC \times AE$	6. In a proportion, the product of the means equals the product of the extremes.

7.7 Segments Intersecting Inside and Outside a Circle

PRINCIPLE 1: *If two chords intersect within a circle, the product of the lengths of the segments of one chord equals the product of the lengths of the segments of the other.*

Thus in Fig. 7-25, $AE \times EB = CE \times ED$.

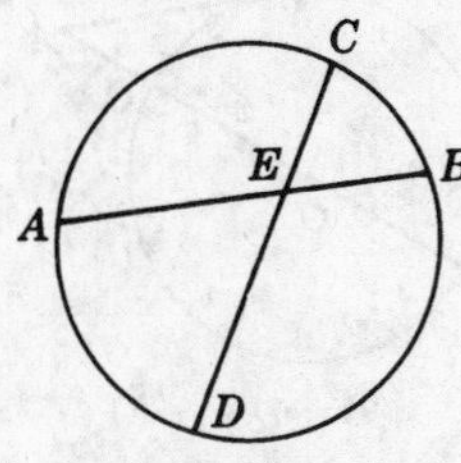

Fig. 7-25

PRINCIPLE 2: *If a tangent and a secant intersect outside a circle, the tangent is the mean proportional between the secant and its external segment.*

Thus in Fig. 7-26, if $\overline{PA}$ is a tangent, then $\dfrac{AB}{AP} = \dfrac{AP}{AC}$.

PRINCIPLE 3: *If two secants intersect outside a circle, the product of the lengths of one of the secants and its external segment equals the product of the lengths of the other secant and its external segment.*

Thus in Fig. 7-27, $AB \times AD = AC \times AE$.

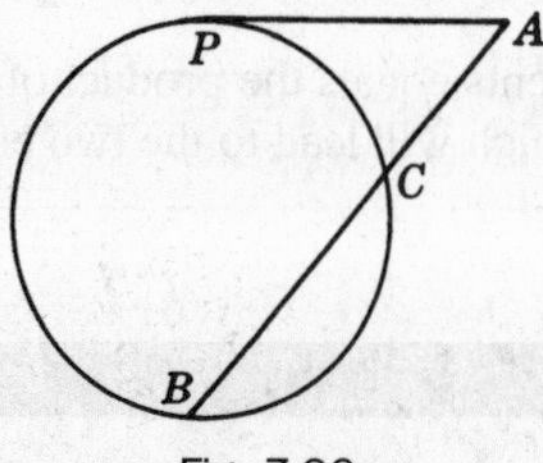

Fig. 7-26

Fig. 7-27

SOLVED PROBLEMS

7.29 Applying principle 1

Find x in each part of Fig. 7-28, if chords $\overline{AB}$ and $\overline{CD}$ intersect in E.

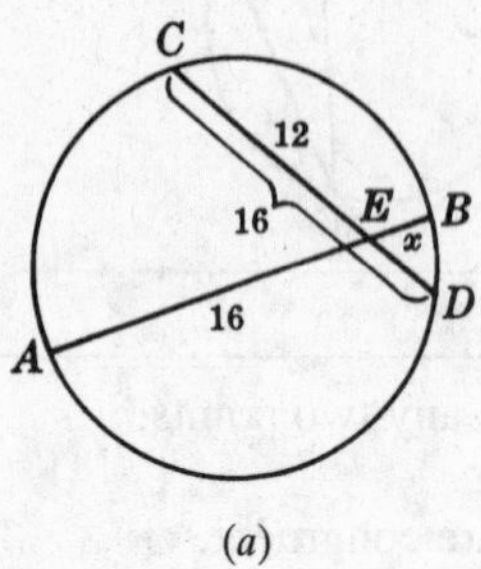

(*a*)

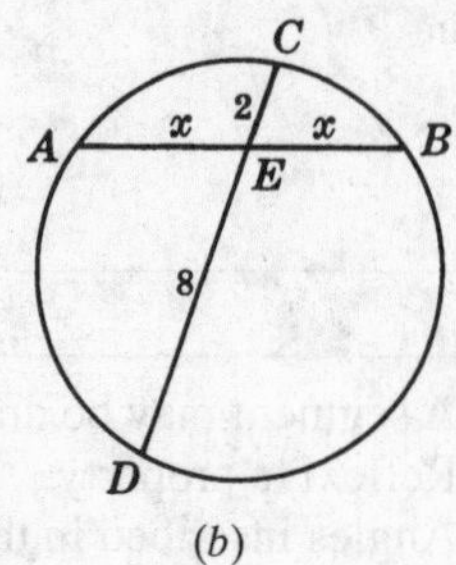

(*b*)

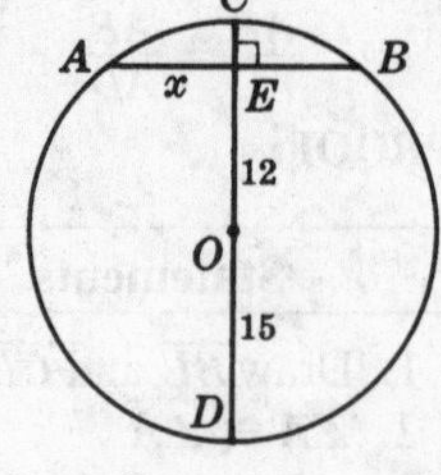

(*c*) Diameter $CD \perp AB$

Fig. 7-28

Solutions

(a) $ED = 4$. Then $16x = 4(12)$, so that $16x = 48$ or $x = 3$.

(b) $AE = EB = x$. Then $x^2 = 8(2)$, so $x^2 = 16$ and $x = 4$.

(c) $CE = 3$ and $AE = EB = x$. Then $x^2 = 27(3)$ or $x^2 = 81$, and $x = 9$.

7.30 Applying principle 2

Find x in each part of Fig. 7-29 if tangent $\overline{AP}$ and $\overline{AB}$ intersect at A.

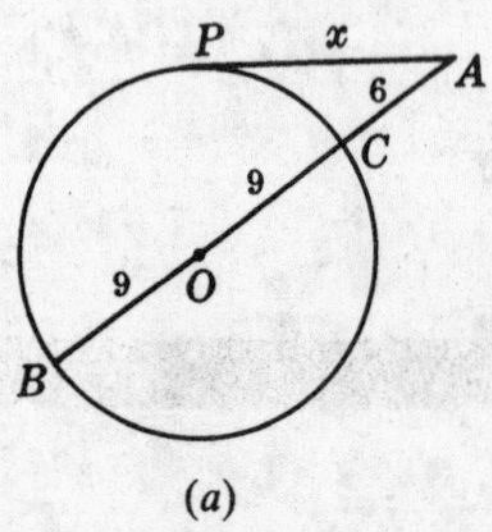

(*a*)

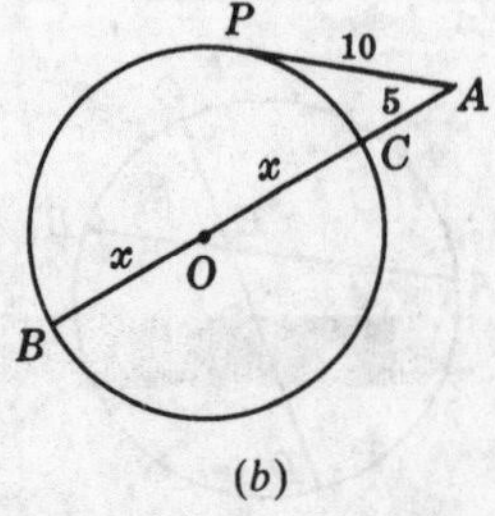

(*b*)

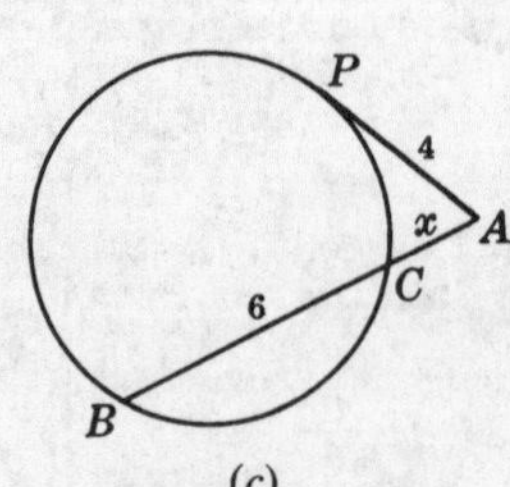

(*c*)

Fig. 7-29

Solutions

(a) $AB = 9 + 9 + 6 = 24$. Then $x^2 = 24(6)$ or $x^2 = 144$, and $x = 12$.

(b) $AB = 2x + 5$. Then $5(2x + 5) = 100$ and $x = 7\frac{1}{2}$.

(c) $AB = x + 6$. Then $x(x + 6) = 16$ or $x^2 + 6x - 16 = 0$. Factoring gives $(x + 8)(x - 2) = 0$ and $x = 2$.

7.31 Applying principle 3

Find x in each part of Fig. 7-30 if secants $\overline{AB}$ and $\overline{AC}$ intersect in A.

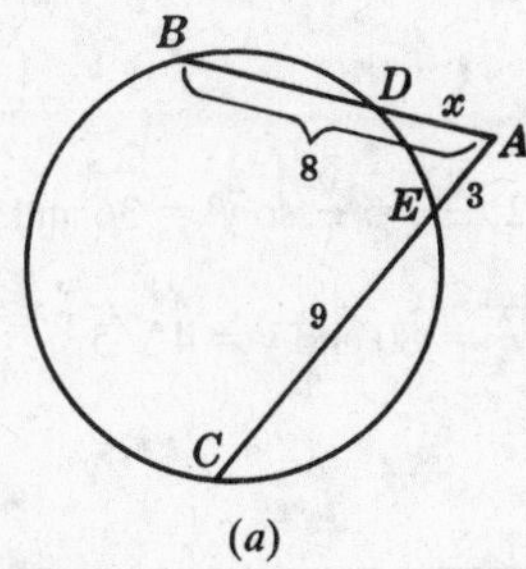

(*a*)

(*b*)

Fig. 7-30

Solutions

(a) $AC = 12$. Then $8x = 12(3)$ and $x = 4\frac{1}{2}$.

(b) $AC = 2x + 2$ and $AB = 12$. Then $2(2x + 2) = 12(5)$ and $x = 14$.

7.8 Mean Proportionals in a Right Triangle

PRINCIPLE 1: *The length of the altitude to the hypotenuse of a right triangle is the mean proportional between the lengths of the segments of the hypotenuse.*

Thus in right $\triangle ABC$ (Fig. 7-31), $\frac{BD}{CD} = \frac{CD}{DA}$.

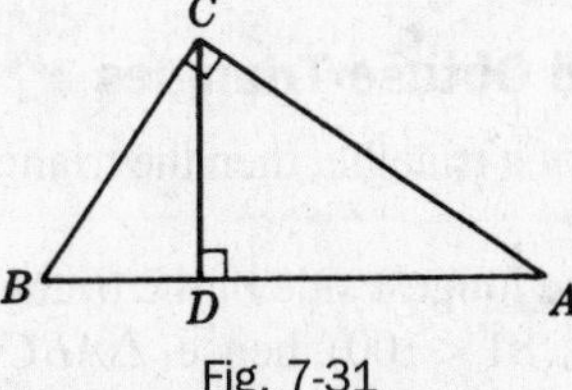

Fig. 7-31

PRINCIPLE 2: *In a right triangle, the length of either leg is the mean proportional between the length of the hypotenuse and the length of the projection of that leg on the hypotenuse.*

Thus in right $\triangle ABC$ (Fig. 7-31), $\frac{AB}{BC} = \frac{BC}{BD}$ and $\frac{AB}{AC} = \frac{AC}{AD}$.

A proof of this principle is given in Chapter 16.

SOLVED PROBLEMS

7.32 Finding mean proportionals in a right triangle

In each triangle in Fig. 7-32, find x and y.

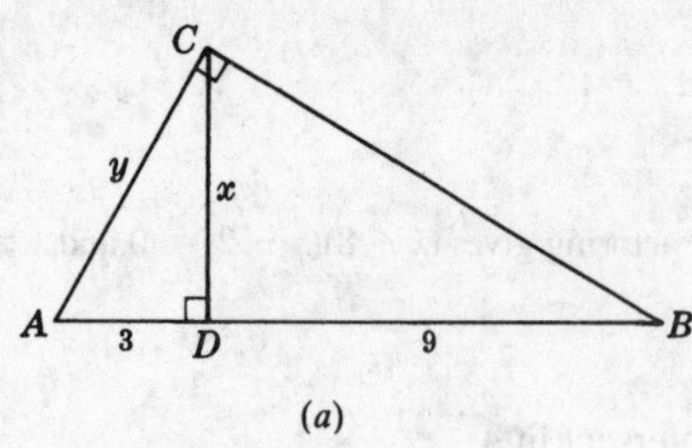

(a)

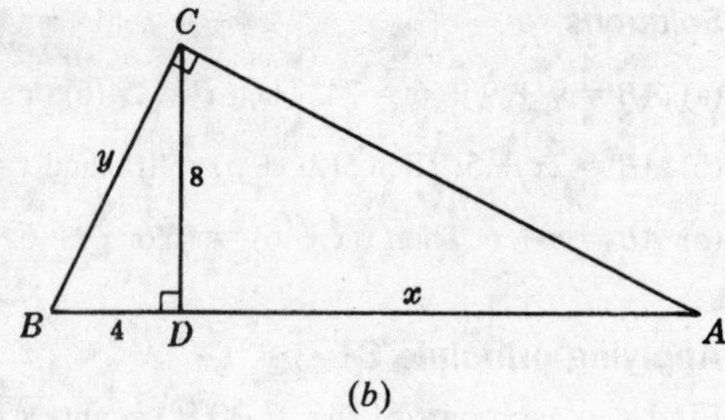

(b)

Fig. 7-32

Solutions

(a) By Principle 1, $\frac{3}{x} = \frac{x}{9}$ or $x^2 = 27$, and $x = 3\sqrt{3}$. By Principle 2, $\frac{12}{y} = \frac{y}{3}$, so $y^2 = 36$ and $y = 6$.

(b) By Principle 1, $\frac{x}{8} = \frac{8}{4}$ and $x = 16$. By Principle 2, $\frac{20}{y} = \frac{y}{4}$, so $y^2 = 80$ and $y = 4\sqrt{5}$.

7.9 Pythagorean Theorem

In a right triangle, the square of the length of the hypotenuse equals the sum of the squares of the lengths of the legs.

Thus in Fig. 7-33, $c^2 = a^2 + b^2$.

A proof of the Pythagorean Theorem is given in Chapter 16.

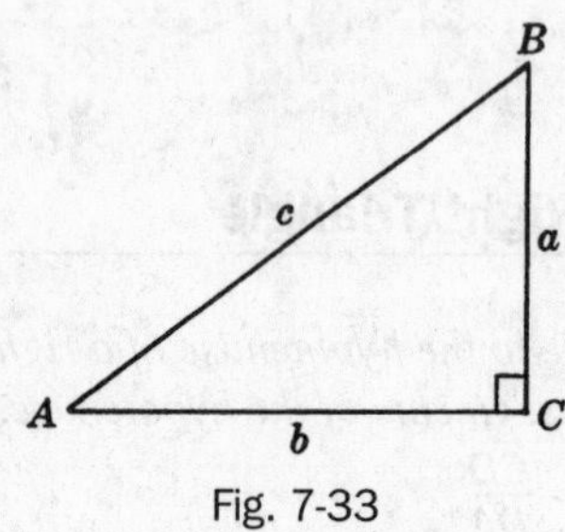

Fig. 7-33

7.9A Tests for Right, Acute, and Obtuse Triangles

If $c^2 = a^2 + b^2$ applies to the three sides of a triangle, then the triangle is a right triangle; but if $c^2 \neq a^2 + b^2$, then the triangle is not a right triangle.

In $\triangle ABC$, if $c^2 < a^2 + b^2$ where c is the longest side of the triangle, then the triangle is an acute triangle. Thus in Fig. 7-34, $9^2 < 6^2 + 8^2$ (that is, $81 < 100$); hence, $\triangle ABC$ is an acute triangle.

In $\triangle ABC$, if $c^2 > a^2 + b^2$ where c is the longest side of the triangle, then the triangle is an obtuse triangle. Thus in Fig. 7-35, $11^2 > 6^2 + 8^2$ (that is, $121 > 100$); hence, $\triangle ABC$ is an obtuse triangle.

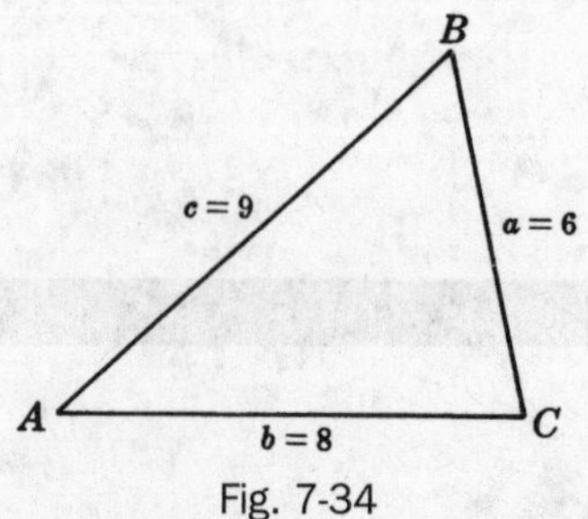

Fig. 7-34

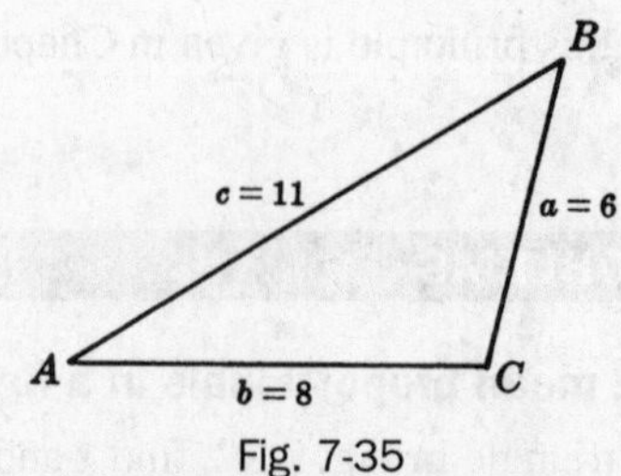

Fig. 7-35

SOLVED PROBLEMS

7.33 Finding the sides of a right triangle

In Fig. 7-36, (a) find the length of hypotenuse c if $a = 12$ and $b = 9$; (b) find a if $b = 6$ and $c = 8$; (c) find b if $a = 4\sqrt{3}$ and $c = 8$.

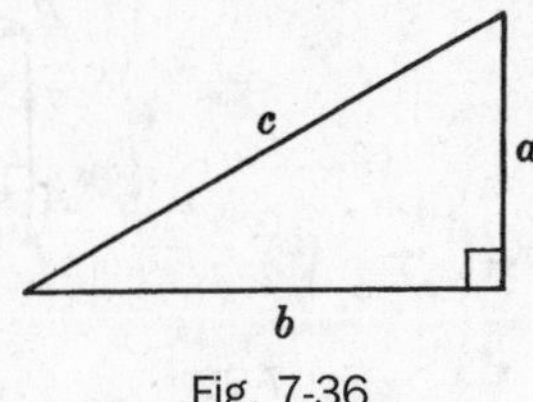

Fig. 7-36

Solutions

(a) $c^2 = a^2 + b^2 = 12^2 + 9^2 = 225$ and $c = 15$.

(b) $a^2 = c^2 - b^2 = 8^2 - 6^2 = 28$ and $a = 2\sqrt{7}$.

(c) $b^2 = c^2 - a^2 = 8^2 - (4\sqrt{3})^2 = 64 - 48 = 16$ and $b = 4$.

7.34 Ratios in a right triangle

In a right triangle, the hypotenuse has length 20 and the ratio of the two arms is 3:4. Find each arm.

Solution

Let the lengths of the two arms be denoted by $3x$ and $4x$. Then $20^2 = (3x)^2 + (4x)^2$.
Multiplying out, we get $400 = 9x^2 + 16x^2$ or $400 = 25x^2$, and $x = 4$; hence, the arms have lengths 12 and 16.

7.35 Applying the pythagorean theorem to an isosceles triangle

Find the length of the altitude to the base of an isosceles triangle if the base is 8 and the equal sides are 12.

Solution

The altitude h of an isosceles triangle bisects the base (Fig. 7-37). Then $h^2 = a^2 - (\frac{1}{2}b)^2 = 12^2 - 4^2 = 128$ and $h = 8\sqrt{2}$.

7.36 Applying the pythagorean theorem to a rhombus

In a rhombus, find (a) the length of a side s if the diagonals are 30 and 40; (b) the length of a diagonal d if a side is 26 and the other diagonal is 20.

Solution

The diagonals of a rhombus are perpendicular bisectors of each other; hence, $s^2 = (\frac{1}{2}d)^2 + (\frac{1}{2}d')^2$ in Fig. 7-38.

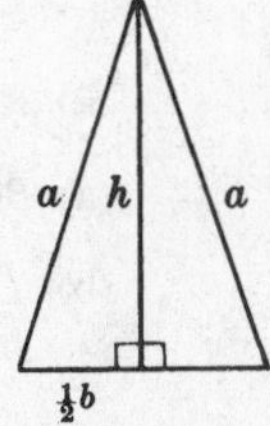

Fig. 7-37

(a) If $d = 30$ and $d' = 40$, then $s^2 = 15^2 + 20^2 = 625$ or $s = 25$.

(b) If $s = 26$ and $d' = 20$, then $26^2 = (\frac{1}{2}d)^2 + 10^2$ or $576 = (\frac{1}{2}d)^2$. Thus, $\frac{1}{2}d = 24$ or $d = 48$.

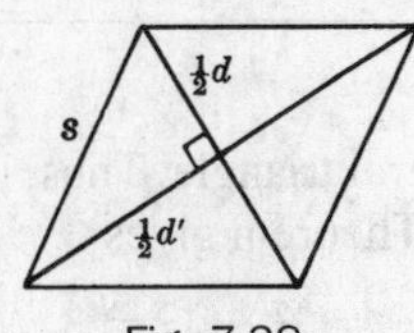

Fig. 7-38

7.37 Applying the pythagorean theorem to a trapezoid

Find x in each part of Fig. 7-39 if $ABCD$ is a trapezoid.

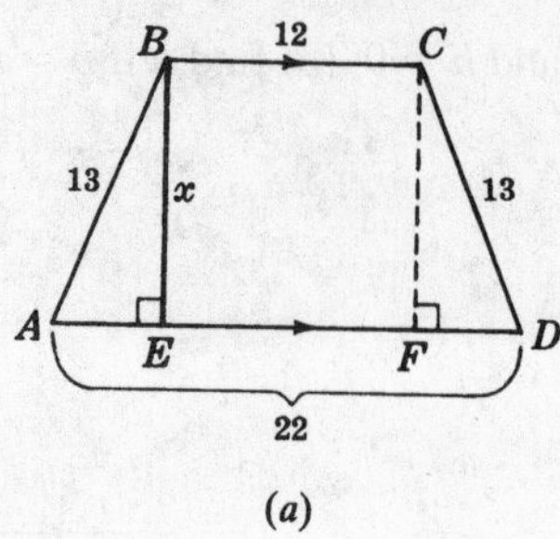

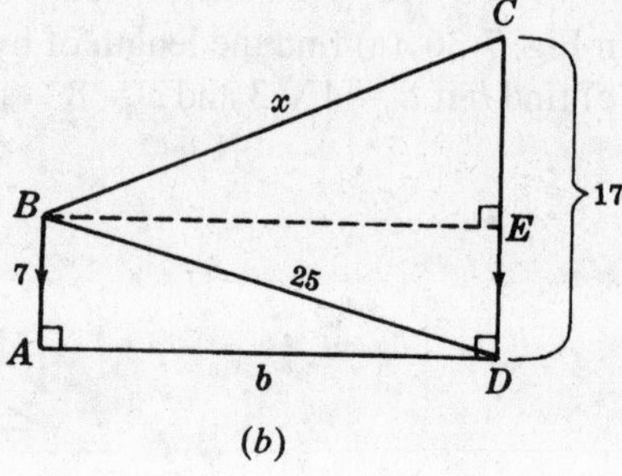

Fig. 7-39

Solutions

The dashed perpendiculars in the diagrams are additional segments needed only for the solutions. Note how rectangles are formed by these added segments.

(a) $EF = BC = 12$ and $AE = \frac{1}{2}(22 - 12) = 5$. Then $x^2 = 13^2 - 5^2 = 144$ or $x = 12$.

(b) $b^2 = 25^2 - 7^2 = 576$ or $b = 24$; also, $BE = b = 24$ and $CE = 17 - 7 = 10$. Then $x^2 = 24^2 + 10^2$ or $x = 26$.

7.38 Applying the pythagorean theorem to a circle

(a) Find the distance d from the center of a circle of radius 17 to a chord whose length is 30 [Fig. 7-40(a)].

(b) Find the length of a common external tangent to two externally tangent circles with radii 4 and 9 [Fig. 7-40(b)].

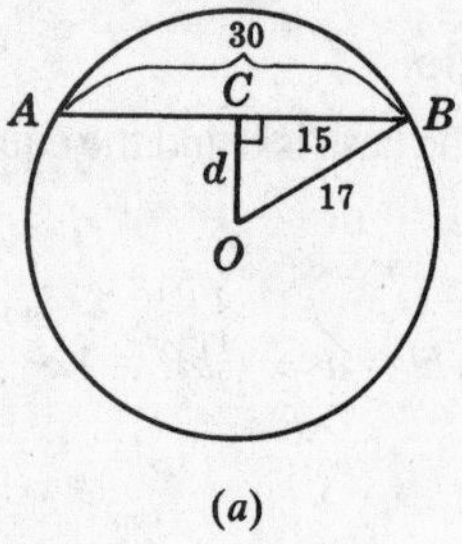

Fig. 7-40

Solutions

(a) $BC = \frac{1}{2}(30) = 15$. Then $d^2 = 17^2 - 15^2 = 64$ and $d = 8$.

(b) $\overline{OS} \cong \overline{PR}$, $RS = 4$, $OQ = 13$, and $SQ = 9 - 4 = 5$. Then in right $\triangle OSQ$, $(OS)^2 = 13^2 - 5^2 = 144$ so $OS = 12$; hence $PR = 12$.

7.10 Special Right Triangles

7.10A The 30°-60°-90° Triangle

A 30°-60°-90° triangle is one-half an equilateral triangle. Thus, in right $\triangle ABC$ (Fig. 7-41), $a = \frac{1}{2}c$. Consider that $c = 2$; then $a = 1$, and the Pythagorean Theorem gives

$$b^2 = c^2 - a^2 = 2^2 - 1^2 = 3 \quad \text{or} \quad b = \sqrt{3}$$

The ratio of the sides is then $a:b:c = 1:\sqrt{3}:2$

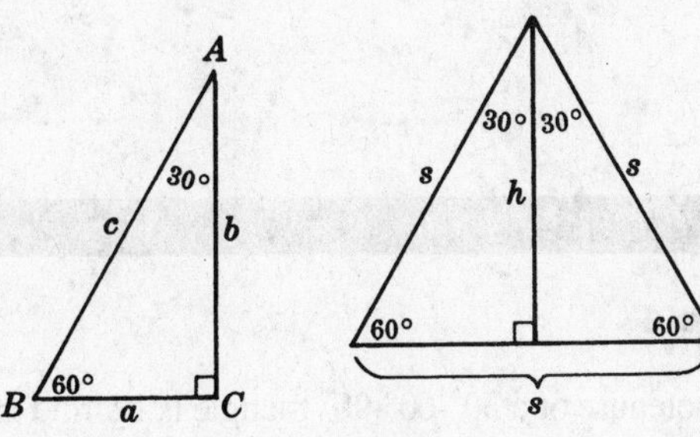

Fig. 7-41

Principles of the 30°-60°-90° Triangle

PRINCIPLE 1: *The length of the leg opposite the* 30° *angle equals one-half the length of the hypotenuse.*

In Fig. 7-41, $a = \frac{1}{2}c$.

PRINCIPLE 2: *The length of the leg opposite the* 60° *angle equals one-half the length of the hypotenuse times the square root of* 3.

In Fig. 7-41, $b = \frac{1}{2}c\sqrt{3}$.

PRINCIPLE 3: *The length of the leg opposite the* 60° *angle equals the length of the leg opposite the* 30° *angle times the square root of* 3.

In Fig. 7-41, $b = a\sqrt{3}$.

Equilateral-Triangle Principle

PRINCIPLE 4: *The length of the altitude of an equilateral triangle equals one-half the length of a side times the square root of* 3. (Principle 4 is a corollary of Principle 2.)

In Fig. 7-41, $h = \frac{1}{2}s\sqrt{3}$.

7.10B The 45°-45°-90° Triangle

A 45°-45°-90° triangle is one-half a square. In right triangle *ABC* (Fig. 7-42), $c^2 = a^2 + a^2$ or $c = a\sqrt{2}$ Hence, the ratio of the sides is $a:a:c = 1:1:\sqrt{2}$.

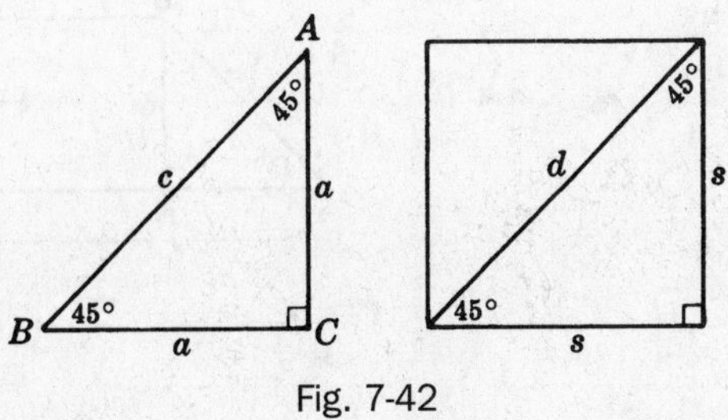

Fig. 7-42

Principles of the 45°-45°-90° Triangle

PRINCIPLE 5: *The length of a leg opposite a* 45° *angle equals one-half the length of the hypotenuse times the square root of* 2.

In Fig. 7-42, $a = \frac{1}{2}c\sqrt{2}$

PRINCIPLE 6: *The length of the hypotenuse equals the length of a side times the square root of* 2.

In Fig. 7-42, $c = a\sqrt{2}$.

Square Principle

PRINCIPLE 7: *In a square, the length of a diagonal equals the length of a side times the square root of 2.*

In Fig. 7-42, $d = s\sqrt{2}$

SOLVED PROBLEMS

7.39 Applying principles 1 to 4

(a) If the length of the hypotenuse of a 30°-60°-90° triangle is 12, find the lengths of its legs [Fig.7-43(a)].

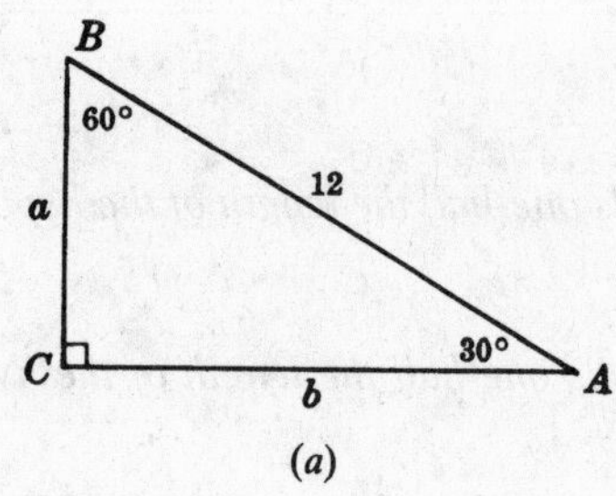

Fig. 7-43

(b) Each leg of an isosceles trapezoid has length 18. If the base angles are 60° and the upper base is 10, find the lengths of the altitude and the lower base [Fig. 7-43(b)].

Solutions

(a) By Principle 1, $a = \frac{1}{2}(12) = 6$. By Principle 2, $b = \frac{1}{2}(12)\sqrt{3} = 6\sqrt{3}$.

(b) By Principle 2, $h = \frac{1}{2}(18)\sqrt{3} = 9\sqrt{3}$. By Principle 1, $AE = FD = \frac{1}{2}(18) = 9$; hence, $b = 9 + 10 + 9 = 28$.

7.40 Applying principles 5 and 6

(a) Find the length of the leg of an isosceles right triangle whose hypotenuse has length 28 [Fig. 7-44(a)].

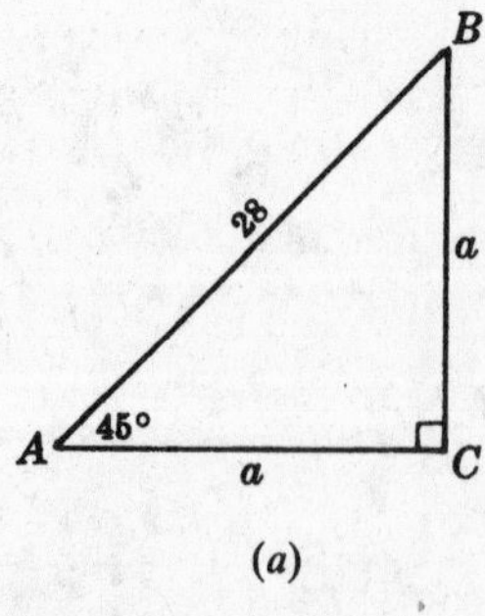

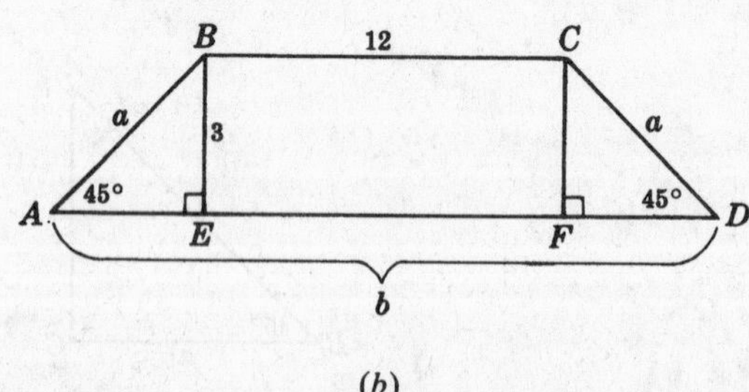

Fig. 7-44

(b) An isosceles trapezoid has base angles measuring 45°. If the upper base has length 12 and the altitude has length 3, find the lengths of the lower base and each leg [Fig. 7-44(b)].

Solutions

(a) By Principle 5, $a = \frac{1}{2}(28)\sqrt{2} = 14\sqrt{2}$.

(b) By Principle 6, $a = 3\sqrt{2}, AE = BE = 3$ and $EF = 12$; hence, $b = 3 + 12 + 3 = 18$.

SUPPLEMENTARY PROBLEMS

7.1. Express each of the following ratios in lowest terms: (7.1)

(a) 20 cents to 5 cents
(b) 5 dimes to 15 dimes
(c) 30 lb to 25 lb
(d) 20° to 14°
(e) 27 min to 21 min
(f) 50% to 25%
(g) 15° to 75°
(h) 33% to 77%
(i) \$2.20 to \$3.30
(j) \$.84 to \$.96
(k) $\frac{1}{2}$lb to $\frac{1}{4}$lb
(l) $2\frac{1}{2}$ days to $3\frac{1}{2}$ days
(m) 5 ft to $\frac{1}{4}$ft
(n) $\frac{1}{2}$ yd to $1\frac{1}{2}$ yd
(o) $16\frac{1}{2}$ m to $5\frac{1}{2}$ m

7.2. Express each of the following ratios in lowest terms: (7.2)

(a) 1 year to 2 months
(b) 2 weeks to 5 days
(c) 3 days to 3 weeks
(d) $2\frac{1}{2}$ h to 20 min
(e) 2 yd to 2 ft
(f) $2\frac{1}{3}$ yd to 2 ft
(g) $1\frac{1}{2}$ft to 9 in
(h) 2 g to 8 mg
(i) 100 lb to 1 ton
(j) \$2 to 25 cents
(k) 2 quarters to 3 dimes
(l) 1 yd² to 2 ft²

7.3. Express each of the following ratios in lowest terms: (7.3)

(a) 20 cents to 30 cents to \$1
(b) \$3 to \$1.50 to 25 cents
(c) 1 quarter to 1 dime to 1 nickel
(d) 1 day to 4 days to 1 week
(e) $\frac{1}{2}$ day to 9 h to 3 h
(f) 2 h to $\frac{1}{2}$ h to 15 min
(g) 1 ton to 200 lb to 40 lb
(h) 3 lb to 1 lb to 8 oz
(i) 1 gal to 1 qt to 1 pt

7.4. Express each of the following ratios in lowest terms: (7.4)

(a) 60 to 70
(b) 84 to 7
(c) 65 to 15
(d) 125 to 500
(e) 630 to 105
(f) 1760 to 990
(g) 0.7 to 2.1
(h) 0.36 to 0.24
(i) 0.002 to 0.007
(j) 0.055 to 0.005
(k) 6.4 to 8
(l) 144 to 2.4
(m) $7\frac{1}{2}$ to $2\frac{1}{2}$
(n) $1\frac{1}{2}$ to 10
(o) $\frac{5}{6}$ to $1\frac{2}{3}$
(p) $\frac{7}{4}$ to $\frac{1}{8}$

7.5. Express each of the following ratios in lowest terms: (7.4)

(a) x to $3x$
(b) $15c$ to 5
(c) $11d$ to 22
(d) $2\pi r$ to πD
(e) πab to πa^2
(f) $4S$ to S^2
(g) S^3 to $6S^2$
(h) $9r^2$ to $6rt$
(i) x to $4x$ to $10x$
(j) $15y$ to $10y$ to $5y$
(k) x^3 to x^2 to x
(l) $12w$ to $10w$ to $8w$ to $2w$

7.6. Use x as the common factor to represent the following numbers and their sum: (7.4)

(a) Two numbers whose ratio is 5:4
(b) Two numbers whose ratio is 9:1
(c) Three numbers whose ratio is 2:5:11
(d) Five numbers whose ratio is 1:2:2:3:7

7.7. If two angles in the ratio of 5:4 are represented by $5x$ and $4x$, express each of the following statements as an equation; then find x and the angles: (7.5)

(a) The angles are adjacent and together form an angle measuring 45°.
(b) The angles are complementary.
(c) The angles are supplementary.
(d) The angles are two angles of a triangle whose third angle is their difference.

7.8. If three angles in the ratio of 7:6:5 are represented by $7x$, $6x$, and $5x$, express each of the following statements as an equation; then find x and the angles: (7.6)

(a) The first and second are adjacent and together form an angle measuring 91°.

(b) The first and third are supplementary.

(c) The first and one-half the second are complementary.

(d) The angles are the three angles of a triangle.

7.9. Solve the following proportions for x: (7.7)

(a) $x:6 = 8:3$ (d) $x:2 = 10:x$ (g) $a:b = c:x$

(b) $5:4 = 20:x$ (e) $(x + 4):3 = 3:(x - 4)$ (h) $x:2y = 18y:x$

(c) $9:x = x:4$ (f) $(2x + 8):(x + 2) = (2x + 5):(x + 1)$

7.10. Solve the following proportions for x: (7.7)

(a) $\frac{5}{7} = \frac{15}{x}$ (c) $\frac{3}{x} = \frac{x}{12}$ (e) $\frac{x+2}{5} = \frac{6}{3}$ (g) $\frac{2x}{x+7} = \frac{3}{5}$

(b) $\frac{7}{x} = \frac{3}{2}$ (d) $\frac{x}{5} = \frac{15}{x}$ (f) $\frac{x-1}{3} = \frac{5}{x+1}$ (h) $\frac{a}{x} = \frac{x}{b}$

7.11. Find the fourth proportional for each of the following sets of numbers: (7.8)

(a) 1, 3, 5 (c) 2, 3, 4 (e) 3, 2, 5 (g) 2, 8, 8

(b) 8, 6, 4 (d) 3, 4, 2 (f) $\frac{1}{3}$, 2, 5 (h) $b, 2a, 3b$

7.12. Find the positive mean proportional between each of the following pairs of numbers: (7.9)

(a) 4 and 9 (c) $\frac{1}{3}$ and 27 (e) 2 and 5 (g) p and q

(b) 12 and 3 (d) $2b$ and $8b$ (f) 3 and 9 (h) a^2 and b

7.13. From each of these equations, form a proportion whose fourth term is x: (a) $cx = bd$; (b) $pq = ax$; (c) $hx = a^2$; (d) $3x = 7$; (e) $x = ab/c$. (7.10)

7.14. In each of the following equations, find the ratio of x to y: (a) $2x = y$; (b) $3y = 4x$; (c) $x = \frac{1}{2}y$; (d) $ax = hy$; (e) $x = by$. (7.10)

7.15. Which of the following is not a proportion? (7.11)

(a) $\frac{4}{3} \stackrel{?}{=} \frac{24}{18}$ (b) $\frac{3}{3} \stackrel{?}{=} \frac{7}{12}$ (c) $\frac{25}{45} \stackrel{?}{=} \frac{10}{18}$ (d) $\frac{.2}{.3} \stackrel{?}{=} \frac{6}{9}$ (e) $\frac{x}{8} \stackrel{?}{=} \frac{3}{4}$ when $x = 6$.

7.16. From each of the following, form a new proportion whose first term is x. Then find x. (7.11)

(a) $\frac{3}{2} = \frac{9}{x}$ (b) $\frac{1}{x} = \frac{5}{4}$ (c) $\frac{a}{x} = \frac{2}{b}$ (d) $\frac{x+5}{5} = \frac{11}{10}$ (e) $\frac{x-20}{20} = \frac{1}{4}$

7.17. Find x in each of these pairs of proportions: (7.11)

(a) $a:b = c:x$ and $a:b = c:d$ (c) $2:3x = 4:5y$ and $2:15 = 4:5y$

(b) $5:7 = x:42$ and $5:7 = 35:42$ (d) $7:5x - 2 = 14:3y$ and $7:18 = 14:3y$

7.18. Find x in each of the following proportions: (7.12)

(a) $\frac{x-7}{8} = \frac{7}{4}$ (b) $\frac{x+y}{6} = \frac{x-y}{3} = \frac{1}{3}$ (c) $\frac{2x-y}{8} = \frac{y-1}{10} = \frac{1}{2}$

7.19. Find x in each part of Fig. 7-45. (7.13)

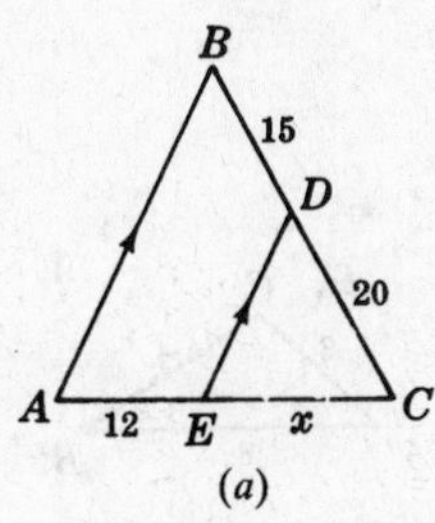

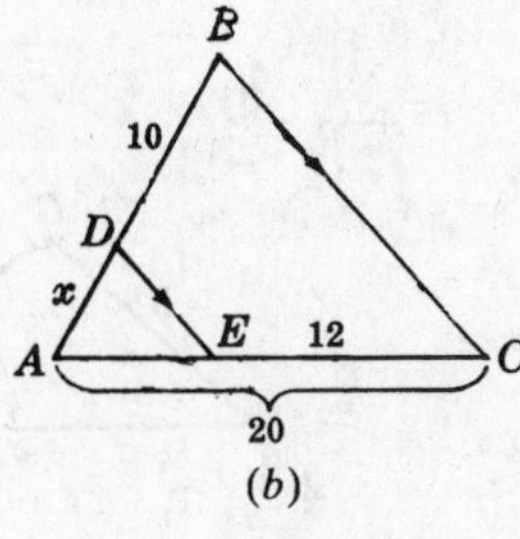

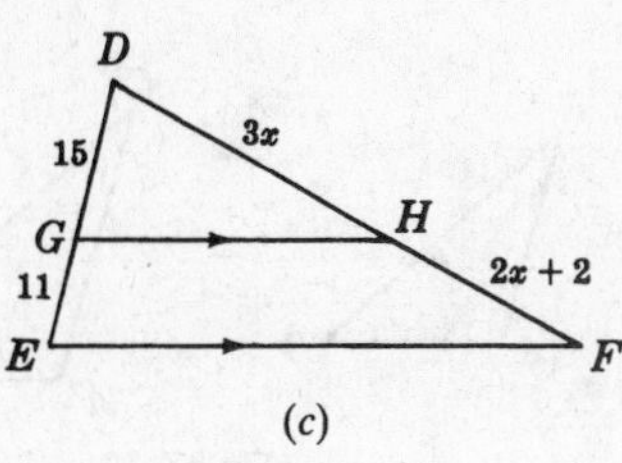

Fig. 7-45

7.20. In which parts of Fig. 7-46 is a line parallel to one side of the triangle? (7.13)

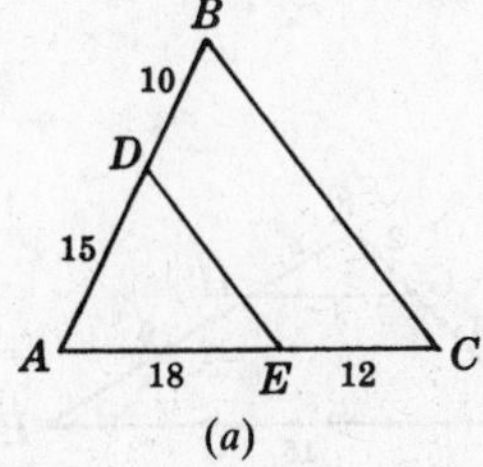

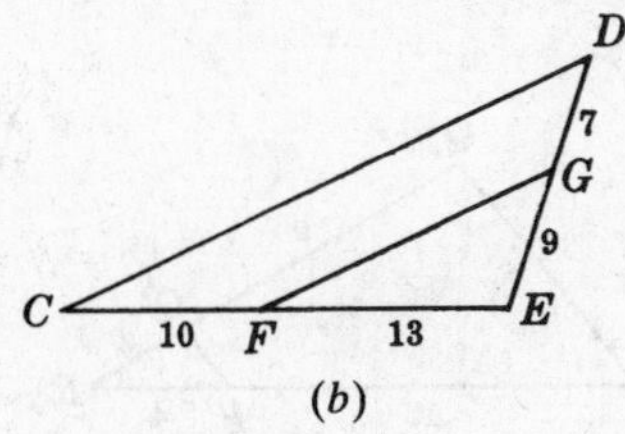

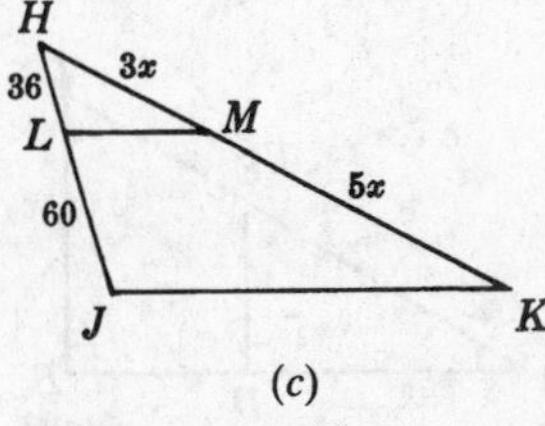

Fig. 7-46

7.21. Find x in each part of Fig. 7-47. (7.14)

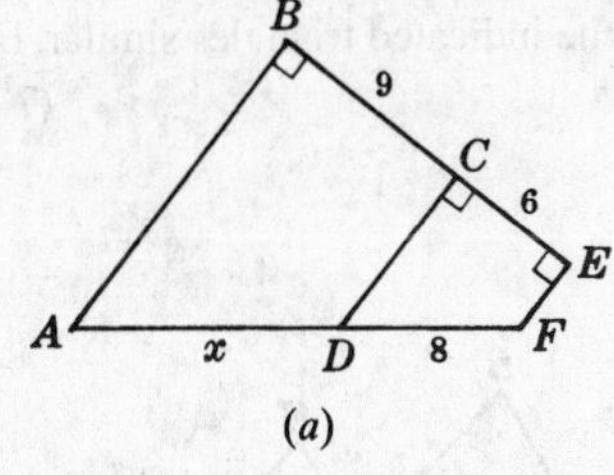

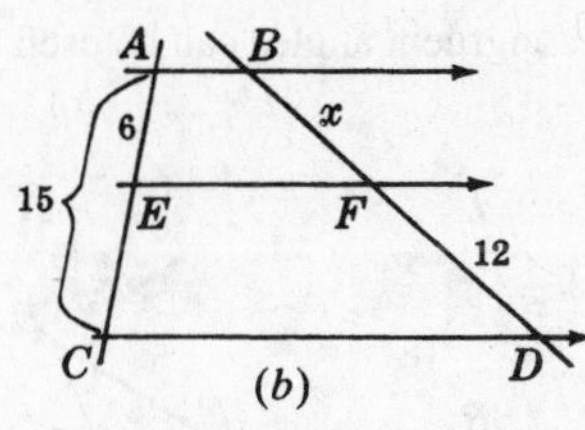

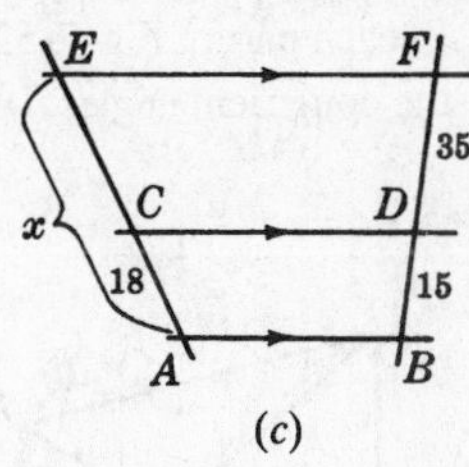

Fig. 7-47

7.22. Find x in each part of Fig. 7-48. (7.15)

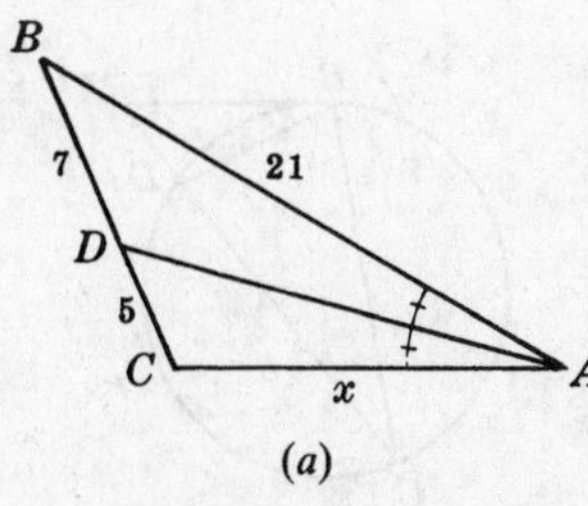

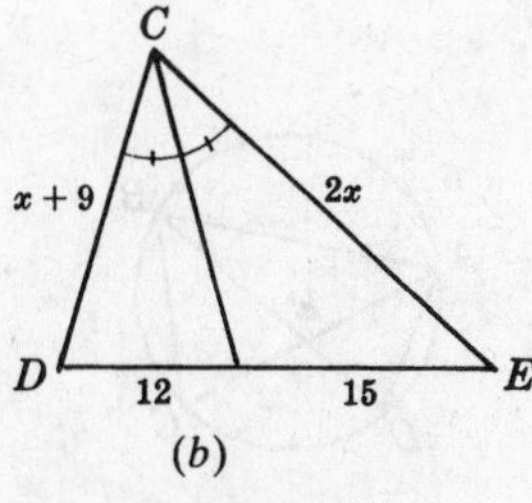

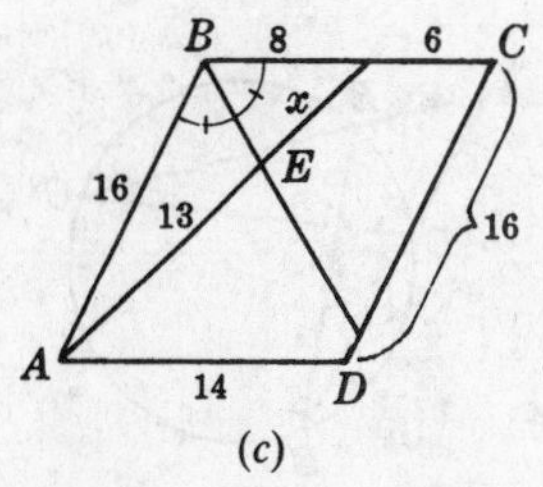

Fig. 7-48

7.23. Prove that three or more parallel lines divide any two transversals proportionately. (7.16)

7.24. In similar triangles ABC and $A'B'C'$ of Fig. 7-49, $\angle B$ and $\angle B'$ are corresponding angles. Find $m\angle B$ if (a) $m\angle A' = 120°$ and $m\angle C' = 25°$; (b) $m\angle A' + m\angle C' = 127°$. (7.17)

7.25. In similar triangles ABC and $A'B'C'$ of Fig. 7-50, $\angle A \cong \angle A'$ and $\angle B \cong \angle B'$. (a) Find a if $c = 24$; (b) find b if $a = 20$; (c) find c if $b = 63$. (7.17)

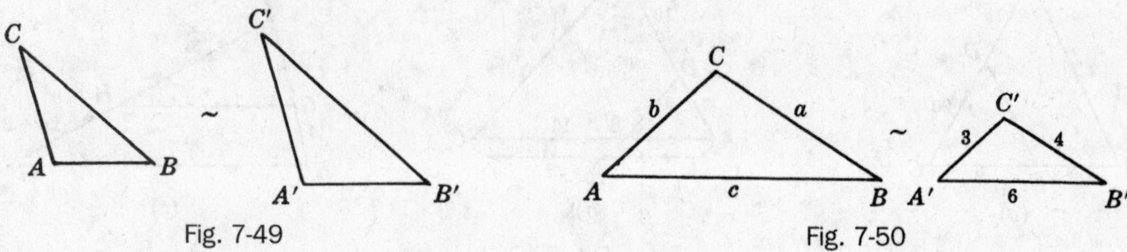

Fig. 7-49

Fig. 7-50

7.26. In each part of Fig. 7-51, show that the indicated triangles are similar. (7.18)

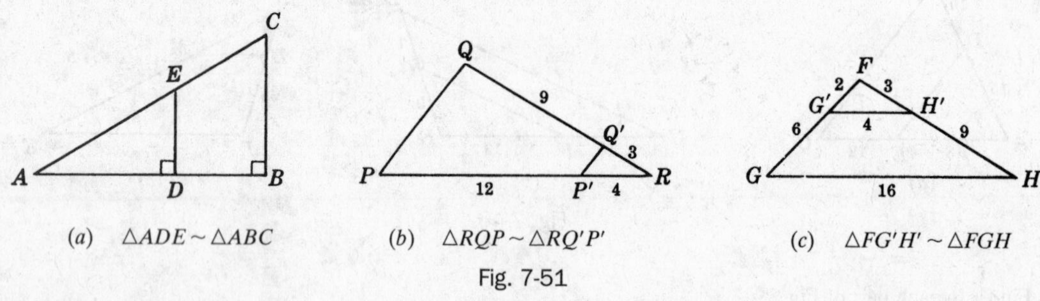

(a) $\triangle ADE \sim \triangle ABC$ (b) $\triangle RQP \sim \triangle RQ'P'$ (c) $\triangle FG'H' \sim \triangle FGH$

Fig. 7-51

7.27. In each part of Fig. 7-52, two pairs of congruent angles can be used to prove the indicated triangles similar. Find the congruent angles. (7.18)

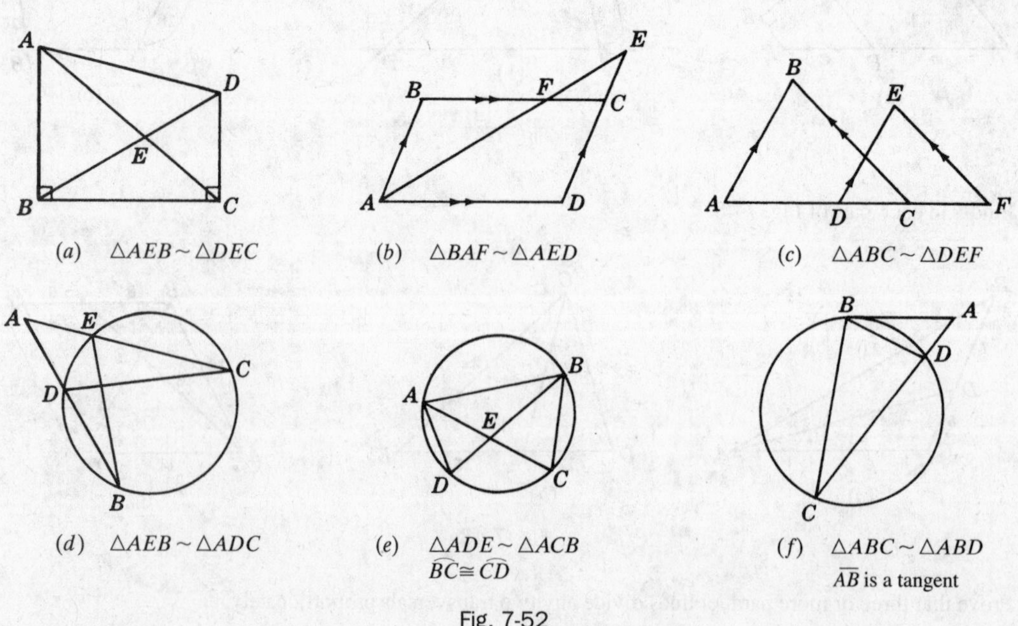

(a) $\triangle AEB \sim \triangle DEC$ (b) $\triangle BAF \sim \triangle AED$ (c) $\triangle ABC \sim \triangle DEF$

(d) $\triangle AEB \sim \triangle ADC$ (e) $\triangle ADE \sim \triangle ACB$, $\overset{\frown}{BC} \cong \overset{\frown}{CD}$ (f) $\triangle ABC \sim \triangle ABD$, $\overline{AB}$ is a tangent

Fig. 7-52

7.28. In each part of Fig. 7-53, determine the angles that can be used to prove the indicated triangles similar. (7.19)

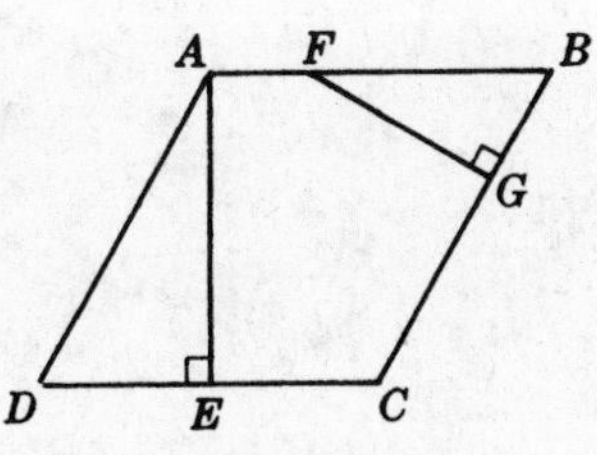

(a) $\triangle AED \sim \triangle FGB$
$ABCD$ is a parallelogram.

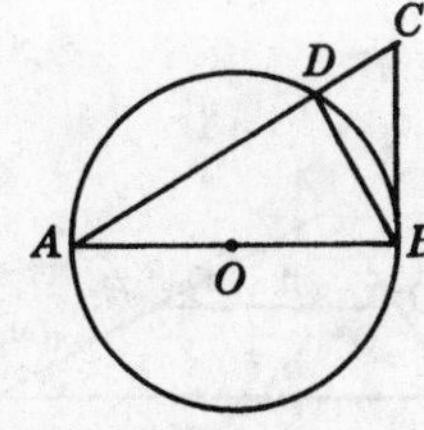

(b) $\triangle ADB \sim \triangle ABC$
$\overline{AB}$ is a diameter.
$\overline{BC}$ is a tangent.

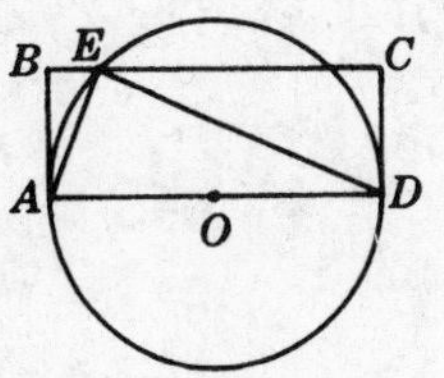

(c) $\triangle ABE \sim \triangle AED$
AD is a diameter.
$ABCD$ is a rectangle.

Fig. 7-53

7.29. In each part of Fig. 7-54, determine the pair of congruent angles and the proportion needed to prove the indicated triangles similar. (7.20)

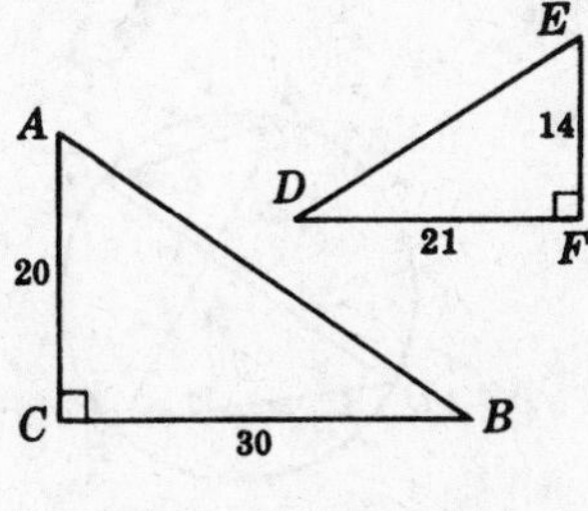

(a) $\triangle ABC \sim \triangle EDF$

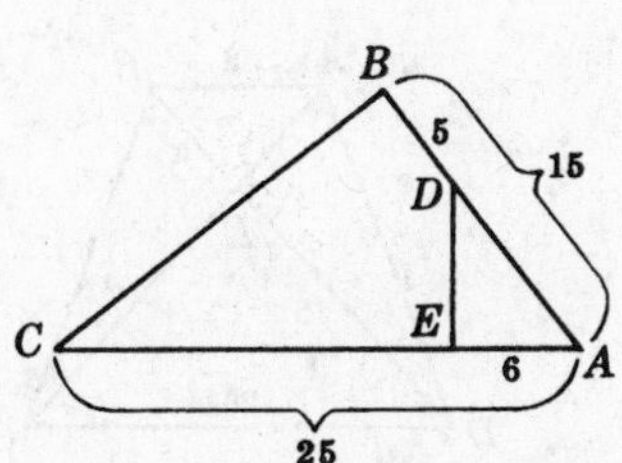

(b) $\triangle ADE \sim \triangle ACB$

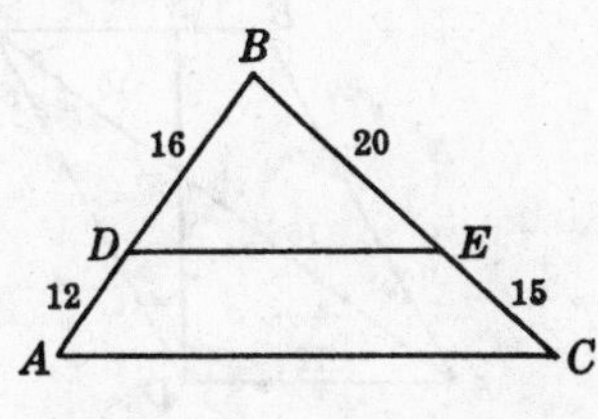

(c) $\triangle BDE \sim \triangle BAC$

Fig. 7-54

7.30. In each part of Fig. 7-55, state the proportion needed to prove the indicated triangles similar. (7.21)

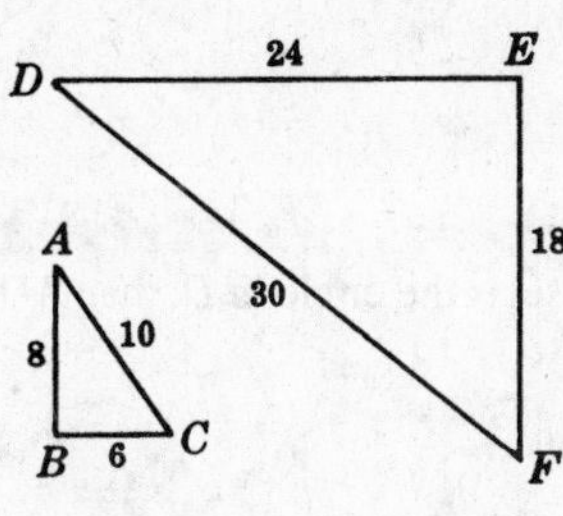

(a) $\triangle ABC \sim \triangle DEF$

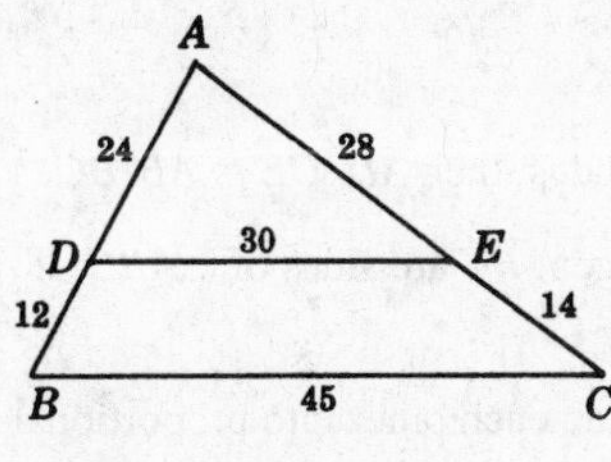

(b) $\triangle ADE \sim \triangle ABC$

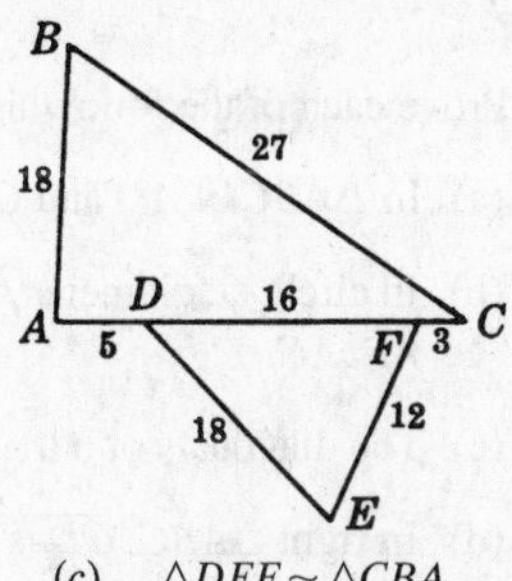

(c) $\triangle DEF \sim \triangle CBA$

Fig. 7-55

7.31. In each part of Fig. 7-56, prove the indicated proportion. (7.25)

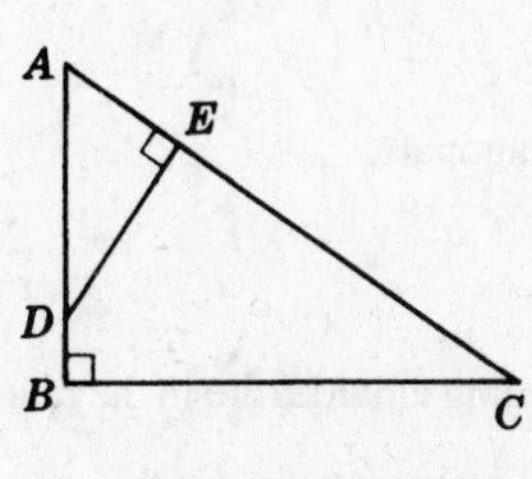

(a) $AD:AC = DE:BC$

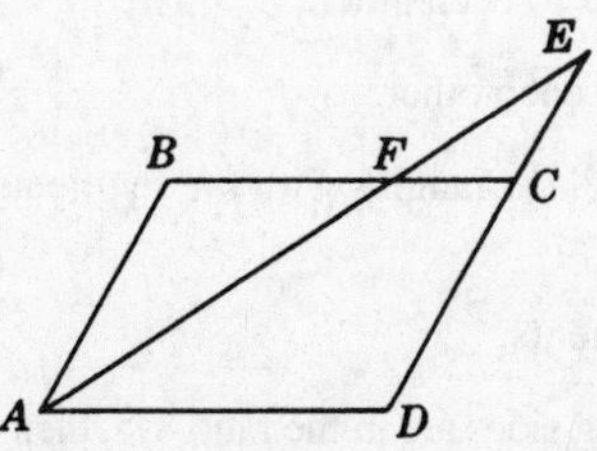

(b) $AB:EC = BF:FC$
$ABCD$ is a parallelogram.

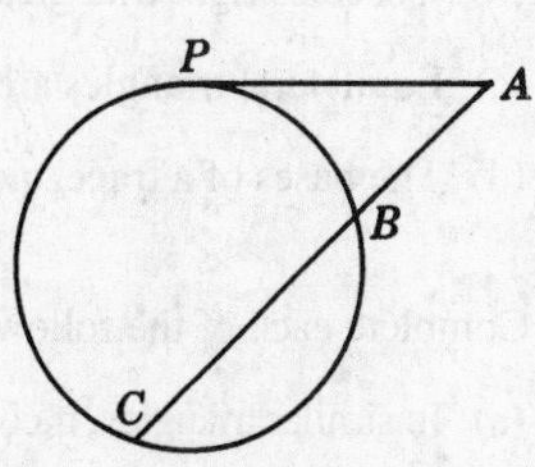

(c) $AC:AP = AP:AB$
$\overline{AP}$ is a tangent.

Fig. 7-56

7.32. In $\triangle ABC$ (Fig. 7-57), $\overline{DE} \parallel \overline{BC}$. (7.22)

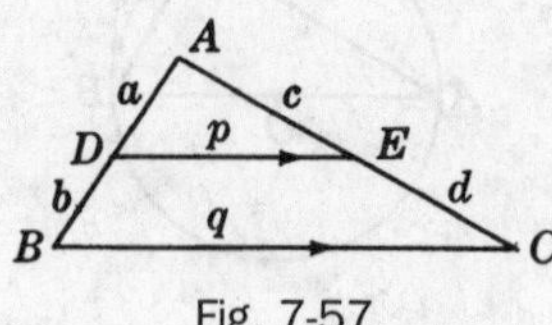

Fig. 7-57

(a) Let $a = 4, AB = 8, p = 10$. Find q. (d) Let $b = 9, p = 20, q = 35$. Find a.

(b) Let $c = 5, AC = 15, q = 24$. Find p. (e) Let $a = 10, p = 24, q = 84$. Find AB.

(c) Let $a = 7, p = 11, q = 22$. Find b. (f) Let $c = 3, p = 4, q = 7$. Find d.

7.33. Find x in each part of Fig. 7-58. (7.22)

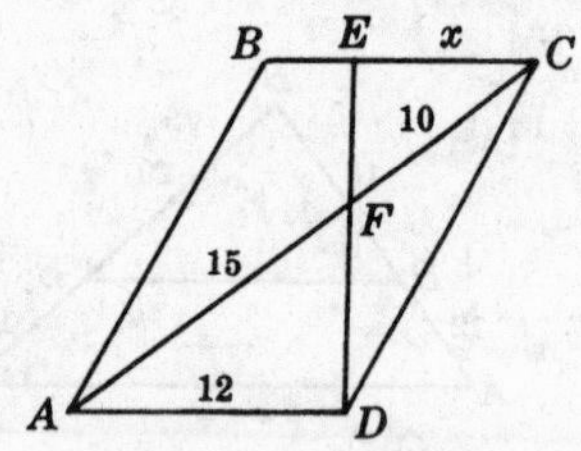

(a) ABCD is a parallelogram.

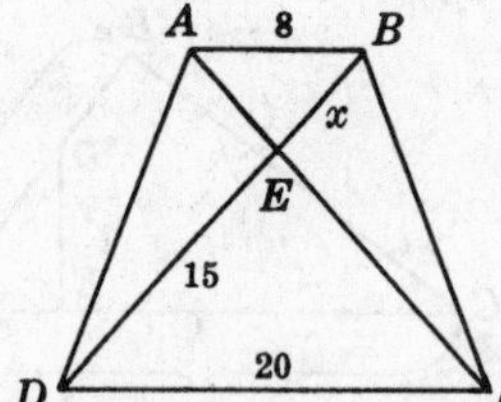

(b) ABCD is a trapezoid.

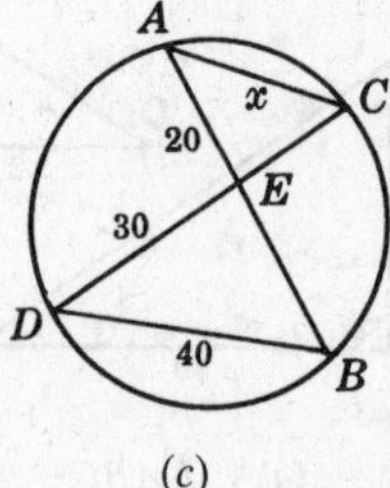

(c)

Fig. 7-58

7.34. A 7-ft upright pole near a vertical tree casts a 6-ft shadow. At that same time, find (a) the height of the tree if its shadow is 36 ft long; (b) the length of the shadow of the tree if its height is 77 ft. (7.23)

7.35. Prove each of the following: (7.23)

(a) In $\triangle ABC$, if $\overline{AD}$ and $\overline{CE}$ are altitudes, then $AD:CE = AB:BC$.

(b) In circle O, diameter $\overline{AB}$ and tangent $\overline{BC}$ are sides of $\triangle ABC$. If $\overline{AC}$ intersects the circle in D, then $AD:AB = AB:AC$.

(c) The diagonals of a trapezoid divide each other into proportional segments.

(d) In right $\triangle ABC$, $\overline{CD}$ is the altitude to the hypotenuse $\overline{AB}$, then $AC:CD = AB:BC$.

7.36. Prove each of the following: (7.24)

(a) A line parallel to one side of a triangle cuts off a triangle similar to the given triangle.

(b) Isosceles right triangles are similar to each other.

(c) Equilateral triangles are similar to each other.

(d) The bases of a trapezoid form similar triangles with the segments of the diagonals.

7.37. Complete each of the following statements: (7.26)

(a) In similar triangles, if corresponding sides are in the ratio 8:5, then corresponding altitudes are in the ratio _?_.

(b) In similar triangles, if corresponding angle bisectors are in the ratio 3:5, then their perimeters are in the ratio _?_.

(c) If the sides of a triangle are halved, then the perimeter is _?_, the angle bisectors are _?_, the medians are _?_, and the radii of the circumscribed circle are _?_.

7.38. (a) Corresponding sides of two similar triangles have lengths 18 and 12. If an altitude of the smaller has length 10, find the length of the corresponding altitude of the larger. (7.26)

(b) Corresponding medians of two similar triangles have lengths 25 and 15. Find the perimeter of the larger if the perimeter of the smaller is 36.

(c) The sides of a triangle have lengths 5, 7, and 8. If the perimeter of a similar triangle is 100, find its sides.

(d) The bases of a trapezoid have lengths 5 and 20, and the altitude has length 12. Find the length of the altitude of the triangle formed by the shorter base and the nonparallel sides extended to meet.

(e) The bases of a trapezoid have lengths 11 and 22. Its altitude has length 9. Find the distance from the point of intersection of the diagonals to each of the bases.

7.39. Complete each of the following statements: (7.27)

(a) If corresponding sides of two similar polygons are in the ratio 3:7, then the ratio of their corresponding altitudes is ?.

(b) If the perimeters of two similar hexagons are in the ratio of 56 to 16, then the ratio of their corresponding diagonals is ?.

(c) If each side of an octagon is quadrupled and the angles remain the same, then its perimeter is ?.

(d) The base of a rectangle is twice that of a similar rectangle. If the radius of the circumscribed circle of the first rectangle is 14, then the radius of the circumscribed circle of the second is ?.

7.40. Prove each of the following: (7.28)

(a) Corresponding angle bisectors of two triangles have the same ratio as a pair of corresponding sides.

(b) Corresponding medians of similar triangles have the same ratio as a pair of corresponding sides.

7.41. Provide the proofs requested in Fig. 7-59. (7.28)

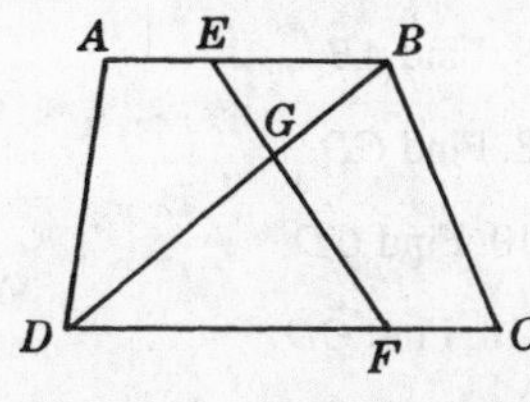

(*a*) **Given:** Trapezoid $ABCD$
To Prove: $GB \times DF = GD \times EB$

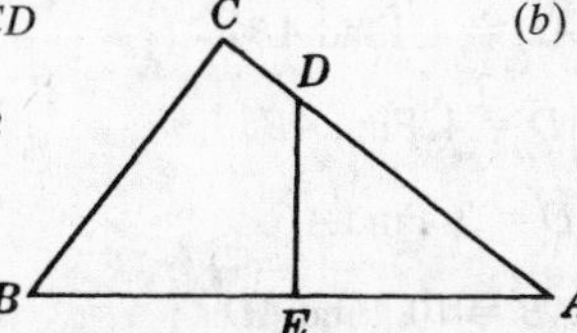

(*b*) **Given:** $\overline{BC} \perp \overline{AC}$, $\overline{DE} \perp \overline{AB}$
To Prove: $DE \times AC = BC \times AE$

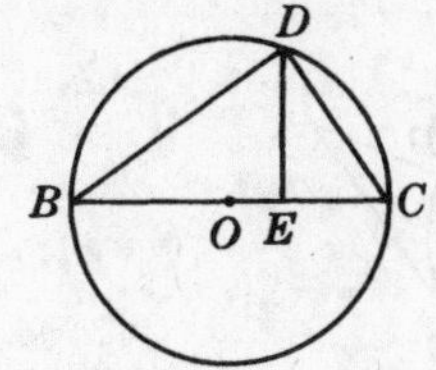

(*c*) **Given:** Diameter $\overline{BC}$, $\overline{DE} \perp \overline{BC}$
To Prove: $(BD^2) = BE \times BC$

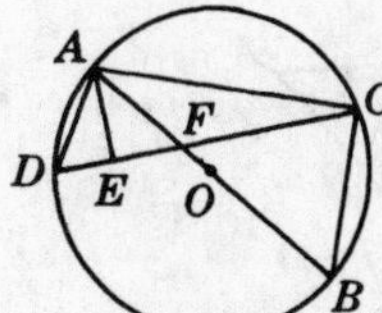

(*d*) **Given:** Circle O, Diameter $\overline{AB}$, $\overline{AE} \perp \overline{CD}$
To Prove: $AD \times BC = AB \times DE$

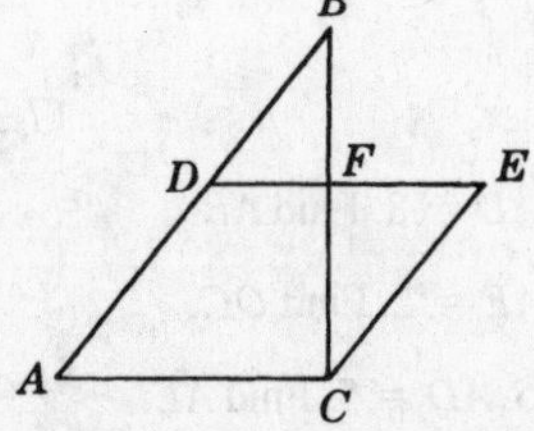

(*e*) **Given:** $\overline{AB} \parallel \overline{CE}$, $\overline{AC} \perp \overline{BC}$, $\overline{DE} \perp \overline{BC}$
To Prove: $AB \times CF = BC \times EC$

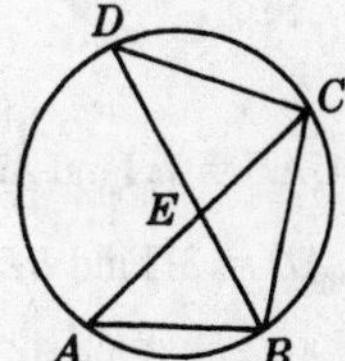

(*f*) **Given:** $\overset{\frown}{BC} \cong \overset{\frown}{CD}$
To Prove: $(BC^2) = AC \times EC$

Fig. 7-59

7.42. Prove each of the following: (7.28)

(a) If two chords intersect in a circle, the product of the lengths of the segments of one chord equals the product of the lengths of the segments of the other.

(b) In a right triangle, the product of the lengths of the hypotenuse and the altitude upon it equals the product of the lengths of the legs.

(c) If in inscribed $\triangle ABC$ the bisector of $\angle A$ intersects $\overline{BC}$ in D and the circle in E, then $BD \times AC = AD \times EC$.

7.43. In Fig. 7-60 (7.29)

(a) Let $AE = 10, EB = 6, CE = 12$. Find ED.

(b) Let $AB = 15, EB = 8, ED = 4$. Find CE.

(c) Let $AE = 6, ED = 4, CD = 13$. Find EB.

(d) Let $ED = 5, EB = 2(AE), CD = 15$. Find AE.

In Fig. 7-61, diameter $\overline{CD} \perp$ chord $\overline{AB}$

(e) Let $OD = 10, OE = 8$. Find AB.

(f) Let $AB = 24, OE = 5$. Find OD.

(g) Let $OD = 25, EC = 18$. Find AB.

(h) Let $AB = 8, OD = 5$. Find EC.

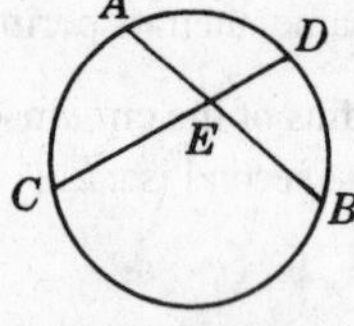

Fig. 7-60

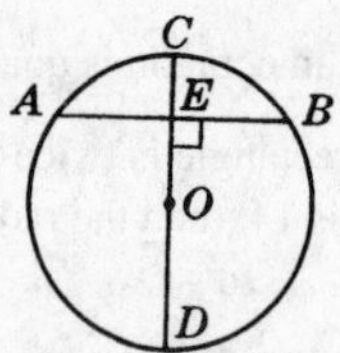

Fig. 7-61

7.44. A point is 12 in from the center of a circle whose radius is 15 in. Find the lengths of the longest and shortest chords that can be drawn through this point. (*Hint*: The longest chord is a diameter, and the shortest chord is perpendicular to this diameter.) (7.29)

7.45. In Fig. 7-62, $\overline{AB}$ is a tangent (7.29)

(a) Let $AC = 16, AD = 4$. Find AB.

(b) Let $CD = 5, AD = 4$. Find AB.

(c) Let $AB = 6, AD = 3$. Find AC.

(d) Let $AC = 20, AB = 10$. Find AD.

(e) Let $AB = 12, AD = 9$. Find CD.

In Fig. 7-63, $\overline{CD}$ is a diameter; $\overline{AB}$ is a tangent

(f) Let $AD = 6, OD = 9$. Find AB.

(g) Let $AD = 2, AB = 8$. Find CD.

(h) Let $AD = 5, AB = 10$. Find OD.

(i) Let $AB = 12, AC = 18$. Find OD.

(j) Let $OD = 5, AB = 12$. Find AD.

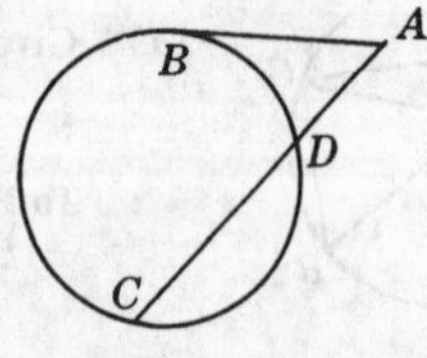

Fig. 7-62

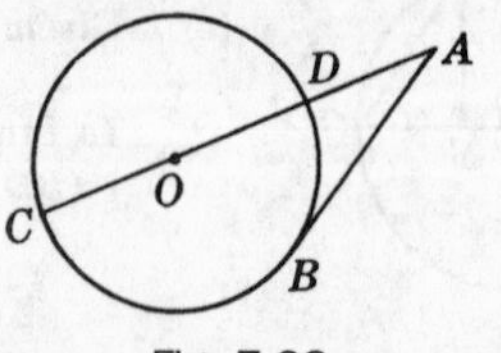

Fig. 7-63

7.46. In Fig. 7-64 (7.31)

(a) Let $AB = 14, AD = 4, AE = 7$. Find AC.

(b) Let $AC = 8, AE = 6, AD = 3$. Find BD.

(c) Let $BD = 5, AD = 7, AE = 4$. Find AC.

(d) Let $AD = DB, EC = 14, AE = 4$. Find AD.

In Fig. 7-65, $\overline{CE}$ is a diameter

(e) Let $OC = 3, AE = 6, AD = 8$. Find AB.

(f) Let $BD = 7, AD = 5, AE = 2$. Find OC.

(g) Let $OC = 11, AB = 15, AD = 5$. Find AE.

(h) Let $OC = 5, AE = 6, BD = 4$. Find AD.

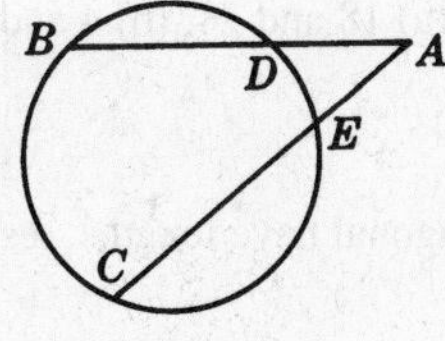

Fig. 7-64

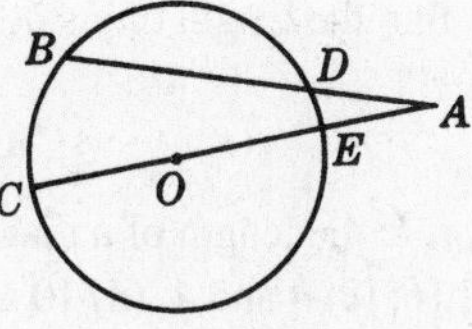

Fig. 7-65

7.47. $\overline{CD}$ is the altitude to hypotenuse $\overline{AB}$ in Fig. 7-66. (7.32)

(a) If $p = 2$ and $q = 6$, find a and h.
(b) If $p = 4$ and $a = 6$, find c and h.
(c) If $p = 16$ and $h = 8$, find q and b.
(d) If $b = 12$ and $q = 6$, find p and h.

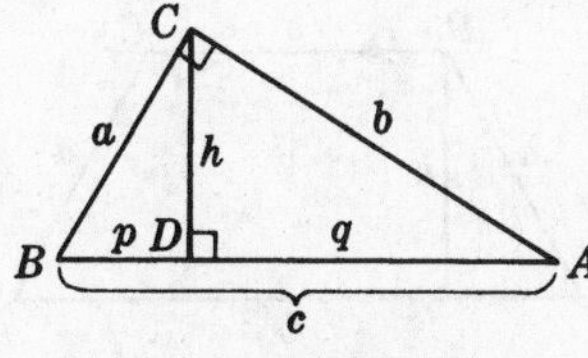

Fig. 7-66

7.48. In a right triangle whose arms have lengths a and b, find the length of the hypotenuse c when (7.33)

(a) $a = 15, b = 20$
(b) $a = 15, b = 36$
(c) $a = 5, b = 4$
(d) $a = 5, b = 5\sqrt{3}$
(e) $a = 7, b = 7$

7.49. In the right triangle in Fig. 7-67, find the length of each missing arm when (7.33)

(a) $a = 12, c = 20$
(b) $b = 6, c = 8$
(c) $b = 15, c = 17$
(d) $a = 2, c = 4$
(e) $a = 5\sqrt{2}, c = 10$
(f) $a = \sqrt{5}, c = 2\sqrt{2}$

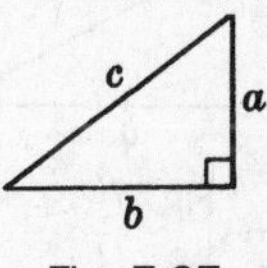

Fig. 7-67

7.50. Find the lengths of the arms of a right triangle whose hypotenuse has length c if these arms have a ratio of (a) 3:4 and $c = 15$; (b) 5:12 and $c = 26$; (c) 8:15 and $c = 170$; (d) 1:2 and $c = 10$. (7.34)

7.51. In a rectangle, find the length of the diagonal if its sides have lengths (a) 9 and 40; (b) 5 and 10. (7.33)

7.52. In a rectangle, find the length of one side if the diagonal has length 15 and the other side has length (a) 9; (b) 5; (c) 10. (7.33)

7.53. Of triangles having sides with lengths as follows, which are right triangles?

(a) 33, 55, 44
(b) 120, 130, 50
(c) $4, 7\frac{1}{2}, 8\frac{1}{2}$
(d) 25, 7, 24
(e) 5 in, 1 ft, 1 ft 1 in
(f) 1 yd, 1 yd 1 ft, 1 yd 2 ft
(g) 11 mi, 60 mi, 61 mi
(h) 5 cm, 5 cm, 7 cm

7.54. Is a triangle a right triangle if its sides have the ratio of (a) 3:4:5; (b) 2:3:4?

7.55. Find the length of the altitude of an isosceles triangle if each of its two congruent sides has length 10 and its base has length (a) 12; (b) 16; (c) 18; (d) 10. (7.35)

7.56. In a rhombus, find the length of a side if the diagonals have lengths (a) 18 and 24; (b) 4 and 8; (c) 6 and $6\sqrt{3}$. (7.36)

7.57. In a rhombus, find the length of a diagonal if a side and the other diagonal have lengths, respectively, (a) 10 and 12; (b) 17 and 16; (c) 4 and 4; (d) 10 and $10\sqrt{3}$. (7.36)

7.58. In isosceles trapezoid *ABCD* in Fig. 7-68 (7.37)

(a) Find a if $b = 32, b' = 20$, and $h = 8$.

(b) Find h if $b = 24, b' = 14$, and $a = 13$.

(c) Find b if $a = 15, b' = 10$, and $h = 12$.

(d) Find b' if $a = 6, b = 21$, and $h = 3\sqrt{3}$.

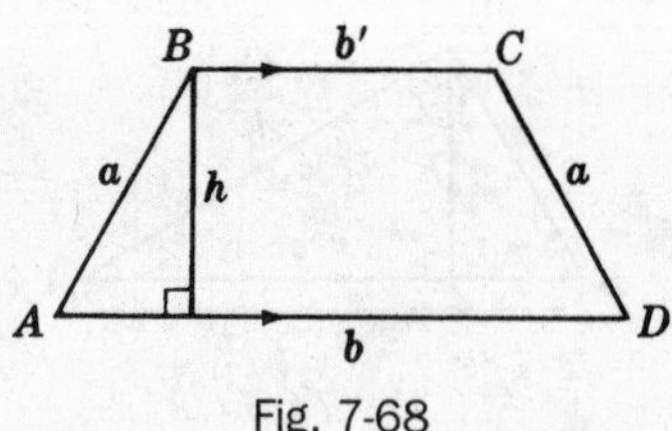

Fig. 7-68

7.59. In a trapezoid *ABCD* in Fig. 7-69 (7.37)

(a) Find d if $a = 11, b = 3$, and $c = 15$.

(b) Find a if $d = 20, b = 12$, and $c = 36$.

(c) Find d if $a = 5, p = 13$, and $c = 14$.

(d) Find p if $a = 20, c = 28$, and $d = 17$.

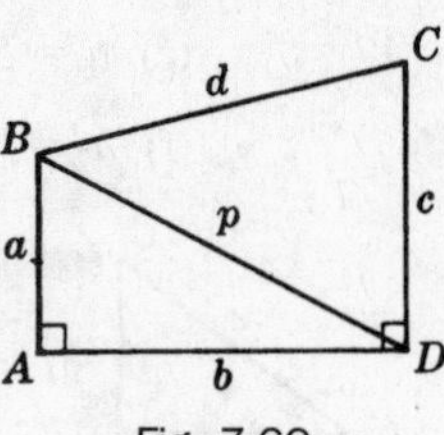

Fig. 7-69

7.60. The radius of a circle is 15. Find (a) the distance from its center to a chord whose length is 18; (b) the length of a chord whose distance from its center is 9. (7.38)

7.61. In a circle, a chord whose length is 16 is at a distance of 6 from the center. Find the length of a chord whose distance from the center is 8. (7.38)

7.62. Two externally tangent circles have radii of 25 and 9. Find the length of a common external tangent. (7.38)

7.63. In a 30°-60°-90° triangle, find the lengths of (a) the legs if the hypotenuse has length 20; (b) the other leg and hypotenuse if the leg opposite 30° has length 7; (c) the other leg and hypotenuse if the leg opposite 60° has length $5\sqrt{3}$. (7.39)

7.64. In an equilateral triangle, find the length of the altitude if the side has length (a) 22; (b) $2a$. Find the side if the altitude has length (c) $24\sqrt{3}$; (d) 24. (7.39)

7.65. In a rhombus which has an angle measuring 60°, find the lengths of (a) the diagonals if a side has length 25; (b) the side and larger diagonal if the smaller diagonal has length 35. (7.39)

7.66. In an isosceles trapezoid which has base angles measuring 60°, find the lengths of (a) the lower base and altitude if the upper base has length 12 and the legs have length 16; (b) the upper base and altitude if the lower base has length 45 and the legs have length 28. (7.39)

7.67. In an isosceles right triangle, find the length of each leg if the hypotenuse has length (a) 34; (b) $2a$. Find the length of the hypotenuse if each leg has length (c) 34; (d) $15\sqrt{2}$. (7.40)

7.68. In a square, find the length of (a) the side if the diagonal has length 40; (b) the diagonal if the side has length 40. (7.40)

7.69. In an isosceles trapezoid which has base angles of measure 45°, find the lengths of (a) the lower base and each leg if the altitude has length 13 and the upper base has length 19; (b) the upper base and each leg if the altitude has length 27 and the lower base has length 65; (c) each leg and the lower base if the upper base has length 25 and the altitude has length 15. (7.40)

7.70. A parallelogram has an angle measuring 45°. Find the distances between its pairs of opposite sides if its sides have lengths 10 and 12. (7.40)

Trigonometry

8.1 Trigonometric Ratios

Trigonometry means "measurement of triangles." Consider its parts: *tri* means "three," *gon* means "angle," and *metry* means "measure." Thus, in trigonometry we study the measurement of triangles.

The following ratios relate the sides and acute angles of a *right triangle*:

1. *Tangent ratio*: The tangent (abbreviated "tan") of an acute angle equals the length of the leg opposite the angle divided by the length of the leg adjacent to the angle.
2. *Sine ratio*: The sine (abbreviated "sin") of an acute angle equals the length of the leg opposite the angle divided by the length of the hypotenuse.
3. *Cosine ratio*: The cosine (abbreviated "cos") of an acute angle equals the length of the leg adjacent to the angle divided by the length of the hypotenuse.

Thus in right triangle *ABC* of Fig. 8-1,

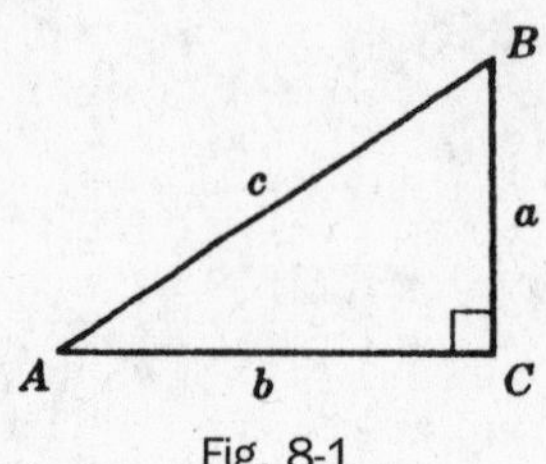

Fig. 8-1

$$\tan A = \frac{\text{length of leg opposite } A}{\text{length of leg adjacent to } A} = \frac{a}{b} \qquad \tan B = \frac{\text{length of leg opposite } B}{\text{length of leg adjacent to } B} = \frac{b}{a}$$

$$\sin A = \frac{\text{length of leg opposite } A}{\text{length of hypotenuse}} = \frac{a}{c} \qquad \sin B = \frac{\text{length of leg opposite } B}{\text{length of hypotenuse}} = \frac{b}{c}$$

$$\cos A = \frac{\text{length of leg adjacent to } A}{\text{length of hypotenuse}} = \frac{b}{c} \qquad \cos B = \frac{\text{length of leg adjacent to } B}{\text{length of hypotenuse}} = \frac{a}{c}$$

If *A* and *B* are the acute angles of a right triangle, then

$$\sin A = \cos B \qquad \cos A = \sin B \qquad \tan A = \frac{1}{\tan B} \qquad \tan B = \frac{1}{\tan A}$$

A scientific calculator can compute the sine, cosine, and tangent of an angle with the "SIN," "COS," and "TAN" buttons, respectively. Make sure the calculator is set to degrees (DEG). For those without a calculator, a table of sines, cosines, and tangents is in the back of this book.

SOLVED PROBLEMS

8.1 Using the table of sines, cosines, and tangents

The following values were taken from a table of sines, cosines, and tangents. State, in equation form, what the values on the first three lines mean. Then use the table at the back of this book to complete the last line.

	Angle	Sine	Cosine	Tangent
(a)	1°	0.0175	0.9998	0.0175
(b)	30°	0.5000	0.8660	0.5774
(c)	60°	0.8660	0.5000	1.7321
(d)	?	?	0.3420	?

Solutions

(a) $\sin 1° = 0.0175$; $\cos 1° = 0.9998$; $\tan 1° = 0.0175$

(b) $\sin 30° = 0.5000$; $\cos 30° = 0.8660$; $\tan 30° = 0.5774$

(c) $\sin 60° = 0.8660$; $\cos 60° = 0.5000$; $\tan 60° = 1.7321$

(d) In the table of trigonometric functions, the cosine value 0.3420 is on the 70° line; hence, the angle measures 70°. Then, from the table, $\sin 70° = 0.9397$ and $\tan 70° = 2.7475$.

8.2 Finding angle measures to the nearest degree

Find the measure of x to the nearest degree if (a) $\sin x = 0.9235$; (b) $\cos x = \frac{21}{25}$ or 0.8400; (c) $\tan x = \sqrt{5}/10$ or 0.2236. Use the table of trigonometric functions.

Solutions

Differences

(a) $\sin 68° = 0.9272$ → 0.0037
$\sin x = 0.9235$
$\sin 67° = 0.9205$ → 0.0030

Since $\sin x$ is nearer to $\sin 67°$, $m\angle x = 67°$ to the nearest degree.

(b) $\cos 32° = 0.8480$ → 0.0080
$\cos x = 0.8400$
$\cos 33° = 0.8387$ → 0.0013

Since $\cos x$ is nearer to $\cos 33°$, $m\angle x = 33°$ to the nearest degree.

(c) $\tan 13° = 0.2309$ → 0.0073
$\tan x = 0.2236$
$\tan 12° = 0.2126$ → 0.0110

Since $\tan x$ is nearer to $\tan 13°$, $m\angle x = 13°$ to the nearest degree.

With a calculator, the above answers can be found with the inverse sine ($\sin^{-1}$), inverse cosine ($\cos^{-1}$), and inverse tangent ($\tan^{-1}$). These usually require first pressing the "2nd" or "INV" button and then "SIN," "COS," or "TAN."

8.3 Finding trigonometric ratios

For each right triangle in Fig. 8-2, find the trigonometric ratios of each acute angle.

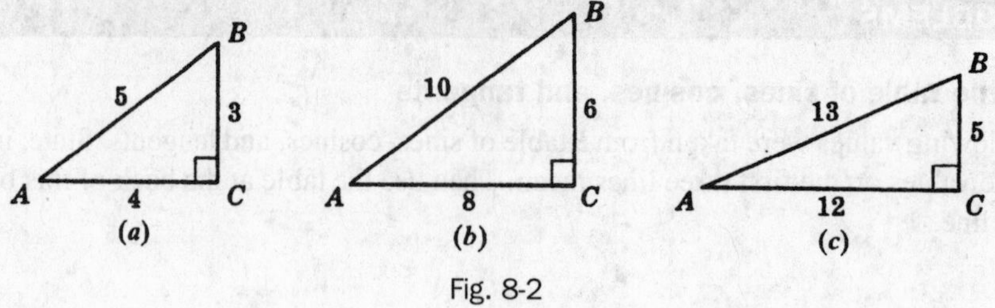

Fig. 8-2

Solutions

Formulas	(a) $a = 3, b = 4, c = 5$	(b) $a = 6, b = 8, c = 10$	(c) $a = 5, b = 12, c = 13$
$\tan A = \frac{a}{b}$ $\tan B = \frac{b}{a}$	$\tan A = \frac{3}{4}$ $\tan B = \frac{4}{3}$	$\tan A = \frac{6}{8} = \frac{3}{4}$ $\tan B = \frac{8}{6} = \frac{4}{3}$	$\tan A = \frac{5}{12}$ $\tan B = \frac{12}{5}$
$\sin A = \frac{a}{c}$ $\sin B = \frac{b}{c}$	$\sin A = \frac{3}{5}$ $\sin B = \frac{4}{5}$	$\sin A = \frac{6}{10} = \frac{3}{5}$ $\sin B = \frac{8}{10} = \frac{4}{5}$	$\sin A = \frac{5}{13}$ $\sin B = \frac{12}{13}$
$\cos A = \frac{b}{c}$ $\cos B = \frac{a}{c}$	$\cos A = \frac{4}{5}$ $\cos B = \frac{3}{5}$	$\cos A = \frac{8}{10} = \frac{4}{5}$ $\cos B = \frac{6}{10} = \frac{3}{5}$	$\cos A = \frac{12}{13}$ $\cos B = \frac{5}{13}$

8.4 Finding measures of angles by trigonometric ratios

Find the measure of angle A, to the nearest degree, in each part of Fig. 8-3.

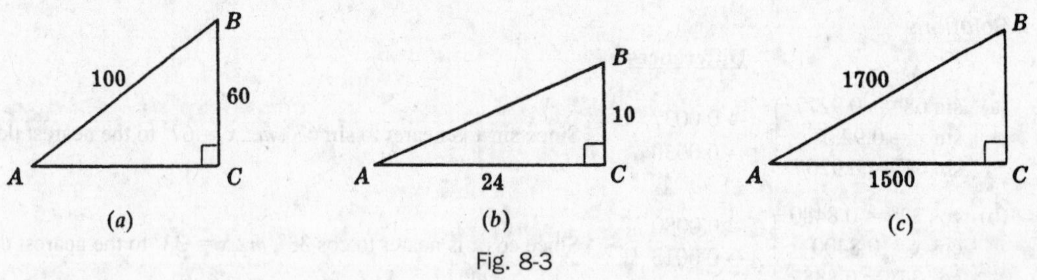

Fig. 8-3

Solutions

(a) $\sin A = \frac{60}{100} = 0.6000$. Since $\sin 37° = 0.6018$ is the nearest-degree sine value, $m\angle A = 37°$.

(b) $\tan A = \frac{10}{24} = 0.4167$. Since $\tan 23° = 0.4245$ is the nearest-degree tangent value, $m\angle A = 23°$.

(c) $\cos A = \frac{1500}{1700} = 0.8824$. Since $\cos 28° = 0.8829$ is the nearest-degree cosine value, $m\angle A = 28°$.

8.5 Trigonometric ratios of 30° and 60°

Show that

(a) $\tan 30° = 0.577$ (c) $\cos 30° = 0.866$ (e) $\sin 60° = 0.866$

(b) $\sin 30° = 0.500$ (d) $\tan 60° = 1.732$ (f) $\cos 60° = 0.500$

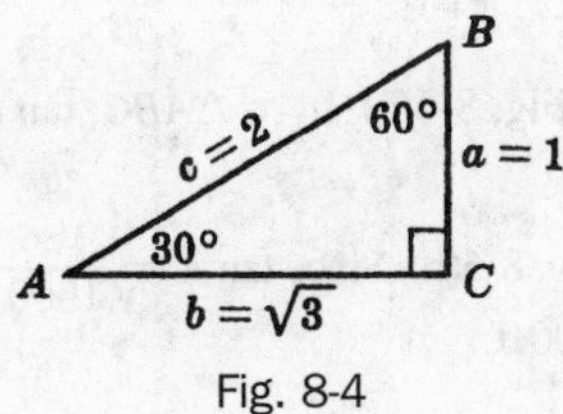

Fig. 8-4

Solutions

The trigonometric ratios for 30° and 60° may be obtained by using a 30°-60°-90° triangle (Fig. 8-4); in such a triangle, the ratio of the sides is $a{:}b{:}c = 1{:}\sqrt{3}{:}2$. Thus:

(a) $\tan 30^\circ = \dfrac{1}{\sqrt{3}} = \dfrac{1}{\sqrt{3}} \cdot \dfrac{\sqrt{3}}{\sqrt{3}} = \dfrac{\sqrt{3}}{3} = 0.577$

(b) $\sin 30^\circ = \dfrac{1}{2} = 0.500$

(c) $\cos 30^\circ = \dfrac{\sqrt{3}}{2} = 0.866$

(d) $\tan 60^\circ = \dfrac{\sqrt{3}}{1} = 1.732$

(e) $\sin 60^\circ = \dfrac{\sqrt{3}}{2} = 0.866$

(f) $\cos 60^\circ = \dfrac{1}{2} = 0.500$

8.6 Finding lengths of sides by trigonometric ratios

In each triangle of Fig. 8-5, solve for x and y to the nearest integer.

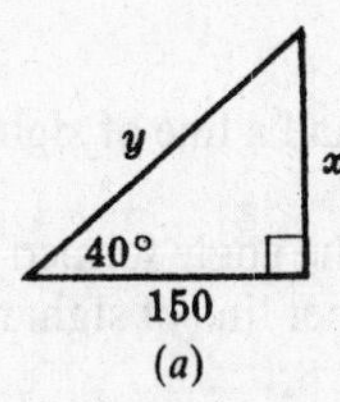

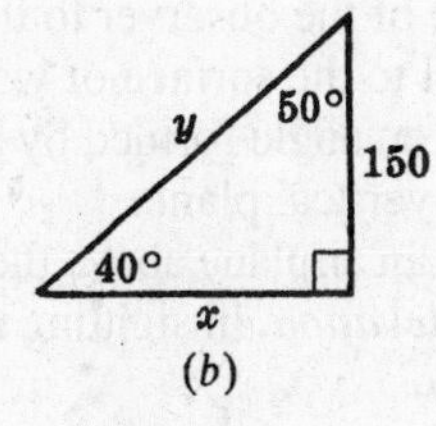

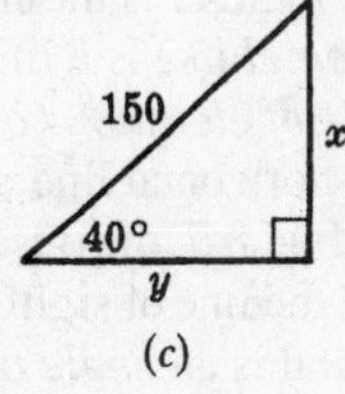

Fig. 8-5

Solutions

(a) Since $\tan 40^\circ = \dfrac{x}{150}$, $x = 150 \tan 40^\circ = 150(0.8391) = 126$.

Since $\cos 40^\circ = \dfrac{150}{y}$, $y = \dfrac{150}{\cos 40^\circ} = \dfrac{150}{0.766} = 196$.

(b) Since $\tan 50^\circ = \dfrac{x}{150}$, $x = 150 \tan 50^\circ = 150(1.1918) = 179$.

Since $\sin 40^\circ = \dfrac{150}{y}$, $y = \dfrac{150}{\sin 40^\circ} = \dfrac{150}{0.6428} = 233$.

(c) Since $\sin 40^\circ = \dfrac{x}{150}$, $x = 150 \sin 40^\circ = 150\,(0.6428) = 96$.

Since $\cos 40^\circ = \dfrac{y}{150}$, $y = 150 \cos 40^\circ = 150(0.776) = 115$.

8.7 Solving trigonometry problems

(a) An aviator flew 70 mi east from A to C. From C, he flew 100 mi north to B. Find the measure of the angle of the turn (to the nearest degree) that must be made at B to return to A.

(b) A road is to be constructed so that it will rise 105 ft for each 1000 ft of horizontal distance. Find the measure of the angle of rise to the nearest degree, and the length of road to the nearest foot for each 1000 ft of horizontal distance.

Solutions

(a) The required angle is $\angle EBA$ in Fig. 8-6(a). In rt. $\triangle ABC$, $\tan B = \frac{70}{100} = 0.7000$; hence, $m\angle B = 35°$ and $m\angle EBA = 180° - 35° = 145°$.

(b) We need to find $m\angle A$ and x in Fig. 8-6(b). Since $\tan A = \frac{105}{1000} = 0.1050$, $m\angle A = 6°$. Then $\cos 6° = \frac{1000}{x}$, so $x = \frac{1000}{\cos 6°} = \frac{1000}{0.9945} = 1006\,\text{ft}$.

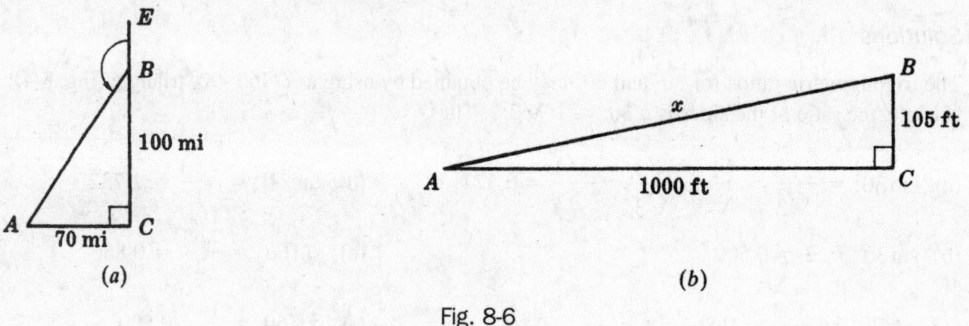

Fig. 8-6

8.2 Angles of Elevation and Depression

Here are some definitions that are involved in observed-angle problems:

The *line of sight* is the line from the eye of the observer to the object sighted.

A *horizontal line* is a line that is parallel to the surface of water.

An *angle of elevation* (*or depression*) is an angle formed by a horizontal line and a line of sight above (or below) the horizontal line and in the same vertical plane.

Thus in Fig. 8-7, the observer is sighting an airplane above the horizontal, and the angle formed by the horizontal and the line of sight is an *angle of elevation*. In sighting the car, the angle her line of sight makes with the horizontal is an *angle of depression*.

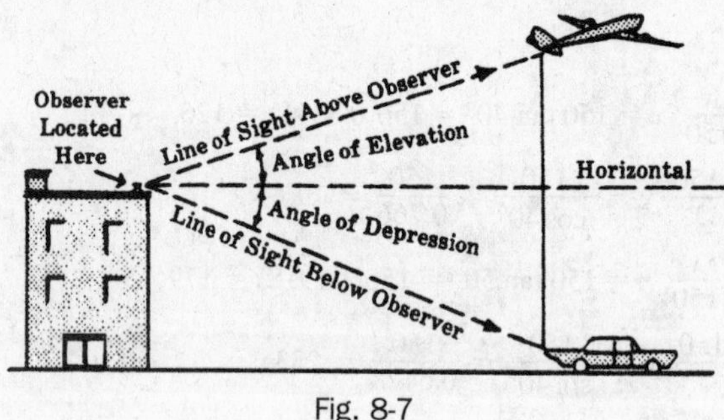

Fig. 8-7

SOLVED PROBLEMS

8.8 Using an angle of elevation

(a) Sighting to the top of a building, Henry found the angle of elevation to measure 21°. The ground is level. The transit is 5 ft above the ground and 200 ft from the building. Find the height of the building to the nearest foot.

(b) If the angle of elevation of the sun at a certain time measures 42°, find to the nearest foot the height of a tree whose shadow is 25 ft long.

Solutions

(a) If x is the height of the part of the building above the transit [Fig. 8-8(a)], then $\tan 21° = \frac{x}{200}$ and $x = 200 \tan 21° = 200(0.3839) = 77$ ft.
Thus, the height of the building is $h = x + 5 = 77 + 5 = 82$ ft.

(b) If h is the height of the tree [Fig. 8-8(b)], then we have $\tan 42° = \frac{h}{25}$ and $h = 25 \tan 42° = 25(0.9004) = 23$ ft.

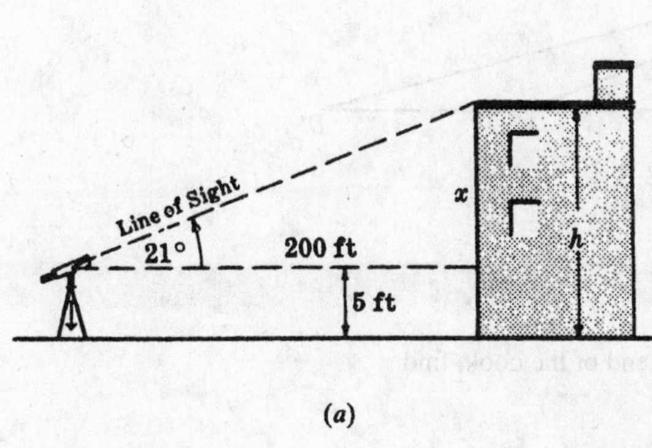

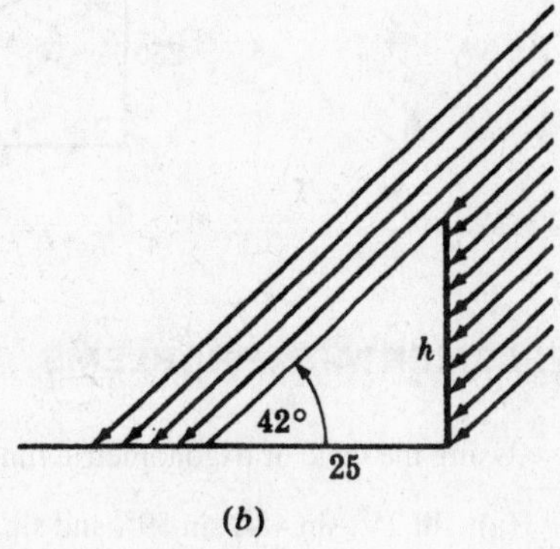

Fig. 8-8

8.9 Using both an angle of elevation and an angle of depression

Standing at the top of a lighthouse 200 ft high, a lighthouse keeper sighted both an airplane and a ship directly beneath the plane. The angle of elevation of the plane measured 25°; the angle of depression of the ship measured 32°. Find (a) the distance d of the boat from the foot of the lighthouse, to the nearest 10 ft; (b) the height of the plane above the water, to the nearest 10 ft.

Solutions

(a) See Fig. 8-9. In $\triangle$III, $\tan 58° = \frac{d}{200}$ and $d = 200 \tan 58° = 200(1.6003) = 320$ ft.

(b) In $\triangle$I, $\tan 25° = \frac{x}{320}$ and $x = 320(0.4663) = 150$ ft. Since the height of the tower is 200 ft, the height of the airplane is $200 + 150 = 350$ ft.

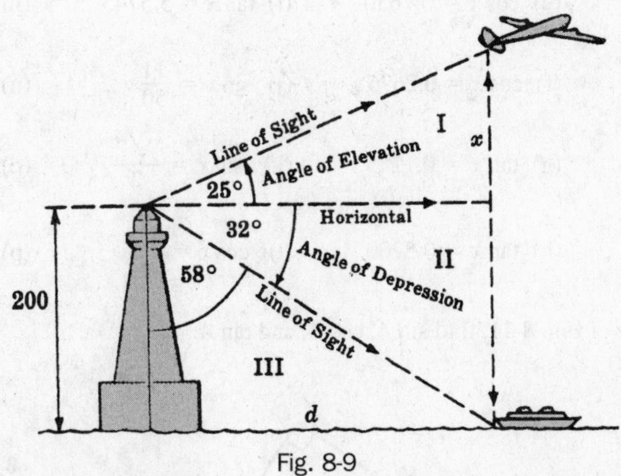

Fig. 8-9

8.10 Using two angles of depression

An observer on the top of a hill 250 ft above the level of a lake sighted two boats directly in line. Find, to the nearest foot, the distance between the boats if the angles of depression noted by the observer measured 11° and 16°.

Solutions

In $\triangle AB'C$ of Fig. 8-10, $m\angle B'AC = 90° - 11° = 79$. Then $CB' = 250 \tan 79°$.

In $\triangle ABC$, $m\angle BAC = 90° - 16° = 74°$. Then $CB = 250 \tan 74°$.

Hence, $BB' = CB' - CB = 250(\tan 79° - \tan 74°) = 250(5.1446 - 3.4874) = 414$ ft.

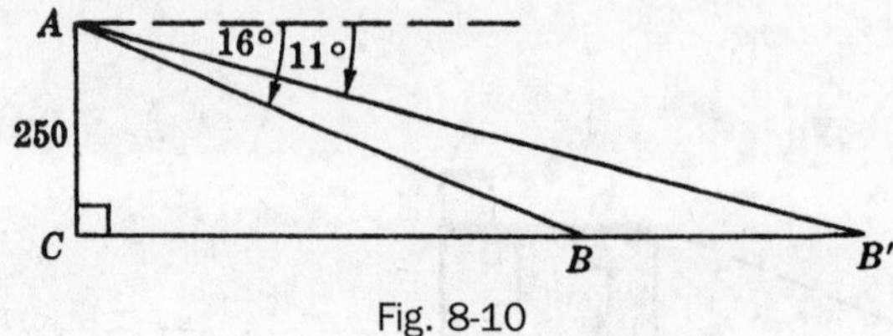

Fig. 8-10

SUPPLEMENTARY PROBLEMS

8.1. Using the table of trigonometric functions at the end of the book, find (8.1)

(a) sin 25°, sin 48°, sin 59°, and sin 89°

(b) cos 15°, cos 52°, cos 74°, and cos 88°

(c) tan 4°, tan 34°, tan 55°, and tan 87°

(d) which trigonometric ratios increase as the measure of the angle increases from 0° to 90°

(e) which trigonometric ratio decreases as the measure of the angle increases from 0° to 90°

(f) which trigonometric ratio has values greater than 1

8.2. Using the table at the end of the book, find the angle for which (8.1)

(a) $\sin x = 0.3420$ (c) $\sin B = 0.9455$ (e) $\cos y = 0.7071$ (g) $\tan W = 0.3443$

(b) $\sin A = 0.4848$ (d) $\cos A' = 0.9336$ (f) $\cos Q = 0.3584$ (h) $\tan B' = 2.3559$

8.3. Using the table at the end of the book, find the measure of x to the nearest degree if (8.2)

(a) $\sin x = 0.4400$ (e) $\cos x = 0.7650$ (i) $\tan x = 5.5745$ (m) $\cos x = \frac{3}{8}$

(b) $\sin x = 0.7280$ (f) $\cos x = 0.2675$ (j) $\sin x = \frac{11}{50}$ (n) $\cos x = \frac{\sqrt{3}}{2}$

(c) $\sin x = 0.9365$ (g) $\tan x = 0.1245$ (k) $\sin x = \frac{\sqrt{2}}{2}$ (o) $\tan x = \frac{2}{7}$

(d) $\cos x = 0.9900$ (h) $\tan x = 0.5200$ (l) $\cos x = \frac{13}{25}$ (p) $\tan x = \frac{\sqrt{3}}{10}$

8.4. In each right triangle of Fig. 8-11, find sin A, cos A, and tan A. (8.3)

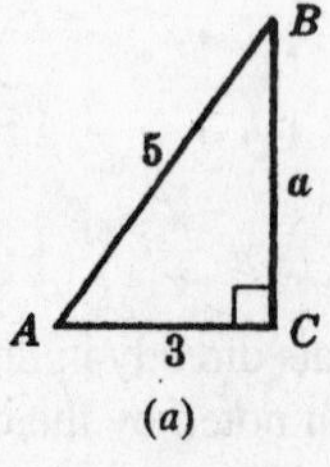

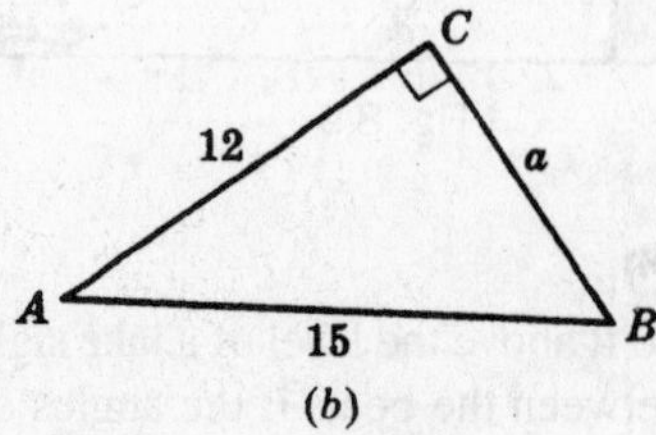

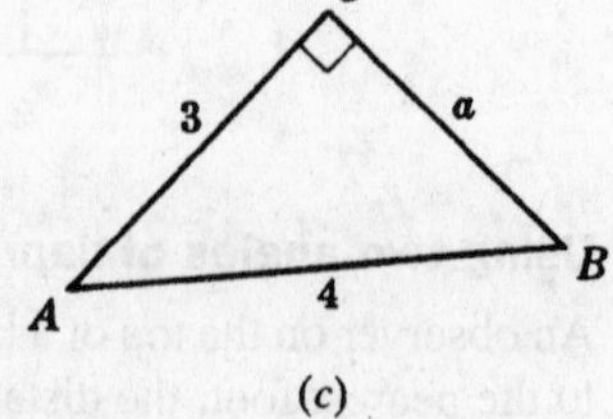

Fig. 8-11

8.5. Find $m\angle A$ to the nearest degree in each part of Fig. 8-12. (8.4)

(a)

(b)

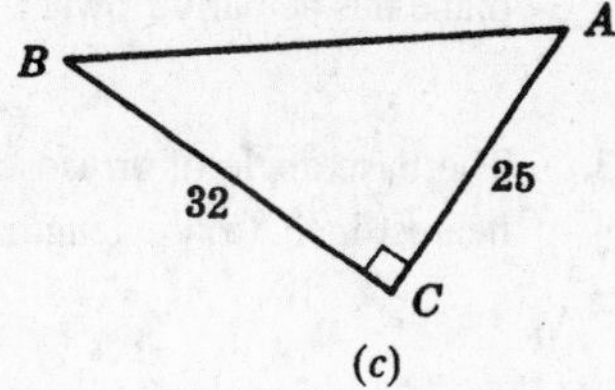

(c)

Fig. 8-12

8.6. Find $m\angle B$ to the nearest degree if (a) $b = 67$ and $c = 100$; (b) $a = 14$ and $c = 50$; (c) $a = 22$ and $b = 55$; (d) $a = 3$ and $b = \sqrt{3}$ in Fig. 8-13. (8.4)

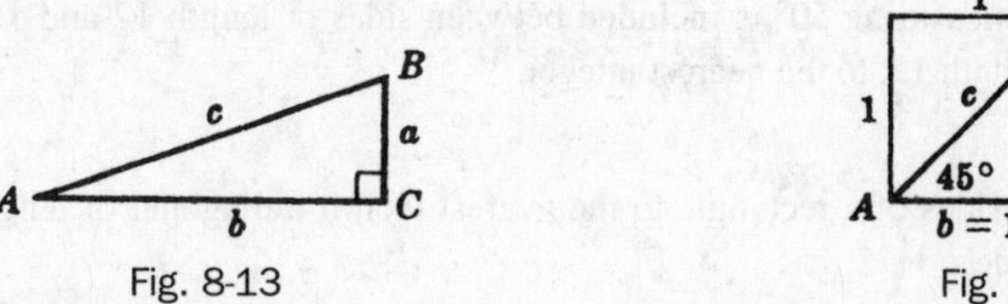

Fig. 8-13 Fig. 8-14

8.7. Making use of a square with a side of 1 (Fig. 8-14), show that (a) the diagonal $c = \sqrt{2}$; (b) $\tan 45° = 1$; (c) $\sin 45° = \cos 45° = 0.707$. (8.5)

8.8. To the nearest degree, find the measure of each acute angle of any right triangle whose sides are in the ratio of (a) 5:12:13; (b) 8:15:17; (c) 7:24:25; (d) 11:60:61. (8.4)

8.9. In each triangle of Fig. 8-15, solve for x and y to the nearest integer. (8.5)

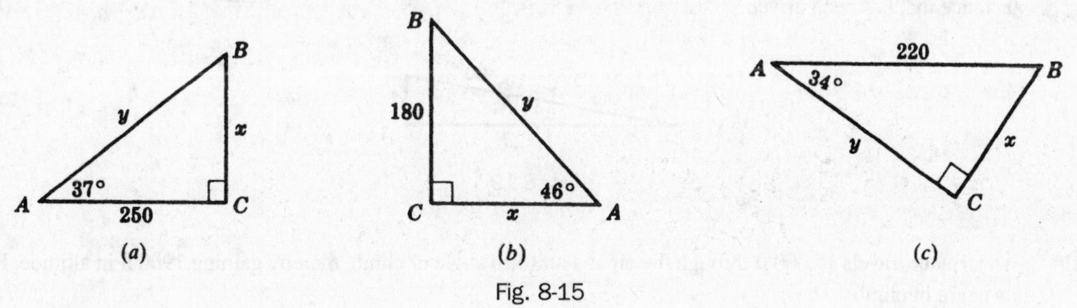

(a) (b) (c)

Fig. 8-15

8.10. A ladder leans against the side of a building and makes an angle measuring 70° with the ground. The foot of the ladder is 30 ft from the building. Find, to the nearest foot, (a) how high up on the building the ladder reaches; (b) the length of the ladder. (8.6)

8.11. To find the distance across a swamp, a surveyor took measurements as shown in Fig. 8-16. $\overline{AC}$ is at right angles to $\overline{BC}$. If $m\angle A = 24°$ and $AC = 350$ ft, find the distance BC across the swamp. (8.6)

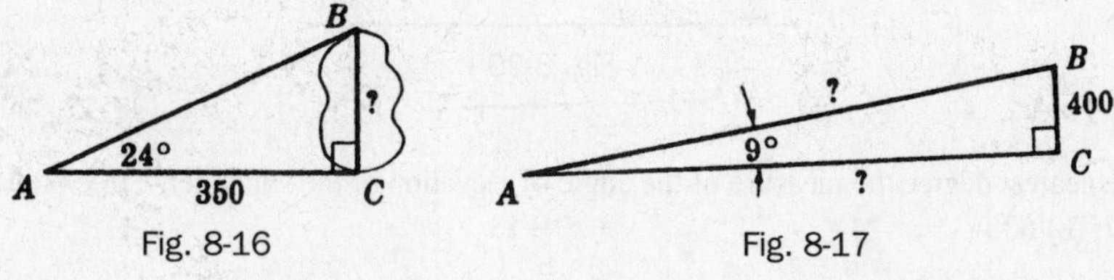

Fig. 8-16 Fig. 8-17

8.12. A plane rises from take-off and flies at a fixed angle measuring 9° with the horizontal ground (Fig. 8-17). When it has gained 400 ft in altitude, find, to the nearest 10 ft, (a) the horizontal distance flown; (b) the distance the plane has actually flown. (8.6)

8.13. The base angle of an isosceles triangle measures 28°, and each leg has length 45 in (Fig. 8-18). Find, to the nearest inch, (a) the length of the altitude drawn to the base; (b) the length of the base. (8.6)

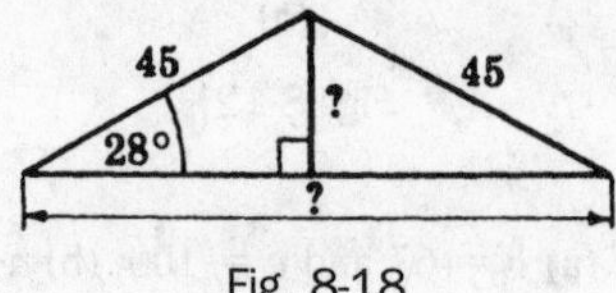

Fig. 8-18

8.14. In a triangle, an angle measuring 50° is included between sides of length 12 and 18. Find the length of the altitude to the side of length 12, to the nearest integer. (8.6)

8.15. Find the lengths of the sides of a rectangle to the nearest inch if a diagonal of length 24 in makes an angle measuring 42° with a side. (8.6)

8.16. A rhombus has an angle measuring 76° and a long diagonal of length 40 ft. Find the length of the short diagonal to the nearest foot. (8.6)

8.17. Find the length of the altitude to the base of an isosceles triangle to the nearest yard if its base has length 40 yd and its vertex angle measures 106°. (8.6)

8.18. A road is inclined uniformly at an angle measuring 6° with the horizontal (Fig. 8-19). After a car is driven 10,000 ft along this road, find, to the nearest 10 ft, the (a) increase in the altitude of the car and driver; (b) horizontal distance that has been driven. (8.7)

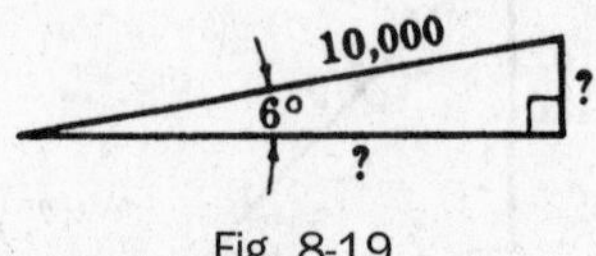

Fig. 8-19

8.19. An airplane travels 15,000 ft through the air at a uniform angle of climb, thereby gaining 1900 ft in altitude. Find its angle of climb. (8.7)

8.20. Sighting to the top of a monument. William found the angle of elevation to measure 16° (Fig. 8-20). The ground is level, and the transit is 5 ft above the ground. If the monument is 86 ft high, find, to the nearest foot, the distance from William to the foot of the monument. (8.8)

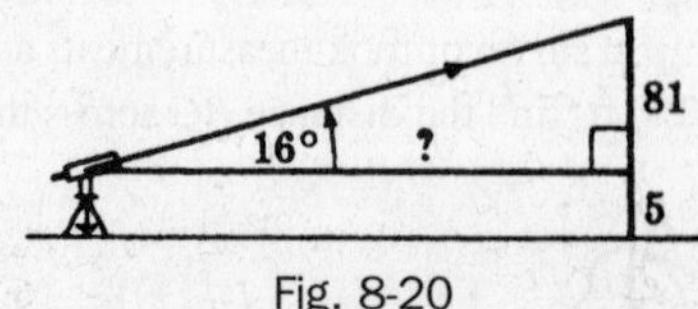

Fig. 8-20

8.21. Find to the nearest degree the measure of the angle of elevation of the sun when a tree 60 ft high casts a shadow of (a) 10 ft; (b) 60 ft.

8.22. At a certain time of day, the angle of elevation of the sun measures 34°. Find, to the nearest foot, the length of the shadow cast by (a) a 15-ft vertical pole; (b) a building 70 ft high. (8.8)

8.23. A light at C is projected vertically to a cloud at B. An observer at A, 1000 ft horizontally from C, notes the angle of elevation of B. Find the height of the cloud, to the nearest foot, if $m\angle A = 37°$. (8.8)

8.24. A lighthouse built at sea level is 180 ft high (Fig. 8-21). From its top, the angle of depression of a buoy measures 24°. Find, to the nearest foot, the distance from the buoy to the foot of the lighthouse. (8.8)

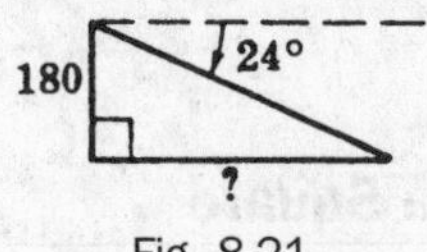

Fig. 8-21

8.25. An observer, on top of a hill 300 ft above the level of a lake, sighted two ships directly in line. Find, to the nearest foot, the distance between the boats if the angles of depression noted by the observer measured (a) 20° and 15°; (b) 35° and 24°; (c) 9° and 6°. (8.10)

8.26. In Fig. 8-22, $m\angle A = 43°$, $m\angle BDC = 54°$, $m\angle C = 90°$, and $DC = 170$ ft. (a) Find the length of $\overline{BC}$. (b) Using the result of (a), find the length of $\overline{AB}$. (8.10)

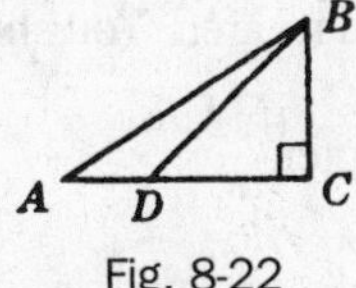

Fig. 8-22

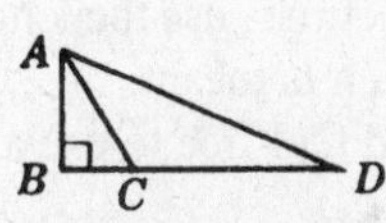

Fig. 8-23

8.27. In Fig. 8-23, $m\angle B = 90°$, $m\angle ACB = 58°$, $m\angle D = 23°$, and $BC = 60$ ft. (a) Find the length of $\overline{AB}$. (b) Using the result of part (a), find the length of $\overline{CD}$.

8.28. Tangents $\overline{PA}$ and $\overline{PB}$ are drawn to a circle from external point P. $m\angle APB = 40°$, and $PA = 25$. Find to the nearest tenth the radius of the circle.

CHAPTER 9

Areas

9.1 Area of a Rectangle and of a Square

A *square unit* is the surface enclosed by a square whose side is 1 unit (Fig. 9-1).

The *area of a closed plane figure*, such as a polygon, is the number of square units contained in its surface. Since a rectangle 5 units long and 4 units wide can be divided into 20 unit squares, its area is 20 square units (Fig. 9-2).

The area of a rectangle equals the product of the length of its base and the length of its altitude (Fig. 9-3). Thus if $b = 8$ in and $h = 3$ in, then $A = 24$ in^2.

The area of a square equals the square of the length of a side (Fig. 9-4). Thus if $s = 6$, then $A = s^2 = 36$.

It follows that the area of a square also equals one-half the square of the length of a diagonal. Since $A = s^2$ and $s = d/\sqrt{2}$, $A = \frac{1}{2}d^2$.

Note that we sometimes use the letter A for both a vertex of a figure and its area. You should have no trouble determining which is meant.

The reader should feel free to use a calculator for the work in this chapter.

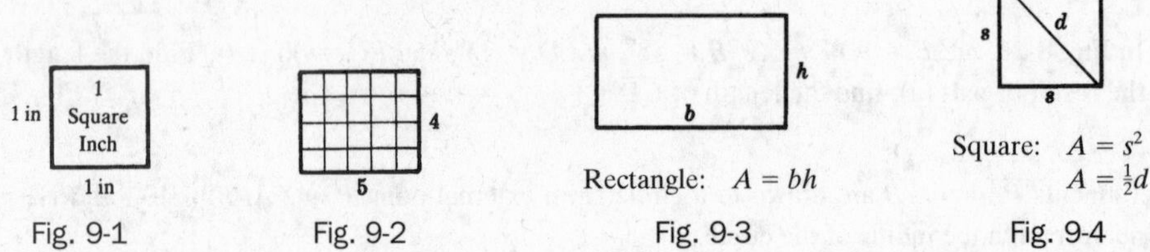

Fig. 9-1 Fig. 9-2 Fig. 9-3 Fig. 9-4

SOLVED PROBLEMS

9.1 Area of a rectangle

(a) Find the area of a rectangle if the base has length 15 and the perimeter is 50.

(b) Find the area of a rectangle if the altitude has length 10 and the diagonal has length 26.

(c) Find the lengths of the base and altitude of a rectangle if its area is 70 and its perimeter is 34.

Solutions

See Fig. 9-5.

(a) Here $p = 50$ and $b = 15$. Since $p = 2b + 2h$, we have $50 = 2(15) + 2h$ so $h = 10$.
Hence, $A = bh = 15(10) = 150$.

(b) Here $d = 26$ and $h = 10$. In right $\triangle ACD$, $d^2 = b^2 + h^2$, so $26^2 = b^2 + 10^2$ or $b = 24$.
Hence, $A = bh = 24(10) = 240$.

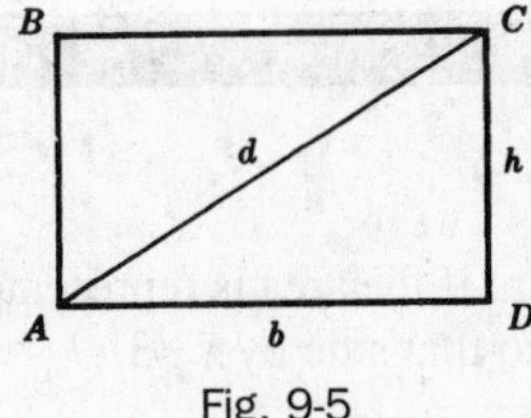

Fig. 9-5

(c) Here $A = 70$ and $p = 34$. Since $p = 2b + 2h$, we have $34 = 2(b + h)$ or $h = 17 - b$. Since $A = bh$, we have $70 = b(17 - b)$, so $b^2 - 17b + 70 = 0$ and $b = 7$ or 10. Then since $h = 17 - b$, we obtain $h = 10$ or 7.
Ans. 10 and 7, or 7 and 10.

9.2 Area of a Square

(a) Find the area of a square whose perimeter is 30.

(b) Find the area of a square if the radius of the circumscribed circle is 10.

(c) Find the side and the perimeter of a square whose area is 20.

(d) Find the number of square inches in a square foot.

Solutions

(a) Since $p = 4s = 30$ in Fig. 9-6(a), $s = 7\frac{1}{2}$. Then $A = s^2 = (7\frac{1}{2})^2 = 56\frac{1}{4}$.

(b) Since $r = 10$ in Fig. 9-6(b), $d = 2r = 20$. Then $A = \frac{1}{2}d^2 = \frac{1}{2}(20)^2 = 200$.

(c) In Fig. 9-6(a), $A = s^2 = 20$; hence, $s = 2\sqrt{5}$. Then perimeter $= 4s = 8\sqrt{5}$.

(d) $A = s^2$. Since 1ft = 12 in, 1 ft^2 = 1ft × 1ft = 12 in × 12 in = 144 in^2.

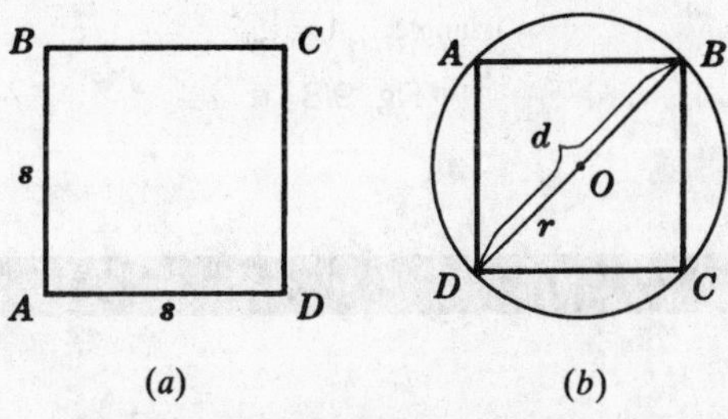

Fig. 9-6

9.2 Area of a Parallelogram

The area of a parallelogram equals the product of the length of a side and the length of the altitude to that side. (A proof of this theorem is given in Chapter 16.) Thus in ▱$ABCD$ (Fig. 9-7), if $b = 10$ and $h = 2.7$, then $A = 10(2.7) = 27$.

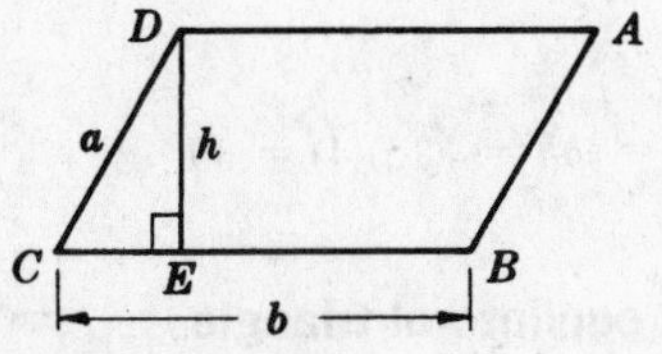

Parallelogram: $A = bh$

Fig. 9-7

SOLVED PROBLEMS

9.3 Area of a parallelogram

(a) Find the area of a parallelogram if the area is represented by x^2-4, the length of a side by $x + 4$, and the length of the altitude to that side by $x-3$.

(b) In a parallelogram, find the length of the altitude if the area is 54 and the ratio of the altitude to the base is 2:3.

Solutions

See Fig. 9-7.

(a) $A = x^2-4$, $b = x + 4$, $h = x-3$. Since $A = bh$, $x^2-4 = (x + 4)(x-3)$ or $x^2-4 = x^2 + x-12$ and $x = 8$. Hence, $A = x^2-4 = 64-4 = 60$.

(b) Let $h = 2x$, $b = 3x$. Then $A = bh$ or $54 = (3x)(2x) = 6x^2$, so $9 = x^2$ and $x = 3$. Hence, $h = 2x = 2(3) = 6$.

9.3 Area of a Triangle

The area of a triangle equals one-half the product of the length of a side and the length of the altitude to that side. (A proof of this theorem is given in Chapter 16.)

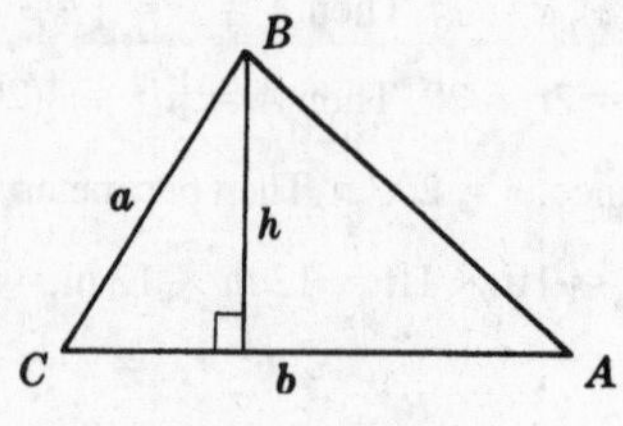

Triangle: $A = \frac{1}{2}bh$

Fig. 9-8

SOLVED PROBLEMS

9.4 Area of a triangle

Find the area of the triangle in Fig. 9-9.

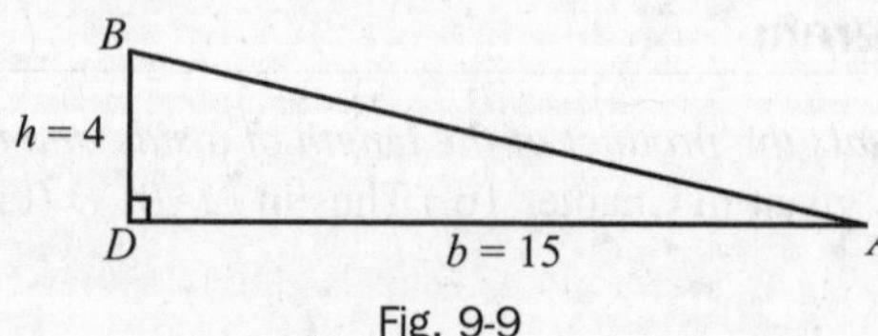

Fig. 9-9

Solution

Here, $b = 15$ and $h = 4$. Thus, $A = \frac{1}{2}bh = \frac{1}{2}(15)(4) = 30$.

9.5 Formulas for the area of an equilateral triangle

Derive the formula for the area of an equilateral triangle (a) whose side has length s; (b) whose altitude has length h.

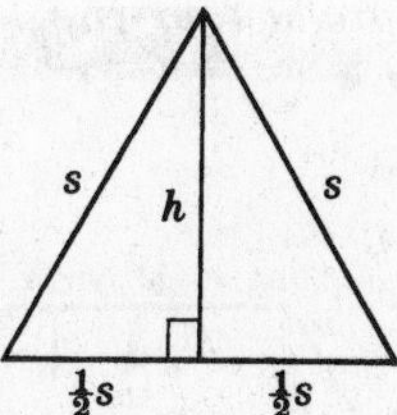

Equilateral Triangle:
$A = \frac{1}{4}s^2\sqrt{3}$
$A = \frac{1}{3}h^2\sqrt{3}$

Fig. 9-10

Solutions

See Fig. 9-10.

(a) Here $A = \frac{1}{2}bh$, where $b = s$ and $h^2 = s^2-(\frac{1}{2}s)^2$ or $h = \frac{1}{2}s\sqrt{3}$.

Then $A = \frac{1}{2}bh = \frac{1}{2}s(\frac{1}{2}s\sqrt{3}) = \frac{1}{4}s^2\sqrt{3}$.

(b) Here $A = \frac{1}{2}bh$, where $b = s$ and $h = \frac{1}{2}s\sqrt{3}$ or $s = \frac{2h}{\sqrt{3}}$.

Then $A = \frac{1}{2}bh = \frac{1}{2}sh = \frac{1}{2}(\frac{2h}{\sqrt{3}})\cdot h$

9.6 Area of an equilateral triangle

In Fig. 9-11, find the area of (a) an equilateral triangle whose perimeter is 24; (b) a rhombus in which the shorter diagonal has length 12 and an angle measures 60°; (c) a regular hexagon with a side of length 6.

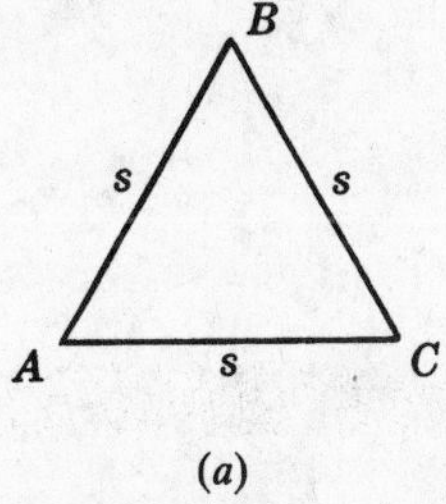

(*a*)

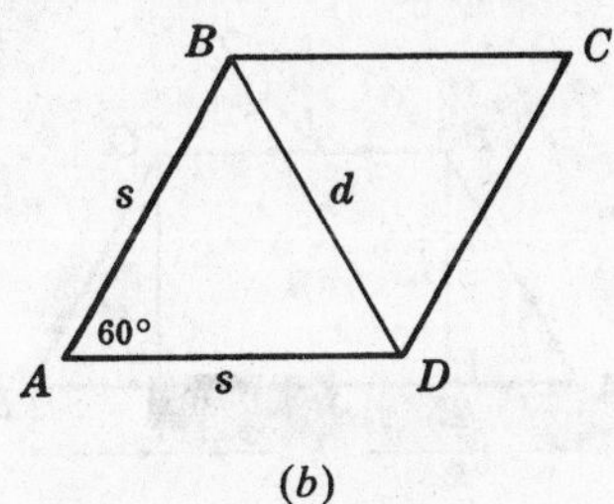

(*b*)

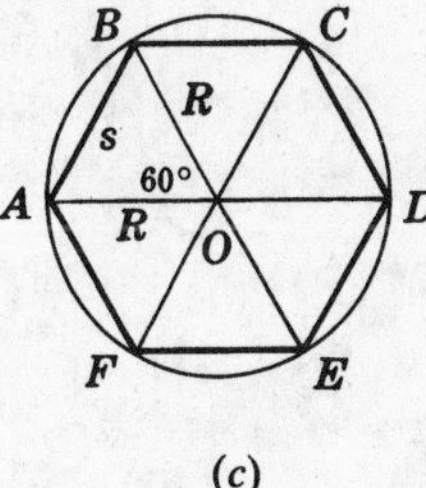

(c)

Fig. 9-11

Solutions

(a) Since $p = 3s = 24$, $s = 8$. Then $A = \frac{1}{4}s^2\sqrt{3} = \frac{1}{4}(64)\sqrt{3} = 16\sqrt{3}$.

(b) Since $m\angle A = 60°$, $\triangle ADB$ is equilateral and $s = d = 12$. The area of the rhombus is twice the area of $\triangle ABD$. Hence $A = 2\left(\frac{1}{4}s^2\sqrt{3}\right) = 2\left(\frac{1}{4}\right)(144)\sqrt{3} = 72\sqrt{3}$.

(c) A side s of the inscribed hexagon subtends a central angle of measure $\frac{1}{6}(360°) = 60°$. Then, since $OA = OB$ = radius R of the circumscribed circle, $m\angle OAB = m\angle OBA = 60°$. Thus $\triangle AOB$ is equilateral.
Area of hexagon = 6(area of $\triangle AOB$) $= 6\left(\frac{1}{4}s^2\sqrt{3}\right) = 6\left(\frac{1}{4}\right)(36\sqrt{3}) = 54\sqrt{3}$.

9.4 Area of a Trapezoid

The area of a trapezoid equals one-half the product of the length of its altitude and the sum of the lengths of its bases. (A proof of this theorem is given in Chapter 16.) Thus if $h = 20$, $b = 27$, and $b' = 23$ in Fig. 9-12, then $A = \frac{1}{2}(20)(27 + 23) = 500$.

The area of a trapezoid equals the product of the lengths of its altitude and median. Since $A = \frac{1}{2}h(b + b')$ and $m = \frac{1}{2}(b + b')$, $A = hm$.

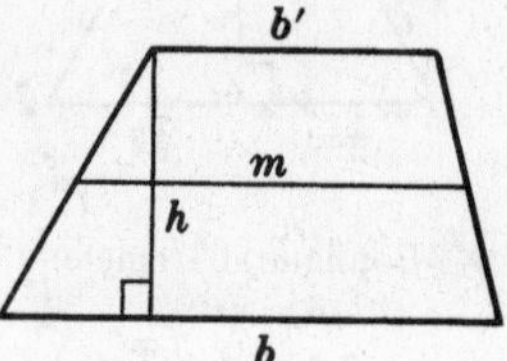

Trapezoid: $A = \frac{1}{2}h(b + b')$
$A = hm$

Fig. 9-12

SOLVED PROBLEMS

9.7 Area of a trapezoid

(a) Find the area of a trapezoid if the bases have lengths 7.3 and 2.7, and the altitude has length 3.8.

(b) Find the area of an isosceles trapezoid if the bases have lengths 22 and 10, and the legs have length 10.

(c) Find the bases of an isosceles trapezoid if the area is $52\sqrt{3}$, the altitude has length $4\sqrt{3}$, and each leg has length 8.

Solutions

See Fig. 9-13.

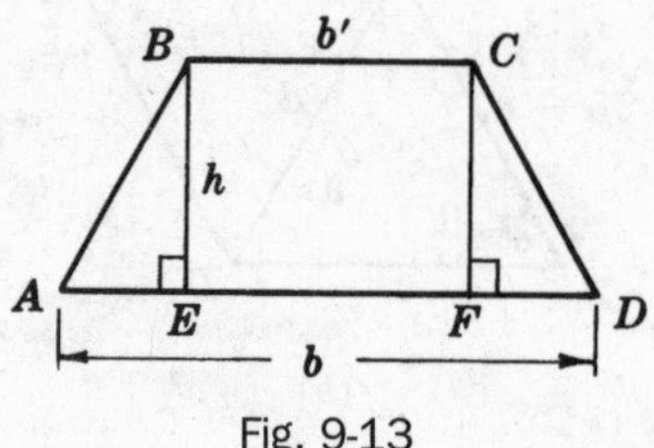

Fig. 9-13

(a) Here $b = 7.3$, $b' = 2.7$, $h = 3.8$. Then $A = \frac{1}{2}h\,(b + b') = \frac{1}{2}\,(3.8 + 2.7) = 19$.

(b) Here $b = 22$, $b' = 10$, $AB = 10$. Also $EF = b' = 10$ and $AE = \frac{1}{2}(22 - 10) = 6$.
In $\triangle BEA$, $h^2 = 10^2 - 6^2 = 64$ so $h = 8$. Then $A = \frac{1}{2}h(b + b') = \frac{1}{2}(8)(22 + 10) = 128$.

(c) $AE = \sqrt{(AB)^2 - h^2} = \sqrt{64 - 48} = 4$. Also $FD = AE = 4$, and $b' = b - (AE + FD) = b - 8$. Then $A = \frac{1}{2}h(b + b') = \frac{1}{2}h(2b - 8)$ or $52\sqrt{3} = \frac{1}{2}(4\sqrt{3})(2b - 8)$, from which $26 = 2b - 8$ or $b = 17$. Then $b' = b - 8 = 17 - 8 = 9$.

9.5 Area of a Rhombus

The area of a rhombus equals one-half the product of the lengths of its diagonals.

Since each diagonal is the perpendicular bisector of the other, the area of triangle I in Fig. 9-14 is $\frac{1}{2}(\frac{1}{2}d)(\frac{1}{2}d') = \frac{1}{8}dd'$. Thus the rhombus, which consists of four triangles congruent to $\triangle$I, has an area of $4(\frac{1}{8}dd')$ or $\frac{1}{2}dd'$.

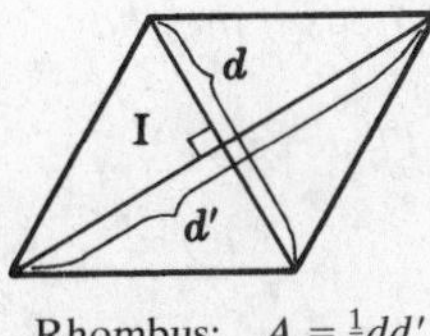

Rhombus: $A = \frac{1}{2}dd'$

Fig. 9-14

SOLVED PROBLEMS

9.8 Area of a rhombus

(a) Find the area of a rhombus if one diagonal has length 30 and a side has length 17.

(b) Find the length of a diagonal of a rhombus if the other diagonal has length 8 and the area of the rhombus is 52.

Solutions

See Fig. 9-15.

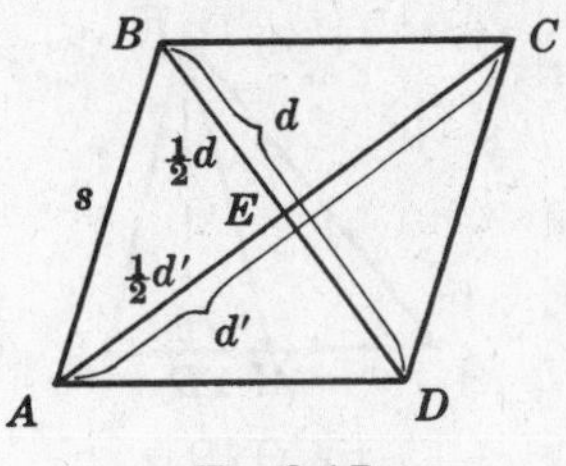

Fig. 9-15

(a) In right $\triangle AEB$, $s^2 = (\frac{1}{2}d)^2 + (\frac{1}{2}d')^2$ or $17^2 = (\frac{1}{2}d)^2 + 15^2$. Then $\frac{1}{2}d = 8$ and $d = 16$. Now $A = \frac{1}{2}dd' = \frac{1}{2}(16)(30) = 240$.

(b) We have $d' = 8$ and $A = 52$. Then $A = \frac{1}{2}dd'$ or $52 = \frac{1}{2}(d)(8)$ and $d = 13$.

9.6 Polygons of the Same Size or Shape

Figure 9-16 shows what we mean when we say that two polygons are of equal area, or are similar, or are congruent.

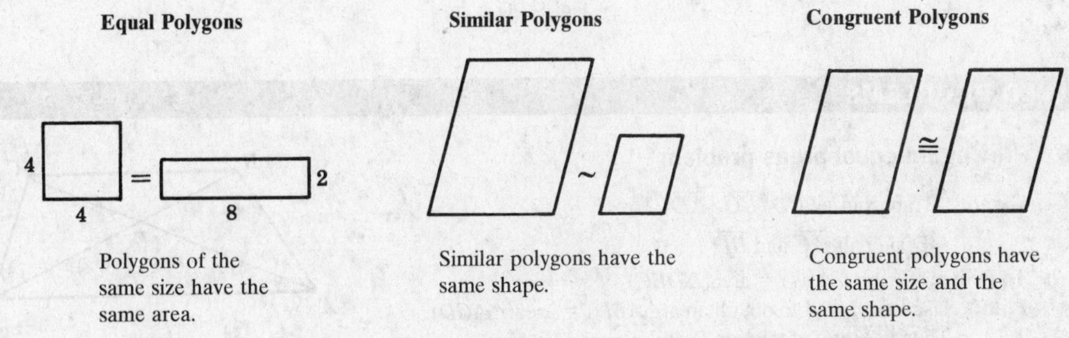

Fig. 9-16

PRINCIPLE 1: *Parallelograms have equal areas if they have congruent bases and congruent altitudes.*

Thus, the two parallelograms shown in Fig. 9-17 are equal.

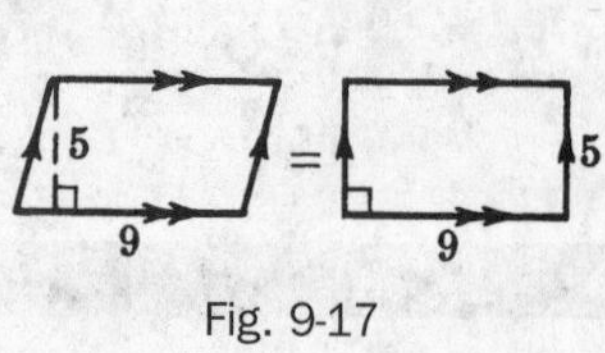

Fig. 9-17

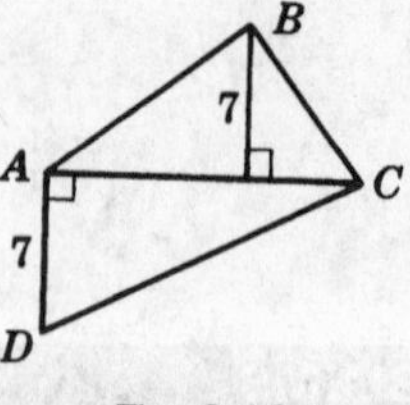

Fig. 9-18

PRINCIPLE 2: *Triangles have equal areas if they have congruent bases and congruent altitudes.*

Thus in Fig. 9-18, the area of $\triangle CAB$ equals the area of $\triangle CAD$.

PRINCIPLE 3: *A median divides a triangle into two triangles with equal areas.*

Thus in Fig. 9-19, where $\overline{BM}$ is a median, the area of $\triangle AMB$ equals the area of $\triangle BMC$ since they have congruent bases ($\overline{AM} \cong \overline{MC}$) and common altitude $\overline{BD}$.

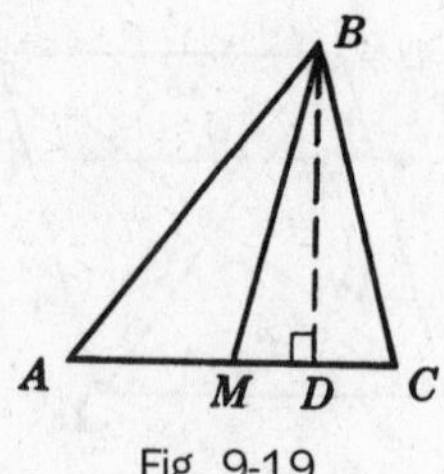

Fig. 9-19

PRINCIPLE 4: *Triangles are equal in area if they have a common base and their vertices lie on a line parallel to the base.*

Thus in Fig. 9-20, the area of $\triangle ABC$ is equal to the area of $\triangle ADC$.

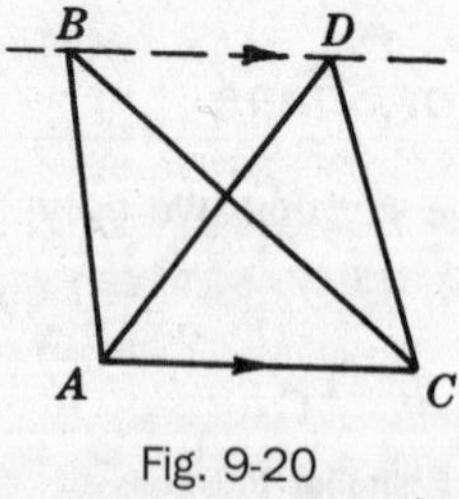

Fig. 9-20

SOLVED PROBLEMS

9.9 Proving an equal-areas problem

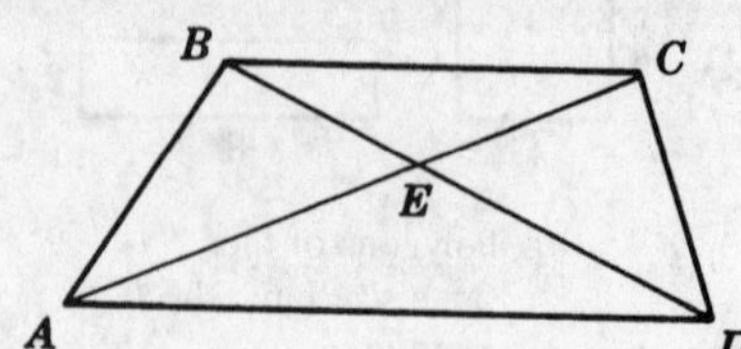

Given: Trapezoid $ABCD$ ($\overline{BC} \parallel \overline{AD}$)
Diagonals $\overline{AC}$ and $\overline{BD}$

To Prove: Area($\triangle AEB$) = area($\triangle DEC$)

Plan: Use Principle 4 to obtain area($\triangle ABD$) = area($\triangle ACD$). Then use the Subtraction Postulate.

PROOF:

Statements	Reasons
1. $\overline{BC} \parallel \overline{AD}$	1. Given
2. Area($\triangle ABD$) = area($\triangle ACD$)	2. Triangles have equal area if they have a common base and their vertices lie on a line parallel to the base.
3. Area($\triangle AED$) = area($\triangle AED$)	3. Identity Postulate
4. Area($\triangle AEB$) = area($\triangle DEC$)	4. Subtraction Postulate

9.10 Proving an equal-areas problem stated in words

Prove that if M is the midpoint of diagonal $\overline{AC}$ in quadrilateral $ABCD$, and $\overline{BM}$ and $\overline{DM}$ are drawn, then the area of quadrilateral $ABMD$ equals the area of quadrilateral $CBMD$.

Solution

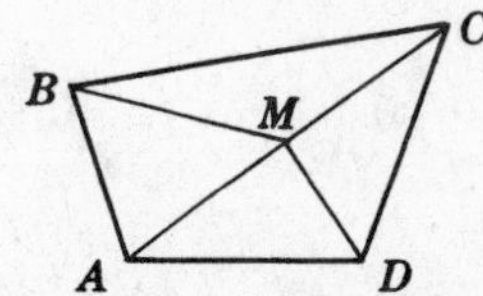

Given: Quadrilateral $ABCD$
M is midpoint of diagonal AC.

To Prove: Area of quadrilateral $ABMD$ equals area of quadrilateral $CBMD$.

Plan: Use Principle 3 to obtain two pairs of triangles which are equal in area. Then use the Addition Postulate.

PROOF:

Statements	Reasons
1. M is the midpoint of $\overline{AC}$.	1. Given
2. $\overline{BM}$ is a median of $\triangle ACB$. $\overline{DM}$ is a median of $\triangle ACD$.	2. A line from a vertex of a triangle to the midpoint of the opposite side is a median.
3. Area ($\triangle AMB$) = area($\triangle BMC$), Area($\triangle AMD$) = area($\triangle DMC$).	3. A median divides a triangle into two triangles of equal area.
4. Area of quadrilateral $ABMD$ equals area of quadrilateral $CBMD$.	4. If equals are added to equals, the results are equal.

9.7 Comparing Areas of Similar Polygons

The areas of similar polygons are to each other as the squares of any two corresponding segments.

Thus if $\triangle ABC \sim \triangle A'B'C'$ and the area of $\triangle ABC$ is 25 times the area of $\triangle A'B'C'$, then the ratio of the lengths of any two corresponding sides, medians, altitudes, radii of inscribed or circumscribed circles, and such is 5:1.

SOLVED PROBLEMS

9.11 Ratio of areas and segments of similar triangles

Find the ratio of the areas of two similar triangles (a) if the ratio of the lengths of two corresponding sides is 3:5; (b) if their perimeters are 12 and 7. Find the ratio of the lengths of a pair of (c) corresponding sides if the ratio of the areas is 4:9; (d) corresponding medians if the areas are 250 and 10.

Solutions

(a) $\frac{A}{A'} = \left(\frac{s}{s'}\right)^2 = \left(\frac{3}{5}\right)^2 = \frac{9}{25}$

(c) $\left(\frac{s}{s'}\right)^2 = \frac{A}{A'} = \frac{4}{9}$ or $\frac{s}{s'} = \frac{2}{3}$

(b) $\frac{A}{A'} = \left(\frac{p}{p'}\right)^2 = \left(\frac{12}{7}\right)^2 = \frac{144}{49}$

(d) $\left(\frac{m}{m'}\right)^2 = \frac{A}{A'} = \frac{250}{10}$ or $\frac{m}{m'} = 5$

9.12 Proportions derived from similar polygons

(a) The areas of two similar polygons are 80 and 5. If a side of the smaller polygon has length 2, find the length of the corresponding side of the larger polygon.

(b) The corresponding diagonals of two similar polygons have lengths 4 and 5. If the area of the larger polygon is 75, find the area of the smaller polygon.

Solutions

(a) $\left(\frac{s}{s'}\right)^2 = \frac{A}{A'}$, so $\left(\frac{s}{2}\right)^2 = \frac{80}{5} = 16$, Then $\frac{s}{2} = 4$ and $s = 8$.

(b) $\frac{A}{A'} = \left(\frac{d}{d'}\right)^2$, so $\frac{A}{75} = \left(\frac{4}{5}\right)^2$. Then $A = 75\left(\frac{16}{25}\right) = 48$.

SUPPLEMENTARY PROBLEMS

9.1. Find the area of a rectangle (9.1)

(a) If the base has length 11 in and the altitude has length 9 in

(b) If the base has length 2 ft and the altitude has length 1 ft 6 in

(c) If the base has length 25 and the perimeter is 90

(d) If the base has length 15 and the diagonal has length 17

(e) If the diagonal has length 12 and the angle between the diagonal and the base measures 60°

(f) If the diagonal has length 20 and the angle between the diagonal and the base measures 30°

(g) If the diagonal has length 25 and the lengths of the sides are in the ratio of 3:4

(h) If the perimeter is 50 and the lengths of the sides are in the ratio of 2:3

9.2. Find the area of a rectangle inscribed in a circle (9.1)

(a) If the radius of the circle is 5 and the base has length 6

(b) If the radius of the circle is 15 and the altitude has length 24

(c) If the radius and the altitude both have length 5

(d) If the diameter has length 26 and the base and altitude are in the ratio of 5:12

9.3. Find the base and altitude of a rectangle (9.1)

(a) If its area is 28 and the base has a length of 3 more than the altitude

(b) If its area is 72 and the base is twice the altitude

(c) If its area is 54 and the ratio of the base to the altitude is 3:2

(d) If its area is 12 and the perimeter is 16

(e) If its area is 70 and the base and altitude are represented by $2x$ and $x + 2$

(f) If its area is 160 and the base and altitude are represented by $3x-4$ and x

9.4. Find the area of (a) a square yard in square inches; (b) a square meter in square decimeters (1 m = 10 dm). (9.2)

9.5. Find the area of a square if (a) a side has length 15; (b) a side has length $3\frac{1}{2}$; (c) a side has length 1.8; (d) a side has length $8a$; (e) the perimeter is 44; (f) the perimeter is 10; (g) the perimeter is $12b$; (h) the diagonal has length 8; (i) the diagonal has length 9; (j) the diagonal has length $8\sqrt{2}$. (9.2)

9.6. Find the area of a square if (a) the radius of the circumscribed circle is 8; (b) the diameter of the circumscribed circle is 12; (c) the diameter of the circumscribed circle is $10\sqrt{2}$; (d) the radius of the inscribed circle is $3\frac{1}{2}$; (e) the diameter of the inscribed circle is 20. (9.2)

9.7. If a floor is 20 m long and 80 m wide, how many tiles are needed to cover it if (a) each tile is 1 m^2; (b) each tile is a square 2 m on a side; (c) each tile is a square 4 m on a side. (9.2)

9.8. If the area of a square is 81, find the length of (a) its side; (b) its perimeter; (c) its diagonal; (d) the radius of the inscribed circle; (e) the radius of the circumscribed circle. (9.2)

9.9. (a) Find the length of the side of a square whose area is $6\frac{1}{4}$. (9.2)

(b) Find the perimeter of a square whose area is 169.

(c) Find the length of the diagonal of a square whose area is 50.

(d) Find the length of the diagonal of a square whose area is 25.

(e) Find the radius of the inscribed circle of a square whose area is 144.

(f) Find the radius of the circumscribed circle of a square whose area is 32.

9.10. Find the area of a parallelogram if the base and altitude have lengths, respectively, of (a) 3 ft and $5\frac{1}{3}$ ft; (b) 4 ft and 1 ft 6 in; (c) 20 and 3.5; (d) 1.8 m and 0.9 m. (9.3)

9.11. Find the area of a parallelogram if the base and altitude have lengths, respectively, of (a) $3x$ and x; (b) $x + 3$ and x; (c) $x-5$ and $x + 5$; (d) $4x + 1$ and $3x + 2$. (9.3)

9.12. Find the area of a parallelogram if

(a) The area is represented by x^2, the base by $x + 3$, and the altitude by $x-2$

(b) The area is represented by x^2-10, the base by x, and the altitude by $x-2$

(c) The area is represented by $2x^2-34$, the base by $x + 3$, and the altitude by $x-3$

9.13. In a parallelogram, find (9.3)

(a) The base if the area is 40 and the altitude has length 15

(b) The length of the altitude if the area is 22 and the base has length 1.1

(c) The length of the base if the area is 27 and the base is three times the altitude

(d) The length of the altitude if the area is 21 and the base has length four more than the altitude

(e) The base if the area is 90 and the ratio of the base to the altitude is 5:2

(f) The length of the altitude to a side of length 20 if the altitude to a side of length 15 is 16

(g) The length of the base if the area is 48, the base is represented by $x + 3$, and the altitude is represented by $x + 1$

(h) The length of the base if the area is represented by $x^2 + 17$, the base by $2x-3$, and the altitude by $x + 1$

9.14. Find the area of a triangle if the lengths of the base and altitude are, respectively, (a) 6 in and $3\frac{2}{3}$ in; (b) 1 yd and 2 ft; (c) 8 and $x-7$; (d) $5x$ and $4x$; (e) $4x$ and $x + 9$; (f) $x + 4$ and $x-4$; (g) $2x-6$ and $x + 3$. (9.4)

9.15. Find the area of (9.4)

(a) A triangle if two sides have lengths 13 and 15 and the altitude to the third side has length 12

(b) A triangle whose sides have lengths 10, 10, and 16

(c) A triangle whose sides have lengths 5, 12, and 13

(d) An isosceles triangle whose base has length 30 and whose legs each have length 17

(e) An isosceles triangle whose base has length 20 and whose vertex angle measures 68°

(f) An isosceles triangle whose base has length 30 and whose base angle measures 62°

(g) A triangle inscribed in a circle of radius 4 if one side is a diameter and another side makes an angle measuring 30° with the diameter

(h) A triangle cut off by a line parallel to the base of a triangle if the base and altitude of the larger triangle have lengths 10 and 5, respectively, and the line parallel to the base is 6

9.16. Find the altitude of a triangle if (9.4)

(a) Its base has length 10 and the triangle is equal in area to a parallelogram whose base and altitude have lengths 15 and 8.

(b) Its base has length 8 and the triangle is equal in area to a square whose diagonal has length 4.

(c) Its base has length 12 and the triangle is equal in area to another triangle whose sides have lengths 6, 8, and 10.

9.17. In a triangle, find the length of (9.4)

(a) A side if the area is 40 and the altitude to that side has length 10

(b) An altitude if the area is 25 and the side to which the altitude is drawn has length 5

(c) A side if the area is 24 and the side has length 2 more than its altitude

(d) A side if the area is 108 and the side and its altitude are in the ratio 3:2

(e) The altitude to a side of length 20, if the sides of the triangle have lengths 12, 16, and 20

(f) The altitude to a side of length 12 if another side and its altitude have lengths 10 and 15

(g) A side represented by $4x$ if the altitude to that side is represented by $x + 7$ and the area is 60

(h) A side if the area is represented by $x^2 - 55$, the side by $2x - 2$, and its altitude by $x - 5$

9.18. Find the area of an equilateral triangle if (a) a side has length 10; (b) the perimeter is 36; (c) an altitude has length 6; (d) an altitude has length $5\sqrt{3}$; (e) a side has length 2b; (f) the perimeter is $12x$; (g) an altitude has length $3r$. (9.6)

9.19. Find the area of a rhombus having an angle of 60° if (a) a side has length 2; (b) the shorter diagonal has length 7; (c) the longer diagonal has length 12; (d) the longer diagonal has length $6\sqrt{3}$. (9.6)

9.20. Find the area of a regular hexagon if (a) a side is 4; (b) the radius of the circumscribed circle is 6; (c) the diameter of the circumscribed circle is 20. (9.6)

9.21. Find the side of an equilateral triangle whose area equals (9.6)

(a) The sum of the areas of two equilateral triangles whose sides have lengths 9 and 12

(b) The difference of the areas of two equilateral triangles whose sides have lengths 17 and 15

(c) The area of a trapezoid whose bases have lengths 6 and 2 and whose altitude has length $9\sqrt{3}$

(d) Twice the area of a right triangle having a hypotenuse of length 5 and an acute angle of measure 30°

9.22. Find the area of trapezoid *ABCD* in Fig. 9-21, if: (9.7)

(a) $b = 25$, $b' = 15$, and $h = 7$

(b) $m = 10$ and $h = 6.9$

(c) $AB = 30$, $m\angle A = 30°$, $b = 24$, and $b' = 6$

(d) $AB = 12$, $m\angle A = 45°$, $b = 13$, and $b' = 7$

(e) $AB = 10$, $m\angle A = 70°$, and $b + b' = 20$

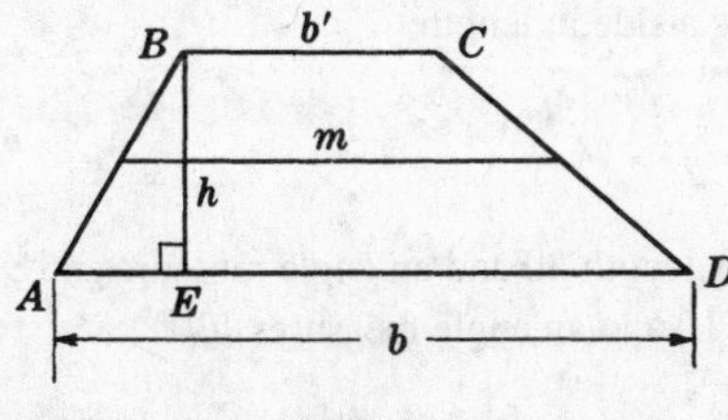

Fig. 9-21

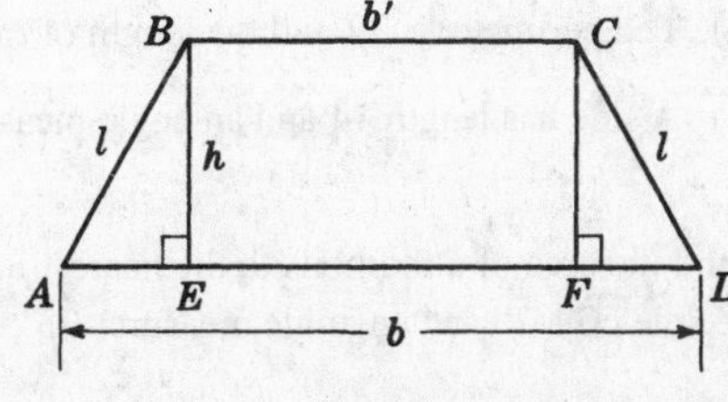

Fig. 9-22

9.23. Find the area of isosceles trapezoid *ABCD* in Fig. 9-22, if (9.7)

(a) $b' = 17$, $l = 10$, and $h = 6$

(b) $b = 22$, $b' = 12$, and $l = 13$

(c) $b = 16$, $b' = 10$, and $m\angle A = 45°$

(d) $b = 20$, $l = 8$, and $m\angle A = 60°$

(e) $b = 40$, $b' = 20$, and $m\angle A = 28°$

9.24. (a) Find the length of the altitude of a trapezoid if the bases have lengths 13 and 7 and the area is 40. (9.7)

(b) Find the length of the altitude of a trapezoid if the sum of the lengths of the bases is twice the length of the altitude and the area is 49.

(c) Find the sum of the lengths of the bases and the median of a trapezoid if the area is 63 and the altitude has length 7.

(d) Find the lengths of the bases of a trapezoid if the upper base has length 3 less than the lower base, the altitude has length 4, and the area is 30.

(e) Find the lengths of the bases of a trapezoid if the lower base has length twice that of the upper base, the altitude has length 6, and the area is 45.

9.25. In an isosceles trapezoid (9.7)

(a) Find the lengths of the bases if each leg has length 5, the altitude has length 3, and the area is 39.

(b) Find the lengths of the bases if the altitude has length 5, each base angle measures 45°, and the area is 90.

(c) Find the lengths of the bases if the area is $42\sqrt{3}$, the altitude has length $3\sqrt{3}$, and each base angle measures 60°.

(d) Find the length of each leg if the bases have lengths 24 and 32 and the area is 84.

(e) Find the length of each leg if the area is 300, the median has length 25, and the lower base has length 30.

9.26. Find the area of a rhombus if (9.8)

(a) The diagonals have lengths 8 and 9.

(b) The diagonals have lengths 11 and 7.

(c) The diagonals have lengths 4 and $6\sqrt{3}$.

(d) The diagonals have lengths $3x$ and $8x$.

(e) One diagonal has length 10 and a side has length 13.

(f) The perimeter is 40 and a diagonal has length 12.

(g) The side has length 6 and an angle measures 30°.

(h) The perimeter is 28 and an angle measures 45°.

(i) The perimeter is 32 and the length of the short diagonal equals a side in length.

(j) A side has length 14 and an angle measures 120°.

9.27. Find the area of a rhombus to the nearest integer if (a) the side has length 30 and an angle measures 55°; (b) the perimeter is 20 and an angle measures 33°; (c) the side has length 10 and an angle measures 130°. (9.8)

9.28. In a rhombus, find the length of (9.8)

(a) A diagonal if the other diagonal has length 7 and the area is 35 (9.8)

(b) The diagonals if their ratio is 4:3 and the area is 54

(c) The diagonals if the longer is twice the shorter and the area is 100

(d) The side if the area is 24 and one diagonal has length 6

(e) The side if the area is 6 and one diagonal has length 4 more than the other

9.29. A rhombus is equal to a trapezoid whose lower base has length 26 and whose other three sides have length 10. Find the length of the altitude of the rhombus if its perimeter is 36. (9.8)

9.30. Provide the proofs requested in Fig. 9-23. (9.9)

(*a*) **Given:** $\triangle ABC$
$DB = \frac{1}{3}AB$
$EC = \frac{1}{3}AC$
To Prove: Area($\triangle DFB$) = area($\triangle FEC$)

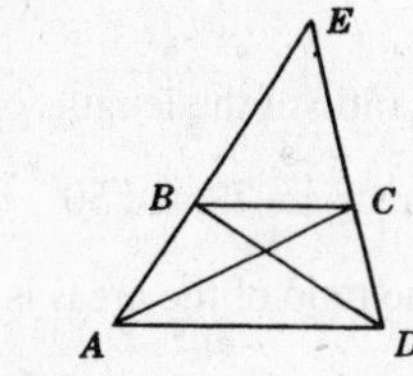

(*b*) **Given:** Trapezoid $ABCD$
$\overline{AB}$ and $\overline{CD}$ extended meet at E.
To Prove: Area($\triangle ECA$) = area($\triangle EBD$)

Fig. 9-23

9.31. Provide the proofs requested in Fig. 9-24. (9.9)

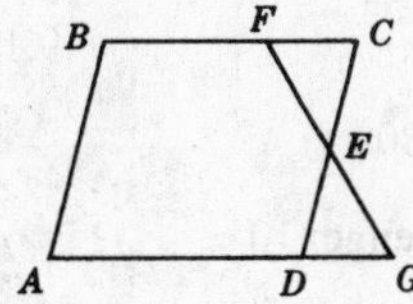

(*a*) **Given:** $\square ABCD$
E is midpoint of $\overline{CD}$.
To Prove:
Area($\square ABCD$) = area(trapezoid $BFGA$)

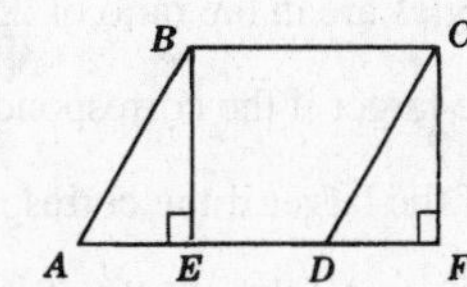

(*b*) **Given:** $\square ABCD$
$\overline{BE}$ and $\overline{CF} \perp \overline{AF}$.
To Prove:
$BCFE$ is a rectangle and equal in area to $\square ABCD$.

Fig. 9-24

9.32. Prove each of the following: (9.10)

(a) A median divides a triangle into two triangles having equal areas.

(b) Triangles are equal in area if they have a common base and their vertices lie in a line parallel to the base.

(c) In a triangle, if lines are drawn from a vertex to the trisection points of the opposite sides, the area of the triangle is trisected.

(d) In trapezoid $ABCD$, base $\overline{AD}$ is twice base $\overline{BC}$. If M is the midpoint of $\overline{AD}$, then $ABCM$ and $BCDM$ are parallelograms which are equal in area.

9.33. (a) In $\triangle ABC$, E is a point on $\overline{BM}$, the median to $\overline{AC}$. Prove that area($\triangle BEA$) = area($\triangle BEC$).

(b) In $\triangle ABC$, Q is a point on $\overline{BC}$, M is the midpoint of $\overline{AB}$, and P is the midpoint of $\overline{AC}$. Prove that area($\triangle BQM$) + area($\triangle PQC$) = area(quadrilateral $APQM$).

(c) In quadrilateral $ABCD$, diagonal $\overline{AC}$ bisects diagonal $\overline{BD}$. Prove that area($\triangle ABC$) = area ($\triangle ACD$).

(d) Prove that the diagonals of a parallelogram divide the parallelogram into four triangles which are equal in area. (9.10)

9.34. Find the ratio of the areas of two similar triangles if the ratio of two corresponding sides is (a) 1:7; (b) 7:2; (c) $1{:}\sqrt{3}$; (d) $a{:}5a$; (e) $9{:}\,x$; (f) $3{:}\sqrt{x}$; (g) $s{:}s\sqrt{2}$. (9.11)

9.35. Find the ratio of the areas of two similar triangles (9.11)

(a) If the ratio of the lengths of two corresponding medians is 7:10

(b) If the length of an altitude of the first is two-thirds of a corresponding altitude of the second

(c) If two corresponding angle bisectors have lengths 10 and 12

(d) If the length of each side of the first is one-third the length of each corresponding side of the second

(e) If the radii of their circumscribed circles are $7\frac{1}{2}$ and 5

(f) If their perimeters are 30 and $30\sqrt{2}$

9.36. Find the ratio of any two corresponding sides of two similar triangles if the ratio of their areas is (a) 100:1; (b) 1:49; (c) 400:81; (d) 25:121; (e) $4{:}y^2$; (f) $9x^2{:}1$; (g) 3:4; (h) 1:2; (i) $x^2{:}5$; (j) $x{:}16$. (9.11)

9.37. In two similar triangles, find the ratio of the lengths of (9.11)

(a) Corresponding sides if the areas are 72 and 50

(b) Corresponding medians if the ratio of the areas is 9:49

(c) Corresponding altitudes if the areas are 18 and 6

(d) The perimeters if the areas are 50 and 40

(e) Radii of the inscribed circles if the ratio of the areas is 1:3

9.38. The areas of two similar triangles are in the ratio of 25:16. Find (9.11)

(a) The length of a side of the larger if the corresponding side of the smaller has length 80

(b) The length of a median of the larger if the corresponding median of the smaller has length 10

(c) The length of an angle bisector of the smaller if the corresponding angle bisector of the larger has length 15

(d) The perimeter of the smaller if the perimeter of the larger is 125

(e) The circumference of the inscribed circle of the larger if the circumference of the inscribed circle of the smaller is 84

(f) The diameter of the circumscribed circle of the smaller if the diameter of the circumscribed circle of the larger is 22.5

(g) The length of an altitude of the larger if the corresponding altitude of the smaller has length $16\sqrt{3}$

9.39. (a) The areas of two similar triangles are 36 and 25. If a median of the smaller triangle has length 10, find the length of the corresponding median of the larger. (9.12)

(b) Corresponding altitudes of two similar triangles have lengths 3 and 4. If the area of the larger triangle is 112, find the area of the smaller.

(c) Two similar polygons have perimeters of 32 and 24. If the area of the smaller is 27, find the area of the larger.

(d) The areas of two similar pentagons are 88 and 22. If a diagonal of the larger has length 5, find the length of the corresponding diagonal of the smaller.

(e) In two similar polygons, the ratio of the lengths of two corresponding sides is $\sqrt{3}{:}1$. If the area of the smaller is 15, find the area of the larger.

Regular Polygons and the Circle

10.1 Regular Polygons

A *regular polygon* is an equilateral and equiangular polygon.

The *center of a regular polygon* is the common center of its inscribed and circumscribed circles.

A *radius of a regular polygon* is a segment joining its center to any vertex. A radius of a regular polygon is also a radius of the circumscribed circle. (Here, as for circles, we may use the word *radius* to mean the number that is "the length of the radius.")

A *central angle of a regular polygon* is an angle included between two radii drawn to successive vertices.

An *apothem of a regular polygon* is a segment from its center perpendicular to one of its sides. An apothem is also a radius of the inscribed circle.

Thus for the regular pentagon shown in Fig. 10-1, $AB = BC = CD = DE = EA$ and $m\angle A = m\angle B = m\angle C = m\angle D = m\angle E$. Also, its center is O, $\overline{OA}$ and $\overline{OB}$ are its radii; $\angle AOB$ is a central angle; and $\overline{OG}$ and $\overline{OF}$ are apothems.

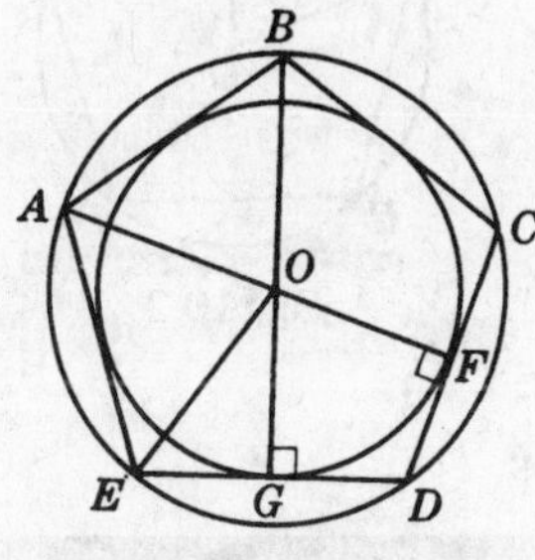

Fig. 10-1

10.1A Regular-Polygon Principles

PRINCIPLE 1: *If a regular polygon of n sides has a side of length s, the perimeter is $p = ns$.*

PRINCIPLE 2: *A circle may be circumscribed about any regular polygon.*

PRINCIPLE 3: *A circle may be inscribed in any regular polygon.*

PRINCIPLE 4: *The center of the circumscribed circle of a regular polygon is also the center of its inscribed circle.*

PRINCIPLE 5: *An equilateral polygon inscribed in a circle is a regular polygon.*

PRINCIPLE 6: *Radii of a regular polygon are congruent.*

PRINCIPLE 7: *A radius of a regular polygon bisects the angle to which it is drawn.*

Thus in Fig. 10-1, $\overline{OB}$ bisects $\angle ABC$.

PRINCIPLE 8: *Apothems of a regular polygon are congruent.*

PRINCIPLE 9: *An apothem of a regular polygon bisects the side to which it is drawn.*

Thus in Fig. 10-1, $\overline{OF}$ bisects $\overline{CD}$, and $\overline{OG}$ bisects $\overline{ED}$.

PRINCIPLE 10: *For a regular polygon of n sides:*

1. *Each central angle c measures* $\frac{360°}{n}$.

2. *Each interior angle i measures* $\frac{(n-2)180°}{n}$.

3. *Each exterior angle e measures* $\frac{360°}{n}$.

Thus for the regular pentagon *ABCDE* of Fig. 10-2,

$$m\angle AOB = m\angle ABS = \frac{360°}{n} = \frac{360°}{5} = 72° \qquad m\angle ABC = \frac{(n-2)\,180°}{n} = \frac{(5-2)\,180°}{5} = 108°$$

and $$m\angle ABC + m\angle ABS = 180°$$

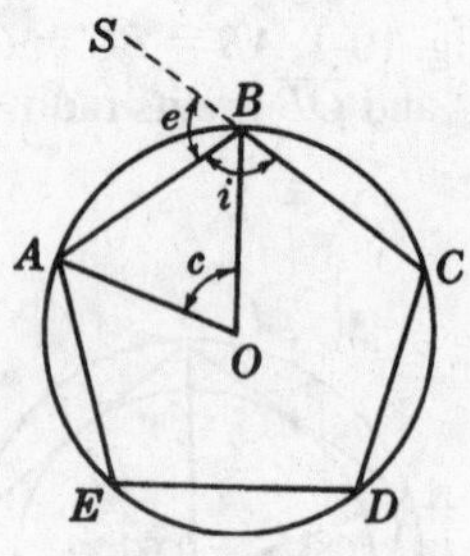

Fig. 10-2

SOLVED PROBLEMS

10.1 Finding measures of lines and angles in a regular polygon

(a) Find the length of a side *s* of a regular pentagon if the perimeter *p* is 35.

(b) Find the length of the apothem *a* of a regular pentagon if the radius of the inscribed circle is 21.

(c) In a regular polygon of five sides, find the measures of the central angle *c*, the exterior angle *e*, and the interior angle *i*.

(d) If an interior angle of a regular polygon measures 108°, find the measures of the exterior angle and the central angle and the number of sides.

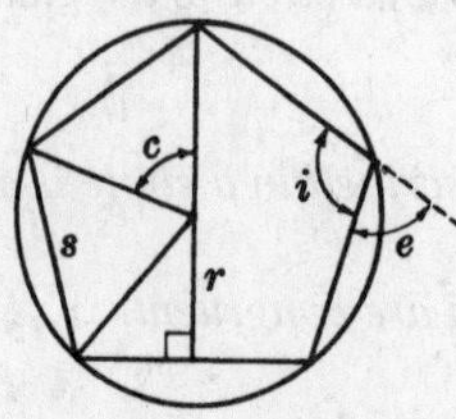

Fig. 10-3

Solutions

(a) $p = 35$. Since $p = 5s$, we have $35 = 5s$ and $s = 7$.

(b) Since an apothem r is a radius of the inscribed circle, it has length 21.

(c) $n = 5$. Then $m\angle c = \dfrac{360°}{n} = \dfrac{360°}{5} = 72°$; $m\angle e = \dfrac{360°}{n} = 72°$; $m\angle i = 180° - m\angle e = 108°$.

(d) $m\angle i = 108°$. Then $m\angle c = 180° - m\angle i = 72°$. Since $m\angle c = \dfrac{360°}{n}$, $n = 5$. (See Fig. 10-3.)

10.2 Proving a regular-polygon problem stated in words

Prove that a vertex angle of a regular pentagon is trisected by diagonals drawn from that vertex.

Solution

Given: Regular pentagon $ABCDE$

Diagonals $\overline{AC}$ and $\overline{AD}$

To Prove: $\overline{AC}$ and $\overline{AD}$ trisect $\angle A$.

Plan: Circumscribe a circle and show that angles BAC, CAD, and DAE are congruent.

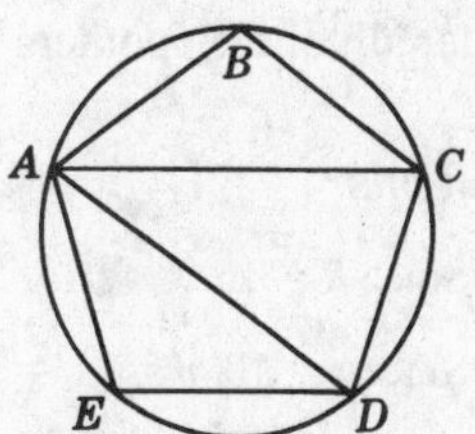

PROOF:

Statements	Reasons
1. $ABCDE$ is a regular pentagon.	1. Given
2. Circumscribe a circle about $ABCDE$.	2. A circle may be circumscribed about any regular polygon
3. $BC = CD = DE$	3. A regular polygon is equilateral.
4. $\overset{\frown}{BC} \cong \overset{\frown}{CD} \cong \overset{\frown}{DE}$	4. In a circle, equal chords have equal arcs.
5. $\angle BAC \cong \angle CAD \cong \angle DAE$	5. In a circle, inscribed angles having congruent arcs are congruent.
6. $\angle A$ is trisected.	6. To divide into three congruent parts is to trisect.

10.2 Relationships of Segments in Regular Polygons of 3, 4, and 6 Sides

In the regular hexagon, square, and equilateral triangle, special right triangles are formed when the apothem r and a radius R terminating in the same side are drawn. In the case of the square we obtain a 45°-45°-90° triangle, while in the other two cases we obtain a 30°-60°-90° triangle. The formulas in Fig. 10-4 relate the lengths of the sides and radii of these regular polygons.

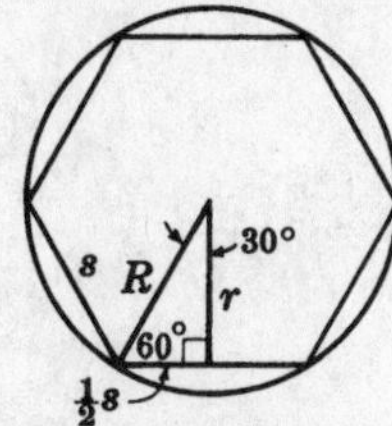

Regular Hexagon
$s = R$
$r = \frac{1}{2}R\sqrt{3}$

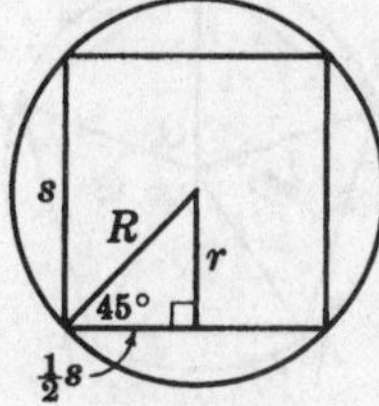

Square
$s = R\sqrt{2}$
$r = \frac{1}{2}s = \frac{1}{2}R\sqrt{2}$

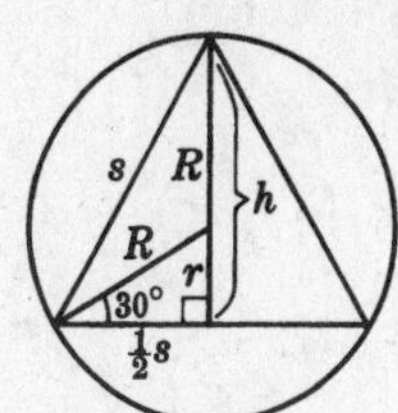

Equilateral Triangle
$s = R\sqrt{3}$, $h = r + R$
$r = \frac{1}{3}h$, $R = \frac{2}{3}h$, $r = \frac{1}{2}R$

Fig. 10-4

SOLVED PROBLEMS

10.3 Applying line relationships in a regular hexagon

In a regular hexagon, (a) find the lengths of the side and apothem if the radius is 12; (b) find the radius and length of the apothem if the side has length 8.

Solutions

(a) Since $R = 12$, $s = R = 12$ and $r = \frac{1}{2}R\sqrt{3} = 6\sqrt{3}$.

(b) Since $s = 8$, $R = s = 8$ and $r = \frac{1}{2}R\sqrt{3} = 4\sqrt{3}$.

10.4 Applying line relationships in a square

In a square, (a) find the lengths of the side and apothem if the radius is 16; (b) find the radius and the length of the apothem if a side has length 10.

Solutions

(a) Since $R = 16$, $s = R\sqrt{2} = 16\sqrt{2}$ and $r = \frac{1}{2}s = 8\sqrt{2}$.

(b) Since $s = 10$, $r = \frac{1}{2}s = 5$ and $R = \dfrac{s}{\sqrt{2}} = \frac{1}{2}s\sqrt{2} = 5\sqrt{2}$.

10.5 Applying line relationships in an equilateral triangle

In an equilateral triangle, (a) find the lengths of the radius, apothem, and side if the altitude has length 6; (b) find the lengths of the side, apothem, and altitude if the radius is 9.

Solutions

(a) Since $h = 6$, we have $r = \frac{1}{3}h = 2$; $R = \frac{2}{3}h = 4$; and $s = R\sqrt{3} = 4\sqrt{3}$.

(b) Since $R = 9$, $s = R\sqrt{3} = 9\sqrt{3}$; $r = \frac{1}{2}R = 4\frac{1}{2}$; and $h = \frac{3}{2}R = 13\frac{1}{2}$.

10.3 Area of a Regular Polygon

The area of a regular polygon equals one-half the product of its perimeter and the length of its apothem.

As shown in Fig. 10-5, by drawing radii we can divide a regular polygon of n sides and perimeter $p = ns$ into n triangles, each of area $\frac{1}{2}rs$. Hence, the area of the regular polygon is $n(\frac{1}{2}rs) = \frac{1}{2}nsr = \frac{1}{2}pr$.

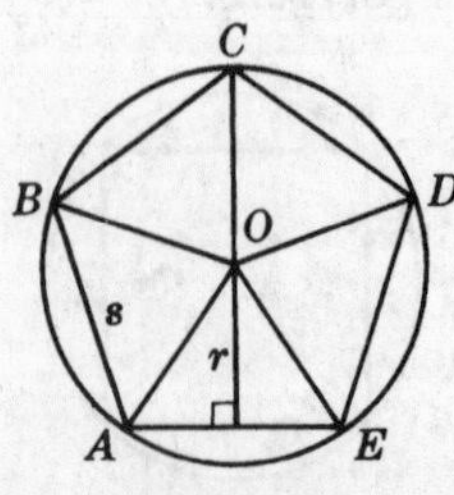

Regular Polygon
$A = \frac{1}{2}nsr = \frac{1}{2}pr$

Fig. 10-5

SOLVED PROBLEMS

10.6 Finding the area of a regular polygon

(a) Find the area of a regular hexagon if the length of the apothem is $5\sqrt{3}$.

(b) Find the area of a regular pentagon to the nearest integer if the length of the apothem is 20.

Solutions

(a) In a regular hexagon, $r = \frac{1}{2}s\sqrt{3}$. Since $r = 5\sqrt{3}$, $s = 10$ and $p = 6(10) = 60$.

Then $A = \frac{1}{2}pr = \frac{1}{2}(60)(5\sqrt{3}) = 150\sqrt{3}$.

(b) In Fig. 10-6, $m\angle AOE = 360°/5 = 72°$ and $m\angle AOF = \frac{1}{2}m\angle AOE = 36°$. Then $\tan 36° = \frac{1}{2}s/20 = s/40$ or $s = 40 \tan 36°$.

Now $A = \frac{1}{2}pr = \frac{1}{2}nsr = \frac{1}{2}(5)(40 \tan 36°)(20) = 1453$.

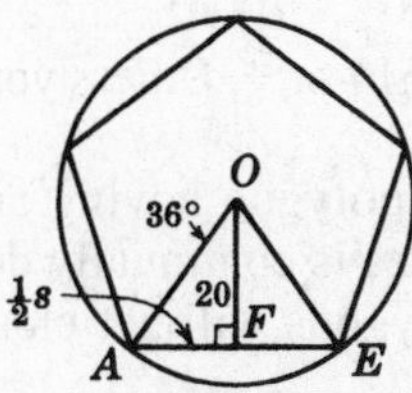

Fig. 10-6

10.4 Ratios of Segments and Areas of Regular Polygons

PRINCIPLE 1: *Regular polygons having the same number of sides are similar.*

PRINCIPLE 2: *Corresponding segments of regular polygons having the same number of sides are in proportion.* "Segments" here includes sides, perimeters, radii or circumferences of circumscribed or inscribed circles, and such.

PRINCIPLE 3: *Areas of regular polygons having the same number of sides are to each other as the squares of the lengths of any two corresponding segments.*

SOLVED PROBLEMS

10.7 Ratios of lines and areas of regular polygons

(a) In two regular polygons having the same number of sides, find the ratio of the lengths of the apothems if the perimeters are in the ratio 5:3.

(b) In two regular polygons having the same number of sides, find the length of a side of the smaller if the lengths of the apothems are 20 and 50 and a side of the larger has length 32.5.

(c) In two regular polygons having the same number of sides, find the ratio of the areas if the lengths of the sides are in the ratio 1:5.

(d) In two regular polygons having the same number of sides, find the area of the smaller if the sides have lengths 4 and 12 and the area of the larger is 10,260.

Solutions

(a) By Principle 2, $r : r' = p : p' = 5:3$.

(b) By Principle 2, $s : s' = r : r'$; thus, $s : 32.5 = 20:50$ and $s = 13$.

(c) By Principle 3, $\frac{A}{A'} = \left(\frac{s}{s'}\right)^2 = \left(\frac{1}{5}\right)^2 = \frac{1}{25}$.

(d) By Principle 3, $\frac{A}{A'} = \left(\frac{s}{s'}\right)^2$. Then $\frac{A}{10{,}260} = \left(\frac{4}{12}\right)^2$ and $A = 1140$.

10.5 Circumference and Area of a Circle

π (pi) is the ratio of the circumference C of any circle to its diameter d; that is, $\pi = C/d$. Hence,

$$C = \pi d \quad \text{or} \quad C = 2\pi r$$

Approximate values for π are 3.1416 or 3.14 or $\frac{22}{7}$. Unless you are told otherwise, we shall use 3.14 for π in solving problems.

A circle may be regarded as a regular polygon having an infinite number of sides. If a square is inscribed in a circle, and the number of sides is continually doubled (to form an octagon, a 16-gon, and so on), the perimeters of the resulting polygons will get closer and closer to the circumference of the circle (Fig. 10-7).

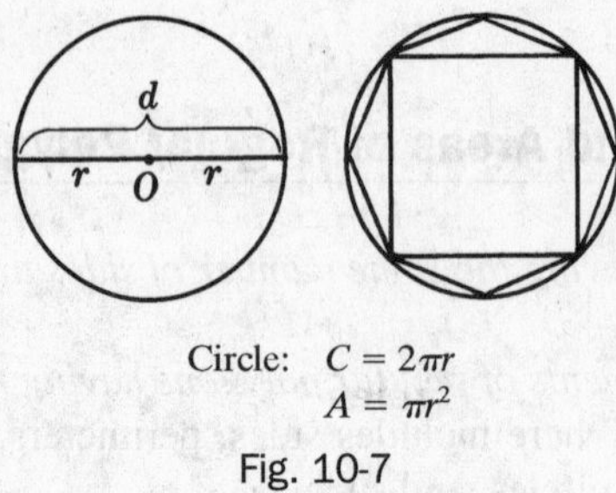

Circle: $C = 2\pi r$
$A = \pi r^2$

Fig. 10-7

Thus to find the area of a circle, the formula $A = \frac{1}{2}pr$ can be used with C substituted for p; doing so, we get

$$A = \tfrac{1}{2}Cr = \tfrac{1}{2}(2\pi r)(r) = \pi r^2$$

All circles are similar figures, since they have the same shape. Because they are similar figures, (1) corresponding segments of circles are in proportion and (2) the areas of two circles are to each other as the squares of their radii or circumferences.

SOLVED PROBLEMS

10.8 Finding the circumference and area of a circle

In a circle, (a) find the circumference and area if the radius is 6; (b) find the radius and area if the circumference is 18π; (c) find the radius and circumference if the area is 144π. (Answer both in terms of π and to the nearest integer.)

Solutions

(a) $r = 6$. Then $C = 2\pi r = 12\pi = 38$ and $A = \pi r^2 = 36\pi = 36(3.14) = 113$.

(b) $C = 18\pi$. Since $C = 2\pi r$, we have $18\pi = 2\pi r$ and $r = 9$. Then $A = \pi r^2 = 81\pi = 254$.

(c) $A = 144\pi$. Since $A = \pi r^2$, we have $144\pi = \pi r^2$ and $r = 12$. Then $C = 2\pi r = 24\pi = 75$.

10.9 Circumference and area of circumscribed and inscribed circles

Find the circumference and area of the circumscribed circle and inscribed circle (a) of a regular hexagon whose side has length 8; (b) of an equilateral triangle whose altitude has length $9\sqrt{3}$. (See Fig. 10-8.)

Solutions

(a) Here $R = s = 8$. Then $C = 2\pi R = 16\pi$ and $A = \pi R^2 = 64\pi$.
Also $r = \frac{1}{2}R\sqrt{3} = 4\sqrt{3}$. Then $C = 2\pi r = 8\pi\sqrt{3}$ and $A = \pi r^2 = 48\pi$.

(b) Here $R = \frac{2}{3}h = 6\sqrt{3}$. Then $C = 2\pi R = 12\pi\sqrt{3}$ and $A = \pi R^2 = 108\pi$.
Also $r = \frac{1}{3}h = 3\sqrt{3}$. Then $C = 2\pi r = 6\pi\sqrt{3}$ and $A = \pi r^2 = 27\pi$.

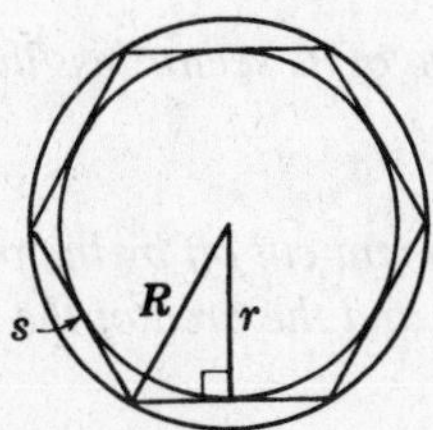

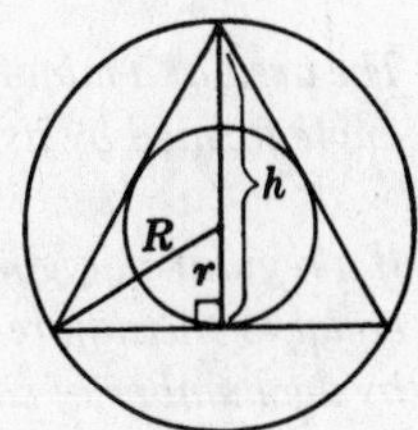

Fig. 10-8

10.10 Ratios of segments and areas in circles

(a) If the circumferences of two circles are in the ratio 2:3, find the ratio of the diameters and the ratio of the areas.

(b) If the areas of two circles are in the ratio 1:25, find the ratio of the diameters and the ratio of the circumferences.

Solutions

(a) $\frac{d}{d'} = \frac{C}{C'} = \frac{2}{3}$ and $\frac{A}{A'} = \left(\frac{C}{C'}\right)^2 = \left(\frac{2}{3}\right)^2 = \frac{4}{9}$.

(b) Since $\frac{A}{A'} = \left(\frac{d}{d'}\right)^2$, $\frac{1}{25} = \left(\frac{d}{d'}\right)^2$ and $\frac{d}{d'} = \frac{1}{5}$. Also, $\frac{C}{C'} = \frac{d}{d'} = \frac{1}{5}$.

10.6 Length of an Arc; Area of a Sector and a Segment

A *sector* of a circle is a part of a circle bounded by two radii and their intercepted arc. Thus in Fig. 10-9, the shaded section of circle O is sector OAB.

A *segment of a circle* is a part of a circle bounded by a chord and its arc. A *minor segment* of a circle is the smaller of the two segments thus formed. Thus in Fig. 10-10, the shaded section of circle Q is minor segment ACB.

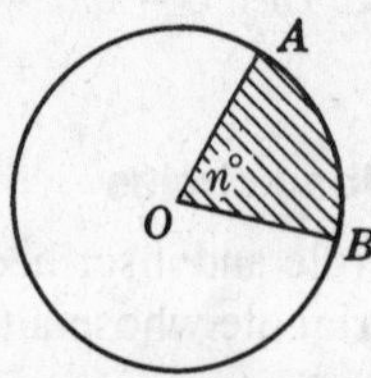

Fig. 10-9

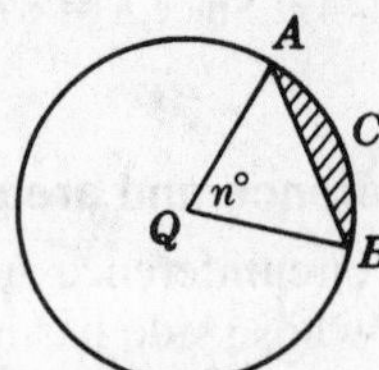

Fig. 10-10

PRINCIPLE 1: *In a circle of radius r, the length l of an arc of measure n° equals* $\frac{n}{360}$ *of the circumference of the circle, or* $l = \frac{n}{360} 2\pi r = \frac{\pi n r}{180}$.

PRINCIPLE 2: *In a circle of radius r, the area K of a sector of measure n° equals* $\frac{n}{360}$ *of the area of the circle, or* $K = \frac{n}{360} \pi r^2$.

PRINCIPLE 3: $\frac{\text{Area of a sector of } n°}{\text{Area of the circle}} = \frac{\text{length of an arc of measure } n°}{\text{circumference of the circle}} = \frac{n}{360}$

PRINCIPLE 4: *The area of a minor segment of a circle equals the area of its sector less the area of the triangle formed by its radii and chord.*

PRINCIPLE 5: *If a regular polygon is inscribed in a circle, each segment cut off by the polygon has area equal to the difference between the area of the circle and the area of the polygon divided by the number of sides.*

SOLVED PROBLEMS

10.11 Length of an arc

(a) Find the length of a 36° arc in a circle whose circumference is 45π.

(b) Find the radius of a circle if a 40° arc has a length of 4π.

Solutions

(a) Here $n° = 36°$ and $C = 2\pi r = 45\pi$. Then $l = \frac{n}{360} 2\pi r = \frac{36}{360} 45\pi = \frac{9}{2}\pi$.

(b) Here $l = 4\pi$ and $n° = 40°$. Then $l = \frac{n}{360} 2\pi r$ yields $4\pi = \frac{40}{360} 2\pi r$, and $r = 18$.

10.12 Area of a sector

(a) Find the area K of a 300° sector of a circle whose radius is 12.

(b) Find the measure of the central angle of a sector whose area is 6π if the area of the circle is 9π.

(c) Find the radius of a circle if an arc of length 2π has a sector of area 10π.

Solutions

(a) $n° = 300°$ and $r = 12$. Then $K = \frac{n}{360}\pi r^2 = \frac{300}{360}144\pi = 120\pi$.

(b) $\frac{\text{Area of sector}}{\text{Area of circle}} = \frac{n}{360}$, so $\frac{6\pi}{9\pi} = \frac{n}{360}$, and $n = 240$. Thus, the central angle measures 240°.

(c) $\frac{\text{Length of arc}}{\text{Circumference}} = \frac{\text{area of sector}}{\text{area of circle}}$, so $\frac{2\pi}{2\pi r} = \frac{10\pi}{\pi r^2}$ and $r = 10$.

10.13 Area of a segment of a circle

(a) Find the area of a segment if its central angle measures 60° and the radius of the circle is 12.

(b) Find the area of a segment if its central angle measures 90° and the radius of the circle is 8.

(c) Find each segment formed by an inscribed equilateral triangle if the radius of the circle is 8.

Solutions

See Fig. 10-11.

(a) $n° = 60°$ and $r = 12$. Then area of sector $OAB = \frac{n}{360}\pi r^2 = \frac{60}{360}144\pi = 24\pi$.

Also, area of equilateral $\triangle OAB = \frac{1}{4}s^2\sqrt{3} = \frac{1}{4}(144)\sqrt{3} = 36\sqrt{3}$.

Hence, area of segment $ACB = 24\pi - 36\sqrt{3}$.

(b) $n° = 90°$ and $r = 8$. Then area of sector $OAB = \frac{n}{360}\pi r^2 = \frac{90}{360}64\pi = 16\pi$.

Also, area of rt. $\triangle OAB = \frac{1}{2}bh = \frac{1}{2}(8)(8) = 32$.

Hence, area of segment $ACB = 16\pi - 32$.

(c) $R = 8$. Since $s = R\sqrt{3} = 8\sqrt{3}$, the area of $\triangle ABC$ is $\frac{1}{4}s^2\sqrt{3} = 48\sqrt{3}$.

Also, area of circle $O = \pi R^2 = 64\pi$.

Hence, area of segment $BDC = \frac{1}{3}(64\pi - 48\sqrt{3})$.

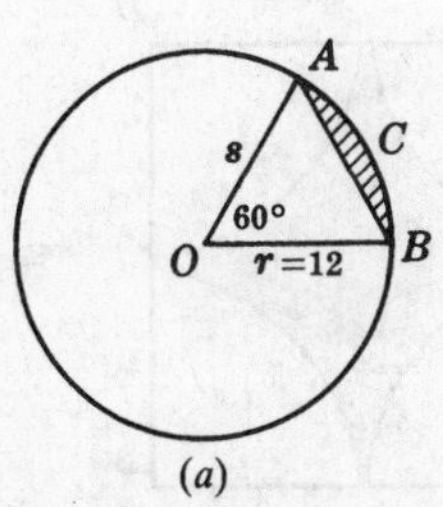

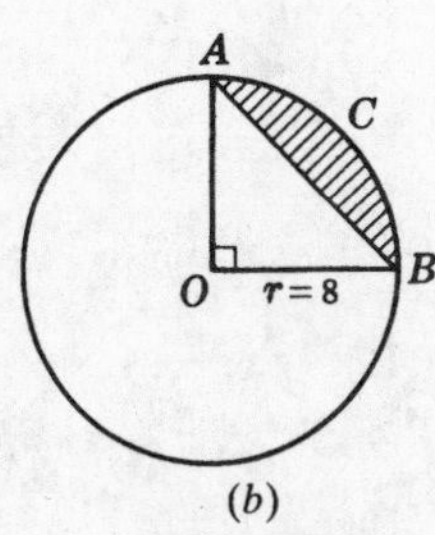

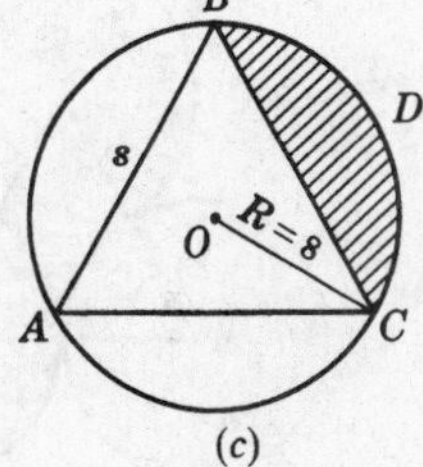

Fig. 10-11

10.14 Area of a segment formed by an inscribed regular polygon

Find the area of each segment formed by an inscribed regular polygon of 12 sides (dodecagon) if the radius of the circle is 12. (See Fig. 10-12.)

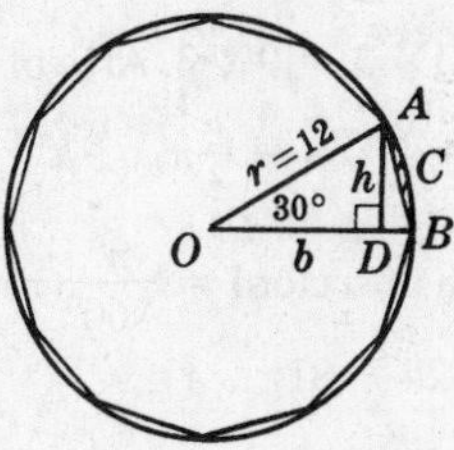

Fig. 10-12

Solution

Area of sector $OAB = \frac{n}{360}\pi r^2 = \frac{30}{360}144\pi = 12\pi$.

To find the area $\triangle OAB$, we draw altitude $\overline{AD}$ to base $\overline{OB}$. Since $m\angle AOB = 30°$, $h = AD = \frac{1}{2}r = 6$.

Then the area of $\triangle OAB$ is $\frac{1}{2}bh = \frac{1}{2}(12)(6) = 36$.

Hence, the area of segment ACB is $12\pi - 36$.

10.7 Areas of Combination Figures

The areas of combination figures like that in Fig. 10-13 may be found by determining individual areas and then adding or substracting as required. Thus, the shaded area in the figure equals the sum of the aeas of the square and the semicircle: $A = 8^2 + \frac{1}{2}(16\pi) = 64 + 8\pi$.

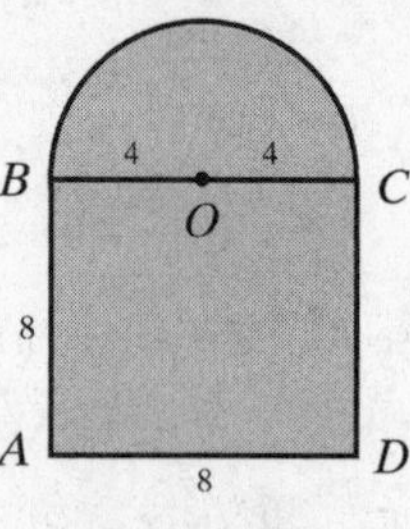

Fig. 10-13

SOLVED PROBLEMS

10.15 Finding areas of combination figures

Find the shaded area in each part of Fig. 10-14. In (a), circles A, B, and C are tangent externally and each has radius 3. In (b), each arc is part of a circle of radius 9.

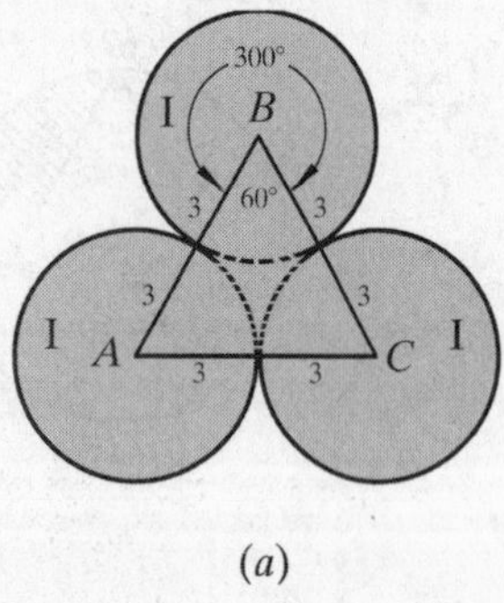

(*a*)

(*b*)

Fig. 10-14

Solutions

(a) Area of $\triangle ABC = \frac{1}{4}s^2\sqrt{3} = \frac{1}{4}(6^2)\sqrt{3} = 9\sqrt{3}$. Area of sector I $= \frac{n°}{360°}(\pi r^2) = \frac{300}{360}(9\pi) = \frac{15}{2}\pi$.

Shaded area $= 9\sqrt{3} + 3\left(\frac{15}{2}\pi\right) = 9\sqrt{3} + \frac{45}{2}\pi$.

(b) Area of square $= 18^2 = 324$. Area of sector I $= \frac{n°}{360°}(\pi r^2) = \frac{90}{360}(81\pi) = \frac{81}{4}\pi$.1

Shaded area $= 324 - 4\left(\frac{81}{4}\pi\right) = 324 - 81\pi$.

SUPPLEMENTARY PROBLEMS

10.1. In a regular polygon, find (10.1)

(a) The perimeter if the length of a side is 8 and the number of sides is 25

(b) The perimeter if the length of a side is 2.45 and the number of sides is 10

(c) The perimeter if the length of a side is $4\frac{2}{3}$ and the number of sides is 24

(d) The number of sides if the perimeter is 325 and the length of a side is 25

(e) The number of sides if the perimeter is $27\sqrt{3}$ and the length of a side is $3\sqrt{3}$

(f) The length of a side if the number of sides is 30 and the perimeter is 100

(g) The length of a side if the perimeter is 67.5 and the number of sides is 15

10.2. In a regular polygon, find (10.1)

(a) The length of the apothem if the diameter of an inscribed circle is 25

(b) The length of the apothem if the radius of the inscribed circle is 23.47

(c) The radius of the inscribed circle if the length of the apothem is $7\sqrt{3}$

(d) The radius of the regular polygon if the diameter of the circumscribed circle is 37

(e) The radius of the circumscribed circle if the radius of the regular polygon is $3\sqrt{2}$

10.3. In a regular polygon of 15 sides, find the measure of (a) the central angle; (b) the exterior angle; (c) the interior angle. (10.1)

10.4. If an exterior angle of a regular polygon measures 40°, find (a) the measure of the central angle; (b) the number of sides; (c) the measure of the interior angle. (10.1)

10.5. If an interior angle of a regular polygon measures 165°, find (a) the measure of the exterior angle; (b) the measure of the central angle; (c) the number of sides. (10.1)

10.6. If a central angle of a regular polygon measures 5°, find (a) the measure of the exterior angle; (b) the number of sides; (c) the measure of the interior angle. (10.1)

10.7. Name the regular polygon whose (10.1)

(a) Central angle measures 45°

(b) Central angle measures 60°

(c) Exterior angle measures 120°

(d) Exterior angle measures 36°

(e) Interior angle is congruent to its central angle

(f) Interior angle measures 150°

10.8. Prove each of the following: (10.2)

(a) The diagonals of a regular pentagon are congruent.

(b) A diagonal of a regular pentagon forms an isosceles trapezoid with three of its sides.

(c) If two diagonals of a regular pentagon intersect, the longer segment of each diagonal is congruent to a side of the regular pentagon.

10.9. In a regular hexagon, find (10.3)

(a) The length of a side if its radius is 9

(b) The perimeter if its radius is 5

(c) The length of the apothem if its radius is 12

(d) Its radius if the length of a side is 6

(e) The length of the apothem if the length of a side is 26

(f) Its radius if the length of the apothem is $3\sqrt{3}$

(g) The length of a side if the length of the apothem is 30

(h) The perimeter if the length of the apothem is $5\sqrt{3}$

10.10. In a square, find (10.4)

(a) The length of a side if the radius is 18

(b) The length of the apothem if the radius is 14

(c) The perimeter if the radius is $5\sqrt{2}$

(d) The radius if the length of a side is 16

(e) The length of a side if the length of the apothem is 1.7

(f) The perimeter if the length of the apothem is $3\frac{1}{2}$

(g) The radius if the perimeter is 40

(h) The length of the apothem if the perimeter is $16\sqrt{2}$

10.11. In an equilateral triangle, find (10.5)

(a) The length of a side if its radius is 30

(b) The length of the apothem if its radius is 28

(c) The length of an altitude if its radius is 18

(d) The perimeter if its radius is $2\sqrt{3}$

(e) Its radius if the length of a side is 24

(f) The length of the apothem if the length of a side is 24

(g) The length of its altitude if the length of a side is 96

(h) Its radius if the length of the apothem is 21

(i) The length of a side if the length of the apothem is $\sqrt{3}$

(j) The length of the altitude if the length of the apothem is $3\frac{1}{3}$

(k) The length of the altitude if the perimeter is 15

(l) The length of the apothem if the perimeter is 54

10.12. (a) Find the area of a regular pentagon to the nearest integer if the length of the apothem is 15. (10.6)

(b) Find the area of a regular decagon to the nearest integer if the length of a side is 20.

10.13. Find the area of a regular hexagon, in radical form, if (a) the length of a side is 6; (b) its radius is 8; (c) the length of the apothem is $10\sqrt{3}$. (10.6)

10.14. Find the area of a square if (a) the length of the apothem is 12; (b) its radius is $9\sqrt{2}$; (c) its perimeter is 40. (10.6)

10.15. Find the area of an equilateral triangle, in radical form, if (10.6)

(a) The length of the apothem is $2\sqrt{3}$.
(b) Its radius is 6.
(c) The length of the altitude is 4.
(d) The length of the altitude is $12\sqrt{3}$.
(e) The perimeter is $6\sqrt{3}$.
(f) The length of the apothem is 4.

10.16. If the area of a regular hexagon is $150\sqrt{3}$, find (a) the length of a side; (b) its radius; (c) the length of the apothem. (10.6)

10.17. If the area of an equilateral triangle is $81\sqrt{3}$, find (a) the length of a side; (b) the length of the altitude; (c) its radius; (d) the length of the apothem. (10.6)

10.18. Find the ratio of the perimeters of two regular polygons having the same number of sides if (10.7)

(a) The ratio of the sides is 1:8.
(b) The ratio of their radii is 4:9.
(c) Their radii are 18 and 20.
(d) Their apothems have lengths 16 and 22.
(e) The length of the larger side is triple that of the smaller.
(f) The length of the smaller apothem is two-fifths that of the larger.
(g) The lengths of the apothems are $20\sqrt{2}$ and 15.
(h) The circumference of the larger circumscribed circle is $2\frac{1}{2}$ times that of the smaller.

10.19. Find the ratio of the perimeters of two equilateral triangles if (a) the sides have lengths 20 and 8; (b) their radii are 12 and 60; (c) their apothems have lengths $2\sqrt{3}$ and $6\sqrt{3}$; (d) the circumferences of their inscribed circles are 120 and 160; (e) their altitudes have lengths $5x$ and x. (10.7)

10.20. Find the ratio of the lengths of the sides of two regular polygons having the same number of sides if the ratio of their areas is (a) 25:1; (b) 16:49; (c) x^2:4; (d) 2:1; (e) 3:y^2; (f) x:18. (10.7)

10.21. Find the ratio of the areas of two regular hexagons if (a) their sides have lengths 14 and 28; (b) their apothems have lengths 3 and 15; (c) their radii are $6\sqrt{3}$ and $\sqrt{3}$; (b) their perimeters are 75 and 250; (e) the circumferences of the circumscribed circles are 28 and 20. (10.7)

10.22. Find the circumference of a circle in terms of π if (a) the radius is 6; (b) the diameter is 14; (c) the area is 25π; (d) the area is 3π. (10.8)

10.23. Find the area of a circle in terms of π if (a) the radius is 3; (b) the diameter is 10; (c) the circumference is 16π; (d) the circumference is π; (e) the circumference is $6\pi\sqrt{2}$. (10.8)

10.24. In a circle, (a) find the circumference and area if the radius is 5; (b) find the radius and area if the circumference is 16π; (c) find the radius and circumference if the area is 16π. (10.8)

10.25. In a regular hexagon, find the circumference of the circumscribed circle if (a) the length of the apothem is $3\sqrt{3}$;(b) the perimeter is 12; (c) the length of a side is $3\frac{1}{2}$. Also find the circumference of its inscribed circle if (d) the length of the apothem is 13; (e) the length of a side is 8; (f) the perimeter is $6\sqrt{3}$. (10.9)

10.26. For a square, find the area in terms of π of the (10.9)

(a) Circumscribed circle if the length of the apothem is 7

(b) Circumscribed circle if the perimeter is 24

(c) Circumscribed circle if the length of a side is 8

(d) Inscribed circle if the length of the apothem is 5

(e) Inscribed circle if the length of a side is $12\sqrt{2}$

(f) Inscribed circle if the perimeter is 80

10.27. Find the circumference and area of the (1) circumscribed circle and (2) inscribed circle of (10.9)

(a) A regular hexagon if the length of a side is 4

(b) A regular hexagon if the length of the apothem is $4\sqrt{3}$

(c) An equilateral triangle if the length of the altitude is 9

(d) An equilateral triangle if the length of the apothem is 4

(e) A square if the length of a side is 20

(f) A square if the length of the apothem is 3

10.28. Find the radius of a pipe having the same capacity as two pipes whose radii are (a) 6 ft and 8 ft; (b) 8 ft and 15 ft; (c) 3 ft and 6 ft. (*Hint*: Find the areas of their circular cross-sections.) (10.10)

10.29. In a circle, find the length of a 90° arc if (10.11)

(a) The radius is 4.

(b) The diameter is 40.

(c) The circumference is 32.

(d) The circumference is 44π.

(e) An inscribed hexagon has a side of length 12.

(f) An inscribed equilateral triangle has an altitude of length 30.

10.30. Find the length of (10.11)

(a) A 90° arc if the radius of the circle is 6

(b) A 180° arc if the circumference is 25

(c) A 30° arc if the circumference is 60π

(d) A 40° arc if the diameter is 18

(e) An arc intercepted by the side of a regular hexagon inscribed in a circle of radius 3

(f) An arc intercepted by a chord of length 12 in a circle of radius 12.

10.31. In a circle, find the area of a 60° sector if (10.12)

(a) The radius is 6.

(b) The diameter is 2.

(c) The circumference is 10π.

(d) The area of the circle is 150π.

(e) The area of the circle is 27.

(f) The area of a 240° sector is 52.

(g) An inscribed hexagon has a side of length 12.

(h) An inscribed hexagon has an area of $24\sqrt{3}$.

10.32. Find the area of a (10.12)

(a) 60° sector if the radius of the circle is 6
(b) 240° sector if the area of the circle is 30
(c) 15° sector if the area of the circle is 72π
(d) 90° sector if its arc length is 4π

10.33. Find the measure of a central angle of an arc whose length is (10.12)

(a) 3 m if the circumference is 9 m
(b) 2 ft if the circumference is 1 yd
(c) 25 if the circumference is 250
(d) 6π if the circumference is 12π
(e) Three-eighths of the circumference
(f) Equal to the radius

10.34. Find the measure of a central angle of a sector whose area is (10.12)

(a) 10 if the area of the circle is 50
(b) 15 cm^2 if the area of the circle is 20 cm^2
(c) 1 ft^2 if the area of the circle is 1 yd^2
(d) 5π if the area of the circle is 12π
(e) Eight-ninths of the area of the circle

10.35. Find the measure of a central angle of (10.11 and 10.12)

(a) An arc whose length is 5π if the area of its sector is 25π
(b) An arc whose length is 12π if the area of its sector is 48π
(c) A sector whose area is 2π if the length of its arc is π
(d) A sector whose area is 10π if the length of its arc is 2π

10.36. Find the radius of a circle if a (10.11 and 10.12)

(a) 120° arc has a length of 8π
(b) 40° arc has a length of 2π
(c) 270° arc has a length of 15π
(d) 30° sector has an area of 3π
(e) 36° sector has an area of $2\frac{1}{2}\pi$
(f) 120° sector has an area of 6π

10.37. Find the radius of a circle if a sector of area (10.12)

(a) 12π has an arc of length 6π
(b) 10π has an arc of length 2π
(c) 25 cm^2 has an arc of length 5 cm
(d) 162 has an arc of length 36

10.38. Find the area of a segment if its central angle is 60° and the radius of the circle is (a) 6; (b) 12; (c) 3; (d) r; (e) $2r$. (10.13)

10.39. Find the area of a segment of a circle if (10.13)

(a) The radius of the circle is 4 and the central angle measures 90°.
(b) The radius of the circle is 30 and the central angle measures 60°.
(c) The radius of the circle and the chord of the segment each have length 12.
(d) The central angle is 90° and the length of the arc is 4π.
(e) Its chord of length 20 units is 10 units from the center of the circle.

10.40. Find the area of a segment of a circle if the radius of the circle is 8 and the central angle measures (a) 120°; (b) 135°; (c) 150°. (10.13)

10.41. If the radius of a circle is 4, find the area of each segment formed by an inscribed (a) equilateral triangle; (b) regular hexagon; (c) square. (10.13 and 10.14)

10.42. Find the area of each segment of a circle if the segments are formed by an inscribed (10.13)

(a) Equilateral triangle and the radius of the circle is 6

(b) Regular hexagon and the radius of the circle is 3

(c) Square and the radius of the circle is 6

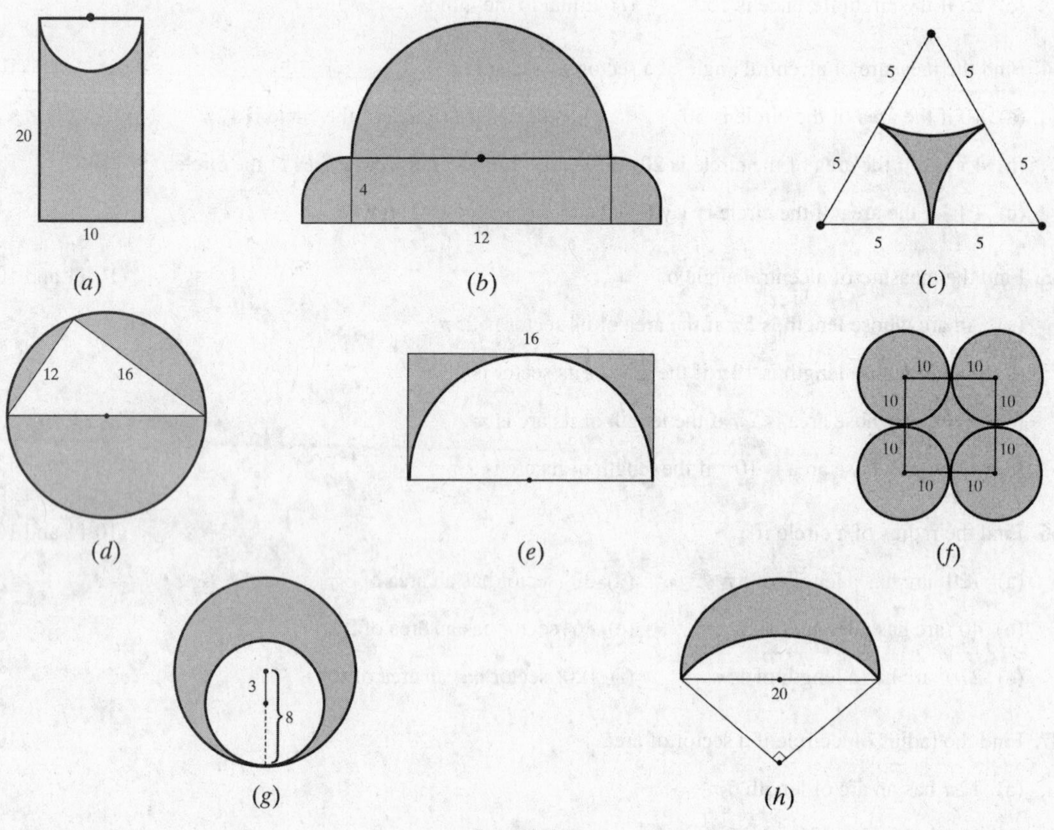

Fig. 10-15

10.43. Find the shaded area in each part of Fig. 10-15. Each heavy dot represents the center of an arc or a circle. (10.15)

10.44. Find the shaded area in each part of Fig. 10-16. Each dot represents the center of an arc or circle. (10.15)

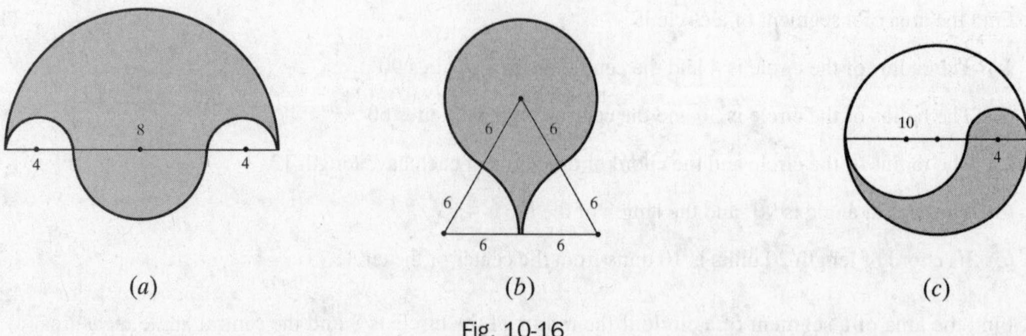

Fig. 10-16

Locus

11.1 Determining a Locus

Locus, in Latin, means *location*. The plural is *loci*. A *locus of points* is the set of points, and only those points, that satisfy given conditions.

Thus, the locus of points that are 1 in from a given point P is the set of points 1 in from P. These points lie on a circle with its center at P and a radius of 1 in, and hence this circle is the required locus (Fig. 11-1). Note that we show loci as long-short dashed figures.

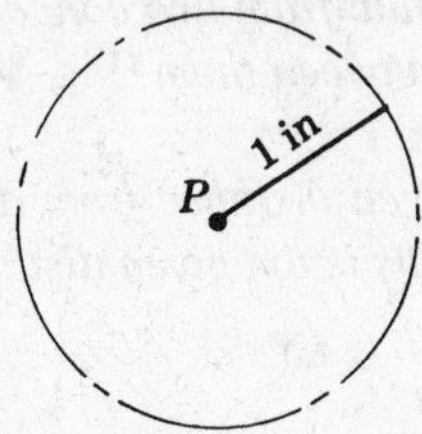

Fig. 11-1

To determine a locus, (1) state what is given and the condition to be satisfied; (2) find several points satisfying the condition which indicate the shape of the locus; then (3) connect the points and describe the locus fully.

All geometric constructions require the use of straightedges and compasses. Hence if a locus is to be *constructed*, such drawing instruments can be used.

11.1A Fundamental Locus Theorems

PRINCIPLE 1: *The locus of points equidistant from two given points is the perpendicular bisector of the line segment joining the two points* (Fig. 11-2).

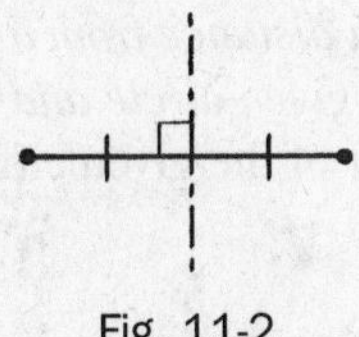

Fig. 11-2

PRINCIPLE 2: *The locus of points equidistant from two given parallel lines is a line parallel to the two lines and midway between them* (Fig. 11-3).

PRINCIPLE 3: *The locus of points equidistant from the sides of a given angle is the bisector of the angle* (Fig. 11-4).

Fig. 11-3 Fig. 11-4

PRINCIPLE 4: *The locus of points equidistant from two given intersecting lines is the bisectors of the angles formed by the lines* (Fig. 11-5).

Fig. 11-5 Fig. 11-6

PRINCIPLE 5: *The locus of points equidistant from two concentric circles is the circle concentric with the given circles and midway between them* (Fig. 11-6).

PRINCIPLE 6: *The locus of points at a given distance from a given point is a circle whose center is the given point and whose radius is the given distance* (Fig. 11-7).

Fig. 11-7 Fig. 11-8

PRINCIPLE 7: *The locus of points at a given distance from a given line is a pair of lines, parallel to the given line and at the given distance from the given line* (Fig. 11-8).

PRINCIPLE 8: *The locus of points at a given distance from a given circle whose radius is greater than that distance is a pair of concentric circles, one on either side of the given circle and at the given distance from it* (Fig. 11-9).

PRINCIPLE 9: *The locus of points at a given distance from a given circle whose radius is less than the distance is a circle, outside the given circle and concentric with it* (Fig. 11-10). (If $r = d$, the locus also includes the center of the given circle.)

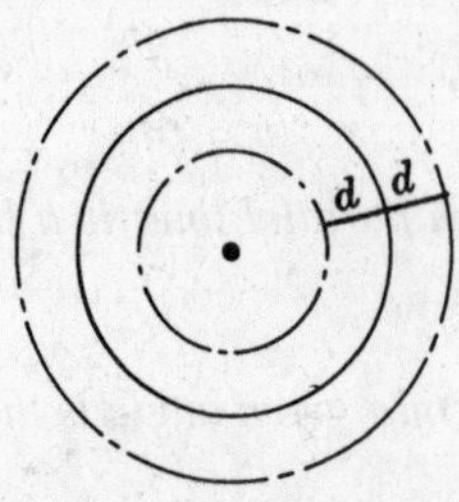

Fig. 11-9

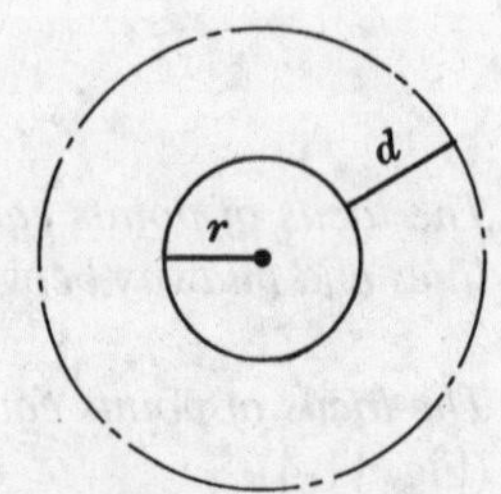

Fig. 11-10

SOLVED PROBLEMS

11.1 Determining loci

Determine the locus of (a) a runner moving equidistant from the sides of a straight track; (b) a plane flying equidistant from two separated aircraft batteries; (c) a satellite 100 mi above the earth; (d) the furthermost point reached by a gun with a range of 10 mi.

Solutions

See Fig. 11-11.

(a) The locus is a line parallel to the two given lines and midway between them.

(b) The locus is the perpendicular bisector of the line joining the two points.

(c) The locus is a circle concentric with the earth and of radius 100 mi greater than that of the earth.

(d) The locus is a circle of radius 10 mi with its center at the gun.

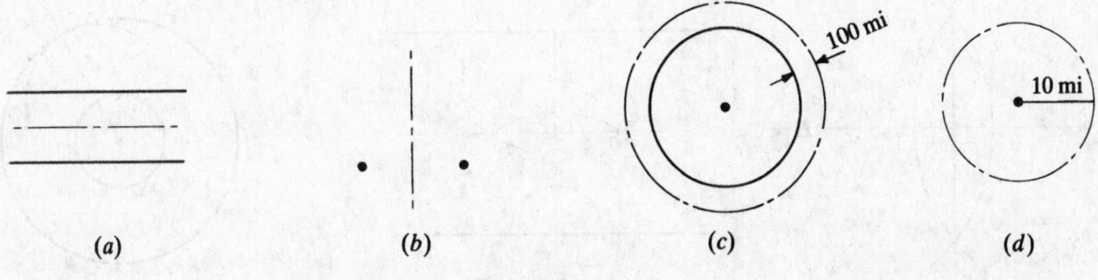

Fig. 11-11

11.2 Determining the locus of the center of a circle

Determine the locus of the center of a circular disk (a) moving so that it touches each of two parallel lines; (b) moving tangentially to two concentric circles; (c) moving so that its rim passes through a fixed point; (d) rolling along a large fixed circular hoop.

Solutions

See Fig. 11-12.

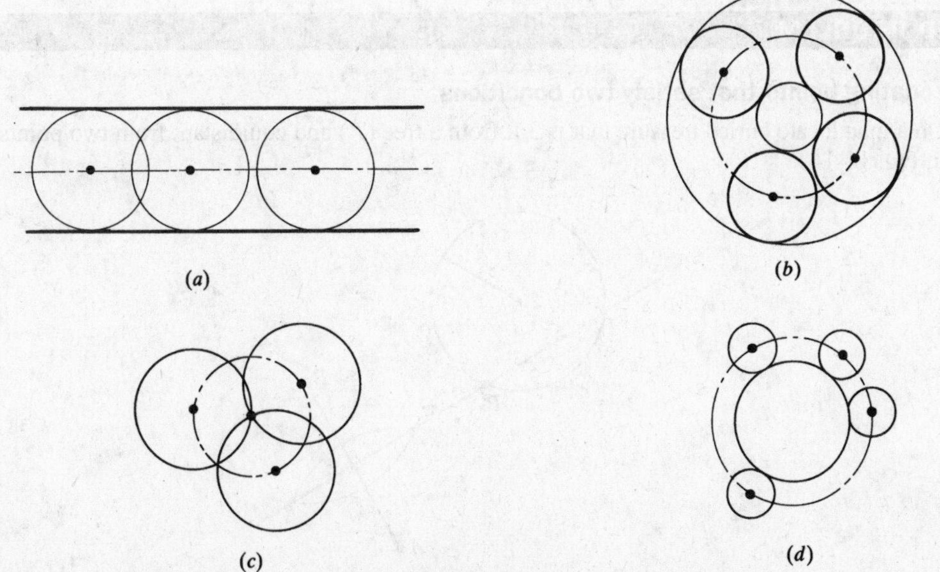

Fig. 11-12

(a) The locus is a line parallel to the two given lines and midway between them.

(b) The locus is a circle concentric with the given circles and midway between them.

(c) The locus is a circle whose center is the given point and whose radius is the radius of the circular disk.

(d) The locus is a circle outside the given circle and concentric to it.

11.3 Constructing loci

Construct (a) the locus of points equidistant from two given points; (b) the locus of points equidistant from two given parallel lines; (c) the locus of points at a given distance from a given circle whose radius is less than that distance.

Solutions

See Fig. 11-13.

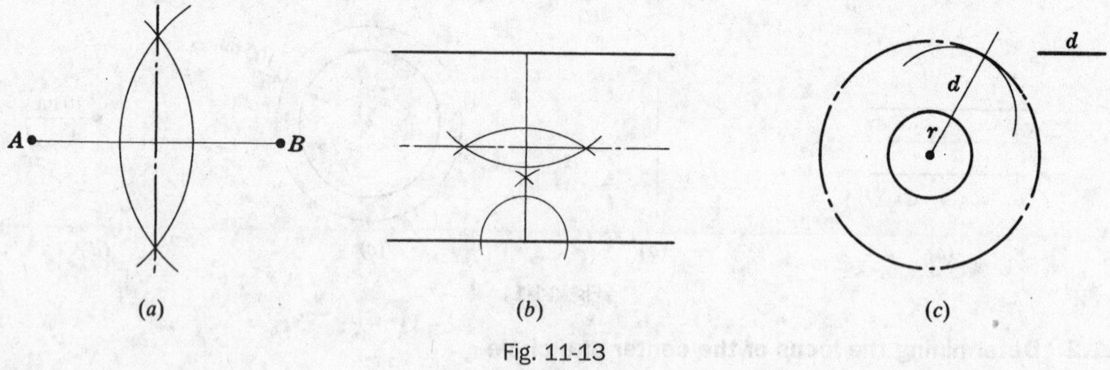

Fig. 11-13

11.2 Locating Points by Means of Intersecting Loci

A point or points which satisfy two conditions may be found by drawing the locus for each condition. The required points are the points of intersection of the two loci.

SOLVED PROBLEM

11.4 Locating points that satisfy two conditions

On a map locate buried treasure that is 3 ft from a tree (T) and equidistant from two points (A and B) in Fig. 11-14.

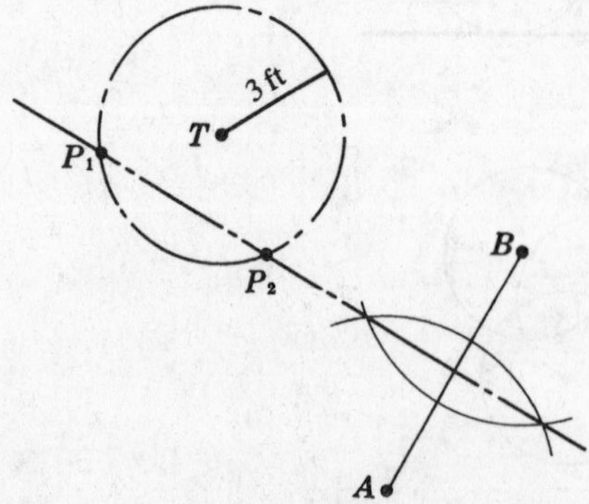

Fig. 11-14

Solution

The required loci are (1) the perpendicular bisector of $\overline{AB}$ and (2) a circle with its center at T and radius 3 ft. As shown, these meet in P_1 and P_2, which are the locations of the treasure.

Note: The diagram shows the two loci intersecting at P_1 and P_2. However, there are three possible kinds of solutions, depending on the location of T with respect to A and B:

1. The solution has two points if the loci intersect.
2. The solution has one point if the perpendicular bisector is tangent to the circle.
3. The solution has no points if the perpendicular bisector does not meet the circle.

11.3 Proving a Locus

To prove that a locus satisfies a given condition, it is necessary to prove the locus theorem *and* its converse or its inverse. Thus to prove that a circle A of radius 2 in is the locus of points 2 in from A, it is necessary to prove either that

1. Any point on circle A is 2 in from A.
2. Any point 2 in from A is on circle A (converse of statement 1).

or that

1. Any point on circle A is 2 in from A.
2. Any point not on circle A is not 2 in from A (inverse of statement 1).

These statements are easily proved using the principle that a point is outside, on, or inside a circle according as its distance from the center is greater than, equal to, or less than the radius of the circle.

SOLVED PROBLEM

11.5 Proving a locus theorem

Prove that the locus of points equidistant from two given points is the perpendicular bisector of the segment joining the two points.

Solution

First prove that any point on the locus satisfies the condition:

Given: Points A and B. $\overline{CD}$ is the $\perp$ bisector of $\overline{AB}$.

To Prove: Any point P on $\overline{CD}$ is equidistant from A and B; that is, $\overline{PA} \cong \overline{PB}$.

Plan: Prove $\triangle PEA \cong \triangle PEB$ to obtain $\overline{PA} \cong \overline{PB}$.

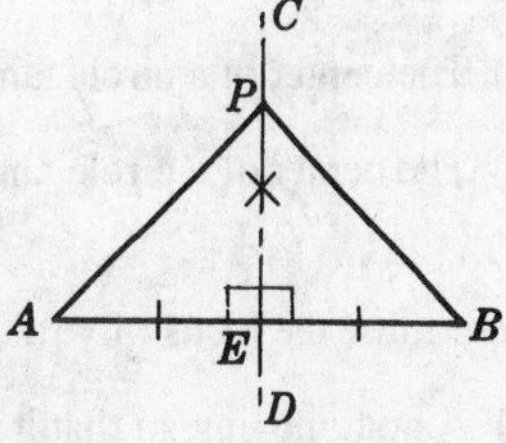

PROOF:

Statements	Reasons
1. $\overline{CD}$ is the $\perp$ bisector of $\overline{AB}$.	1. Given
2. $\angle PEA \cong \angle PEB$	2. Perpendiculars form right angles; all right angles are congruent.
3. $\overline{AE} \cong \overline{EB}$	3. To bisect is to divide into congruent parts.
4. $\overline{PE} \cong \overline{PE}$	4. Reflexive property
5. $\triangle PEA \cong \triangle PEB$	5. SAS
6. $\overline{PA} \cong \overline{PB}$	6. Corresponding parts of $\cong$ triangles are $\cong$.

Then prove that any point satisfying the condition is on the locus:

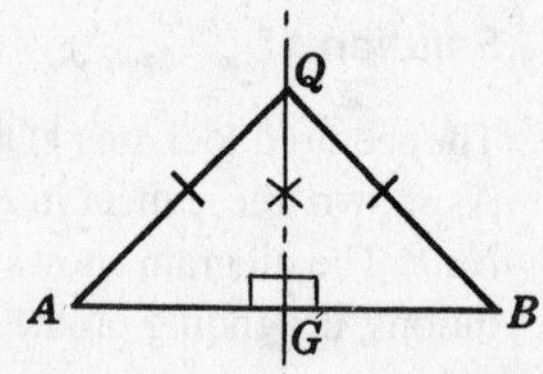

Given: Any point Q which is equidistant from points A and B ($\overline{QA} \cong \overline{QB}$).
To Prove: Q is on the perpendicular bisector of $\overline{AB}$.
Plan: Draw $\overline{QG}$ perpendicular to $\overline{AB}$ and prove by congruent triangles that $\overline{QG}$ bisects $\overline{AB}$.

PROOF:

Statements	Reasons
1. Draw $\overline{QG} \perp \overline{AB}$.	1. Through an external point, a line can be drawn perpendicular to a given line.
2. $\overline{QA} \cong \overline{QB}$	2. Given
3. $\angle QGA$ and $\angle QGB$ are rt. $\angle$s; $\triangle QGA$ and $\triangle QGB$ are rt. $\triangle$s.	3. Perpendiculars form right angles; $\triangle$s with a rt.$\angle$ are rt. $\triangle$s.
4. $\overline{QG} \cong \overline{QG}$	4. Reflexive property
5. $\triangle QGA \cong \triangle QGB$	5. Hy-leg
6. $\overline{AG} \cong \overline{GB}$	6. Corresponding parts of $\cong$ triangles are $\cong$.
7. $\overline{QG}$ bisects $\overline{AB}$.	7. To bisect is to divide into two congruent parts
8. $\overline{QG}$ is $\perp$ bisector of $\overline{AB}$.	8. A line perpendicular to a segment and bisecting it is its perpendicular bisector.

SUPPLEMENTARY PROBLEMS

11.1. Determine the locus of (11.1)

(a) The midpoints of the radii of a given circle

(b) The midpoints of chords of a given circle parallel to a given line

(c) The midpoints of chords of fixed length in a given circle

(d) The vertex of the right angle of a triangle having a given hypotenuse

(e) The vertex of an isosceles triangle having a given base

(f) The center of a circle which passes through two given points

(g) The center of a circle tangent to a given line at a given point on that line

(h) The center of a circle tangent to the sides of a given angle

11.2. Determine the locus of (11.1)

(a) A boat moving so that it is equidistant from the parallel banks of a stream

(b) A swimmer maintaining the same distance from two floats

(c) A police helicopter in pursuit of a car which has just passed the junction of two straight roads and which may be on either one of them

(d) A treasure buried at the same distance from two intersecting straight roads

11.3. Determine the locus of (a) a planet moving at a fixed distance from its sun; (b) a boat moving at a fixed distance from the coast of a circular island; (c) plants laid at a distance of 20 ft from a straight row of other plants; (d) the outer extremity of a clock hand. (11.1)

11.4. Excluding points lying outside rectangle *ABCD* in Fig. 11-15, find the locus of points which are (11.1)

(a) Equidistant from $\overline{AD}$ and $\overline{BC}$

(b) Equidistant from $\overline{AB}$ and $\overline{CD}$

(c) Equidistant from *A* and *B*

(d) Equidistant from *B* and *C*

(e) 5 units from $\overline{BC}$

(f) 10 units from $\overline{AB}$

(g) 20 units from $\overline{CD}$

(h) 10 units from *B*

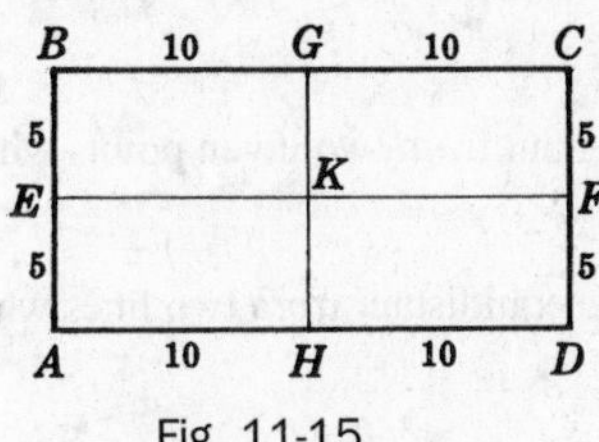

Fig. 11-15

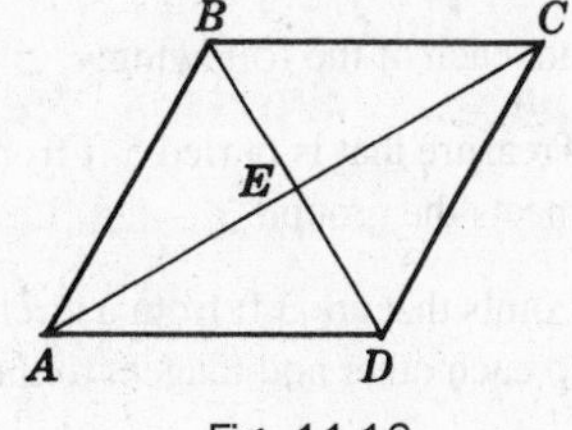

Fig. 11-16

11.5. Find the locus of points in rhombus *ABCD* in Fig. 11-16, which are equidistant from (a) $\overline{AB}$ and $\overline{AD}$; (b) $\overline{AB}$ and $\overline{BC}$; (c) *A* and *C*; (d) *B* and *D*; (e) each of the four sides. (11.1)

11.6. In Fig. 11-17, find the locus of points which are on or inside circle *C* and (11.1 and 11.2)

(a) 5 units from *O*

(b) 15 units from *O*

(c) Equidistant from circles *A* and *C*

(d) 10 units from circle *C*

(e) 10 units from circle *A*

(f) 5 units from circle *B*

(g) The center of a circle tangent to circles *A* and *C*

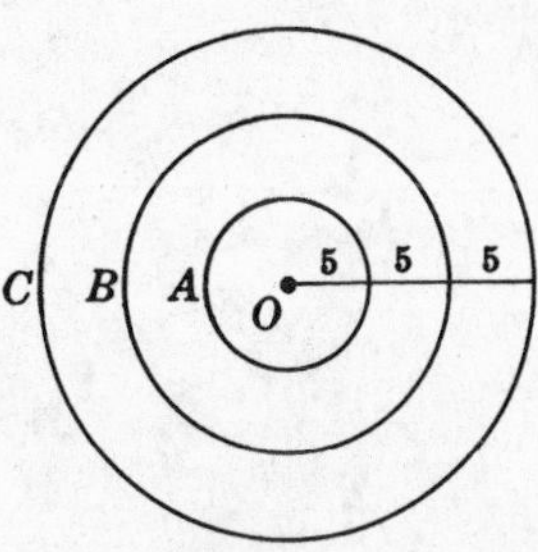

Fig. 11-17

11.7. Determine the locus of the center of (a) a coin rolling around and touching a smaller coin; (b) a coin rolling around and touching a larger coin; (c) a wheel moving between two parallel bars and touching both of them; (d) a wheel moving along a straight metal bar and touching it. (11.2)

11.8. Find the locus of points that are in rectangle *ABCD* of Fig. 11-18 and the center of a circle (11.2)

(a) Tangent to $\overline{AD}$ and $\overline{BC}$

(b) Tangent to $\overline{AB}$ and $\overline{CD}$

(c) Tangent to $\overline{AD}$ and $\overline{EF}$

(d) Of radius 10, tangent to $\overline{BC}$

(e) Of radius 20, tangent to $\overline{AD}$

(f) Tangent to $\overline{BC}$ at *G*

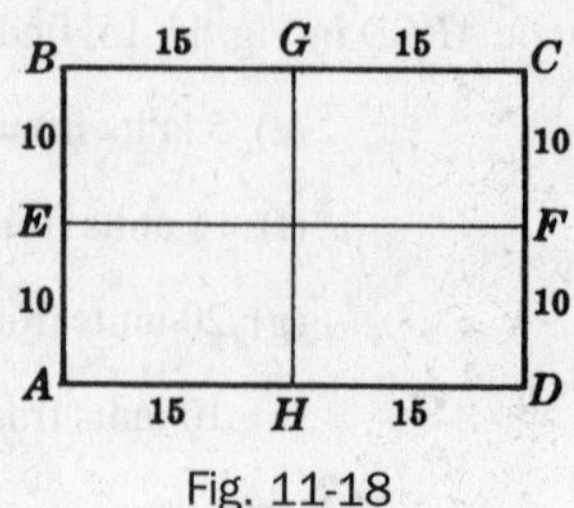

Fig. 11-18

11.9. Locate each of the following: (11.4)

(a) Treasure that is buried 5 ft from a straight fence and equidistant from two given points where the fence meets the ground

(b) Points that are 3 ft from a circle whose radius is 2 ft and are equidistant from two lines which are parallel to each other and tangent to the circle

(c) A point equidistant from the three vertices of a given triangle

(d) A point equidistant from two given points and equidistant from two given parallels

(e) Points equidistant from two given intersecting lines and 5 ft from their intersection

(f) A point that is equidistant from the sides of an angle and $\frac{1}{2}$ in from their intersection

11.10. Locate the point or points which satisfy the following conditions with respect to $\triangle ABC$ in Fig. 11-19: (11.4)

(a) Equidistant from its sides

(b) Equidistant from its vertices

(c) Equidistant from A and B and from $\overline{AB}$ and $\overline{BC}$

(d) Equidistant from $\overline{BC}$ and $\overline{AC}$ and 5 units from C

(e) 5 units from B and 10 units from A

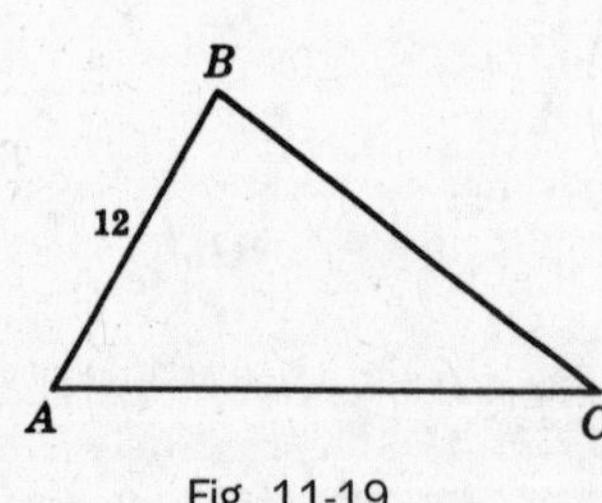

Fig. 11-19

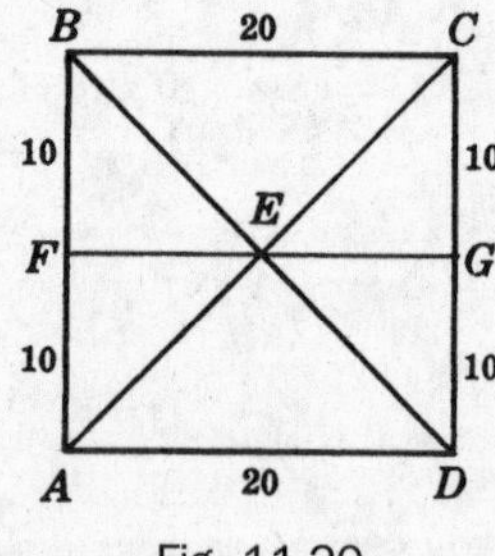

Fig. 11-20

11.11. Excluding points lying outside square $ABCD$ in Fig. 11-20, how many points are there that are (11.4)

(a) Equidistant from its vertices

(b) Equidistant from its sides

(c) 5 units from E and on one of the diagonals

(d) 5 units from E and equidistant from $\overline{AD}$ and $\overline{BC}$

(e) 5 units from $\overline{FG}$ and equidistant from $\overline{AB}$ and $\overline{CD}$

(f) 20 units from A and 10 units from B

11.12. Prove that the locus of points equidistant from the sides of an angle is the bisector of the angle. (11.5)

CHAPTER 12

Analytic Geometry

12.1 Graphs

A *number line* is a line on which distances from a point are marked off in equal units, positively in one direction and negatively in the other. The *origin* is the zero point from which distances are measured. Figure 12-1 shows a horizontal number line.

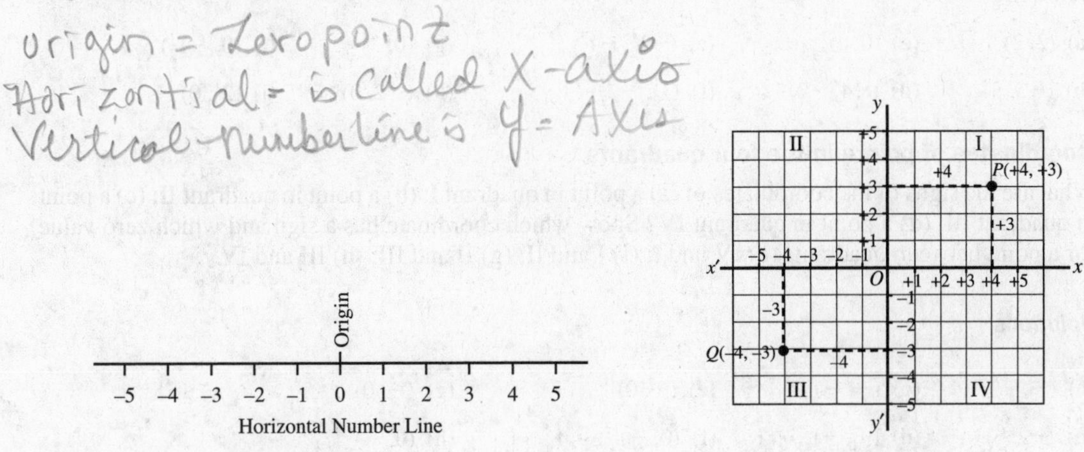

Fig. 12-1

Fig. 12-2

The *graph* shown in Fig. 12-2 is formed by combining two number lines at right angles to each other so that their zero points coincide. The horizontal number line is called the *x-axis*, and the vertical number line is the *y*-axis. The point where the two lines cross each other is, again, called the *origin*.

A *point is located on a graph by its coordinates,* which are its distances from the axes. The *abscissa* or *x-coordinate* of a point is its distance from the *y*-axis. The *ordinate* or *y-coordinate* of a point is its distance from the *x*-axis.

When the coordinates of a point are stated, the *x*-coordinate precedes the *y*-coordinate. Thus, the coordinates of point *P* in Fig. 12-2 are written (4, 3); those for *Q* are (−4, −3). Note the parentheses.

The *quadrants* of a graph are the four parts cut off by the axes. These are numbered I, II, III, and IV in a counterclockwise direction, as shown in Fig. 12-2.

SOLVED PROBLEMS

12.1 Locating points on a graph

Give the coordinates of the following points in Fig. 12-3:

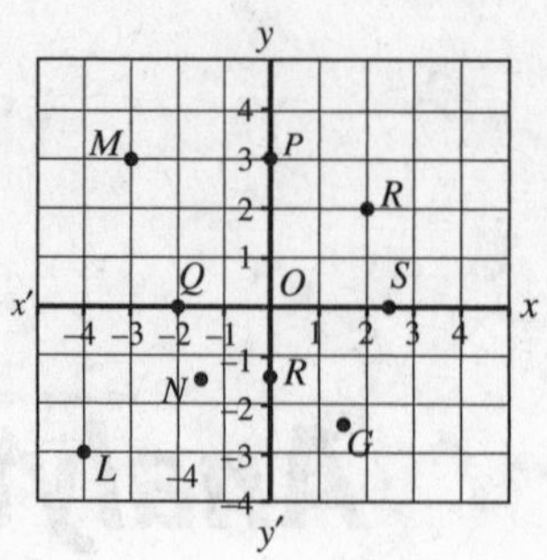

Fig. 12-3

(a) B (c) O (e) N (g) P (i) R

(b) M (d) L (f) G (h) Q (j) S

Solutions

(a) $(2, 2)$ (c) $(0, 0)$ (e) $(-1\frac{1}{2}, -1\frac{1}{2})$ (g) $(0, 3)$ (i) $(0, -1\frac{1}{2})$

(b) $(-3, 3)$ (d) $(-4, -3)$ (f) $(1\frac{1}{2}, -2\frac{1}{2})$ (h) $(-2, 0)$ (j) $(2\frac{1}{2}, 0)$

12.2 Coordinates of points in the four quadrants

What are the signs of the coordinates of (a) a point in quadrant I; (b) a point in quadrant II; (c) a point in quadrant III; (d) a point in quadrant IV? Show which coordinate has a sign and which zero value for a point between quadrants (e) IV and I; (f) I and II; (g) II and III; (h) III and IV.

Solutions

(a) $(+, +)$ (c) $(-, -)$ (e) $(+, 0)$ (g) $(-, 0)$

(b) $(-, +)$ (d) $(+, -)$ (f) $(0, +)$ (h) $(0, -)$

12.3 Graphing a quadrilateral

If the vertices of a rectangle have the coordinates $A(3, 1)$, $B(-5, 1)$, $C(-5, -3)$, and $D(3-3)$, find its perimeter and area.

Solution

The base and height of the rectangle are 8 and 4 (see Fig. 12-4). Hence, the perimeter is $2b + 2h = 2(8) + 2(4) = 24$, and the area is $bh = (8)(4) = 32$.

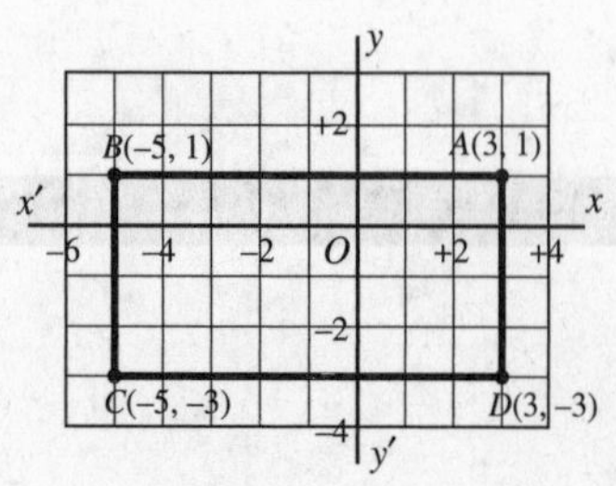

Fig. 12-4

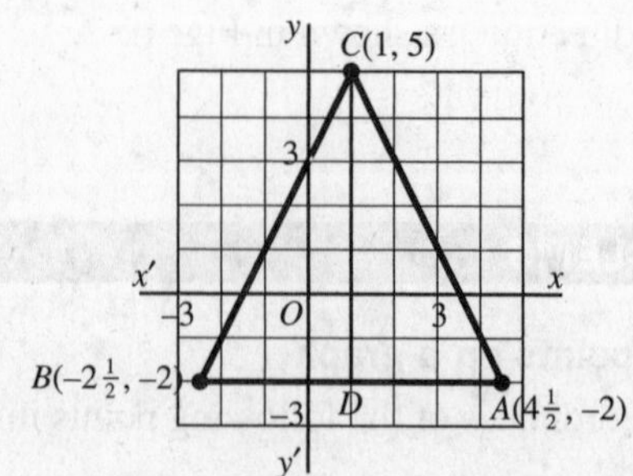

Fig. 12-5

12.4 Graphing a triangle

If the vertices of a triangle have the coordinates $A\left(4\frac{1}{2},-2\right)$, $B\left(-2\frac{1}{2},-2\right)$ and $C(1, 5)$, find its area.

Solution

The length of the base is $BA = 7$ (see Fig. 12-5). The height is $CD = 7$. Then $A = \frac{1}{2}bh = \frac{1}{2}(7)(7) = 24\frac{1}{2}$.

12.2 Midpoint of a Segment

The coordinates (x_m, y_m) of the midpoint M of the line segment joining $P(x_1, y_1)$ to $Q(x_2, y_2)$ are

$$x_m = \frac{1}{2}(x_1 + x_2) \qquad \text{and} \qquad y_m = \frac{1}{2}(y_1 + y_2)$$

In Fig. 12-6, segment y_m is the median of trapezoid $CPQD$, whose bases are y_1 and y_2. Since the length of a median is one-half the sum of the bases, $y_m = \frac{1}{2}(y_1 + y_2)$. Similarly, segment x_m is the median of trapezoid $ABQP$, whose bases are x_1 and x_2; hence, $x_m = \frac{1}{2}(x_1 + x_2)$.

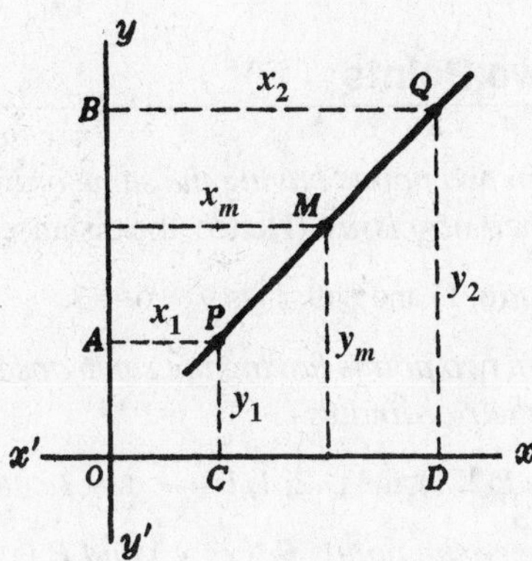

Fig. 12-6

SOLVED PROBLEMS

12.5 Applying the midpoint formula

If M is the midpoint of $\overline{PQ}$, find the coordinates of (a) M if the coordinates of P and Q are $P(3, 4)$ and $Q(5, 8)$; (b) Q if the coordinates of P and M are $P(1, 5)$ and $M(3, 4)$.

Solutions

(a) $x_m = \frac{1}{2}(x_1 + x_2) = \frac{1}{2}(3 + 5) = 4$; $y_m = \frac{1}{2}(y_1 + y_2) = \frac{1}{2}(4 + 8) = 6$.

(b) $x_m = \frac{1}{2}(x_1 + x_2)$, so $3 = \frac{1}{2}(1 + x_2)$ and $x_2 = 5$; $y_m = \frac{1}{2}(y_1 + y_2)$, so $4 = \frac{1}{2}(5 + y_2)$ and $y_2 = 3$.

12.6 Determining if segments bisect each other

The vertices of a quadrilateral are $A(0, 0)$, $B(0, 3)$, $C(4, 3)$, and $D(4, 0)$.

(a) Show that $ABCD$ is a rectangle.

(b) Show that the midpoint of $\overline{AC}$ is also the midpoint of $\overline{BD}$.

(c) Do the diagonals bisect each other? Why?

Solutions

(a) From Fig. 12-7, $AB = CD = 3$ and $BC = AD = 4$; hence, $ABCD$ is a parallelogram. Since $\angle BAD$ is a right angle, $ABCD$ is a rectangle.

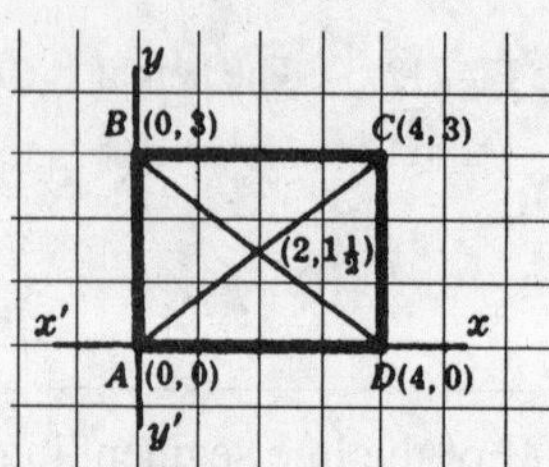

Fig. 12-7

(b) The coordinates of the midpoint of $\overline{AC}$ are $x = \frac{1}{2}(0 + 4) = 2$, $y = \frac{1}{2}(0 + 3) = 1\frac{1}{2}$. The coordinates of the midpoint of $\overline{BD}$ are $x = \frac{1}{2}(0 + 4) = 2$, $y = \frac{1}{2}(3 + 0) = 1\frac{1}{2}$. Hence, $(2, 1\frac{1}{2})$ is the midpoint of both $\overline{AC}$ and $\overline{BD}$.

(c) Yes, since the midpoints of both diagonals are the same point.

12.3 Distance Between Two Points

PRINCIPLE 1: *The distance between two points having the same ordinate (or y-value) is the absolute value of the difference of their abscissas.* (Hence, the distance between two points must be *positive.*)

Thus, the distance between the point $P(6, 1)$ and $Q(9, 1)$ is $9 - 6 = 3$.

PRINCIPLE 2: *The distance between two points having the same abscissa (or x-value) is the absolute value of the difference of their ordinates.*

Thus, the distance between the points $P(2, 1)$ and $Q(2, 4)$ is $4 - 1 = 3$.

PRINCIPLE 3: *The distance d between the points $P_1(x_1, y_1)$ and $P_2(x_2, y_2)$ is*

$$d = \sqrt{(x_2 - x_1)^2 + (y_2 - y_1)^2} \qquad \text{or} \qquad d = \sqrt{(\Delta x)^2 + (\Delta y)^2}$$

The difference $x_2 - x_1$ is denoted by the symbol Δx; the difference $y_2 - y_1$ is denoted by Δy. Delta (Δ) is the fourth letter of the Greek alphabet, corresponding to our *d*. The difference Δx and Δy may be positive or negative.

SOLVED PROBLEMS

12.7 Providing and using the distance formula

(a) Prove the distance formula (Principle 3) algebraically.

(b) Use it to find the distance between $A(2, 5)$ and $B(6, 8)$.

Solutions

(a) See Fig. 12-8. By Principle 1, $P_1S = x_2 - x_1 = \Delta x$. By Principle 2, $P_2S = y_2 - y_1 = \Delta y$. Also, in right triangle P_1SP_2,

$$(P_1P_2)^2 = (P_1S)^2 + (P_2S)^2$$

or

$$d^2 = (x_2 - x_1)^2 + (y_2 - y_1)^2$$

and

$$d = \sqrt{(x_2 - x_1)^2 + (y_2 - y_1)^2}$$

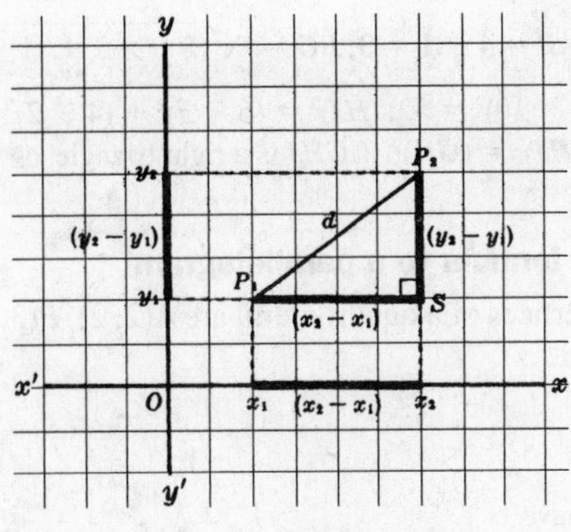

Fig. 12-8

(b) The distance from $A(2, 5)$ to $B(6, 8)$ is found as follows:

$$\underline{(x, y)}$$
$$B(6, 8) \rightarrow x_2 = 6, y_2 = 8$$
$$\underline{A(2, 5) \rightarrow x_1 = 2, y_1 = 5}$$
$$d^2 = (x_2 - x_1)^2 + (y_2 - y_1)^2$$
$$d^2 = (6 - 2)^2 + (8 - 5)^2 = 4^2 + 3^2 = 25 \quad \text{and} \quad d = 5$$

12.8 Finding the distance between two points

Find the distance between the points (a) $(-3, 5)$ and $(1, 5)$; (b) $(3, -2)$ and $(3, 4)$; (c) $(3, 4)$ and $(6, 8)$; $(-3, 2)$ and $(9, -3)$.

Solutions

(a) Since both points have the same ordinate (or y-value), $d = x_2 - x_1 = 1 - (-3) = 4$

(b) Since both points have the same abscissa (or x-value), $d = y_2 - y_1 = 4 - (-2) = 6$

(c) $d = \sqrt{(x_2 - x_1)^2 + (y_2 - y_1)^2} = \sqrt{(6 - 3)^2 + (8 - 4)^2} = \sqrt{3^2 + 4^2} = 5$

(d) $d = \sqrt{(x_2 - x_1)^2 + (y_2 - y_1)^2} = \sqrt{[9 - (-3)]^2 + (-3 - 2)^2} = \sqrt{12^2 + (-5)^2} = 13$

12.9 Applying the distance formula to a triangle

(a) Find the lengths of the sides of a triangle whose vertices are $A(1, 1)$, $B(1, 4)$, and $C(5, 1)$.

(b) Show that the triangle whose vertices are $G(2, 10)$, $H(3, 2)$, and $J(6, 4)$ is a right triangle.

Solutions

See Fig. 12-9.

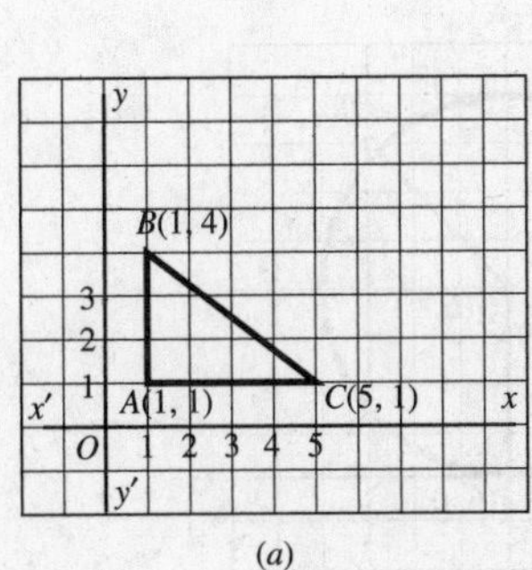

(a)

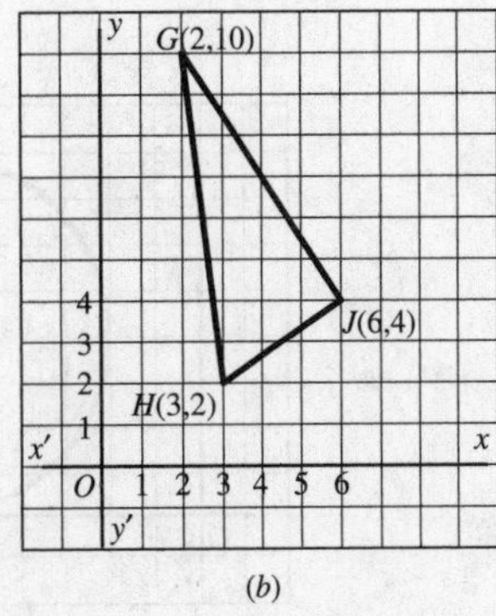

(b)

Fig. 12-9

(a) $AC = 5 - 1 = 4$ and $AB = 4 - 1 = 3$; $BC = \sqrt{(5-1)^2 + (1-4)^2} = \sqrt{4^2 + (-3)^2} = 5$.

(b) $(GJ)^2 = (6-2)^2 + (4-10)^2 = 52$; $(HJ)^2 = (6-3)^2 + (4-2)^2 = 13$; $(GH)^2 = (2-3)^2 + (10-2)^2 = 65$. Since $(GJ)^2 + (HJ)^2 = (GH)^2$, $\triangle GHJ$ is a right triangle.

12.10 Applying the distance formula to a parallelogram

The coordinates of the vertices of a quadrilateral are $A(2, 2)$, $B(3, 5)$, $C(6, 7)$, and $D(5, 4)$. Show that $ABCD$ is a parallelogram.

Solution

See Fig. 12-10, where we have

$$\begin{aligned} AB &= \sqrt{(3-2)^2 + (5-2)^2} = \sqrt{1^2 + 3^2} = \sqrt{10} \\ CD &= \sqrt{(6-5)^2 + (7-4)^2} = \sqrt{1^2 + 3^2} = \sqrt{10} \\ BC &= \sqrt{(6-3)^2 + (7-5)^2} = \sqrt{3^2 + 2^2} = \sqrt{13} \\ AD &= \sqrt{(5-2)^2 + (4-2)^2} = \sqrt{3^2 + 2^2} = \sqrt{13} \end{aligned}$$

Thus, $\overline{AB} \cong \overline{CD}$ and $\overline{BC} \cong \overline{AD}$. Since opposite sides are congruent, $ABCD$ is a parallelogram.

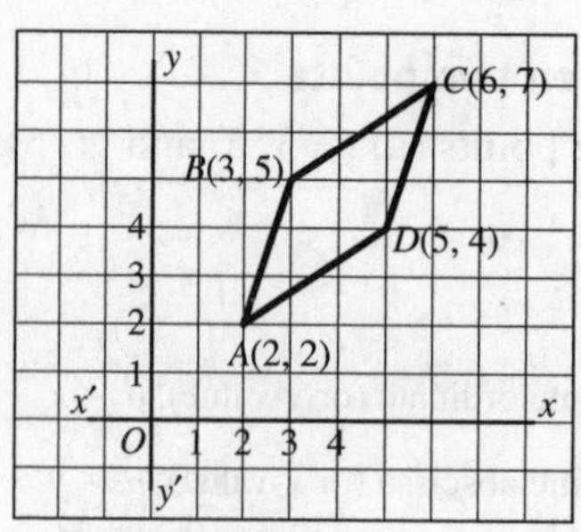

Fig. 12-10

12.11 Applying the distance formula to a circle

A circle is tangent to the x-axis and has its center at (6, 4). Where is the point (9, 7) with respect to the circle?

Solution

Since the circle is tangent to the x-axis, $\overline{AQ}$ in Fig. 12-11 is a radius. By Principle 2, $AQ = 4$.

By Principle 3, $BQ = \sqrt{(9-6)^2 + (7-4)^2} = \sqrt{3^2 + 3^2} = \sqrt{18}$. Since $\sqrt{18}$ is greater than 4, $\overline{BQ}$ is greater than a radius so B is outside the circle.

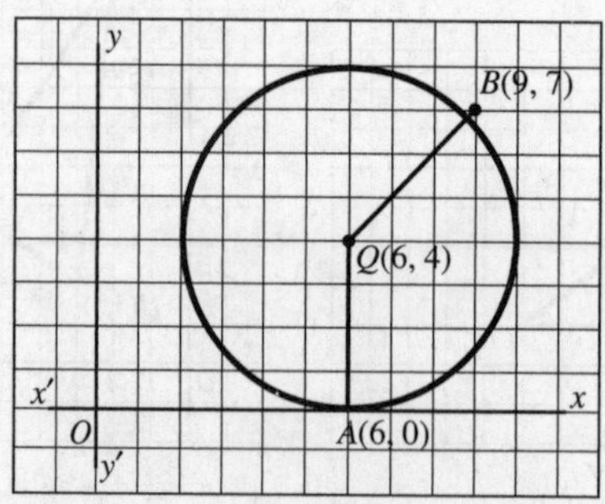

Fig. 12-11

12.4 Slope of a Line

PRINCIPLE 1: *If a line passes through the points $P_1(x_1, y_1)$ and $P_2(x_2, y_2)$, then*

$$\text{Slope of } \overleftrightarrow{P_1P_2} = \frac{y_2 - y_1}{x_2 - x_1} = \frac{\Delta y}{\Delta x}$$

PRINCIPLE 2: *The line whose equation is $y = mx + b$ has slope m.*

PRINCIPLE 3: *The slope of a line equals the tangent of its inclination.*

The inclination i of a line is the angle above the x-axis that is included between the line and the positive direction of the x-axis (see Fig. 12-12). In the figure,

$$\text{Slope of } \overleftrightarrow{P_1P_2} = \frac{y_2 - y_1}{x_2 - x_1} = \frac{\Delta y}{\Delta x} = m = \tan i$$

The slope is independent of the order in which the end points are selected. Thus,

$$\text{Slope of } \overleftrightarrow{P_1P_2} = \frac{y_2 - y_1}{x_2 - x_1} = \frac{y_1 - y_2}{x_1 - x_2} = \text{slope of } \overleftrightarrow{P_2P_1}$$

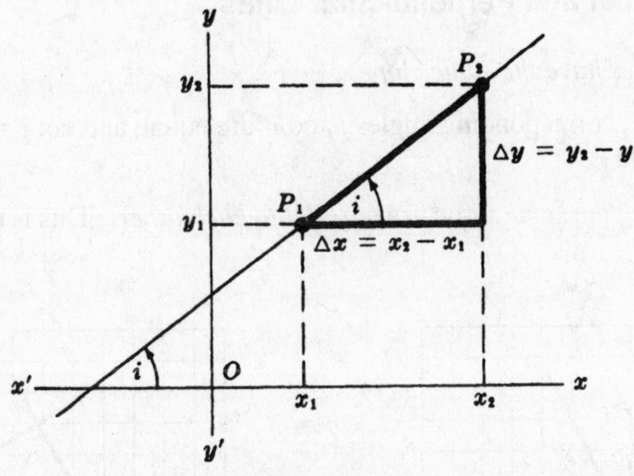

Fig. 12-12

12.4A Positive and Negative Slopes

PRINCIPLE 4: *If a line slants upward from left to right, its inclination i is an acute angle and its slope is positive* (Fig. 12-13).

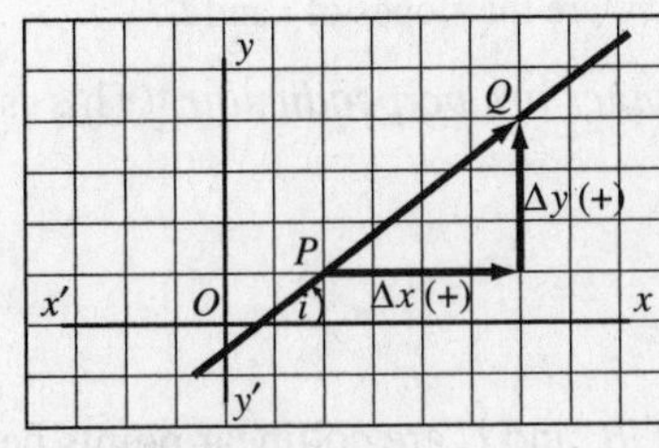

Slope of $PQ = \frac{\Delta y}{\Delta x} = \frac{+}{+} = +$

Fig. 12-13

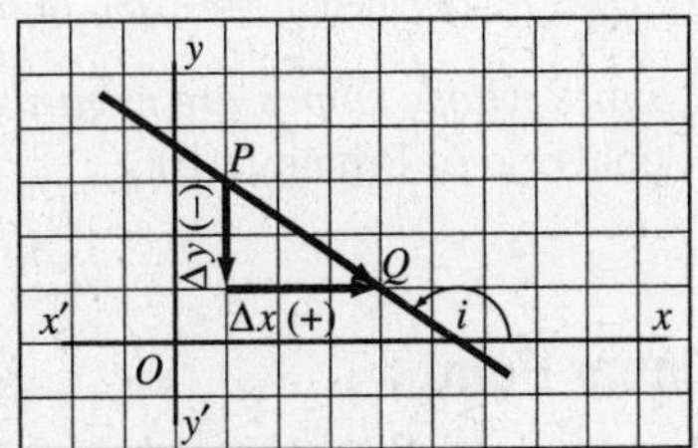

Slope of $PQ = \frac{\Delta y}{\Delta x} = \frac{-}{+} = -$

Fig. 12-14

PRINCIPLE 5: *If a line slants downward from left to right, its inclination is an obtuse angle and its slope is negative* (Fig. 12-14).

PRINCIPLE 6: *If a line is parallel to the x-axis, its inclination is 0° and its slope is* 0 (Fig. 12-15).

PRINCIPLE 7: *If a line is perpendicular to the x-axis, its inclination is 90° and it has no slope* (Fig. 12-16).

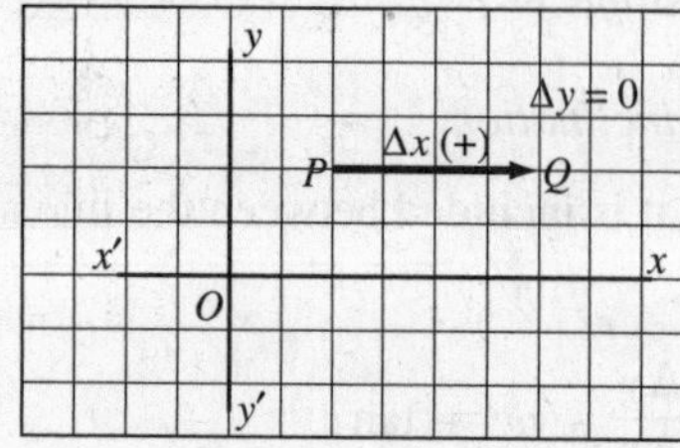

Slope of $PQ = \frac{\Delta y}{\Delta x} = \frac{0}{+} = 0$

Fig. 12-15

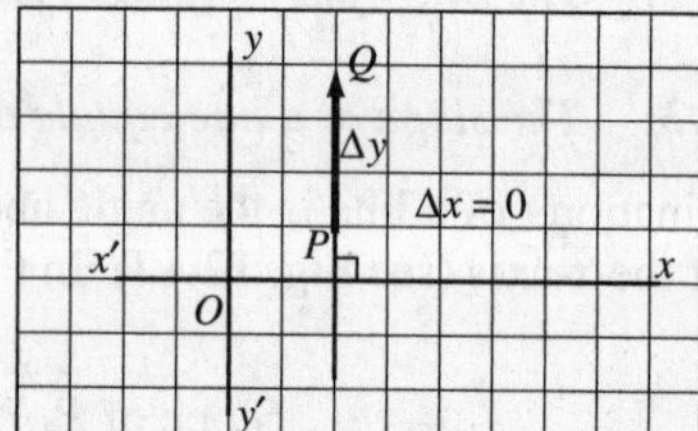

Slope of $PQ = \frac{\Delta y}{\Delta x} = \frac{+}{0}$ (meaningless)

Fig. 12-16

12-4B Slopes of Parallel and Perpendicular Lines

PRINCIPLE 8: *Parallel lines have the same slope.*

In Fig. 12-17, $l \parallel l'$; hence, corresponding angles i and i' are equal, and $\tan i = \tan i'$ or $m = m'$, where m and m' are the slopes of l and l'.

PRINCIPLE 9: *Lines having the same slope are parallel to each other.* (This is the converse of Principle 8.)

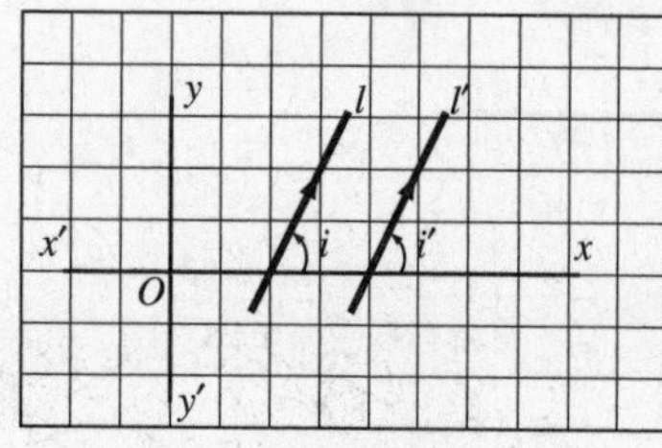

Fig. 12-17

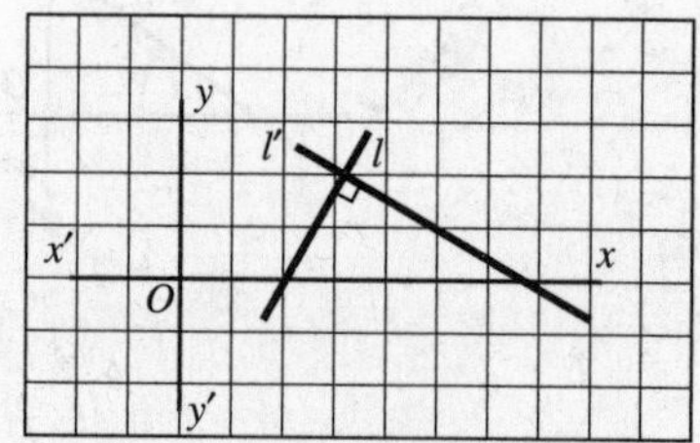

Fig. 12-18

PRINCIPLE 10: *Perpendicular lines have slopes that are negative reciprocals of each other.* (Negative reciprocals are numbers, such as $\frac{2}{5}$ and $-\frac{5}{2}$, whose product is -1.)

Thus in Fig. 12-18, if $l \perp l'$, then $m = -1/m'$ or $mm' = -1$, where m and m' are the slopes of l and l'.

PRINCIPLE 11: *Lines whose slopes are negative reciprocals of each other are perpendicular.* (This is the converse of Principle 10.)

12.4C Collinear Points

Collinear points are points which lie on the same straight line. Thus, A, B, and C are collinear points here:

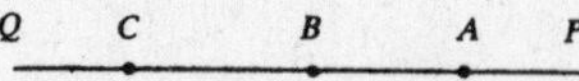

PRINCIPLE 12: *The slope of a straight line is constant all along the line.*

Thus if $\overleftrightarrow{PQ}$ above is a straight line, the slope of the segment from A to B equals the slope of the segment from C to Q.

PRINCIPLE 13: *If the slope of a segment between a first point and a second equals the slope of the segment between either point and a third, then the points are collinear.*

SOLVED PROBLEMS

12.12 Slope and inclination of a line

(a) Find the slope of the line through $(-2, -1)$ and $(4, 3)$.

(b) Find the slope of the line whose equation is $3y - 4x = 15$.

(c) Find the inclination of the line whose equation is $y = x + 4$.

Solutions

(a) By principle 1, $m = \dfrac{y_2 - y_1}{x_2 - x_1} = \dfrac{3 - (-1)}{4 - (-2)} = \dfrac{4}{6} = \dfrac{2}{3}$

(b) We may rewrite $3y - 4x = 15$ as $y = \frac{4}{3}x + 5$, from which $m = \frac{4}{3}$.

(c) Since $y = x + 4$, we have $m = 1$; thus, $\tan i = 1$ and $i = 45°$.

12.13 Slopes of parallel or perpendicular lines

Find the slope of $\overleftrightarrow{CD}$ if (a) $\overleftrightarrow{AB} \parallel \overleftrightarrow{CD}$ and the slope of $\overleftrightarrow{AB}$ is $\frac{2}{3}$; (b) $\overleftrightarrow{AB} \perp \overleftrightarrow{CD}$ and the slope of $\overleftrightarrow{AB}$ is $\frac{3}{4}$.

Solutions

(a) By Principle 8, slope of $\overleftrightarrow{CD}$ = slope of $\overleftrightarrow{AB} = \frac{2}{3}$.

(b) By Principle 10, slope of $\overleftrightarrow{CD} = -\dfrac{1}{\text{slope of } AB} = -\dfrac{1}{3/4} = -\dfrac{4}{3}$.

12.14 Applying principles 9 and 11 to triangles and quadrilaterals

Complete each of the following statements:

(a) In quadrilateral $ABCD$, if the slopes of $\overleftrightarrow{AB}$, $\overleftrightarrow{BC}$, $\overleftrightarrow{CD}$, and $\overleftrightarrow{DA}$ are $\frac{1}{2}$, -2, $\frac{1}{2}$, and -2, respectively, the quadrilateral is a __?__.

(b) In triangle LMP, if the slopes of $\overleftrightarrow{LM}$ and $\overleftrightarrow{MP}$ are 5 and $-\frac{1}{5}$, then LMP is a __?__ triangle.

Solutions

(a) Since the slopes of the opposite sides are equal, $ABCD$ is a parallelogram. In addition, the slopes of adjacent sides are negative reciprocals; hence, those sides are $\perp$ and $ABCD$ is a rectangle.

(b) Since the slopes of $\overleftrightarrow{LM}$ and $\overleftrightarrow{MP}$ are negative reciprocals, $\overline{LM} \perp \overline{MP}$ and the triangle is a right triangle.

12.15 Applying principle 12

(a) $\overleftrightarrow{AB}$ has a slope of 2 and points A, B, and C are collinear. What are the slopes of $\overline{AC}$ and $\overline{BC}$?

(b) Find y if $G(1, 4)$, $H(3, 2)$, and $J(9, y)$ are collinear.

Solutions

(a) By Principle 12, $\overline{AC}$ and $\overline{BC}$ have a slope of 2.

(b) By Principle 12, slope of $\overleftrightarrow{GJ}$ = slope of $\overleftrightarrow{GH}$. Hence $\frac{y-4}{9-1} = \frac{2-4}{3-1}$, so that $\frac{y-4}{8} = \frac{-2}{2} = -1$ and $y = -4$.

12.5 Locus in Analytic Geometry

A locus of points is the set of points, and only those points, satisfying a given condition. In geometry, a line or curve (or set of lines or curves) on a graph is the locus of analytic points that satisfy the equation of the line or curve.

Think of the locus as the path of a point moving according to a given condition or as the set of points satisfying a given condition.

PRINCIPLE 1: *The locus of points whose abscissa is a constant k is a line parallel to the y-axis; its equation is $x = k$.* (See Fig. 12-19.)

PRINCIPLE 2: *The locus of points whose ordinate is a constant k is a line parallel to the x-axis; its equation is $y = k$.* (See Fig. 12-19.)

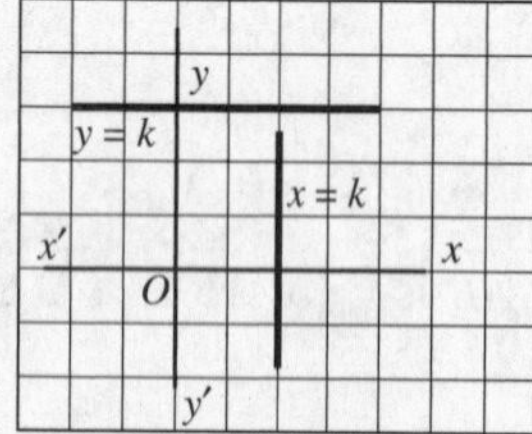

Fig. 12-19

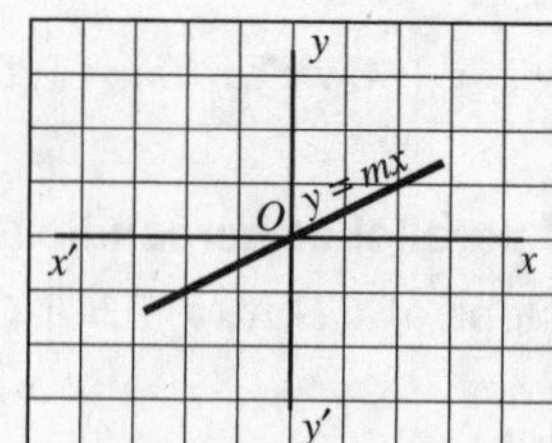

Fig. 12-20

PRINCIPLE 3: *The locus of points whose ordinate equals the product of a constant m and its abscissa is a straight line passing through the origin; its equation is $y = mx$.*

The constant m is the slope of the line. (See Fig. 12-20.)

PRINCIPLE 4: *The locus of points whose ordinate and abscissa are related by either of the equations*

$$y = mx + b \quad \text{or} \quad \frac{y - y_1}{x - x_1} = m$$

where m and b are constants, is a line (Fig. 12-21).

In the equation $y = mx + b$, m is the slope and b is the *y-intercept*. The equation $\frac{y - y_1}{x - x_1} = m$ tells us that the line passes through the fixed point (x_1, y_1) and has a slope of m.

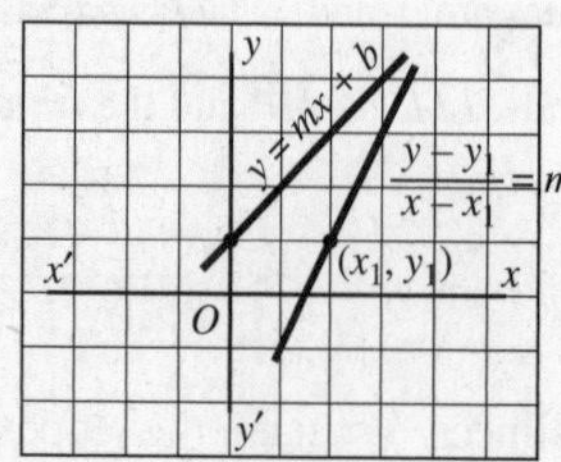

Fig. 12-21

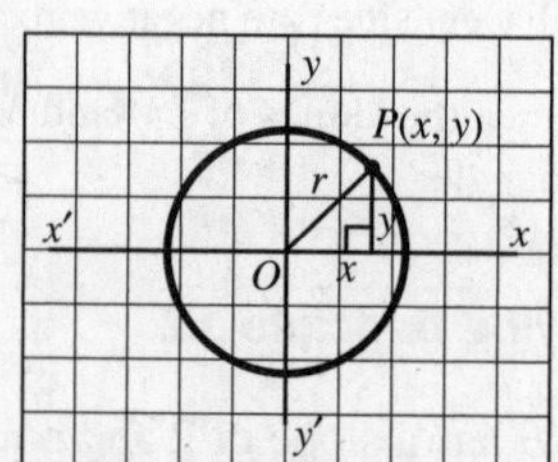

Fig. 12-22

PRINCIPLE 5: *The locus of points such that the sum of the squares of the coordinates is a constant is a circle whose center is the origin.*

The constant is the square of the radius, and the equation of the circle is

$$x^2 + y^2 = r^2$$

(see Fig. 12-22). Note that for any point $P(x, y)$ on the circle, $x^2 + y^2 = r^2$.

SOLVED PROBLEMS

12.16 Applying principles 1 and 2

Graph and give the equation of the locus of points (a) whose ordinate is -2; (b) that are 3 units from the y-axis; (c) that are equidistant from the points (3, 0) and (5, 0).

Solutions

(a) From Principle 2, the equation is $y = -2$; see Fig. 12-23(a).

(b) From Principle 1, the equation is $x = 3$ and $x = -3$; see Fig. 12-23(b).

(c) The equation is $x = 4$; see Fig. 12-23(c).

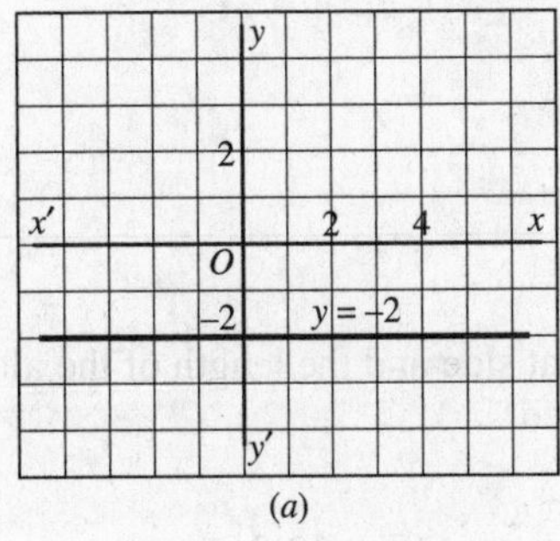

(a)

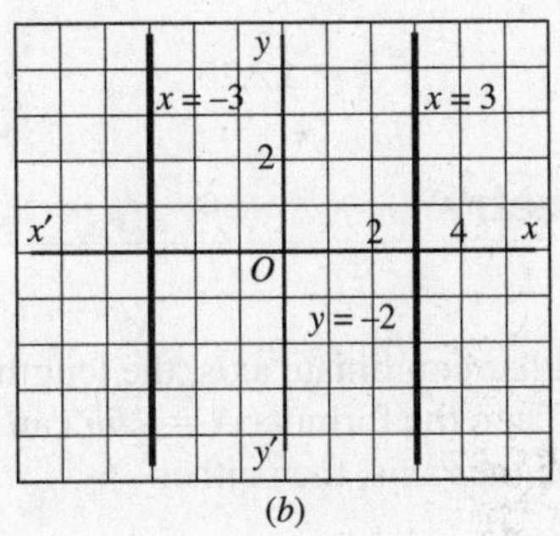

(b)

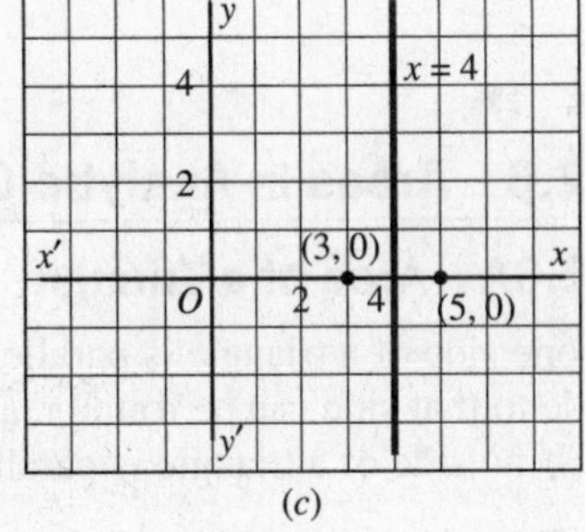

(c)

Fig. 12-23

12.17 Applying principles 3 and 4

Graph and describe the locus whose equation is (a) $y = \frac{1}{3}x + 1$; (b) $y = \frac{3}{2}x$; (c) $\dfrac{y-1}{x-1} = \dfrac{3}{4}$.

Solutions

(a) The locus is a line whose y-intercept is 1 and whose slope equals $\frac{1}{3}$. See Fig. 12-24(a).

(b) The locus is a line which passes through the origin and has slope $\frac{3}{2}$. See Fig. 12-24(b).

(c) The locus is a line which passes through the point (1, 1) and has slope $\frac{3}{4}$. See Fig. 12-24(c).

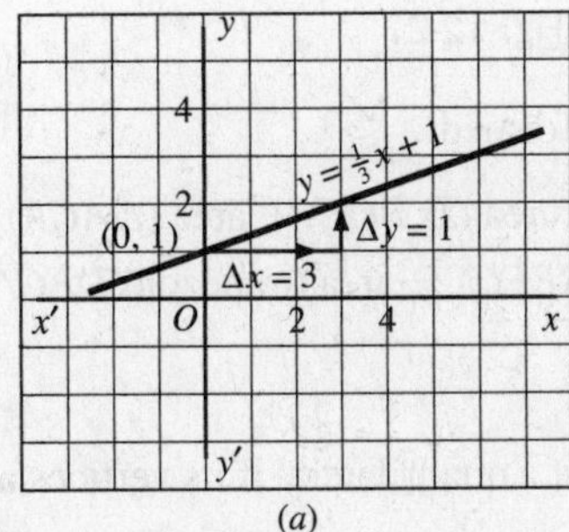

(a)

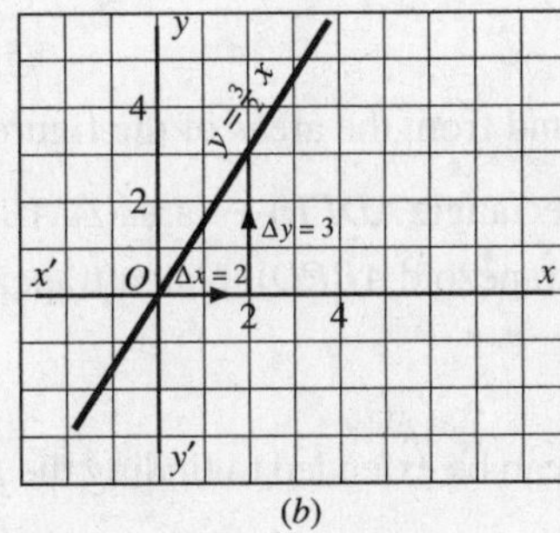

(b)

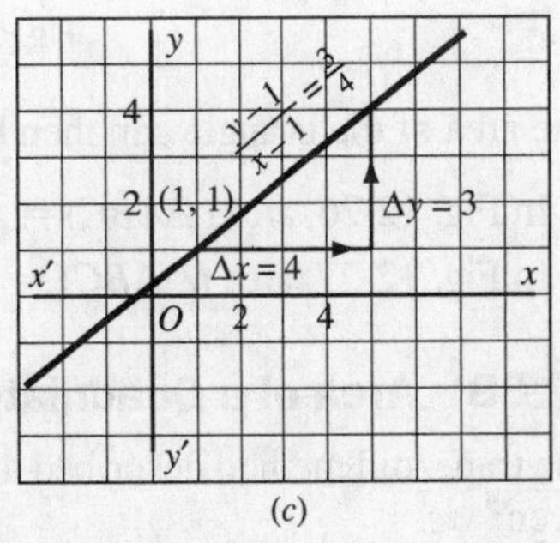

(c)

Fig. 12-24

12.18 Applying principle 5

Graph and give the equation of the locus of points (a) 2 units from the origin; (b) 2 units from the locus of $x^2 + y^2 = 9$.

Solutions

(a) The locus is a circle whose equation is $x^2 + y^2 = 4$. See Fig. 12-25(a).

(b) The locus is a pair of circles, each 2 units from the circle with center at O and radius 3. Their equations are $x^2 + y^2 = 25$ and $x^2 + y^2 = 1$. See Fig. 12-25(b).

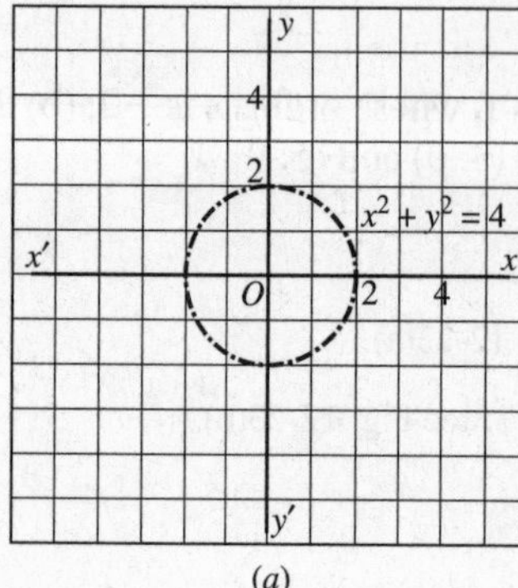

(a)

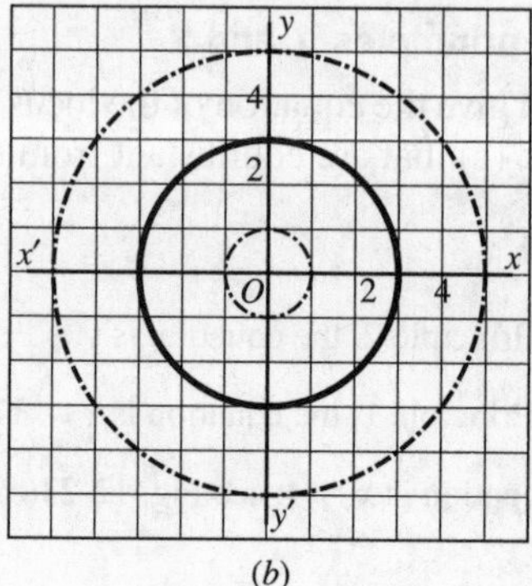

(b)

Fig. 12-25

12.6 Areas in Analytic Geometry

12.6A Area of a Triangle

If one side of a triangle is parallel to either coordinate axis, the length of that side and the length of the altitude to that side can be found readily. Then the formula $A = \frac{1}{2}bh$ can be used.

If no side of a triangle is parallel to either axis, then either

1. The triangle can be enclosed in a rectangle whose sides are parallel to the axes (Fig. 12-26), or
2. Trapezoids whose bases are parallel to the y-axis can be formed by dropping perpendiculars to the x-axis (Fig. 12-27).

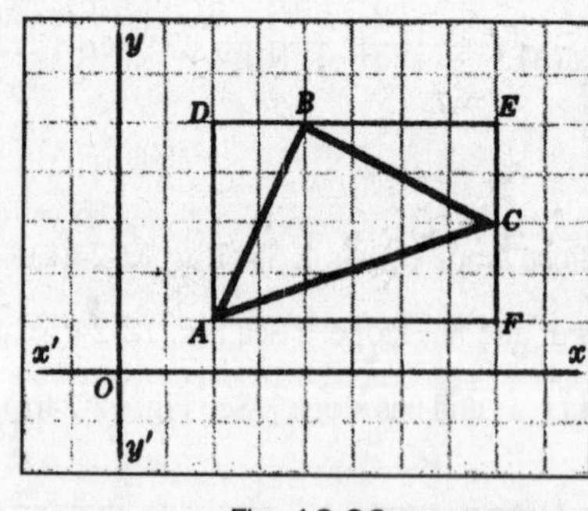

Fig. 12-26

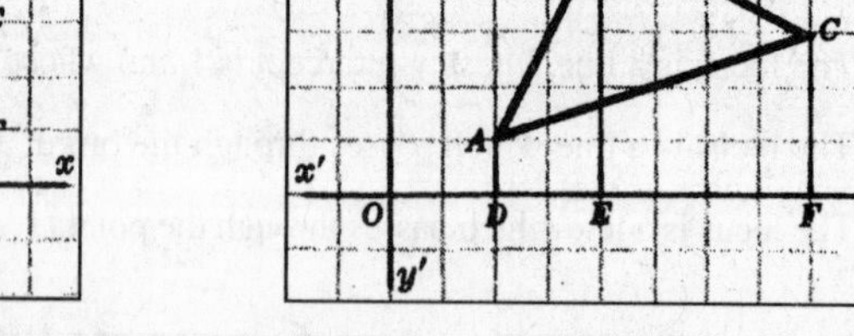

Fig. 12-27

The area of the triangle can then be found from the areas of the figures so formed:

1. In Fig. 12-26, area ($\triangle ABC$) = area(rectangle $ADEF$) − [area($\triangle ABD$) + area ($\triangle BCE$) + area ($\triangle ACF$)].
2. In Fig. 12-27, area ($\triangle ABC$) = area(trapezoid $ABED$) + area(trapezoid $BEFC$) − area (trapezoid $DFCA$).

12.6B Area of a Quadrilateral

The trapezoid method described above can be extended to finding the area of a quadrilateral if its vertices are given.

SOLVED PROBLEMS

12.19 Area of a triangle having a side parallel to an axis

Find the area of the triangle whose vertices are $A(1, 2)$, $B(7, 2)$, and $C(5, 4)$.

Solution

From the graph of the triangle (Fig. 12-28), we see that $b = 7 - 1 = 6$ and $h = 4 - 2 = 2$. Then $A = \frac{1}{2}bh = \frac{1}{2}(6)(2) = 6$.

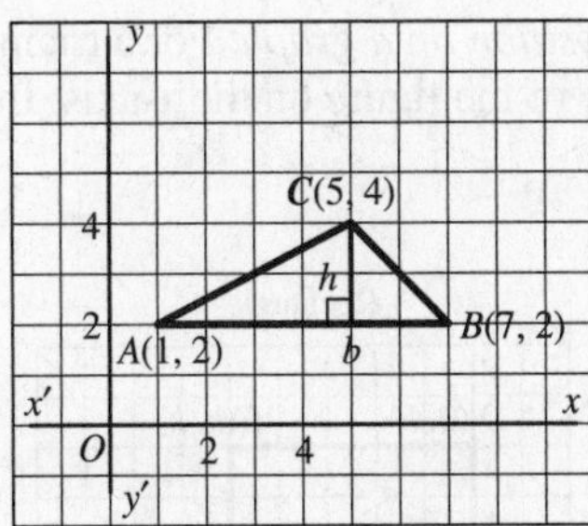

Fig. 12-28

12.20 Area of a triangle having no side parallel to an axis

Find the area of $\triangle ABC$ whose vertices are $A(2, 4)$, $B(5, 8)$ and $C(8, 2)$ (a) using the rectangle method; (b) using the trapezoid method.

Solutions

See Fig. 12-29.

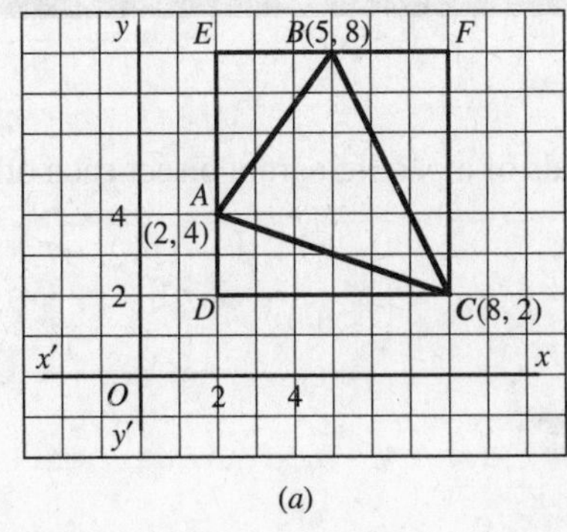

(*a*)

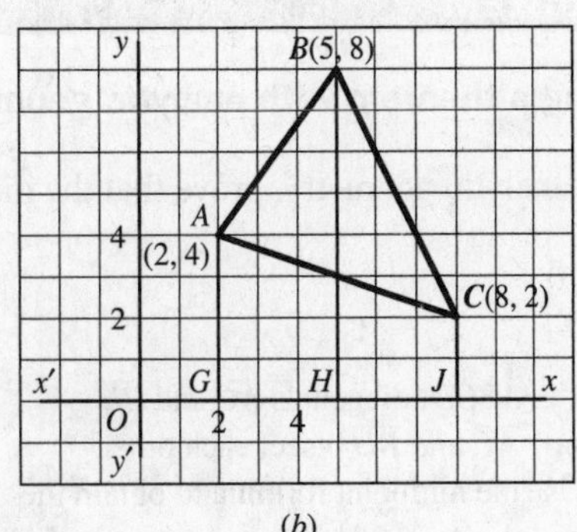

(*b*)

Fig. 12-29

(a) Area of rectangle $DEFC = bh = 6(6) = 36$. Then:

Area of $\triangle DAC = \frac{1}{2}bh = \frac{1}{2}(2)(6) = 6$.

Area of $\triangle ABE = \frac{1}{2}bh = \frac{1}{2}(3)(4) = 6$.

Area of $\triangle BCF = \frac{1}{2}bh = \frac{1}{2}(3)(6) = 9$.

So area($\triangle ABC$) = area($DEFC$) − area($\triangle DAC + \triangle ABE + \triangle BCF$) = $36 - (6 + 6 + 9) = 15$.

(b) Area of trapezoid $ABHG = \frac{1}{2}h(b + b') = \frac{1}{2}(3)(4 + 8) = 18$.
Area of trapezoid $BCJH = \frac{1}{2}(3)(2 + 8) = 15$.
Area of trapezoid $ACJG = \frac{1}{2}(6)(2 + 4) = 18$.
Then area($\triangle ABC$) = area($ABHG$) + area($BCJH$) − area($ACJG$) = 18 + 15 − 18 = 15.

12.7 Proving Theorems with Analytic Geometry

Many theorems of plane geometry can be proved with analytic geometry. The procedure for proving a theorem has two major steps, as follows:

1. *Place each figure in a convenient position on a graph.* For a triangle, rectangle, or parallelogram, place one vertex at the origin and one side of the figure on the x-axis. Indicate the coordinates of each vertex (Fig. 12-30).

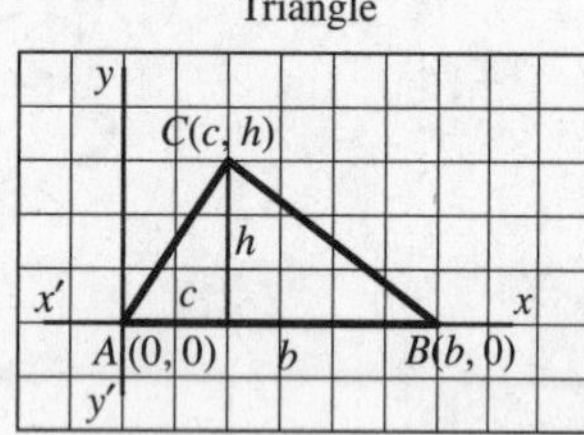

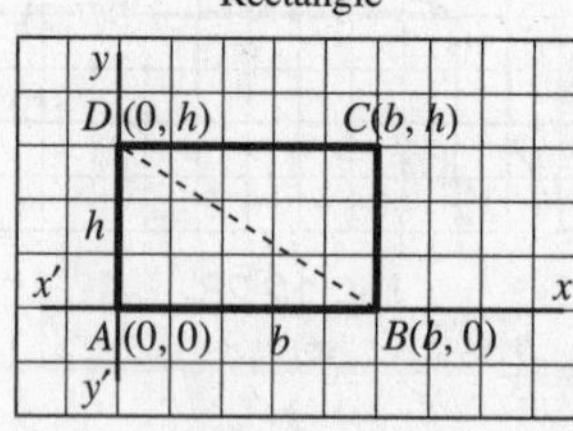

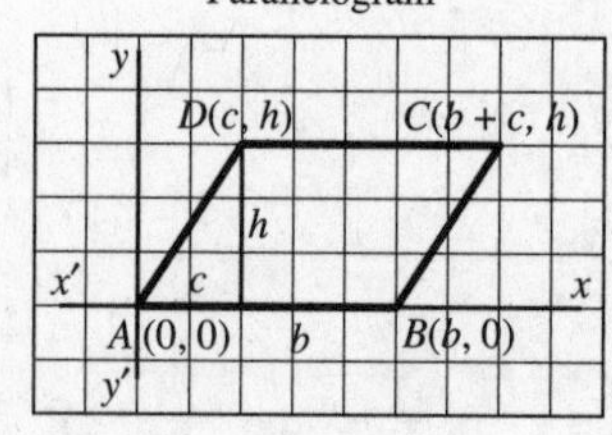

Fig. 12-30

2. *Apply the principles of analytic geometry.* For example, prove that lines are parallel by showing that their slopes are equal; or that lines are perpendicular by showing that their slopes are negative reciprocals of each other. Use the midpoint formula when the midpoint of a segment is involved, and use the distance formula to obtain the lengths of segments.

SOLVED PROBLEM

12.21 Proving a theorem with analytic geometry

Using analytic geometry, prove that the diagonals of a parallelogram bisect each other.

Solution

Given: $\square ABCD$, diagonals $\overline{AC}$ and $\overline{BD}$.
To Prove: $\overline{AC}$ and $\overline{BD}$ bisect each other.
Plan: Use the midpoint formula to obtain the coordinates of the midpoints of the diagonals

Place $\square$ $ABCD$ with vertex A at the origin and side $\overline{AD}$ along the x-axis (Fig. 12-31). Then the vertices have the coordinates $A(0, 0)$, $B(a, b)$, $C(a + c, b)$, and $D(c, 0)$.

By the midpoint formula, the midpoint of $\overline{AC}$ has the coordinates $\left(\frac{a + c}{2}, \frac{b}{2}\right)$, and the midpoint of $\overline{BD}$ has the coordinates $\left(\frac{a + c}{2}, \frac{b}{2}\right)$. Then the diagonals bisect each other, since the midpoints of both diagonals are the same point.

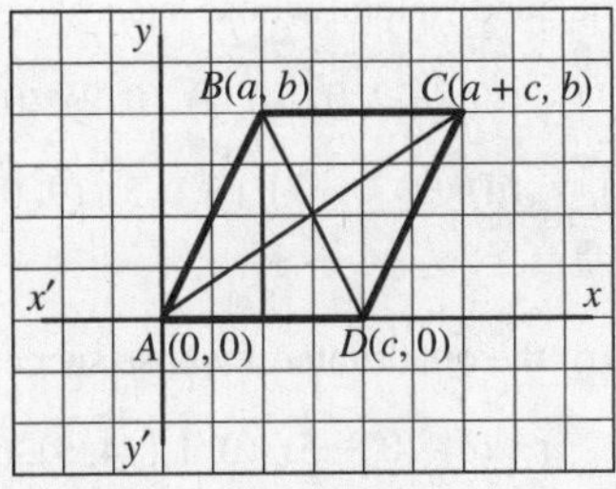

Fig. 12-31

SUPPLEMENTARY PROBLEMS

12.1. State the coordinates of each lettered point in Fig. 12-32. (12.1)

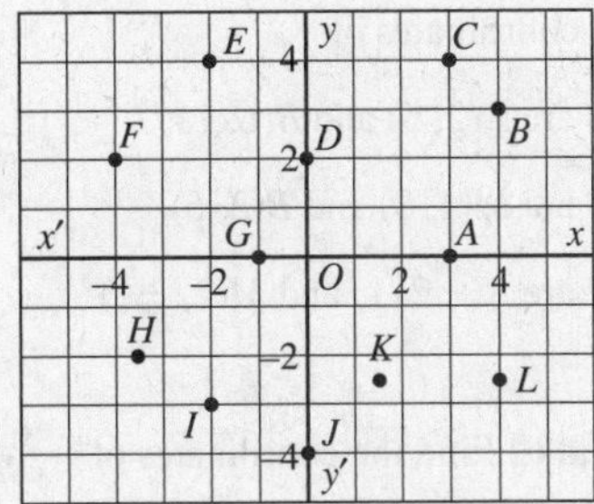

Fig. 12-32

12.2. Plot each of the following points: (12.2)

$A(-2, -3)$	$C(0, -1)$	$E(3, -4)$	$G(0, 3)$
$B(-3, 2)$	$D(-3, 0)$	$F(1\frac{1}{2}, 2\frac{1}{2})$	$H(3\frac{1}{2}, 0)$

12.3. Plot the following points: $A(2, 3)$, $B(-3, 3)$, $C(-3, -2)$, $D(2, -2)$. Then find the perimeter and area of square $ABCD$.

12.4. Plot the following points: $A(4, 3)$, $B(-1, 3)$, $C(-3, -3)$, $D(2, -3)$. Then find the area of parallelogram $ABCD$ and triangle BCD. (12.3, 12.4)

12.5. Find the midpoint of the segment joining (12.5)

(a) (0, 0) and (8, 6)
(b) (0, 0) and (5, 7)
(c) (0, 0) and (−8, 12)
(d) (14, 10) and (0, 0)
(e) (−20, −5) and (0, 0)
(f) (0, 4) and (0, 16)
(g) (8, 0) and (0, −2)
(h) (−10, 0) and (0, −5)
(i) (3, 4) and (7, 6)
(j) (−2, −8) and (−4, −12)
(k) (7, 9) and (3, 3)
(l) (2, −1) and (−2, −5)

12.6. Find the midpoints of the sides of a triangle whose vertices are (12.5)

(a) (0, 0), (8, 0), (0, 6)
(b) (−6, 0), (0, 0), (0, 10)
(c) (12, 0), (0, −4), (0, 0)
(d) (3, 5), (5, 7), (3, 11)
(e) (4, 0), (0, −6), (−4, 10)
(f) (−1, −2), (0, 2), (1, −1)

12.7. Find the midpoints of the sides of the quadrilateral whose successive vertices are (12.5)

(a) (0, 0), (0, 4), (2, 10), (6, 0)
(b) (−3, 5), (−1, 9), (7, 3), (5, −1)
(c) (−2, 0), (0, 4), (6, 2), (0, −10)
(d) (−3, −7), (−1, 5), (9, 0), (5, −8)

12.8. Find the midpoints of the diagonals of the quadrilateral whose successive vertices are (12.5)

(a) (0, 0), (0, 5), (4, 12), (8, 1)
(b) (−4, −1), (−2, 3), (6, 1), (2, −8)
(c) (0, −5), (0, 1), (4, 9), (4, 3)

12.9. Find the center of a circle if the end points of a diameter are (12.5)

(a) (0, 0) and (−4, 6)
(b) (−1, 0) and (−5, −12)
(c) (−3, 1) and (0, −5)
(d) (0, 0) and $(2a, 2b)$
(e) (a, b) and $(3a, 5b)$
(f) $(a, 2b)$ and $(a, 2c)$

12.10. If M is the midpoint of $\overline{AB}$, find the coordinates of (12.5)

(a) M if the coordinates of A and B are $A(2, 5)$ and $B(6, 11)$
(b) A if the coordinates of M and B are $M(1, 3)$ and $B(3, 6)$
(c) B if the coordinates of A and M are $A(-2, 1)$ and $M(2, -1)$

12.11. The trisection points of $\overline{AD}$ are B and C. Find the coordinates of (12.5)

(a) B if the coordinates of A and C are $A(1, 2)$ and $C(3, 5)$
(b) D if the coordinates of B and C are $B(0, 5)$ and $C(1\frac{1}{2}, 4)$
(c) A if the coordinates of B and C are $B(0, 6)$ and $C(2, 3)$

12.12. $A(0, 0)$, $B(0, 5)$, $C(6, 5)$, and $D(6, 0)$ are the vertices of quadrilateral $ABCD$. (12.6)

(a) Prove that $ABCD$ is a rectangle.
(b) Show that the midpoints of $\overline{AC}$ and $\overline{BD}$ have the same coordinates.
(c) Do the diagonals bisect each other? Why?

12.13. The vertices of $\triangle ABC$ are $A(0, 0)$, $B(0, 4)$, and $C(6, 0)$. (12.6)

(a) If $\overline{AD}$ is the median to $\overline{BC}$, find the coordinates of D and the midpoint of $\overline{AD}$.
(b) If $\overline{CE}$ is the median to $\overline{AB}$, find the coordinates of E and the midpoint of $\overline{CE}$.
(c) Do the medians, $\overline{AD}$ and $\overline{CE}$, bisect each other? Why?

12.14. Find the distance between each of the following pairs of points: (12.8)

(a) (0, 0) and (0, 5)
(b) (4, 0) and (−2, 0)
(c) (0, −3) and (0, 7)
(d) (−6, −1) and (−6, 11)
(e) (5, 3) and (5, 8.4)
(f) (−1.5, 7) and (6, 7)
(g) $(-3, -4\frac{1}{2})$ and $(-3, 4\frac{1}{2})$
(h) (a, b) and $(2a, b)$

12.15. Find the distances separating pairs of the following collinear points: (12.8)

(a) $(5, -2), (5, 1), (5, 4)$ (c) $(-4, 2), (-3, 2), (0, 2)$

(b) $(0, -6), (0, -2), (0, 12)$ (d) $(0, b), (a, b), (3a, b)$

12.16. Find the distance between each of the following pairs of points: (12.8)

(a) $(0, 0)$ and $(5, 12)$ (e) $(-3, -6)$, and $(3, 2)$ (i) $(3, 4)$ and $(4, 7)$

(b) $(-3, -4)$ and $(0, 0)$ (f) $(2, 3)$ and $(-10, 12)$ (j) $(-1, -1)$ and $(1, 3)$

(c) $(0, -6)$ and $(9, 6)$ (g) $(2, 2)$ and $(5, 5)$ (k) $(-3, 0)$ and $(0, \sqrt{7})$

(d) $(4, 1)$ and $(7, 5)$ (h) $(0, 5)$ and $(-5, 0)$ (l) $(a, 0)$ and $(0, a)$

12.17. Show that the triangles having the following vertices are isosceles triangles: (12.9)

(a) $A(3, 5)$, $B(6, 9)$, and $C(2, 6)$ (c) $G(5, -5)$, $H(-2, -2)$, and $J(8, 2)$

(b) $D(2, 0)$, $E(6, 0)$, and $F(4, 4)$ (d) $K(7, 0)$, $L(3, 4)$, and $M(2, -1)$

12.18. Which of the triangles having the following vertices are right triangles? (12.9)

(a) $A(7, 0)$, $B(6, 3)$, and $C(12, 5)$ (c) $G(1, -1)$, $H(5, 0)$, and $J(3, 8)$

(b) $D(2, 0)$, $E(5, 2)$, and $F(1, 8)$ (d) $K(-4, 0)$, $L(-2, 4)$, and $M(4, -1)$

12.19. The vertices of $\triangle ABC$ are $A(-2, 2)$, $B(4, 4)$, and $C(8, 2)$. Find the length of the median to (a) $\overline{AB}$; (b) $\overline{AC}$; (c) $\overline{BC}$. (12.9)

12.20. (a) The vertices of quadrilateral $ABCD$ are $A(0, 0)$, $B(3, 2)$, $C(7, 7)$, and $D(4, 5)$. Show that $ABCD$ is a parallelogram. (12.10)

(b) The vertices of quadrilateral $DEFG$ are $D(3, 5)$, $E(1, 1)$, $F(5, 3)$, and $G(7, 7)$. Show that $DEFG$ is a rhombus.

(c) The vertices of quadrilateral $HJKL$ are $H(0, 0)$, $J(4, 4)$, $K(0, 8)$, and $L(-4, 4)$. Show that $HJKL$ is a square.

12.21. Find the radius of a circle that has its center at (12.11)

(a) $(0, 0)$ and passes through $(-6, 8)$ (d) $(2, 0)$ and passes through $(7, -12)$

(b) $(0, 0)$ and passes through $(3, -4)$ (e) $(4, 3)$ and is tangent to the y-axis

(c) $(0, 0)$ and passes through $(-5, 5)$ (f) $(-1, 7)$ and is tangent to the line $x = -4$

12.22. A circle has its center at the origin and a radius of 10. State whether each of the following points is on, inside, or outside of this circle: (a) $(6, 8)$; (b) $(-6, 8)$; (c) $(0, 11)$; (d) $(-10, 0)$; (e) $(7, 7)$; (f) $(-9, 4)$; (g) $(9, \sqrt{19})$. (12.11)

12.23. Find the slope of the line through each of the following pairs of points: (12.12)

(a) $(0, 0)$ and $(5, 9)$ (e) $(-2, -3)$ and $(7, 15)$ (i) $(3, -9)$ and $(0, 0)$

(b) $(0, 0)$ and $(9, 5)$ (f) $(-2, -3)$ and $(2, 1)$ (j) $(0, -2)$ and $(8, 10)$

(c) $(0, 0)$ and $(6, 15)$ (g) $(3, -4)$ and $(5, 6)$ (k) $(-1, -5)$ and $(1, -7)$

(d) $(2, 3)$ and $(6, 15)$ (h) $(0, 0)$ and $(-4, 8)$ (l) $(-3, -4)$ and $(-1, -2)$

12.24. Find the slope of the line whose equation is (12.12)

(a) $y = 3x - 4$ (e) $y = 5x$ (i) $3y = -12x + 6$ (m) $\frac{1}{5}y = x - 3$

(b) $y = 4x - 3$ (f) $y = 5$ (j) $3y = 12 - 2x$ (n) $\frac{1}{3}y = 2x - 6$

(c) $y = -\frac{1}{2}x + 5$ (g) $2y = 6x - 10$ (k) $y + x = 21$ (o) $\frac{1}{4}y = 7 - x$

(d) $y = 8 - 7x$ (h) $2y = 10x - 6$ (l) $2x = 12 - y$ (p) $\frac{1}{4}y + 2x = 1$

12.25. Find the inclination, to the nearest degree, of each of the following lines: (12.12)

(a) $y = 3x - 1$ (c) $2y = 5x + 10$ (e) $5y = 5x - 3$

(b) $y = \frac{1}{3}x - 1$ (d) $y = \frac{2}{5}x + 5$ (f) $y = -3$

12.26. Find the slope of a line whose inclination is (a) 5°; (b) 17°; (c) 20°; (d) 35°; (e) 45°; (f) 73°; (g) 85°. (12.12)

12.27. Find the inclination, to the nearest degree, of a line whose slope is (a) 0; (b) 0.4663; (c) 1, (d) 1.4281; (e) $\frac{1}{8}$; (f) $\frac{1}{2}$; (g) $\frac{3}{4}$; (h) $1\frac{1}{3}$; (i) $2\frac{1}{5}$. (12.12)

12.28. In hexagon *ABCDEF* of Fig. 12-33, $\overline{CD} \parallel \overline{AF}$. Which sides or diagonals have (a) positive slope; (b) negative slope; (c) zero slope; (d) no slope?

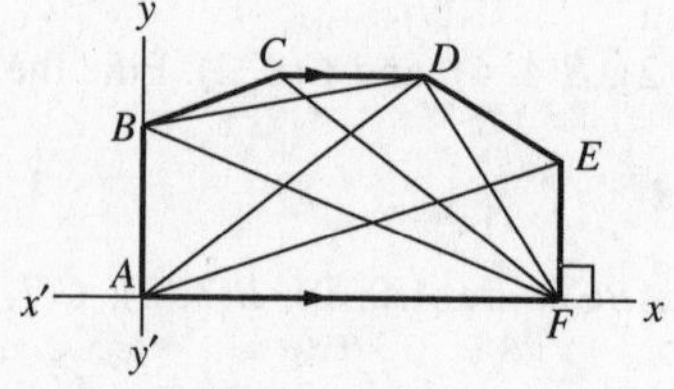

Fig. 12-33

12.29. Find the slope of a line that is parallel to a line whose slope is (a) 0; (b) has no slope; (c) 5; (d) −5; (e) 0.5; (f) −0.0005. (12.13)

12.30. Find the slope of a line that is parallel to the line whose equation is (12.13)

(a) $y = 0$ (c) $x = 7$ (e) $y = 5x - 2$ (g) $3y - 6x = 12$

(b) $x = 0$ (d) $y = 7$ (f) $x + y = 5$

12.31. Find the slope of a line that is parallel to a line which passes through (a) (0, 0) and (2, 3); (b) (2, −1) and (5, 6); (c) (3, 4) and (5, 2); (d) (1, 2) and (0, −4). (12.13)

12.32. Find the slope of a line that is perpendicular to a line whose slope is (12.13)

(a) $\frac{1}{2}$ (c) 3 (e) 0.1 (g) $-\frac{4}{5}$ (i) 0

(b) 1 (d) $2\frac{1}{2}$ (f) −1 (h) $-3\frac{1}{4}$ (j) has no slope

12.33. Find the slope of a line that is perpendicular to a line which passes through (a) (0, 0) and (0, 5); (b) (0, 0) and (2, 1); (c) (0, 0) and (3, −1); (d) (1, 1) and (3, 3). (12.13)

12.34. In rectangle $DEFG$, the slope of $\overline{DE}$ is $\frac{2}{3}$. What is the slope of (a) $\overline{EF}$; (b) $\overline{FG}$; (c) $\overline{DG}$? (12.14)

12.35. In $\square ABCD$ the slope of $\overline{AB}$ is 1 and the slope of $\overline{BC}$ is $-\frac{1}{2}$. What is the slope of (a) $\overline{AD}$; (b) $\overline{CD}$; (c) the altitude of $\overline{AD}$; (d) the altitude to $\overline{CD}$? (12.14)

12.36. The vertices of $\triangle ABC$ are $A(0, 5)$, $B(3, 7)$, and $C(5, -1)$. What is the slope of the altitude to (a) $\overline{AB}$; (b) $\overline{BC}$; (c) $\overline{AC}$? (12.14)

12.37. Which of the following sets of points are collinear: (a) (2, 1), (4, 4), (8, 10); (b) (−1, 1), (2, 4), (6, 8); (c) (1, −1), (3, 4), (5, 8)? (12.15)

12.38. What values of k will make the following trios of points collinear (a) $A(0, 1)$, $B(2, 7)$, $C(6, k)$; (b) $D(-1, 5)$, $E(3, k)$, $F(5, 11)$; (c) $G(0, k)$, $H(1, 1)$, $I(3, -1)$? (12.15)

12.39. State the equation of the line or pair of lines which is the locus of points (12.16)

(a) Whose abscissa is −5
(b) Whose ordinate is $3\frac{1}{2}$
(c) 3 units from the x-axis
(d) 5 units below the x-axis
(e) 4 units from the y-axis
(f) 3 units from the line $x = 2$
(g) 6 units above the line $y = -2$
(h) 1 unit to the right of the y-axis
(i) Equidistant from the lines $x = 5$ and $x = 13$

12.40. State the equation of the locus of the center of a circle that (12.18)

(a) Is tangent to the x-axis at (6, 0)
(b) Is tangent to the y-axis at (0, 5)
(c) Is tangent to the lines $x = 4$ and $x = 8$
(d) Passes through the origin and (10, 0)
(e) Passes through (3, 7) and (9, 7)
(f) Passes through (3, −2) and (3, 8)

12.41. State the equation of the line or pair of lines which is the locus of points (12.16)

(a) Whose coordinates are equal
(b) Whose ordinate is 5 more than the abscissa
(c) Whose abscissa is 4 less than the ordinate
(d) Whose ordinate exceeds the abscissa by 10
(e) The sum of whose coordinates is 12
(f) The difference of whose coordinates is 2
(g) Equidistant from the x-axis and y-axis
(h) Equidistant from $x + y = 3$ and $x + y = 7$

12.42. Describe the locus of each of the following equations: (12.17)

(a) $y = 2x + 5$
(b) $\dfrac{y - 3}{x - 2} = 4$
(c) $\dfrac{y + 3}{x + 2} = \dfrac{5}{4}$
(d) $y = \frac{1}{2}x$
(e) $x + y = 7$
(f) $3y = x$

12.43. State the equation of a line which passes through the origin and has a slope of (a) 4; (b) −2; (c) $\frac{3}{2}$; (d) $-\frac{2}{5}$; (e) 0. (12.17)

12.44. State the equation of a line which has a y- intercept of (12.17)

(a) 5 and a slope of 4
(b) 2 and a slope of -3
(c) -1 and a slope of $\frac{1}{3}$
(d) 8 and is parallel to $y = 3x - 2$
(e) -3 and is parallel to $y = 7 - 4x$
(f) 0 and is parallel to $y - 2x = 8$

12.45. State the equation of a line which has a slope of 2 and passes through (a) (1, 4); (b) $(-2, 3)$; (c) $(-4, 0)$; (d) $(0, -7)$. (12.17)

12.46. State the equation of a line (12.17)

(a) Which passes through the origin and has a slope of 4
(b) Which passes through (0, 3) and has a slope of $\frac{1}{2}$
(c) Which passes through (1, 2) and has a slope of 3
(d) Which passes through $(-1, -2)$ and has a slope of $\frac{1}{3}$
(e) Which passes through the origin and is parallel to a line that has a slope of 2

12.47. (a) Describe the locus of the equation $x^2 + y^2 = 49$. (12.18)

(b) State the equation of the locus of points 4 units from the origin.
(c) State the equations of the locus of points 3 units from the locus of $x^2 + y^2 = 25$.

12.48. State the equation of the locus of point 5 units from (a) the origin; (b) the circle $x^2 + y^2 = 16$; (c) the circle $x^2 + y^2 = 49$. (12.18)

12.49. What is the radius of the circle whose equation is (a) $x^2 + y^2 = 9$; (b) $x^2 + y^2 = \frac{16}{9}$; (c) $9x^2 + 9y^2 = 36$; (d) $x^2 + y^2 = 3$? (12.18)

12.50. What is the equation of a circle whose center is the origin and whose radius is (a) 4; (b) 11; (c) $\frac{2}{3}$; (d) $1\frac{1}{2}$; (e) $\sqrt{5}$; (f) $\frac{1}{3}\sqrt{3}$? (12.18)

12.51. Find the area of $\triangle ABC$, whose vertices are $A(0, 0)$ and (12.19)

(a) $B(0, 5)$ and $C(4, 5)$
(b) $B(0, 5)$ and $C(4, 2)$
(c) $B(0, 8)$ and $C(-5, 8)$
(d) $B(0, 8)$ and $C(-5, 12)$
(e) $B(6, 2)$ and $C(7, 0)$
(f) $B(6, -5)$ and $C(10, 0)$

12.52. Find the area of a (12.19)

(a) Triangle whose vertices are $A(0, 0)$, $B(3, 4)$, and $C(8, 0)$
(b) Triangle whose vertices are $D(1, 1)$, $E(5, 6)$, and $F(1, 7)$
(c) Rectangle three of whose vertices are $H(2, 2)$, $J(2, 6)$, and $K(7, 2)$
(d) Parallelogram three of whose vertices are $L(3, 1)$, $M(9, 1)$, and $P(5, 5)$

12.53. Find the area of $\triangle DEF$, whose vertices are $D(0, 0)$ and (a) $E(6, 4)$ and $F(8, 2)$; (b) $E(3, 2)$ and $F(6, -4)$; (c) $E(-2, 3)$ and $F(10, 7)$. (12.20)

12.54. Find the area of a triangle whose vertices are (a) (0, 0), (2, 3), and (4, 1); (b) (1, 1), (7, 3), and (3, 6); (c) (−1, 2), (0, −2), and (3, 1). (12.20)

12.55. The vertices of $\triangle ABC$ are $A(2, 1)$, $B(8, 9)$, and $C(5, 7)$. (a) Find the area of $\triangle ABC$. (b) Find the length of $\overline{AB}$. (c) Find the length of the altitude to $\overline{AB}$. (12.20)

12.56. Find the area of a quadrilateral whose vertices are (a) (3, 3), (10, 4), (8, 7), and (5, 6); (b) (0, 4), (5, 8), (10, 6), and (14, 0); (c) (0, 1), (2, 4), (8, 10), and (12, 2).

12.57. Find the area of the quadrilateral formed by the lines (12.19)

(a) $x = 0, x = 5, y = 0$, and $y = 6$

(b) $x = 0, x = 7, y = -2$, and $y = 5$

(c) $x = -3, x = 5, y = 3$, and $y = -8$

(d) $x = 0, x = 6, y = 0$, and $y = x + 1$

(e) $y = 0, y = 4, y = x$, and $y = x + 4$

(f) $y = 0, y = 6, y = 2x$, and $y = 2x + 6$

12.58. Prove each of the following with analytic geometry: (12.21)

(a) A line segment joining the midpoints of two sides of a triangle is parallel to the third side.

(b) The diagonals of a rhombus are perpendicular to each other.

(c) The median to the base of an isosceles triangle is perpendicular to the base.

(d) The length of a line segment joining the midpoints of two sides of a triangle equals one-half the length of the third side.

(e) If the midpoints of the sides of a rectangle taken in succession are joined, the quadrilateral formed is a rhombus.

(f) In a right triangle, the length of the median to the hypotenuse is one-half the length of the hypotenuse.

Inequalities and Indirect Reasoning

13.1 Inequalities

An inequality is a statement that quantities are not equal. If two quantities are unequal, the first is either greater than or less than the other. The inequality symbols are: $\neq$, meaning unequal to; $>$, meaning greater than; and $<$, meaning less than. Thus, $4 \neq 3$ is read "four is unequal to three"; $7 > 2$ is read "seven is greater than two"; and $1 < 5$ is read "one is less than five."

Two inequalities may be of the same order or of opposite order. In inequalities of the same order, the same inequality symbol is used; in inequalities of the opposite order, opposite inequality symbols are used. Thus, $5 > 3$ and $10 > 7$ are inequalities of the same order; $5 > 3$ and $7 < 10$ are inequalities of opposite order.

Inequalities of the same order may be combined, as follows. The inequalities $x < y$ and $y < z$ may be combined into $x < y < z$, which states that y is greater than x and less than z. The inequalities $a > b$ and $b > c$ may be combined into $a > b > c$, which states b is less than a and greater than c.

13.1A Inequality Axioms

Axioms are statements that are accepted as true without proof and are used in the same way as theorems.

AXIOM 1: *A quantity may be substituted for its equal in any inequality.*

Thus if $x > y$ and $y = 10$, then $x > 10$.

AXIOM 2: *If the first of three quantities is greater than the second, and the second is greater than the third, then the first is greater than the third.*

Thus if $x > y$ and $y > z$, then $x > z$.

AXIOM 3: *The whole is greater than any of its parts.*

Thus, $AB > AM$ and $m\angle BAD > m\angle BAC$ in Fig. 13-1.

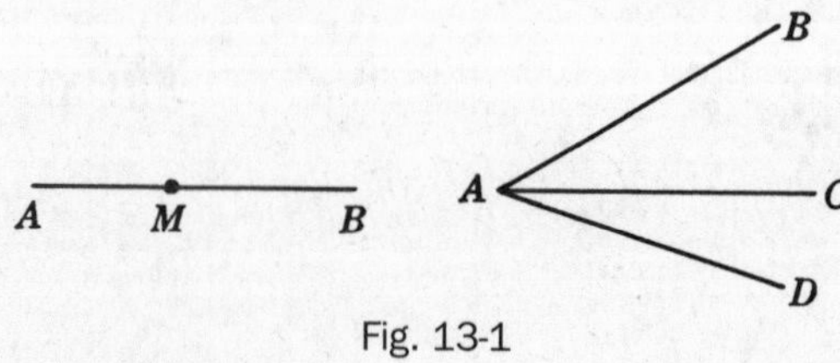

Fig. 13-1

13.1B Inequality Axioms of Operation

AXIOM 4: *If equals are added to unequals, the sums are unequal in the same order.*

Since $5 > 4$ and $4 = 4$, we know that $5 + 4 > 4 + 4$ (or $9 > 8$). If $x - 4 < 5$, then $x - 4 + 4 < 5 + 4$ or $x < 9$.

AXIOM 5: *If unequals are added to unequals of the same order, the sums are unequal in the same order.*

Since $5 > 3$ and $4 > 1$, we have $5 + 4 > 3 + 1$ (or $9 > 4$). If $2x - 4 < 5$ and $x + 4 < 8$, then $2x - 4 + x + 4 < 5 + 8$ or $3x < 13$.

AXIOM 6: *If equals are subtracted from unequals, the differences are unequal in the same order.*

Since $10 > 5$ and $3 = 3$, we have $10 - 3 > 5 - 3$ (or $7 > 2$). If $x + 6 < 9$ and $6 = 6$, then $x + 6 - 6 < 9 - 6$ or $x < 3$.

AXIOM 7: *If unequals are subtracted from equals, the differences are unequal in the opposite order.*

Since $10 = 10$ and $5 > 3$, we have $10 - 5 < 10 - 3$ (or $5 < 7$). If $x + y = 12$ and $y > 5$, then $x + y - y < 12 - 5$ or $x < 7$.

AXIOM 8: *If unequals are multiplied by the same positive number, the products are unequal in the same order.*

Thus if $\frac{1}{4}x < 5$, then $4(\frac{1}{4}x) < 4(5)$ or $x < 20$.

AXIOM 9: *If unequals are multiplied by the same negative number, the results are unequal in the opposite order.*

Thus if $\frac{1}{2}x < 5$, then $(-2)(\frac{1}{2}x) > (-2)(5)$ or $-x > -10$ or $x < 10$.

AXIOM 10: *If unequals are divided by the same positive number, the results are unequal in the same order.*

Thus if $4x > 20$, then $\frac{4x}{4} > \frac{20}{4}$ or $x > 5$.

AXIOM 11: *If unequals are divided by the same negative number, the results are unequal in the opposite order.*

Thus if $-7x < 42$, then $\frac{-7x}{-7} > \frac{42}{-7}$ or $x > -6$.

13.1C Inequality Postulate

POSTULATE 1: *The length of a line segment is the shortest distance between two points.*

13.1D Triangle Inequality Theorems

PRINCIPLE 1: *The sum of the lengths of two sides of a triangle is greater than the length of the third side.* (Corollary: *The length of the longest side of a triangle is less than the sum of the lengths of the other two sides and greater than their difference.*)

Thus in Fig. 13-2, $BC + CA > AB$ and $AB > BC - AC$.

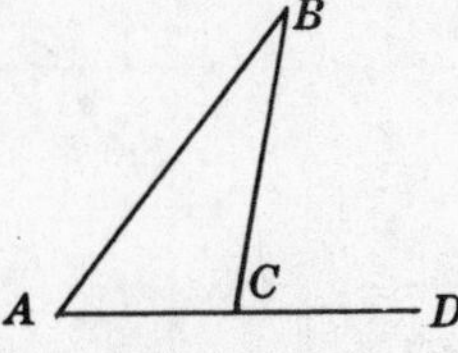

Fig. 13-2

PRINCIPLE 2: *In a triangle, the measure of an exterior angle is larger than the measure of either nonadjacent interior angle.*

Thus in Fig. 13-2, $m\angle BCD > m\angle BAC$ and $m\angle BCD > m\angle ABC$.

PRINCIPLE 3: *If the lengths of two sides of a triangle are unequal, the measures of the angles opposite these sides are unequal, the larger angle being opposite the longer side.* (Corollary: *The largest angle of a triangle is opposite the longest side.*)

Thus in Fig. 13-2, if $BC > AC$, then $m\angle A > m\angle B$.

PRINCIPLE 4: *If the measures of two angles of a triangle are unequal, the lengths of the sides opposite these angles are unequal, the longer side being opposite the larger angle.* (Corollary: *The longest side of a triangle is opposite the largest angle.*)

Thus in Fig. 13-2, if $m\angle A > m\angle B$, then $BC > AC$.

PRINCIPLE 5: *The perpendicular from a point to a line is the shortest segment from the point to the line.*

Thus in Fig. 13-3, if $\overline{PC} \perp \overline{AB}$ and $\overline{PD}$ is any other line from P to $\overline{AB}$, then $PC < PD$.

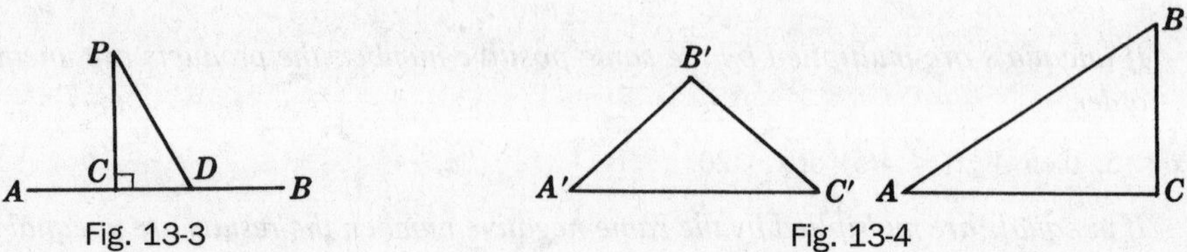

Fig. 13-3 Fig. 13-4

PRINCIPLE 6: *If two sides of a triangle are congruent to two sides of another triangle, the triangle having the greater included angle has the greater third side.*

Thus in Fig. 13-4, if $BC = B'C', AC = A'C'$, and $m\angle C > m\angle C'$, then $AB > A'B'$.

PRINCIPLE 7: *If two sides of a triangle are congruent to two sides of another triangle, the triangle having the greater third side has the greater angle opposite this side.*

Thus in Fig. 13-4, if $BC = B'C', AC = A'C'$, and $AB > A'B'$, then $m\angle C > m\angle C'$.

13.1E Circle Inequality Theorems

PRINCIPLE 8: *In the same or equal circles, the greater central angle has the greater arc.*

Thus in Fig. 13-5, if $m\angle AOB > m\angle COD$, then $m\widehat{AB} > m\widehat{CD}$.

PRINCIPLE 9: *In the same or equal circles, the greater arc has the greater central angle.* (This is the converse of Principle 8.)

Thus in Fig. 13-5, if $m\widehat{AB} > m\widehat{CD}$, then $m\angle AOB > m\angle COD$.

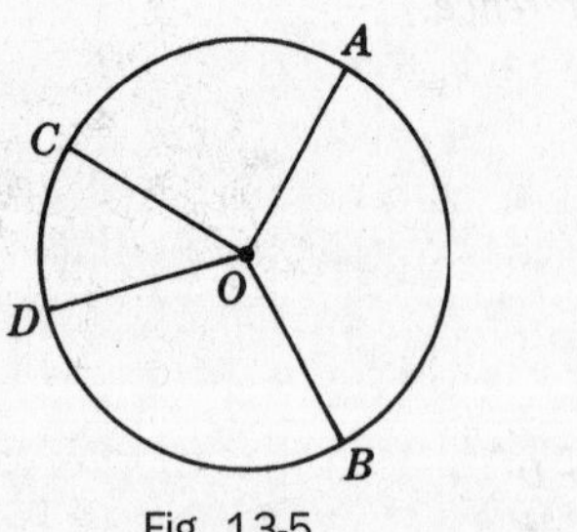

Fig. 13-5

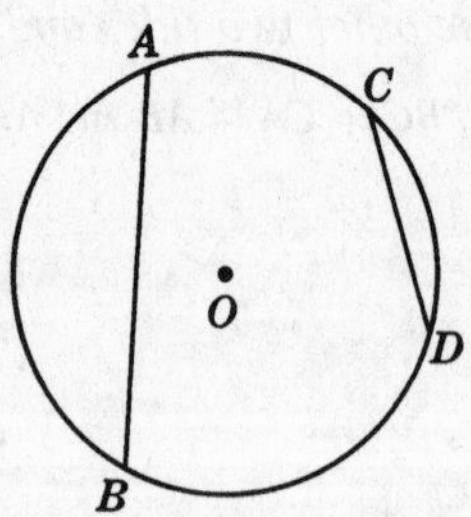

Fig. 13-6

PRINCIPLE 10: *In the same or equal circles, the greater chord has the greater minor arc.*

Thus in Fig. 13-6, if $AB > CD$, then $m\widehat{AB} > m\widehat{CD}$.

PRINCIPLE 11: *In the same or equal circles, the greater minor arc has the greater chord.* (This is the converse of Principle 10.)

Thus in Fig. 13-6, if $m\widehat{AB} > m\widehat{CD}$, then $AB > CD$.

PRINCIPLE 12: *In the same or equal circles, the greater chord is at a smaller distance from the center.*

Thus in Fig. 13-7, if $AB > CD$, then $OE < OF$.

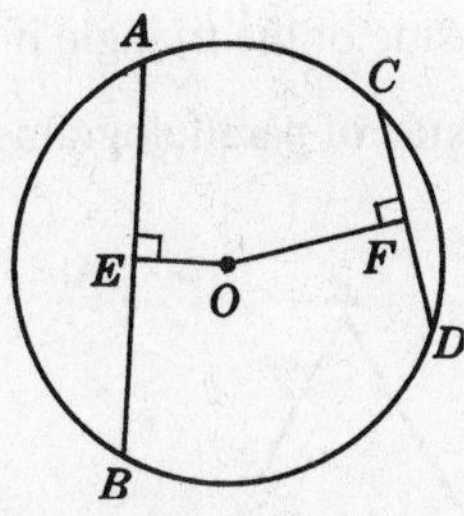

Fig. 13-7

PRINCIPLE 13: *In the same or equal circles, the chord at the smaller distance from the center is the greater chord.* (This is the converse of Principle 12.)

Thus in Fig. 13-7, if $OE < OF$, then $AB > CD$.

SOLVED PROBLEMS

13.1 Selecting inequality symbols

Determine which inequality symbol, > or <, makes each of the following true:

(a) 5 _?_ 3 (c) −5 _?_ 3 (e) If $x = 3$, then x^2 _?_ x.

(b) 6 _?_ 9 (d) −5 _?_ −3 (f) If $x > 10$, then 10 _?_ x.

Solutions

(a) > (b) < (c) < (d) < (e) > (f) <

13.2 Applying inequality axioms

Complete each of the following statements:

(a) If $a > b$ and $b > 8$, then a _?_ 8.

(b) If $x > y$ and $y = 15$, then x _?_ 15.

(c) If $c < 20$ and $d < 5$, then $c + d$ _?_ 25.

(d) If $x > y$ and $y > 6$, then x _?_ y _?_ 6.

(e) If $x > y$, then $\frac{1}{2}x$ _?_ $\frac{1}{2}y$.

(f) If $e < \frac{1}{4}f$, then $4e$ _?_ f.

(g) If $-y < z$ then y _?_ $-z$.

(h) If $-4x > p$, then x _?_ $-\frac{1}{4}p$.

(i) If Paul and Jack have equal amounts of money and Paul spends more than Jack, then Paul will have _?_ than Jack.

(j) If Anne is now older than Helen, then 10 years ago, Anne was _?_ than Helen.

Solutions

(a) > (c) < (e) > (g) > (i) less

(b) > (d) >, > (f) < (h) < (j) older

13.3 Applying triangle inequality theorems (Fig. 13-8)

(a) Determine the integer values that the length of side a of the triangle can have if the other two sides have lengths 3 and 7.

(b) Determine which is the longest side of the triangle if two angles have measures 59° and 60°.

(c) Determine which is the longest side of parallelogram $ABCD$ if E is the midpoint of the diagonals and $m\angle AEB > m\angle AED$.

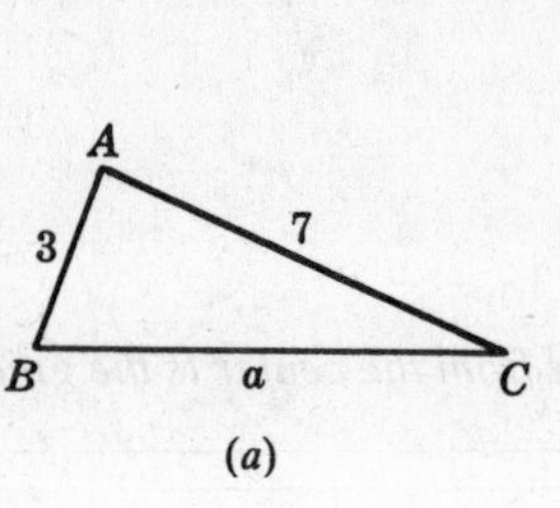

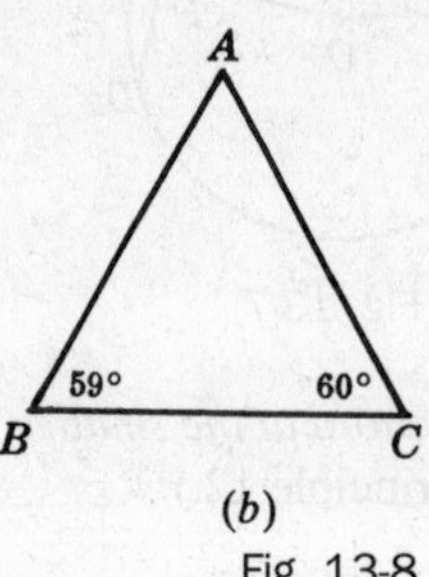

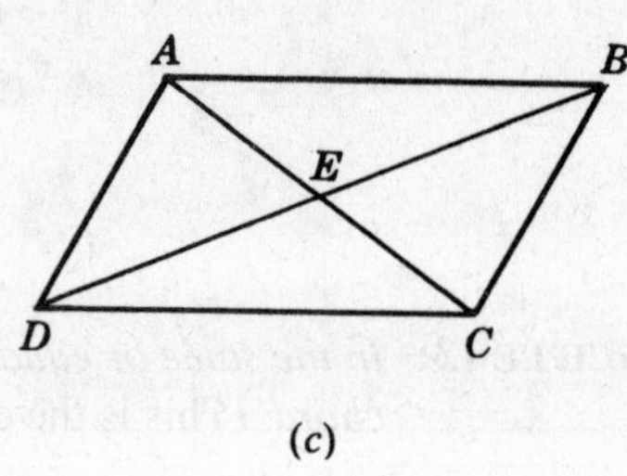

Fig. 13-8

Solutions

(a) Since a must be less than $3 + 7 = 10$ and greater than $7 - 3 = 4$, a can have the integer values of 5, 6, 7, 8, 9.

(b) Since $m\angle B = 59°$ and $m\angle C = 60°$, $m\angle A = 180° - (59° + 60°) = 61°$. Then the longest side is opposite the largest angle, $\angle A$, so the longest side is BC.

(c) In $\square ABCD$, $AE = CE$ and $DE = EB$. Since $m\angle AEB > m\angle AED$, $AB > AD$ or AB ($= DC$) is the longest side (Principle 6).

13.4 Applying circle inequality theorems

In Fig. 13-9, compare

(a) OD and OF if $\angle C$ is the largest angle of $\triangle ABC$

(b) AC and BC if $m\overset{\frown}{AC} > m\overset{\frown}{BC}$

(c) $m\overset{\frown}{BC}$ and $m\overset{\frown}{AC}$ if $OF > OE$

(d) $m\angle AOB$ and $m\angle BOC$ if $m\overset{\frown}{AB} > m\overset{\frown}{BC}$

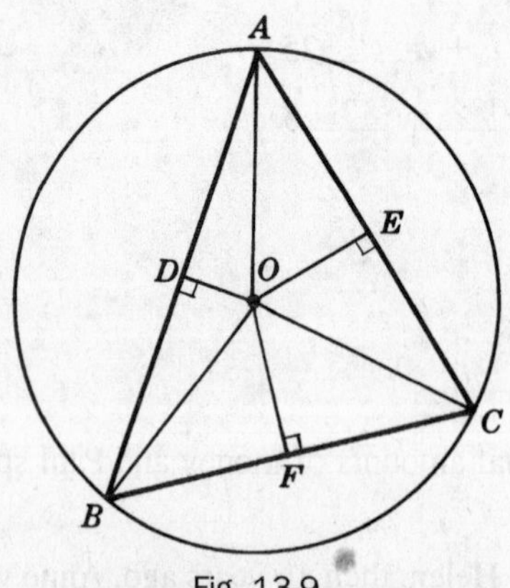

Fig. 13-9

Solutions

(a) Since $\angle C$ is the largest angle of the triangle, $\overline{AB}$ is the longest side, or $AB > BC$; hence, $OD < OF$ by Principle 12.

(b) Since $m\overset{\frown}{AC} > m\overset{\frown}{BC}$, $AC > BC$, the greater arc having the greater chord.

(c) Since $OF > OE$, $BC < AC$ by Principle 13; hence, $m\overset{\frown}{BC} < m\overset{\frown}{AC}$ by Principle 10.

(d) Since $m\overset{\frown}{AB} > m\overset{\frown}{BC}$, $m\angle AOB > m\angle BOC$, the greater arc having the greater central angle.

13.5 Proving an inequality problem

Prove that in $\triangle ABC$, if M is the midpoint of $\overline{AC}$ and $BM > AM$, then $m\angle A + m\angle C > m\angle B$.

Given: $\triangle ABC$, M is midpoint of AC.
$BM > AM$

To Prove: $m\angle A + m\angle C > m\angle B$

Plan: Prove $m\angle A > m\angle 1$ and $m\angle C > m\angle 2$ and then add unequals.

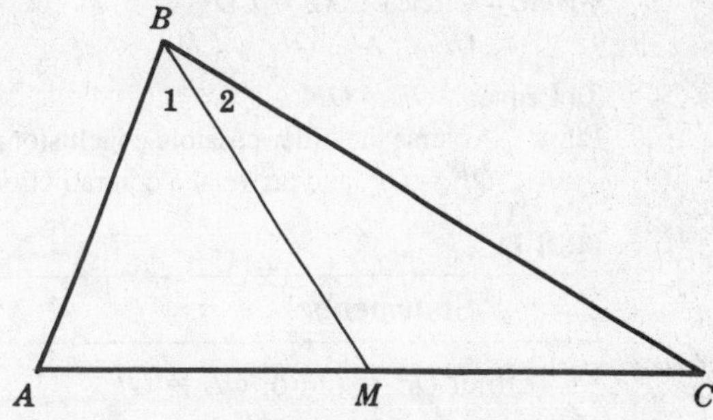

PROOF:

Statements	Reasons
1. M is the midpoint of $\overline{AC}$.	1. Given
2. $\overline{AM} \cong \overline{MC}$	2. A midpoint divides a line into two congruent parts.
3. $BM > AM$	3. Given
4. $BM > MC$	4. A quantity may be substituted for its equal in any inequality. Definition of congruent segments.
5. In $\triangle AMB$, $m\angle A > \angle 1$. In $\triangle BMC$, $m\angle C > \angle 2$.	5. In a triangle, the larger angle lies opposite the longer side.
6. $m\angle A + m\angle C > m\angle B$	6. If unequals are added to unequals, the sums are unequal in the same order.

13.2 Indirect Reasoning

We often arrive at a correct conclusion by *indirect* reasoning. In this form of reasoning, the correct conclusion is reached by eliminating all possible conclusions except one. The remaining possibility must be the correct one. Suppose we are given the years 1492, 1809, and 1960 and are assured that one of these years is the year in which a president of the United States was born. By eliminating 1492 and 1960 as impossibilities, we know by indirect reasoning that 1809 is the correct answer. (Had we known that 1809 was the year in which Lincoln was born, the reasoning would have been direct.)

In proving a theorem by indirect reasoning, a possible conclusion may be eliminated if we assume it is true and that assumption results in a contradiction of some given or known fact.

SOLVED PROBLEMS

13.6 Applying indirect reasoning in life situations

Explain how indirect reasoning is used in each of the following situations:

(a) A detective determines the murderer of a slain person.

(b) A librarian determines which volume of a set of books is in use.

Solutions

(a) The detective, using a list of all those who could have been a murderer in the case, eliminates all except one. He or she concludes that the remaining one is the murderer.

(b) The librarian finds all the books of the set except one by looking on the shelf and checking the records. He or she concludes that the missing one is the one in use.

13.7 Proving an inequality theorem by the indirect method

Prove that in the same or equal circles, unequal chords are unequally distant from the center.

Given: Circle O, $AB \neq CD$
$\overline{OE} \perp \overline{AB}$, $\overline{OF} \perp \overline{CD}$

To Prove: $OE \neq OF$

Plan: Assume the other possible conclusion, $OE = OF$, and arrive at a contradiction.

PROOF:

Statements	Reasons
1. Either $OE = OF$ or $OE \neq OF$.	1. Two quantities are either equal or unequal.
2. Assume $OE = OF$.	2. This is one of the possible conclusions.
3. If $OE = OF$, then $AB = CD$.	3. In the same or equal circles, chords equally distant from the center are equal.
4. But $AB \neq CD$.	4. Given
5. The assumption $OE = OF$ is not valid.	5. It leads to a contradiction.
6. Hence, $OE \neq OF$.	6. This is the only remaining possibility.

SUPPLEMENTARY PROBLEMS

13.1. Determine which inequality symbol, > or <, makes each of the following true: (13.1)

(a) If $y > 15$, then 15 __?__ y.

(b) If $x = 2$, then $3x - 1$ __?__ 4.

(c) If $x = 2$ and $y = 3$, then xy __?__ 5.

(d) If $a = 4$ and $b = \frac{1}{4}$, then a/b __?__ 15.

(e) If $a = 5$, then a^2 __?__ $4a$.

(f) If $b = \frac{1}{2}$, then b^2 __?__ b.

13.2. Complete each of the following statements: (13.2)

(a) If $y > x$ and $x = z$, then y __?__ z.

(b) If $a + b > c$ and $b = d$, then $a + d$ __?__ c.

(c) If $a < b$ and $b < 15$, then a __?__ 15.

(d) If $z > y$, $y > x$, and $x = 10$, then z __?__ 10.

13.3. Complete each of the following statements about Fig. 13-10: (13.2)

(a) BC __?__ BD

(b) $m\angle BAD$ __?__ $m\angle BAC$

(c) $\triangle ADC$ __?__ $\triangle ABC$

(d) If $m\angle A = m\angle C$, then AB __?__ BD.

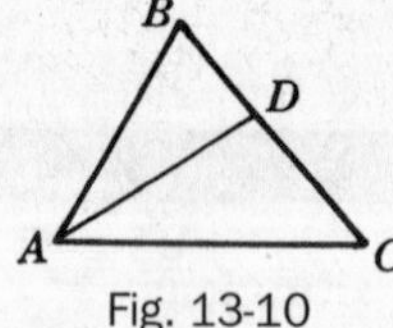

Fig. 13-10

13.4. Complete each of the following statements: (13.2)

(a) If Mary and Ann earn the same weekly wage and Mary is to receive a larger increase than Ann, then Mary will earn __?__ Ann earns.

(b) If Bernice, who is the same weight as Helen, loses more weight than Helen, then Bernice will weigh __?__ Helen weighs.

13.5. Complete each of the following statements: (13.2)

(a) If $a > 3$, then $4a$ __?__ 12.

(b) If $x - 3 > 15$, then x __?__ 18.

(c) If $3x < 18$, then x __?__ 6.

(d) If $f > 8$, then $f + 7$ __?__ 15.

(e) If $x = y$, then $x + 5$ __?__ $y + 6$.

(f) If $g = h$, then $g - 10$ __?__ $h - 9$.

13.6. Which of the following sets of numbers can be the lengths of the sides of a triangle? (13.3)

(a) 3, 4, 8 (b) 5, 7, 12 (c) 3, 4, 6 (d) 2, 7, 8 (e) 50, 50, 5

13.7. What integer values can the length of the third side of a triangle have if the two sides have lengths (a) 2 and 6; (b) 3 and 8; (c) 4 and 7; (d) 4 and 6; (e) 4 and 5; (f) 7 and 7 ? (13.3)

13.8. In Fig. 13-11, arrange, in descending order of size, (a) the angles of $\triangle ABC$; (b) the sides of $\triangle DEF$; (c) the angles 1, 2, and 3. (13.3)

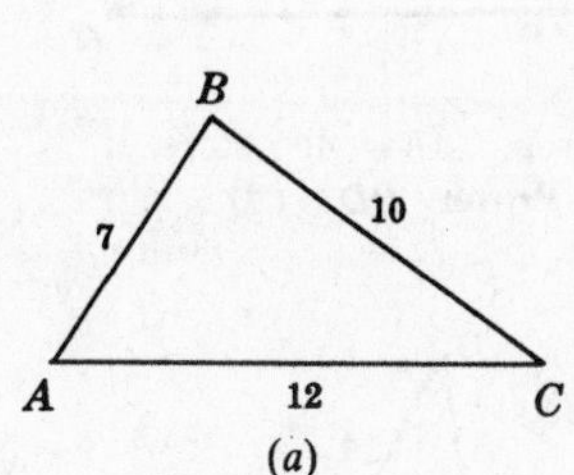

(*a*)

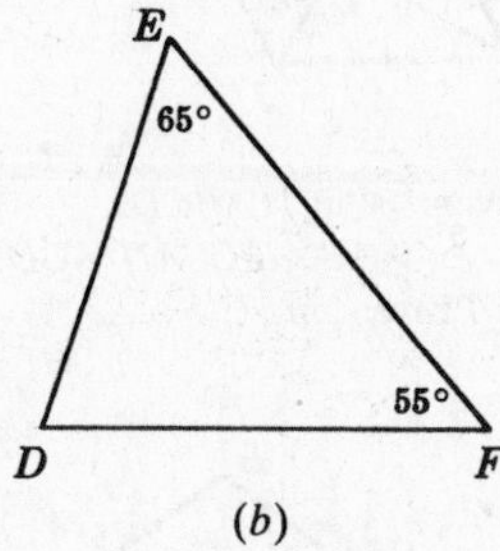

(*b*)

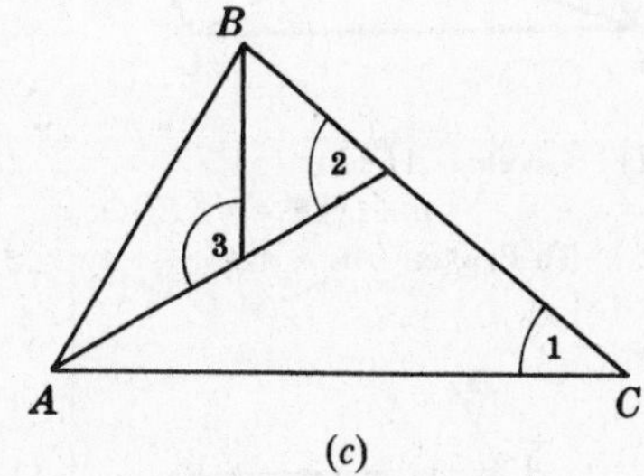

(c)

Fig. 13-11

13.9. (a) In quadrilateral *ABCD* of Fig. 13-12, compare $m\angle BAC$ and $m\angle ACD$ if $AB = CD$ and $BC > AD$.

(b) In $\triangle ABC$ of Fig. 13-13, compare *AB* and *BC* if $\overline{BM}$ is the median to $\overline{AC}$ and $m\angle AMB > m\angle BMC$. (13.3)

13.10. Arrange, in descending order of magnitude, (13.4)

(a) The sides of $\triangle ABC$ in Fig. 13-14

(b) The central angles *AOB*, *BOC*, and *AOC* in Fig. 13-14

(c) The sides of trapezoid *ABCD* in Fig. 13-15

(d) The distances of the sides of $\triangle DEF$ from the center in Fig. 13-16

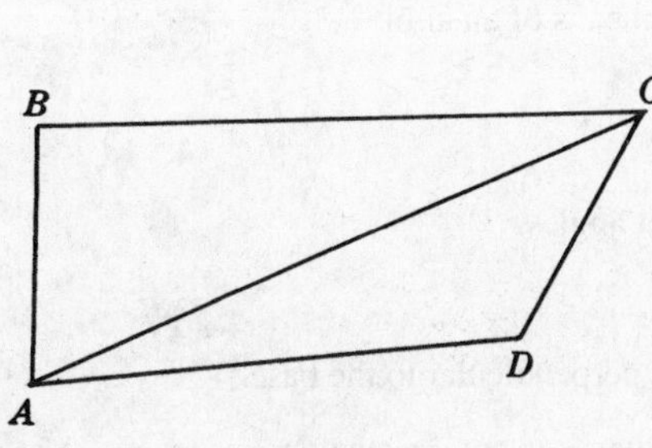

Fig. 13-12

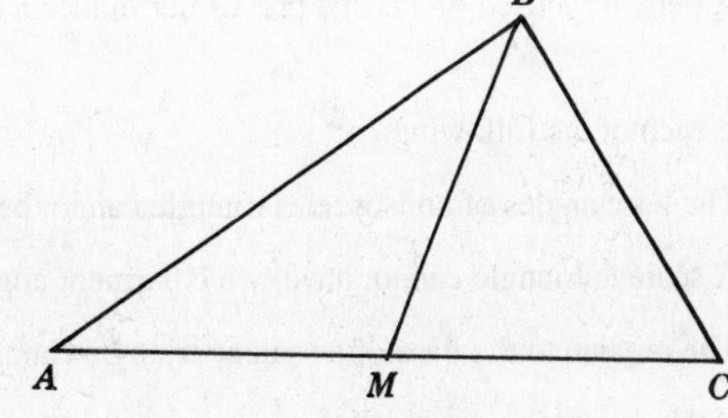

Fig. 13-13

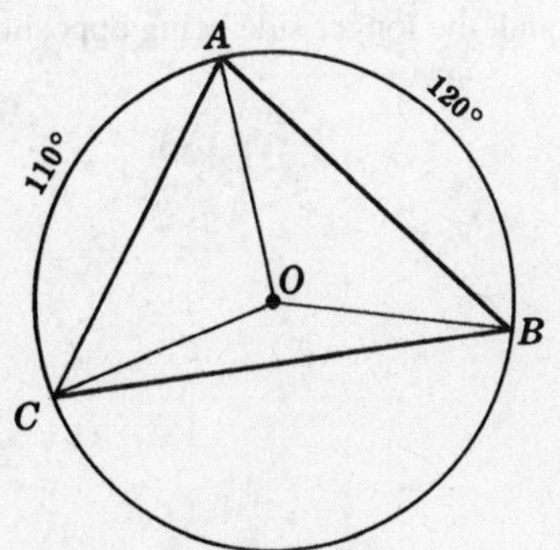

Fig. 13-14

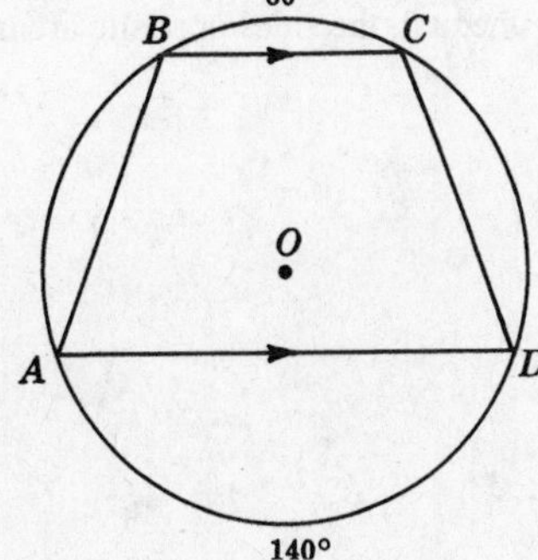

Fig. 13-15

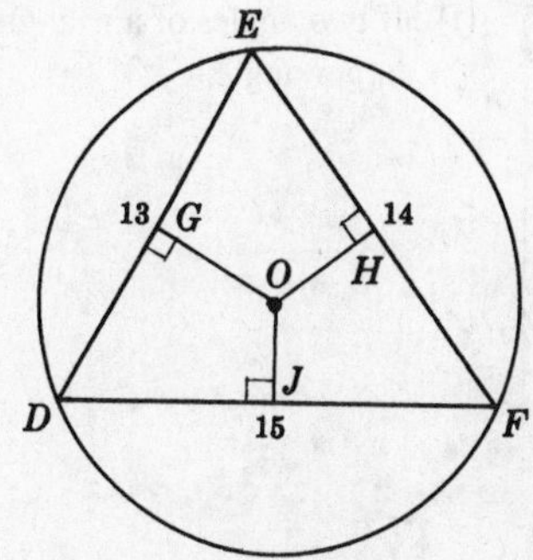

Fig. 13-16

13.11. Provide the proofs requested in Fig. 13-17. (13.5)

(*a*) **Given:** Parallelogram $ABCD$
$AC > BD$
To Prove: $m\angle BDA > m\angle CAD$

(*b*) **Given:** Rhombus $FGHJ$
$m\angle G > m\angle F$
To Prove: $FL > GL$

(*c*) **Given:** $\overline{AD}$ bisects $\angle A$
$\overline{CD}$ bisects $\angle C$, $AB > BC$
To Prove: $AD > CD$

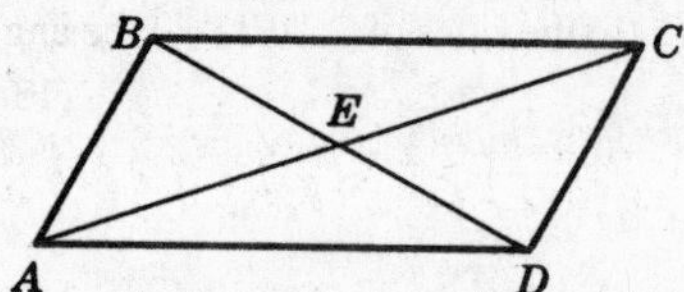

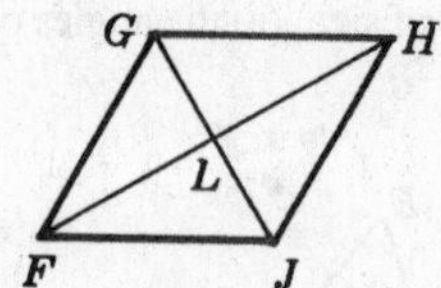

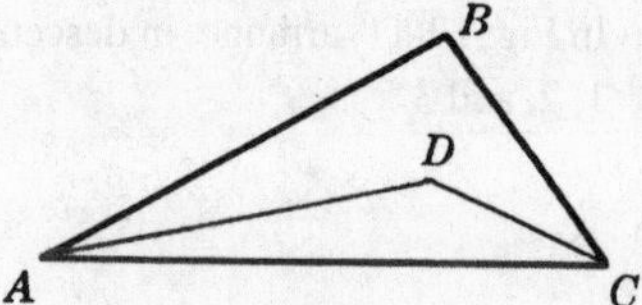

(*d*) **Given:** $\overrightarrow{AE} \parallel \overrightarrow{BC}$
$m\angle DAE > m\angle EAC$
To Prove: $AC > AB$

(*e*) **Given:** Quad. $ABCD$
$AB > BC$, $AD > CD$
To Prove: $m\angle C > m\angle A$

(*f*) **Given:** $AB = AC$
To Prove: $BD > CD$

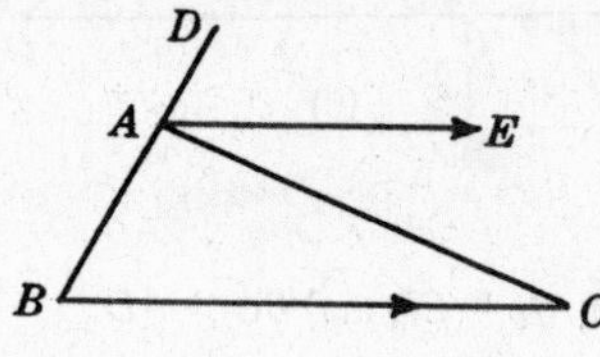

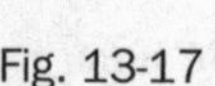

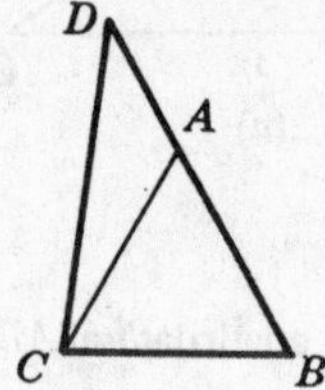

Fig. 13-17

13.12. Explain how indirect reasoning is used in each of the following situations: (13.6)

(a) A person determines which of his ties has been borrowed by his roommate.

(b) A girl determines that the electric motor in her train set is not defective even though her toy trains do not run.

(c) A teacher finds which of his students did not do their assigned homework.

(d) A mechanic finds the reason why the battery in a car does not work.

(e) A person accused of a crime proves her innocence by means of an alibi.

13.13. Prove each of the following: (13.7)

(a) The base angles of an isosceles triangle cannot be right angles.

(b) A scalene triangle cannot have two congruent angles.

(c) The median to the base of a scalene triangle cannot be perpendicular to the base.

(d) If the diagonals of a parallelogram are not congruent, then it is not a rectangle.

(e) If a diagonal of a parallelogram does not bisect a vertex angle, then the parallelogram is not a rhombus.

(f) If two angles of a triangle are unequal, the sides opposite are unequal, the longer side being opposite the larger angle.

Improvement of Reasoning

14.1 Definitions

"Was Lincoln an educated man?" is a question that cannot be properly answered unless we agree upon the meaning of "an educated man." Understanding cannot exist and progress cannot be made in any discussion or problem unless the terms involved are properly defined or, by agreement, are to be undefined.

14.1A Requirements of a Good Definition

PRINCIPLE 1: *All terms in a definition must have been previously defined (or be those that, by agreement, are left undefined).*

Thus, if we are to define a regular polygon as an equilateral and equiangular polygon, it is necessary that equilateral, equiangular, and polygon be previously defined.

PRINCIPLE 2: *The term being defined should be placed in the next larger set or class to which it belongs.*

Thus, the terms polygon, quadrilateral, parallelogram, and rectangle should be defined in that order. Once the term polygon has been defined, the term quadrilateral is then defined as a kind of polygon. Then the term parallelogram is defined as a kind of quadrilateral and, lastly, the term rectangle is defined as a kind of parallelogram.

Proper sequence in definition can be understood by using a circle to represent a set of objects. In Fig. 14-1, the set of rectangles is in the next larger set of parallelograms. In turn, the set of parallelograms is in the next larger set of quadrilaterals, and, finally, the set of quadrilaterals is in the next larger set of polygons.

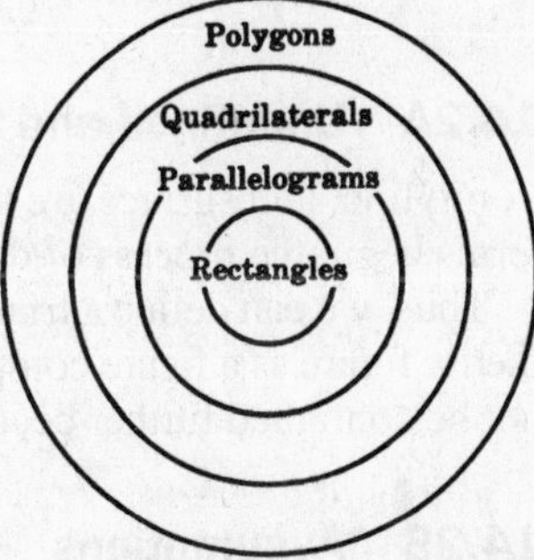

Fig. 14-1

PRINCIPLE 3: *The term being defined should be distinguished from all other members of its class.*

Thus, the definition of a triangle as a polygon with three sides is a good one since it shows how the triangle differs from all other polygons.

PRINCIPLE 4: *The distinguishing characteristics of a defined term should be as few as possible.*

Thus, a right triangle should be defined as a triangle having a right angle, and not as a triangle having a right angle and two acute angles.

SOLVED PROBLEMS

14.1 Observing proper sequence in definition

In which order should the terms in each of the following sets be defined: (a) Englishman, European, Londoner; (b) quadrilateral, square, rectangle, parallelogram.

Solutions

(a) European, Englishman, Londoner

(b) Quadrilateral, parallelogram, rectangle, square

14.2 Correcting faulty definitions

Correct the following definition: A trapezoid is a quadrilateral having two parallel sides.

Solution

The given definition is incomplete. The correct definition is "a trapezoid is a quadrilateral having only two parallel sides." This definition distinguishes a trapezoid from a parallelogram.

14.2 Deductive Reasoning in Geometry

The kinds of terms and statements discussed in this section comprise the deductive structure of geometry, which can be visualized as in Fig. 14-2.

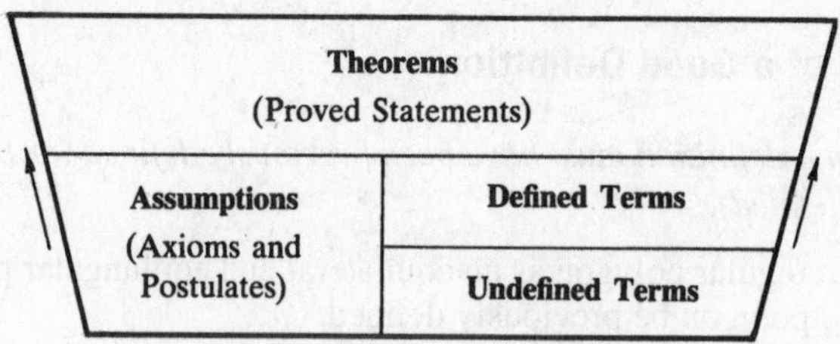

DEDUCTIVE STRUCTURE OF GEOMETRY

Fig. 14-2

14.2A Undefined and Defined Terms

Point, line, and surface are the terms in geometry which are, by agreement, not defined. These undefined terms begin the process of definition in geometry and underlie the definitions of all other geometric terms.

Thus, we can define a triangle in terms of a polygon, a polygon in terms of a geometric figure, and a geometric figure as a figure composed of line segments, or parts of lines. However, the process of definition cannot be continued further because the term "line" is undefined.

14.2B Assumptions

Postulates and axioms are the statements which are not proved in geometry. They are called assumptions because we willingly accept them as true. These assumptions enable us to begin the process of proof in the same way that undefined terms enable us to begin the process of definition.

Thus, when we draw a line segment between two points, we justify this by using as a reason the postulate "two points determine one and only one straight line." This reason is an assumption since we assume it to be true without requiring further justification.

14.2C Theorems

Theorems are the statements which are proved in geometry. By using definitions and assumptions as reasons, we deduce or prove the basic theorems. As we use each new theorem to prove still more theorems, the process of deduction grows. However, if a new theorem is used to prove a previous one, the logical sequence is violated.

For example, the theorem "the sum of the measures of the angles of a triangle equals 180°" is used to prove that "the sum of the measures of the angles of a pentagon is 540°." This, in turn, enables us to prove that "each angle of a regular pentagon measures 108°." However, it would be violating logical sequence if we tried to use the last theorem to prove either of the first two.

14.3 Converse, Inverse, and Contrapositive of a Statement

DEFINITION 1: *The converse of a statement* is the statement that is formed by interchanging the hypothesis and conclusion.

Thus, the converse of the statement "lions are wild animals" is "wild animals are lions." Note that the converse is not necessarily true.

DEFINITION 2: *The negative of a statement is the denial of the statement.*

Thus, the negative of the statement "a burglar is a criminal" is "a burglar is not a criminal."

DEFINITION 3: *The inverse of a statement is formed by denying both the hypothesis and the conclusion.*

Thus, the inverse of the statement "a burglar is a criminal" is "a person who is not a burglar is not a criminal." Note that the inverse is not necessarily true.

DEFINITION 4: *The contrapositive of a statement is formed by interchanging the negative of the hypothesis with the negative of the conclusion. Hence, the contrapositive is the converse of the inverse and the inverse of the converse.*

Thus, the contrapositive of the statement "if you live in New York City, then you will live in New York State" is "if you do not live in New York State, then you do not live in New York City." Note that both statements are true.

14.3A Converse, Inverse, and Contrapositive Principles

PRINCIPLE 1: *A statement is considered false if one false instance of the statement exists.*

PRINCIPLE 2: *The converse of a definition is true.*

Thus, the definition "a quadrilateral is a four-sided polygon" and its converse "a four-sided polygon is a quadrilateral" are both true.

PRINCIPLE 3: *The converse of a true statement other than a definition is not necessarily true.*

The statement "vertical angles are congruent angles" is true, but its converse, "congruent angles are vertical angles" is not necessarily true.

PRINCIPLE 4: *The inverse of a true statement is not necessarily true.*

The statement "a square is a quadrilateral" is true, but its inverse, "a non-square is not a quadrilateral," is not necessarily true.

PRINCIPLE 5: *The contrapositive of a true statement is true, and the contrapositive of a false statement is false.*

The statement "a triangle is a square" is false, and its contrapositive, "a non-square is not a triangle," is also false.

The statement "right angles are congruent angles" is true, and its contrapositive, "angles that are not congruent are not right angles," is also true.

14.3B Logically Equivalent Statements

Logically equivalent statements are pairs of related statements that are either both true or both false. Thus according to Principle 5, a statement and its contrapositive are logically equivalent statements. Also, the converse and inverse of a statement are logically equivalent, since each is the contrapositive of the other.

The relationships among a statement and its inverse, converse, and contrapositive are summed up in the rectangle of logical equivalency in Fig. 14-3:

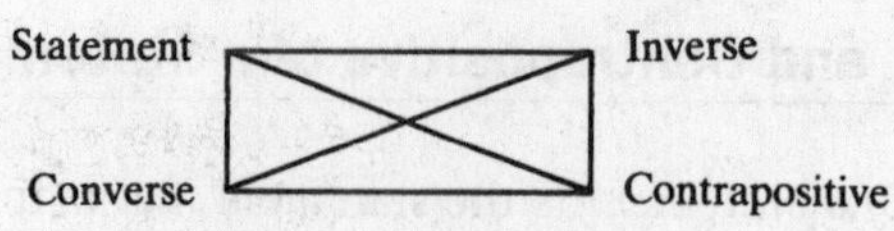

Rectangle of Logical Equivalency

Fig. 14-3

1. Logically equivalent statements are at diagonally opposite vertices. Thus, the logically equivalent pairs of statements are (a) a statement and its contrapositive, and (b) the inverse and converse of the same statement.
2. Statements that are not logically equivalent are at adjacent vertices. Thus, pairs of statements that are not logically equivalent are (a) a statement and its inverse, (b) a statement and its converse, (c) the converse and contrapositive of the same statement, and (d) the inverse and contrapositive of the same statement.

SOLVED PROBLEMS

14.3 Converse of a statement

State the converse of each of the following statements, and indicate whether or not it is true.

(a) Supplementary angles are two angles the sum of whose measures is 180°.

(b) A square is a parallelogram with a right angle.

(c) A regular polygon is an equilateral and equiangular polygon.

Solutions

(a) Two angles the sum of whose measures is 180° are supplementary. (True)

(b) A parallelogram with a right angle is a square. (False)

(c) An equilateral and equiangular polygon is a regular polygon. (True)

14.4 Negative of a statement

State the negative of (a) $a = b$; (b) $m\angle B \neq m\angle C$; (c) $\angle C$ is the complement of $\angle D$; (d) "the point does not lie on the line."

Solutions

(a) $a \neq b$

(b) $m\angle B = m\angle C$

(c) $\angle C$ is not the complement of $\angle D$.

(d) The point lies on the line.

14.5 Inverse of a statement

State the inverse of each of the following statements, and indicate whether or not it is true.

(a) A person born in the United States is a citizen of the United States.

(b) A sculptor is a talented person.

(c) A triangle is a polygon.

Solutions

(a) A person who is not born in the United States is not a citizen of the United States. (False, since there are naturalized citizens)

(b) One who is not a sculptor is not a talented person. (False, since one may be a fine musician, etc.)

(c) A figure that is not a triangle is not a polygon. (False, since the figure may be a quadrilateral, etc.)

14.6 Forming the converse, inverse, and contrapositive

State the converse, inverse, and contrapositive of the statement "a square is a rectangle." Determine the truth or falsity of each, and check the logical equivalence of the statement and its contrapositive, and of the converse and inverse.

Solutions

Statement: A square is a rectangle. (True)
Converse: A rectangle is a square. (False)
Inverse: A figure that is not a square is not a rectangle. (False)
Contrapositive: A figure that is not a rectangle is not a square. (True)
Thus, the statement and its contrapositive are true and the converse and inverse are false.

14.4 Partial Converse and Partial Inverse of a Theorem

A partial converse of a theorem is formed by interchanging any one condition in the hypothesis with one consequence in the conclusion.

A partial inverse of a theorem is formed by denying one condition in the hypothesis and one consequence in the conclusion.

Thus from the theorem "if a line bisects the vertex angle of an isosceles triangle, then it is an altitude to the base," we can form a partial inverse or partial converse as shown in Fig. 14-4.

In forming a partial converse or inverse, the basic figure, such as the triangle in Fig. 14-4, is kept and not interchanged or denied.

(*a*) Theorem

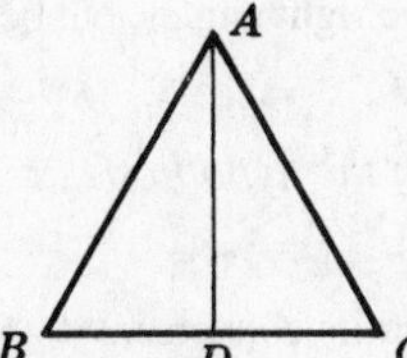

Given: $\triangle ABC$
(*1*) $\overline{AB} \cong \overline{AC}$
(2) $\overline{AD}$ bisects $\angle A$.
To Prove: (*3*) $\overline{AD}$ is the altitude to $\overline{BC}$.

(*b*) Partial Converse

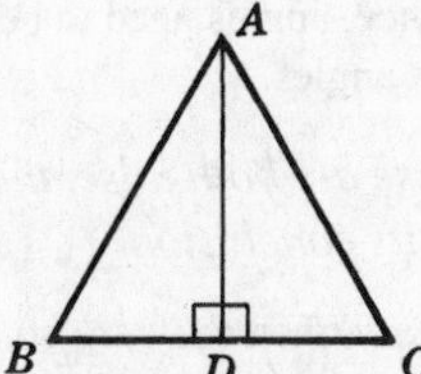

Given: $\triangle ABC$
(2) $\overline{AD}$ bisects $\angle A$.
(3) $\overline{AD}$ is the altitude to $\overline{BC}$.
To Prove: (*1*) $\overline{AB} \cong \overline{AC}$.

(*c*) Partial Inverse

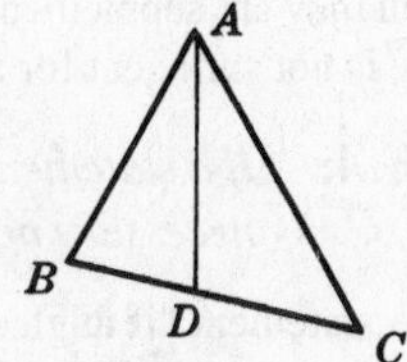

Given: $\triangle ABC$
(*1′*) $\overline{AB} \ncong AC$
(2) $\overline{AD}$ bisects $\angle A$.
To Prove: (*3′*) $\overline{AD}$ is not the altitude to $\overline{BC}$.

Fig. 14-4

In Fig. 14-4(b), the partial converse is formed by interchanging statements (1) and (3). Stated in words, the partial converse is: "If the bisector of an angle of a triangle is an altitude, then the triangle is isosceles." Another partial converse may be formed by interchanging (2) and (3).

In Fig. 14-4(c), the partial inverse is formed by replacing statements (1) and (3) with their negatives, (1′) and (3′). Stated in words, the partial inverse is: "If two sides of a triangle are not congruent, the line segment that bisects their included angle is not an altitude to the third side." Another partial inverse may be formed by negating (2) and (3).

SOLVED PROBLEMS

14.7 Forming partial converses with partial inverses of a theorem

Form (a) partial converses and (b) partial inverses of the statement "congruent supplementary angles are right angles."

Solutions

(a) Partial converses: (1) Congruent right angles are supplementary.
(2) Supplementary right angles are congruent.

(b) Partial inverses: (1) Congruent angles that are not supplementary are not right angles.
(2) Supplementary angles that are not congruent are not right angles.

14.5 Necessary and Sufficient Conditions

In logic and in geometry, it is often important to determine whether the conditions in the hypothesis of a statement are necessary or sufficient to justify its conclusion. This is done by ascertaining the truth or falsity of the statement and its converse, and then applying the following principles.

PRINCIPLE 1: *If a statement and its converse are both true, then the conditions in the hypothesis of the statement are necessary and sufficient for its conclusion.*

For example, the statement "if angles are right angles, then they are congruent and supplementary" is true, and its converse, "if angles are congruent and supplementary, then they are right angles" is also true. Hence, being right angles is necessary and sufficient for the angles to be congruent and supplementary.

PRINCIPLE 2: *If a statement is true and its converse is false, then the conditions in the hypothesis of the statement are sufficient but not necessary for its conclusion.*

The statement "if angles are right angles, then they are congruent" is true, and its converse, "if angles are congruent, then they are right angles," is false. Hence, being right angles is sufficient for the angles to be congruent. However, the angles need not be right angles to be congruent.

PRINCIPLE 3: *If a statement is false and its converse is true, then the conditions in the hypothesis are necessary but not sufficient for its conclusion.*

The statement "if angles are supplementary, then they are right angles" is false, and its converse, "if angles are right angles, then they are supplementary," is true. Hence, angles need to be supplementary to be right angles, but being supplementary is not sufficient for angles to be right angles.

PRINCIPLE 4: *If a statement and its converse are both false, then the conditions in the hypothesis are neither necessary nor sufficient for its conclusion.*

Thus the statement "if angles are supplementary, then they are congruent" is false, and its converse, "if angles are congruent, then they are supplementary," is false. Hence, being supplementary is neither necessary nor sufficient for the angles to be congruent.

These principles are summarized in the table that follows.

When the Conditions in the Hypothesis of a Statement are Necessary or Sufficient to Justify its Conclusion

Principle	Statement	Converse	Sufficient?	Necessary?
1	True	True	Yes	Yes
2	True	False	Yes	No
3	False	True	No	Yes
4	False	False	No	No

SOLVED PROBLEMS

14.8 Determining necessary and sufficient conditions

For each of the following statements, determine whether the conditions in the hypothesis are necessary or sufficient to justify the conclusion.

(a) A regular polygon is equilateral and equiangular.

(b) An equiangular polygon is regular.

(c) A regular polygon is equilateral.

(d) An equilateral polygon is equiangular.

Solutions

(a) Since the statement and its converse are both true, the conditions are necessary and sufficient.

(b) Since the statement is false and its converse is true, the conditions are necessary but not sufficient.

(c) Since the statement is true and its converse is false, the conditions are sufficient but not necessary.

(d) Since both the statement and its converse are false, the conditions are neither necessary nor sufficient.

SUPPLEMENTARY PROBLEMS

14.1. State the order in which the terms in each of the following sets should be defined: (14.1)

(a) Jewelry, wedding ring, ornament, ring

(b) Automobile, vehicle, commercial automobile, taxi

(c) Quadrilateral, rhombus, polygon, parallelogram

(d) Obtuse triangle, obtuse angle, angle, isosceles obtuse triangle

14.2. Correct each of the following definitions: (14.2)

(a) A regular polygon is an equilateral polygon.

(b) An isosceles triangle is a triangle having at least two congruent sides and angles.

(c) A pentagon is a geometric figure having five sides.

(d) A rectangle is a parallelogram whose angles are right angles.

(e) An inscribed angle is an angle formed by two chords.

(f) A parallelogram is a quadrilateral whose opposite sides are congruent and parallel.

(g) An obtuse angle is an angle larger than a right angle.

14.3. State the negative of each of the following statements: (14.4)

(a) $x + 2 = 4$

(b) $3y \neq 15$

(c) She loves you.

(d) His mark was more than 65.

(e) Joe is heavier than Dick.

(f) $a + b \neq c$

14.4. State the inverse of each of the following statements, and indicate whether or not it is true. (14.5)

(a) A square has congruent diagonals.

(b) An equiangular triangle is equilateral.

(c) A bachelor is an unmarried person.

(d) Zero is not a positive number.

14.5. State the converse, inverse, and contrapositive of each of the following statements. Indicate the truth or falsity of each, and check the logical equivalence of the statement and its contrapositive, and of the converse and inverse. (14.6)

(a) If two sides of a triangle are congruent, the angles opposite these sides are congruent.

(b) Congruent triangles are similar triangles.

(c) If two lines intersect, then they are not parallel.

(d) A senator of the United States is a member of its Congress.

14.6. Form partial converses and partial inverses of the theorems given in Fig. 14-5. (14.7)

(a)

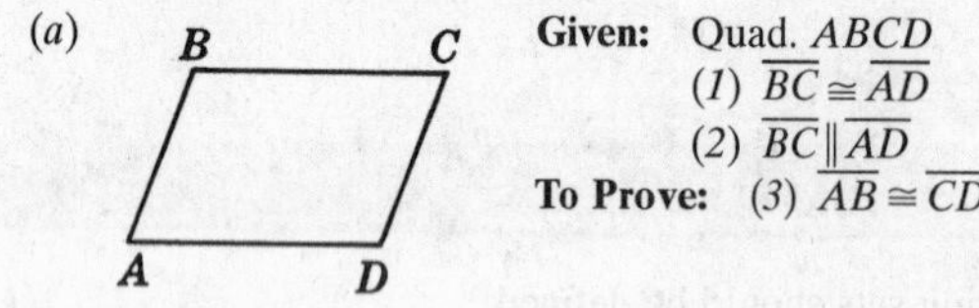

Given: Quad. $ABCD$
(1) $\overline{BC} \cong \overline{AD}$
(2) $\overline{BC} \parallel \overline{AD}$
To Prove: (3) $\overline{AB} \cong \overline{CD}$

(b)

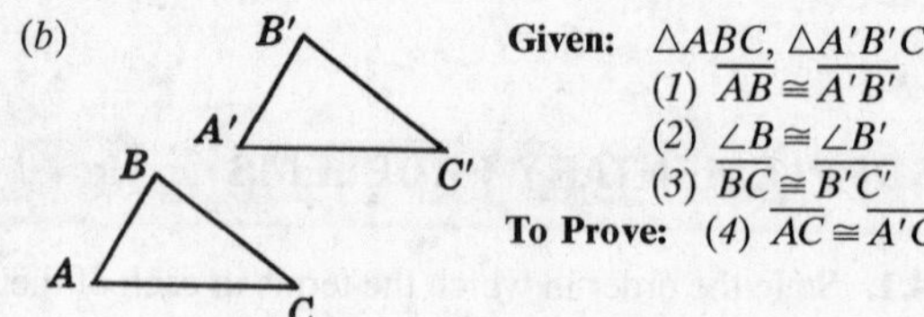

Given: $\triangle ABC$, $\triangle A'B'C'$
(1) $\overline{AB} \cong \overline{A'B'}$
(2) $\angle B \cong \angle B'$
(3) $\overline{BC} \cong \overline{B'C'}$
To Prove: (4) $\overline{AC} \cong \overline{A'C'}$

Fig. 14-5

14.7. For each of the following statements, determine whether the conditions in the hypothesis are necessary or sufficient to justify the conclusion. (14.8)

(a) Senators of the United States are elected members of Congress, two from each state.

(b) Elected members of Congress are senators of the United States.

(c) Elected persons are government officials.

(d) If a woman lives in New York City, then she lives in New York State.

(e) A bachelor is an unmarried man.

(f) A bachelor is an unmarried person.

(g) A quadrilateral having two pairs of congruent sides is a parallelogram.

Constructions

15.1 Introduction

Geometric figures are constructed with straightedge and compass. Since constructions are based on deductive reasoning, measuring instruments such as the ruler and protractor are not permitted. However, a ruler may be used as a straightedge if its markings are disregarded.

In constructions, it is advisable to plan ahead by making a sketch of the situation; such a sketch will usually reveal the needed construction steps. Construction lines should be made light to distinguish them from the required figure.

The following constructions are detailed in this chapter:

1. To construct a line segment congruent to a given line segment
2. To construct an angle congruent to a given angle
3. To bisect a given angle
4. To construct a line perpendicular to a given line through a given point on the line
5. To bisect a given line segment
6. To construct a line perpendicular to a given line through a given external point
7. To construct a triangle given its three sides
8. To construct an angle of measure 60°
9. To construct a triangle given two sides and the included angle
10. To construct a triangle given two angles and the included side
11. To construct a triangle given two angles and a side not included
12. To construct a right triangle given its hypotenuse and a leg
13. To construct a line parallel to a given line through a given external point
14. To construct a tangent to a given circle through a given point on the circle
15. To construct a tangent to a given circle through a given point outside the circle
16. To circumscribe a circle about a triangle
17. To locate the center of a given circle
18. To inscribe a circle in a given triangle
19. To inscribe a square in a given circle
20. To inscribe a regular octagon in a given circle
21. To inscribe a regular hexagon in a given circle
22. To inscribe an equilateral triangle in a given circle
23. To construct a triangle similar to a given triangle on a given line segment as base

15.2 Duplicating Segments and Angles

CONSTRUCTION 1: *To construct a line segment congruent to a given line segment*

Given: Line segment $\overline{AB}$ (Fig. 15-1)
To construct: A line segment congruent to $\overline{AB}$
Construction: On a working line w, with any point C as a center and a radius equal to AB, construct an arc intersecting w at D. Then $\overline{CD}$ is the required line segment.

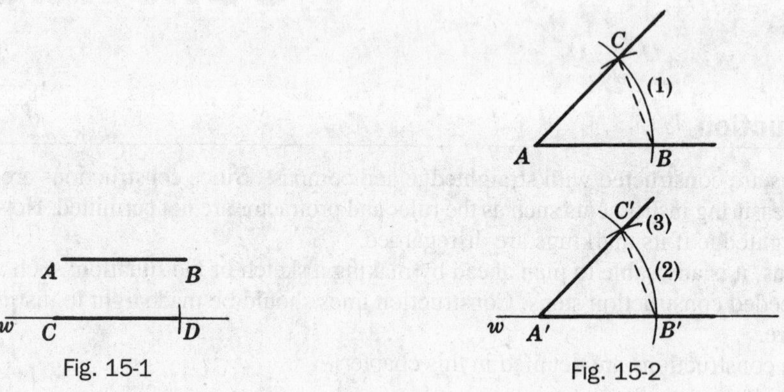

Fig. 15-1

Fig. 15-2

CONSTRUCTION 2: *To construct an angle congruent to a given angle*

Given: $\angle A$ (Fig. 15-2)
To construct: An angle congruent to $\angle A$
Construction: With A as center and a convenient radius, construct an arc (1) intersecting the sides of $\angle A$ at B and C. With A', a point on a working line w, as center and the same radius, construct arc (2) intersecting w at B'. With B' as center and a radius equal to BC, construct arc (3) intersecting arc (2) at C'. Draw $A'C'$. Then $\angle A'$ is the required angle. ($\triangle ABC \cong \triangle A'B'C'$ by SSS; hence $\angle A \cong \angle A'$.)

SOLVED PROBLEMS

15.1 Combining line segments

Given line segments with lengths a and b (Fig. 15-3), construct line segments with lengths equal to (a) $a + 2b$; (b) $2(a + b)$; (c) $b - a$.

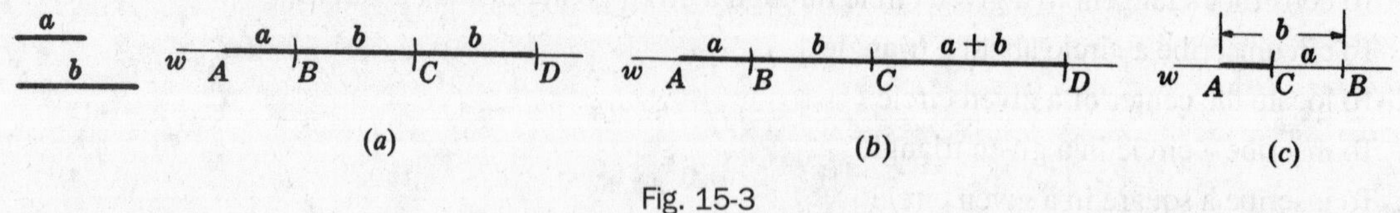

Fig. 15-3

Solutions

Use construction 1.

(a) On a working line w, construct a line segment $\overline{AB}$ with length a. From B, construct a line segment with length equal to b, to point C; and from C construct a line segment with length b, to point D. Then $\overline{AD}$ is the required line segment.

(b) Similar to (a). $AD = a + b + (a + b)$.

(c) Similar to (a). First construct $\overline{AB}$ with length b, then $\overline{BC}$ with length a. $AC = b - a$.

15.2 Combining angles

Given $\triangle ABC$ in Fig. 15-4, construct angles whose measures are equal to (a) $2A$; (b) $A + B + C$; (c) $B - A$.

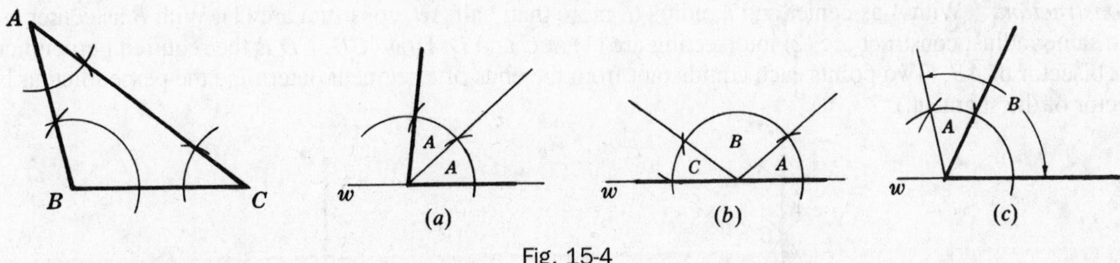

Fig. 15-4

Solutions

Use construction 2.

(a) Using a working line w as one side, duplicate $\angle A$. Construct another duplicate of $\angle A$ adjacent to $\angle A$, as shown. The exterior sides of the copied angles form the required angle.

(b) Using a working line w as one side, duplicate $\angle A$. Construct $\angle B$ adjacent to $\angle A$. Then construct $\angle C$ adjacent to $\angle B$. The exterior sides of the copied angles A and C form the required angle. Note that the angle is a straight angle.

(c) Using a working line w as one side, duplicate $\angle B$. Then duplicate $\angle A$ from the new side of $\angle B$ as shown. The difference is the required angle.

15.3 Constructing Bisectors and Perpendiculars

CONSTRUCTION 3: *To bisect a given angle*

Given: $\angle A$ (Fig. 15-5)
To construct: The bisector of $\angle A$
Construction: With A as center and a convenient radius, construct an arc intersecting the sides of $\angle A$ at B and C. With B and C as centers and equal radii, construct arcs intersecting in D. Draw $\overrightarrow{AD}$. Then $\overrightarrow{AD}$ is the required bisector. ($\triangle ABD \cong \triangle ACD$ by SSS; hence, $\angle 1 \cong \angle 2$.)

Fig. 15-5

Fig. 15-6

CONSTRUCTION 4: *To construct a line perpendicular to a given line through a given point on the line*

Given: Line w and point P on w (Fig. 15-6)
To construct: A perpendicular to w at P
Construction: Using construction 3, bisect the straight angle at P. Then $\overrightarrow{DP}$ is the required perpendicular; $\overleftrightarrow{DP}$ is the required line.

CONSTRUCTION 5: *To bisect a given line segment* (to construct the perpendicular bisector of a given line segment)

Given: Line segment $\overline{AB}$ (Fig. 15-7)
To construct: The perpendicular bisector of $\overline{AB}$
Construction: With *A* as center and a radius of more than half $\overline{AB}$, construct arc (1). With *B* as center and the same radius, construct arc (2) intersecting arc (1) at *C* and *D*. Draw $\overleftrightarrow{CD}$. $\overleftrightarrow{CD}$ is the required perpendicular bisector of $\overline{AB}$. (Two points each equidistant from the ends of a segment determine the perpendicular bisector of the segment.)

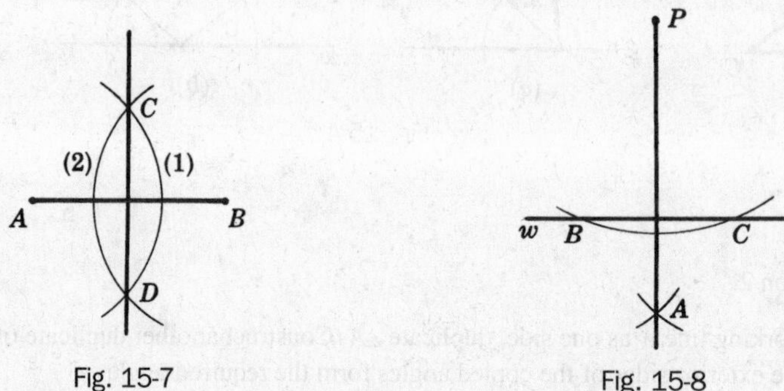

Fig. 15-7 Fig. 15-8

CONSTRUCTION 6: *To construct a line perpendicular to a given line through a given external point*

Given: Line *w* and point *P* outside of *w* (Fig. 15-8)
To construct: A perpendicular to *w* through *P*
Construction: With *P* as center and a sufficiently long radius, construct an arc intersecting *w* at *B* and *C*. With *B* and *C* as centers and equal radii of more than half $\overline{BC}$, construct arcs intersecting at *A*. Draw $\overleftrightarrow{PA}$. Then $\overleftrightarrow{PA}$ is the required perpendicular. (Points *P* and *A* are each equidistant from *B* and *C*.)

SOLVED PROBLEMS

15.3 Constructing special lines in a triangle

In scalene $\triangle ABC$ [Fig. 15-9(a)], construct (a) a perpendicular bisector of $\overline{AB}$ and (b) a median to $\overline{AB}$. In $\triangle DEF$ [Fig. 15-9(b)], *D* is an obtuse angle; construct (c) the altitude to $\overline{DF}$ and (d) the bisector of $\angle E$.

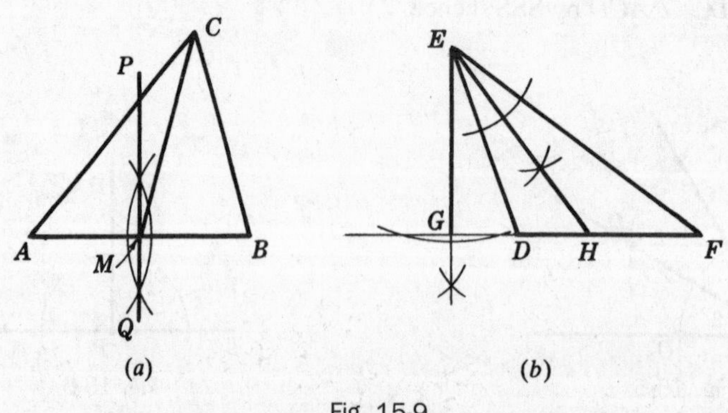

Fig. 15-9

Solutions

(a) Use construction 5 to obtain $\overleftrightarrow{PQ}$ the perpendicular bisector of $\overline{AB}$.

(b) Point *M* is the midpoint of $\overline{AB}$. Draw $\overline{CM}$, the median to $\overline{AB}$.

(c) Use construction 6 to obtain $\overline{EG}$, the altitude to $\overline{DF}$ (extended).

(d) Use construction 3 to bisect $\angle E$. $\overrightarrow{EH}$ is the required bisector.

15.4 Constructing bisectors and perpendiculars to obtain required angles

(a) Construct angles measuring 90°, 45°, and 135°.

(b) Given an angle with measure A (Fig. 15-10), construct an angle whose measure is $90° + A$.

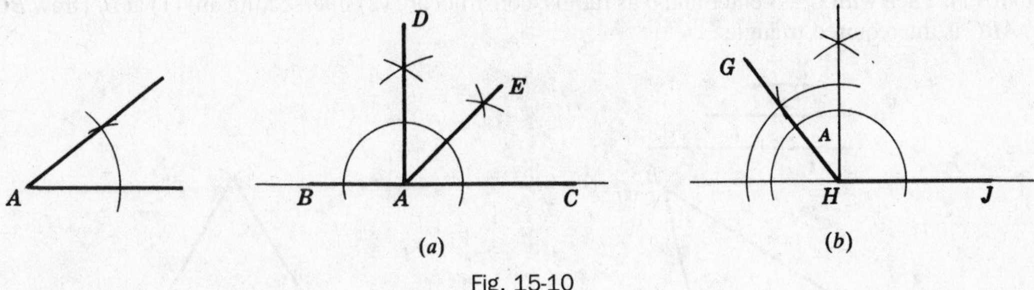

Fig. 15-10

Solutions

(a) In Fig. 15-10(a), $m\angle DAB = 90°$, $m\angle CAE = 45°$, $m\angle BAE = 135°$

(b) In Fig. 15-10(b), $m\angle GHJ = 90° + A$.

15.4 Constructing a Triangle

15.4A Determining a Triangle

A triangle is determined when a set of given data fix its size and shape. Since the parts needed to prove congruent triangles fix the size and shape of the triangles, a triangle is determined when the given data consist of three sides, or two sides and the angle included by those sides, or two angles and a side included by those angles, or two angles and a side not included by those angles, or the hypotenuse and either leg of a right triangle.

15.4B Sketching Triangles to be Constructed

Before doing the actual construction, it is very helpful to make a preliminary sketch of the required triangle. In this sketch:

1. Show the position of each of the given parts of the triangle.
2. Draw the given parts heavy, the remaining parts light.
3. Approximate the sizes of the given parts.
4. Use small letters for sides to agree with the capital letters for the angles opposite them.

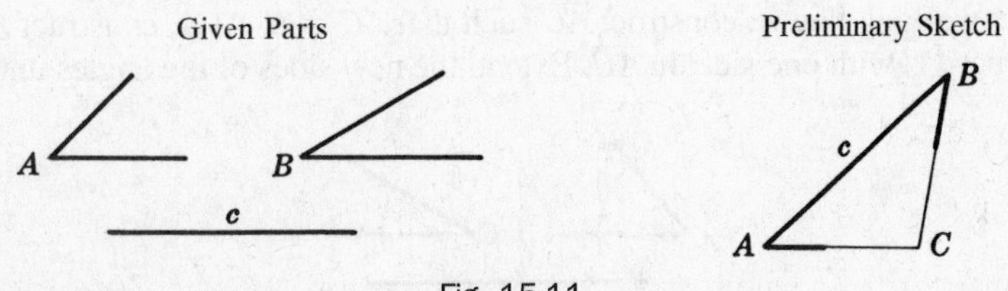

Fig. 15-11

As an example, you might make a sketch like that in Fig. 15-11 before constructing a triangle given two angles and an included side.

15.4C Triangle Constructions

CONSTRUCTION 7: *To construct a triangle given its three sides*

Given: Sides of lengths a, b, and c (Fig. 15-12)
To construct: $\triangle ABC$
Construction: On a working line w, construct $\overline{AC}$ such that $AC = b$. With A as center and c as radius, construct arc (1). Then with C as center and a as radius, construct arc (2) intersecting arc (1) at B. Draw $\overline{BC}$ and $\overline{AB}$. $\triangle ABC$ is the required triangle.

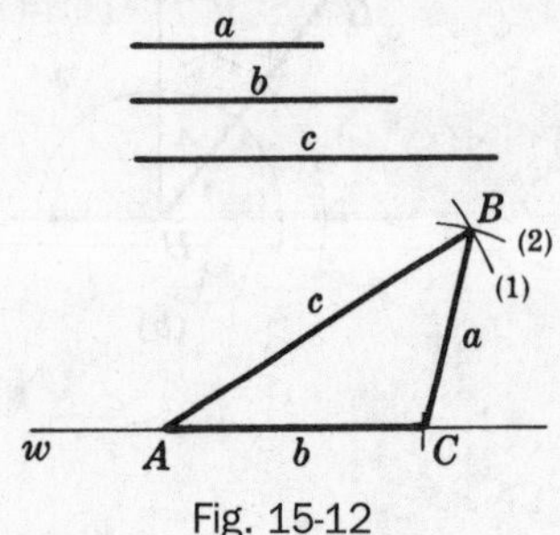

Fig. 15-12

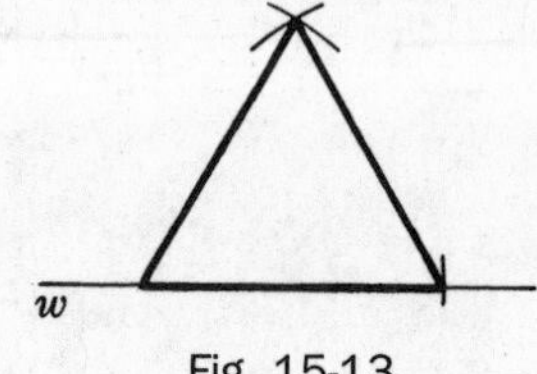

Fig. 15-13

CONSTRUCTION 8: *To construct an angle of measure 60°*

Given: Line w (Fig. 15-13)
To construct: An angle of measure 60°
Construction: Using a convenient length as a side, construct an equilateral triangle using construction 7. Then any angle of the equilateral triangle is the required angle.

CONSTRUCTION 9: *To construct a triangle given two sides and the included angle*

Given: $\angle A$, segments of lengths b and c (Fig. 15-14)
To construct: $\triangle ABC$
Construction: On a working line w, construct $\overline{AC}$ such that $AC = b$. At A, construct $\angle A$ with one side $\overline{AC}$. On the other side of $\angle A$, construct $\overline{AB}$ such that $AB = c$. Draw $\overline{BC}$. Then the required triangle is $\triangle ABC$.

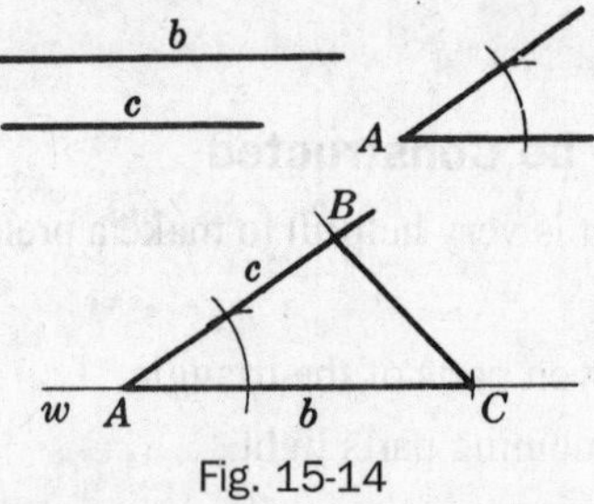

Fig. 15-14

CONSTRUCTION 10: *To construct a triangle given two angles and the included side*

Given: $\angle A$, $\angle C$, and a segment of length b (Fig. 15-15)
To construct: $\triangle ABC$
Construction: On a working line w, construct $\overline{AC}$ such that $AC = b$. At A, construct $\angle A$ with one side on $\overline{AC}$, and at C, construct $\angle C$ with one side on $\overline{AC}$. Extend the new sides of the angles until they meet at B.

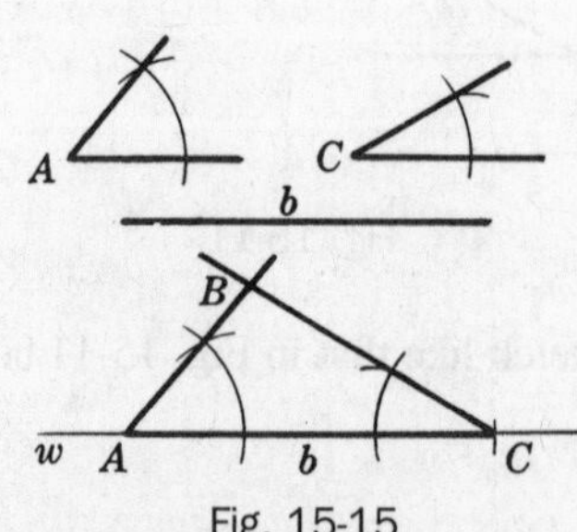

Fig. 15-15

CONSTRUCTION 11: *To construct a triangle given two angles and a side not included*

Given: $\angle A$, $\angle B$, and a segment of length b (Fig. 15-16)
To construct: $\triangle ABC$
Construction: On a working line w, construct $\overline{AC}$ such that $AC = b$. At C, construct an angle with measure equal to $m\angle A + m\angle B$ so that the extension of $\overline{AC}$ will be one side of the angle. The remainder of the straight angle at C will be $\angle C$. At A, construct $\angle A$ with one side on $\overline{AC}$. The intersection of the new sides of the angles is B.

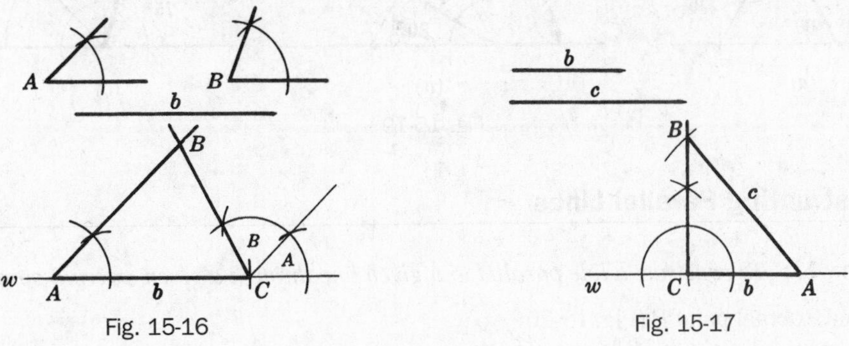

Fig. 15-16

Fig. 15-17

CONSTRUCTION 12: *To construct a right triangle given its hypotenuse and a leg*

Given: Hypotenuse with length c and leg with length b of right triangle ABC (Fig. 15-17)
To construct: Right triangle ABC
Construction: On a working line w, construct $\overline{AC}$ such that $AC = b$. At C construct a perpendicular to $\overline{AC}$. With A as center and a radius of c, construct an arc intersecting the perpendicular at B.

SOLVED PROBLEMS

15.5 Constructing a triangle

Construct an isosceles triangle, given the lengths of the base and an arm (Fig. 15-18).

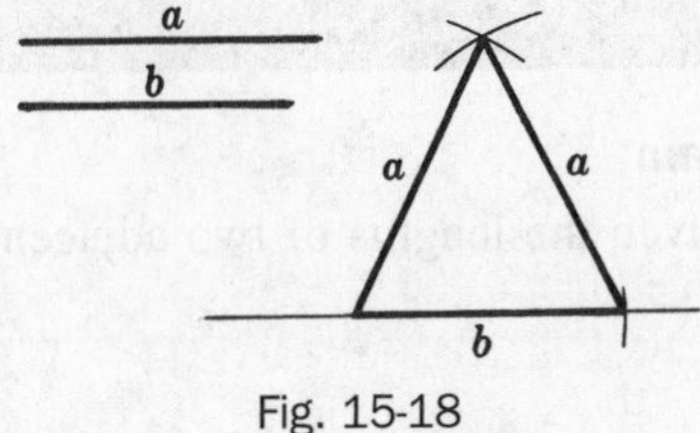

Fig. 15-18

Solution

Use construction 7, since all three sides of the triangle are known.

15.6 Constructing angles based on the construction of the 60° angle

Construct an angle of measure (a) 120°; (b) 30°; (c) 150°; (d) 105°; (e) 75°.

Solutions

(a) Use construction 8 [Fig. 15-19(a)] to construct 120° as 180°−60°.

(b) Use constructions 8 and 3 to construct 30° as $\frac{1}{2}(60°)$ [Fig. 15-19(b)].

(c) Use (b) to construct 150° as 180°–30° [Fig. 15-19(b)].

(d) Use constructions 3, 4, and 8 to construct 105° as $60° + \frac{1}{2}(90°)$ [Fig. 15-19(c)].

(e) Use (d) to construct 75° as 180° − 105° [Fig. 15-19(c)].

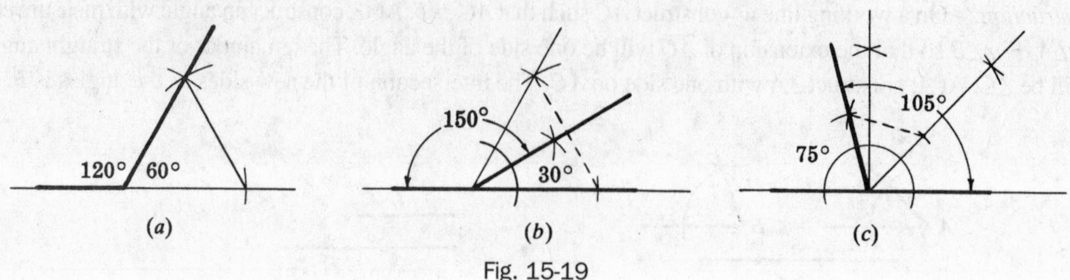

Fig. 15-19

15.5 Constructing Parallel Lines

CONSTRUCTION 13: *To construct a line parallel to a given line through a given external point*

Given: $\overleftrightarrow{AB}$ and external point P (Fig. 15-20)

To construct: A line through P parallel to $\overleftrightarrow{AB}$

Construction: Draw a line $\overleftrightarrow{RS}$ through P intersecting $\overleftrightarrow{AB}$ in Q. Construct $\angle SPD \cong \angle PQB$. Then $\overleftrightarrow{CD}$ is the required parallel. (If two corresponding angles are congruent, the lines cut by the transversal are parallel.)

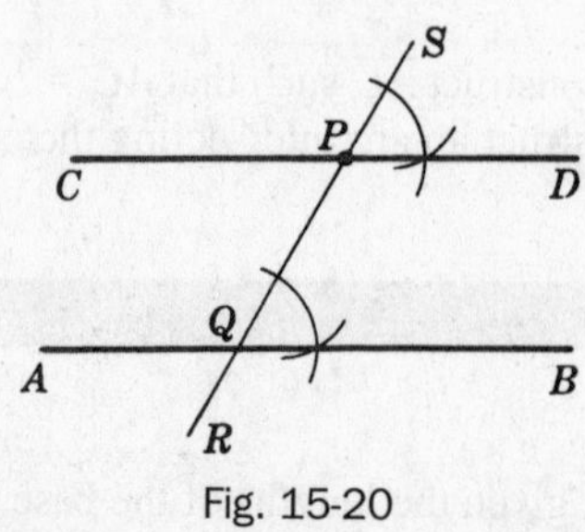

Fig. 15-20

SOLVED PROBLEM

15.7 Constructing a parallelogram

Construct a parallelogram given the lengths of two adjacent sides a and b and of a diagonal d (Fig. 15-21).

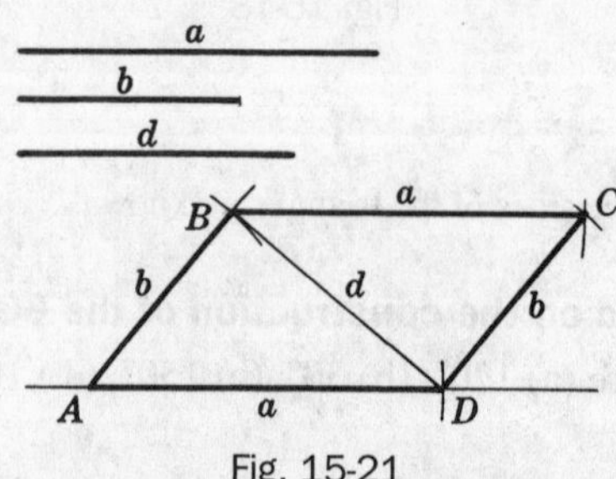

Fig. 15-21

Solution

Three vertices of the parallelogram are obtained by constructing $\triangle ABD$ by construction 7. The fourth vertex, C, is obtained by constructing $\triangle BCD$ upon diagonal $\overline{BD}$ by construction 7. Vertex C may also be obtained by constructing $\overline{BC} \parallel \overline{AD}$ and $\overline{DC} \parallel \overline{AB}$.

15.6 Circle constructions

CONSTRUCTION 14: *To construct a tangent to a given circle through a given point on the circle*

Given: Circle O and point P on the circle (Fig. 15-22)
To construct: A tangent to circle O at P
Construction: Draw radius $\overline{OP}$ and extend it outside the circle. Construct $\overleftrightarrow{AB} \perp \overleftrightarrow{OP}$ at P. $\overleftrightarrow{AB}$ is the required tangent. (A line perpendicular to a radius at its outer extremity is a tangent to the circle.)

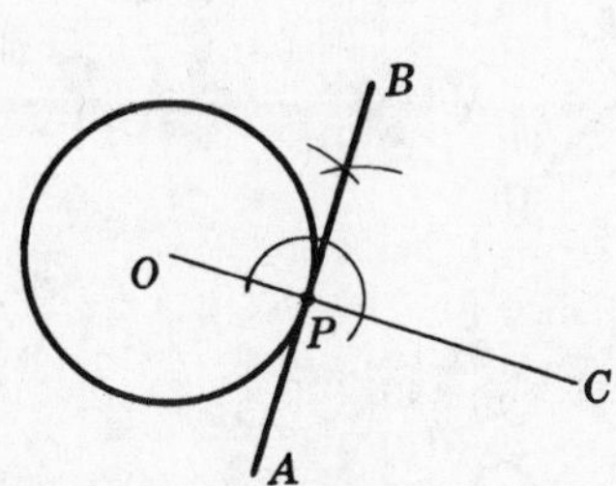

Fig. 15-22

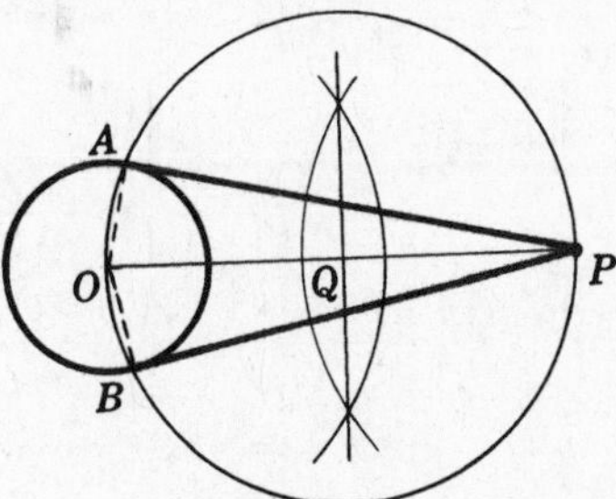

Fig. 15-23

CONSTRUCTION 15: *To construct a tangent to a given circle through a given point outside the circle*

Given: Circle O and point P outside the circle (Fig. 15-23)
To construct: A tangent to circle O from P
Construction: Draw $\overline{OP}$, and make $\overline{OP}$ the diameter of a new circle Q. Connect P to A and B, the intersections of circles O and Q. Then $\overline{PA}$ and $\overline{PB}$ are tangents. ($\angle OAP$ and $\angle OBP$ are right angles, since angles inscribed in semicircles are right angles.)

CONSTRUCTION 16: *To circumscribe a circle about a triangle*

Given: $\triangle ABC$ (Fig. 15-24)
To construct: The circumscribed circle of $\triangle ABC$
Construction: Construct the perpendicular bisectors of two sides of the triangle. Their intersection is the center of the required circle, and the distance to any vertex is the radius. (Any point on the perpendicular bisector of a segment is equidistant from the ends of the segment.)

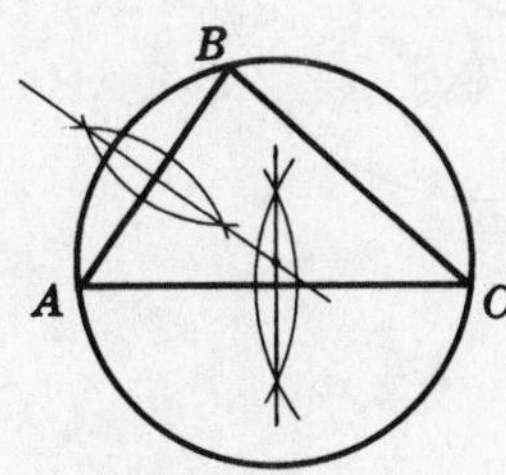

Fig. 15-24

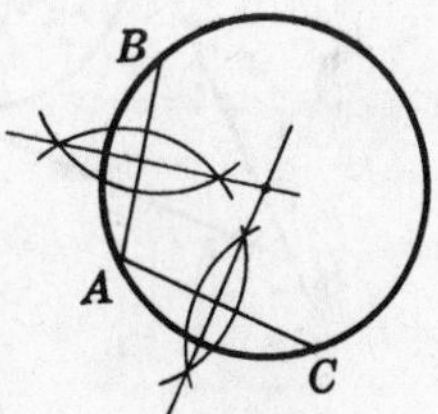

Fig. 15-25

CONSTRUCTION 17: *To locate the center of a given circle*

Given: A circle (Fig. 15-25)
To construct: The center of the given circle
Construction: Select any three points A, B, and C on the circle. Construct the perpendicular bisectors of line segments $\overline{AB}$ and $\overline{AC}$. The intersection of these perpendicular bisectors is the center of the circle.

CONSTRUCTION 18: *To inscribe a circle in a given triangle*

Given: $\triangle ABC$ (Fig. 15-26)
To construct: The circle inscribed in $\triangle ABC$
Construction: Construct the bisectors of two of the angles of $\triangle ABC$. Their intersection is the center of the required circle, and the distance (perpendicular) to any side is the radius. (Any point on the bisector of an angle is equidistant from the sides of the angle.)

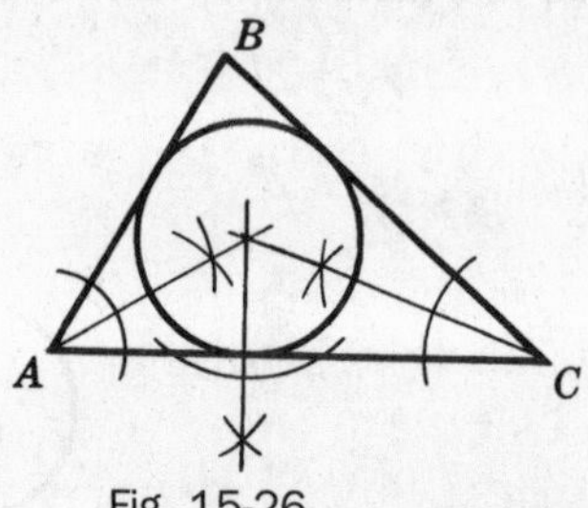

Fig. 15-26

SOLVED PROBLEMS

15.8 Constructing tangents

A secant from a point P outside circle O in Fig. 15-27 meets the circle in B and A. Construct a triangle circumscribed about the circle so that two of its sides meet in P and the third side is tangent to the circle at A.

Solution

Use constructions 14 and 15: At A construct a tangent to circle O. From P construct tangents to circle O intersecting the first tangent in C and D. The required triangle is $\triangle PCD$.

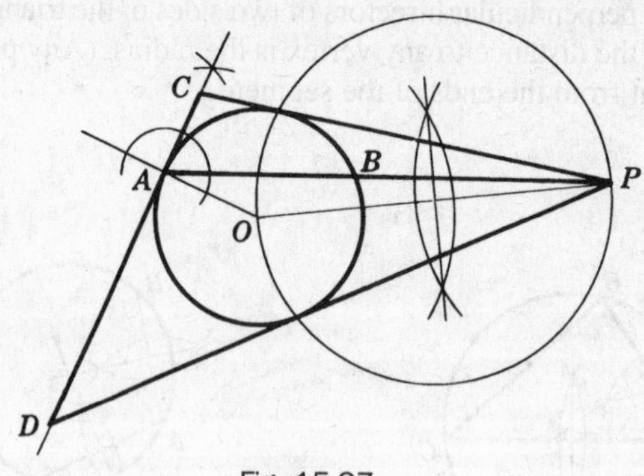

Fig. 15-27

15.9 Constructing circles

Construct the circumscribed and inscribed circles of isosceles triangle DEF in Fig. 15-28.

Solution

Use constructions 16 and 18. In doing so, note that the bisector of $\angle E$ is also the perpendicular bisector of $\overline{DF}$. Then the center of each circle is on $\overline{EG}$. I, the center of the inscribed circle, is found by constructing the bisector of $\angle D$ or $\angle F$. C, the center of the circumscribed circle, is found by constructing the perpendicular bisector of $\overline{DE}$ or $\overline{EF}$.

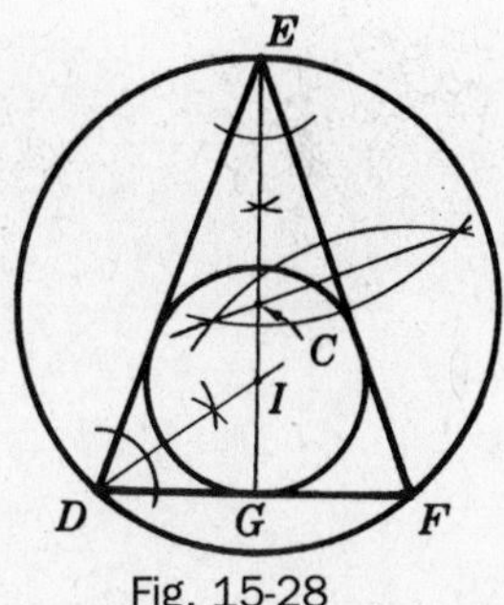

Fig. 15-28

15.7 Inscribing and Circumscribing Regular Polygons

CONSTRUCTION 19: *To inscribe a square in a given circle*

Given: Circle O (Fig. 15-29)
To construct: A square inscribed in circle O
Construction: Draw a diameter, and construct another diameter perpendicular to it. Join the end points of the diameters to form the required square.

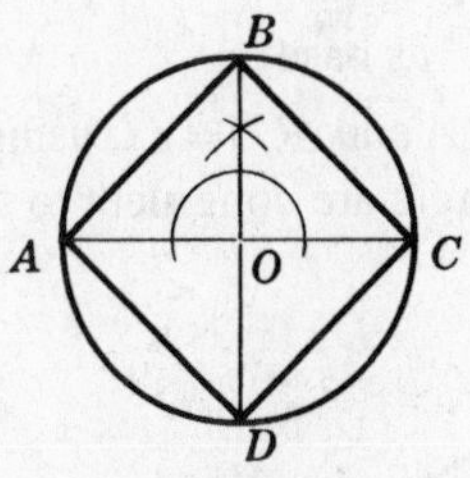

Fig. 15-29

CONSTRUCTION 20: *To inscribe a regular octagon in a given circle*

Given: Circle O (Fig. 15-30)
To construct: A regular octagon inscribed in circle O
Construction: As in construction 19, construct perpendicular diameters. Then bisect the angles formed by these diameters, dividing the circle into eight congruent arcs. The chords of these arcs are the sides of the required regular octagon.

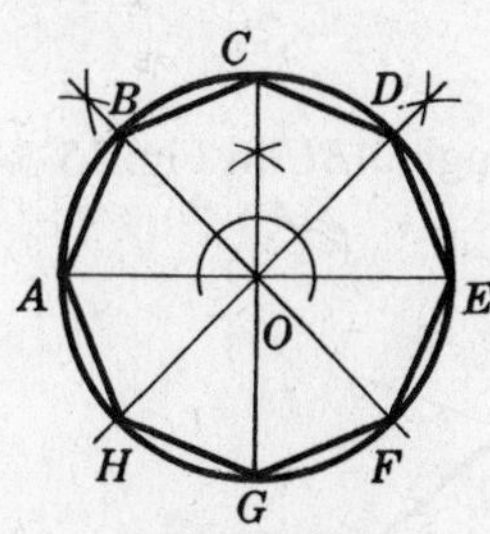

Fig. 15-30

CONSTRUCTION 21: *To inscribe a regular hexagon in a given circle*

Given: Circle O (Fig. 15-31)
To construct: A regular hexagon inscribed in circle O
Construction: Draw diameter $\overline{AD}$ and, using A and D as centers, construct four arcs having the same radius as circle O and intersecting the circle. Construct the required regular hexagon by joining consecutive points in which these arcs intersect the circle.

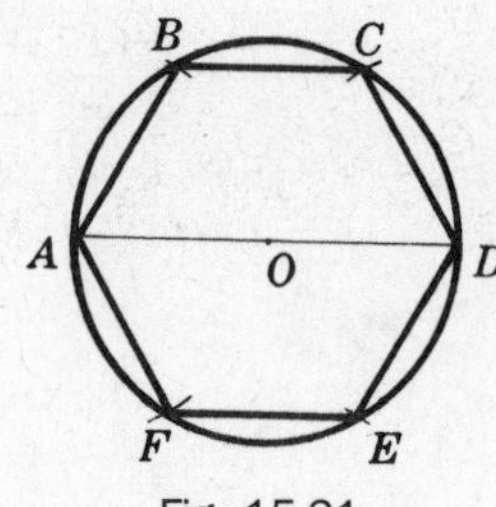

Fig. 15-31

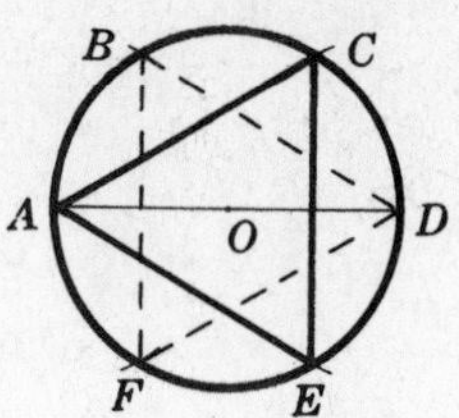

Fig. 15-32

CONSTRUCTION 22: *To inscribe an equilateral triangle in a given circle*

Given: Circle O (Fig. 15-32)
To construct: An equilateral triangle inscribed in circle O
Construction: Inscribed equilateral triangles are obtained by joining alternately the six points of division obtained in construction 21.

15.8 Constructing Similar Triangles

CONSTRUCTION 23: *To construct a triangle similar to a given triangle on a given line segment as base*

Given: $\triangle ABC$ and line segment $\overline{A'C'}$ (Fig. 15-33)
To construct: $\triangle A'B'C' \sim \triangle ABC$ on $\overline{A'C'}$ as base
Construction: On $\overline{A'C'}$ construct $\angle A' \cong \angle A$ and $\angle C' \cong \angle C$ using construction 2. Extend the other sides until they meet, at B. (If two angles of one triangle are congruent to two angles of another triangle, the triangles are similar.)

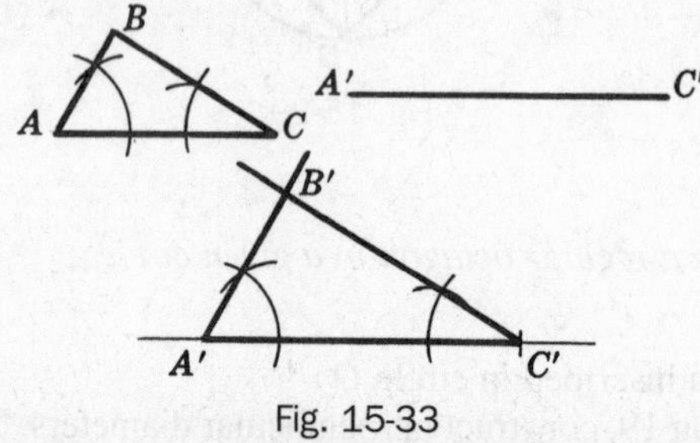

Fig. 15-33

SOLVED PROBLEM

15.10 Constructing similar triangles

Construct a triangle similar to triangle ABC in Fig. 15-34, with a base twice as long as the base of the given triangle.

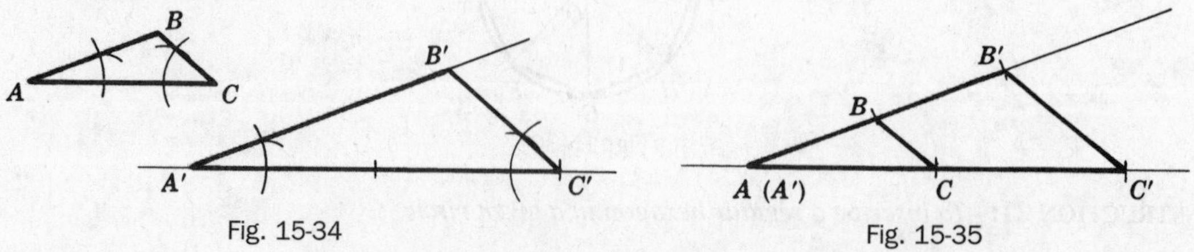

Fig. 15-34

Fig. 15-35

Solution

Construct $\overline{A'C'}$ twice as long as $\overline{AC}$, and then use construction 23.
Alternative method (Fig. 15-35): Extend two sides of $\triangle ABC$ to twice their lengths and join the endpoints.

SUPPLEMENTARY PROBLEMS

15.1. Given line segments with lengths a and b as follows: $\underline{\quad a \quad}$ $\underline{\quad b \quad}$. Construct a line segment whose length equals (a) $a + b$; (b) $a - b$; (c) $2a + b$; (d) $a + 3b$; (e) $2(a + b)$; (f) $2(3b - a)$. (15.1)

15.2. Given line segments with lengths a, b, and c: $\underline{\quad a \quad}$ $\underline{\quad b \quad}$ $\underline{\quad c \quad}$. Construct a line segment whose length equals (a) $a + b + c$; (b) $a + c - b$; (c) $a + 2(b + c)$; (d) $b + 2(a - c)$; (e) $3(b + c - a)$. (15.1)

15.3. Given angles with measures A and B (Fig. 15-36). Construct an angle with measure (a) $A + B$; (b) $A - B$; (c) $2B - A$; (d) $2A - B$; (e) $2(A - B)$. (15.2)

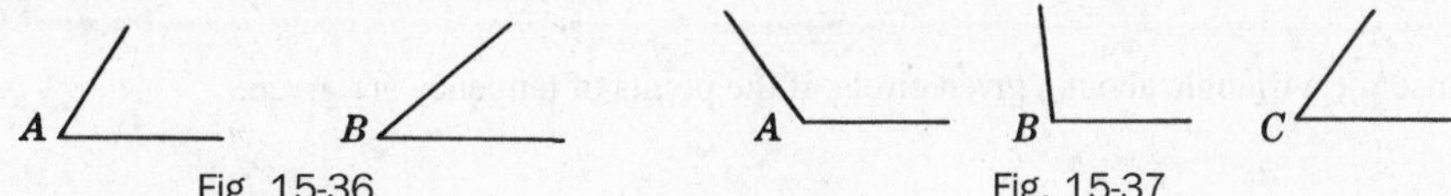

Fig. 15-36 Fig. 15-37

15.4. Given angles with measures A, B, and C (Fig. 15-37). Construct an angle with measure (a) $A + C$; (b) $B + C - A$; (c) $2C$; (d) $B - C$; (e) $2(A - B)$. (15.2)

15.5. In a right triangle, construct (a) the bisector of the right angle; (b) the perpendicular bisector of the hypotenuse; (c) the median to the hypotenuse. (15.3)

15.6. For each kind of triangle (acute, right, and obtuse), show that the following sets of rays and segments are concurrent, that is, they intersect in one point: (a) the angle bisectors; (b) the medians; (c) the altitudes; (d) the perpendicular bisectors. (15.3)

15.7. Given $\triangle ABC$ in Fig. 15-38, construct (a) the supplement of $\angle A$; (b) the complement of $\angle B$; (c) the complement of $\frac{1}{2}\angle C$. (15.4)

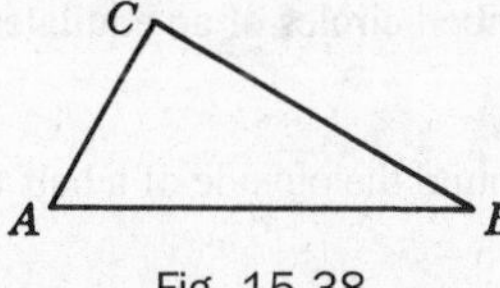

Fig. 15.38

15.8. Construct an angle with measure equal to (a) $22\frac{1}{2}^\circ$; (b) $67\frac{1}{2}^\circ$; (c) $112\frac{1}{2}^\circ$. (15.4)

15.9. Given an acute angle, construct (a) its supplement; (b) its complement; (c) half its supplement; (d) half its complement. (15.4)

15.10. By actual construction, illustrate that the difference between the measures of the supplement and complement of an acute angle equals 90°. (15.4)

15.11. Construct a right triangle given its (a) legs; (b) hypotenuse and a leg; (c) leg and an acute angle adjacent to the leg; (d) leg and an acute angle opposite the leg; (e) hypotenuse and an acute angle. (15.5)

15.12. Construct an isosceles triangle given (a) an arm and a vertex angle; (b) an arm and a base angle; (c) an arm and the altitude to the base; (d) the base and the altitude to the base. (15.5)

15.13. Construct an isosceles right triangle given (a) a leg; (b) the hypotenuse; (c) the altitude to the hypotenuse. (15.5)

15.14. Construct a triangle given (a) two sides and the median to one of them; (b) two sides and the altitude to one of them; (c) an angle, the angle bisector of the given angle, and a side adjacent to the given angle. (15.5)

15.15. Construct angles of measure 15° and 165°. (15.6)

15.16. Given an angle with measure A, construct angles with measure (a) $A + 60°$; (b) $A + 30°$; (c) $A + 120°$. (15.6)

15.17. Construct a parallelogram, given (a) two adjacent sides and an angle; (b) the diagonals and the acute angle at their intersection; (c) the diagonals and a side; (d) two adjacent sides and the altitude to one of them; (e) a side, an angle, and the altitude to the given side. (15.7)

15.18. Circumscribe a triangle about a given circle, if the points of tangency are given. (15.8)

15.19. Secant $\overleftrightarrow{AB}$ passes through the center of circle O in Fig. 15-39. Circumscribe a quadrilateral about the circle so that A and B are opposite vertices. (15.8)

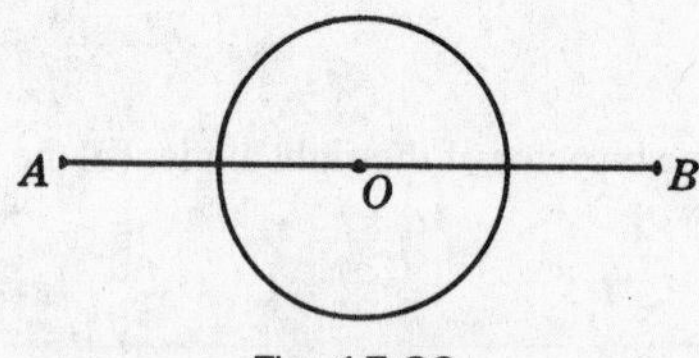

Fig. 15-39

15.20. Circumscribe and inscribe circles about (a) an acute triangle; (b) an obtuse triangle. (15.9)

15.21. Circumscribe a circle about (a) a right triangle; (b) a rectangle; (c) a square. (15.9)

15.22. Construct the inscribed and circumscribed circles of an equilateral triangle. (15.9)

15.23. Locate the center of a circle drawn around the outside of a half-dollar piece. (15.9)

15.24. In a given circle, inscribe (a) a square; (b) a regular octagon; (c) a regular 16-gon; (d) a regular hexagon; (e) an equilateral triangle; (f) a regular dodecagon.

15.25. Construct a triangle similar to a given triangle with a base (a) three times as long; (b) half as long; (c) one and one-half times as long. (15.10)

Proofs of Important Theorems

16.1 Introduction

The theorems proved in this chapter are considered the most important in the logical sequence of geometry. They are as follows:

1. If two sides of a triangle are congruent, the angles opposite these sides are congruent. (Base angles of an isosceles triangle are congruent.)
2. The sum of the measures of the angles in a triangle equals 180°.
3. If two angles of a triangle are congruent, the sides opposite these angles are congruent.
4. Two right triangles are congruent if the hypotenuse and a leg of one are congruent to the corresponding parts of the other.
5. A diameter perpendicular to a chord bisects the chord and its arcs.
6. An angle inscribed in a circle is measured by one-half its intercepted arc.
7. An angle formed by two chords intersecting inside a circle is measured by one-half the sum of the intercepted arcs.

8a. An angle formed by two secants intersecting outside a circle is measured by one-half the difference of its intercepted arcs.

8b. An angle formed by a tangent and a secant intersecting outside a circle is measured by one-half the difference of its intercepted arcs.

8c. An angle formed by two tangents intersecting outside a circle is measured by one-half the difference of its intercepted arcs.

9. If three angles of one triangle are congruent to three angles of another triangle, the triangles are similar.
10. If the altitude is drawn to the hypotenuse of a right triangle, then (a) the two triangles thus formed are similar to the given triangle and to each other, and (b) each leg of the given triangle is the mean proportional between the hypotenuse and the projection of that leg upon the hypotenuse.
11. The square of the length of the hypotenuse of a right triangle equals the sum of the squares of the lengths of the other two sides.
12. The area of a parallelogram equals the product of the length of one side and the length of the altitude to that side.
13. The area of a triangle is equal to one-half the product of the length of one side and the length of the altitude to that side.
14. The area of a trapezoid is equal to one-half the product of the length of the altitude and the sum of the lengths of the bases.
15. The area of a regular polygon is equal to one-half the product of its perimeter and the length of its apothem.

16.2 The Proofs

1. If two sides of a triangle are congruent, the angles opposite these sides are congruent. (Base angles of an isosceles triangle are congruent.)

Given: $\triangle ABC, \overline{AB} \cong \overline{BC}$

To Prove: $\angle A \cong \angle C$

Plan: When the bisector of the vertex angle is drawn, the angles to be proved congruent become corresponding angles of congruent triangles.

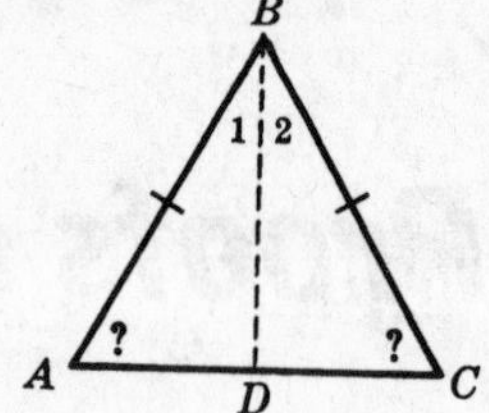

PROOF:

Statements	Reasons
1. Draw $\overline{BD}$ bisecting $\angle B$.	1. An angle may be bisected.
2. $\angle 1 \cong \angle 2$	2. To bisect is to divide into two congruent parts.
3. $\overline{AB} \cong \overline{BC}$	3. Given
4. $\overline{BD} \cong \overline{BD}$	4. Reflexive property
5. $\triangle ADB \cong \triangle CDB$	5. SAS
6. $\angle A \cong \angle C$	6. Corresponding parts of congruent triangles are congruent.

2. The sum of the measures of the angles in a triangle equals 180°.

Given: $\triangle ABC$

To Prove: $m\angle A + m\angle B + m\angle C = 180°$

Plan: When a line is drawn through one vertex parallel to the opposite side, a straight angle is formed whose parts can be proved congruent to the angles of the triangle.

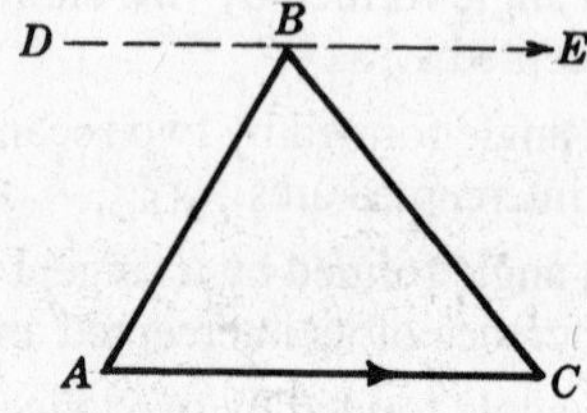

PROOF:

Statements	Reasons
1. Through B, draw $\overleftrightarrow{DE} \parallel \overleftrightarrow{AC}$.	1. Through an external point, a line can be drawn parallel to a given line.
2. $m\angle DBE = 180°$	2. A straight angle is an angle whose measure is 180°.
3. $m\angle DBA + m\angle ABC + m\angle CBE = 180°$	3. The whole equals the sum of its parts.
4. $\angle A \cong \angle DBA$, $\angle C \cong \angle CBE$	4. Alternate interior angles of parallel lines are congruent.
5. $m\angle A + m\angle B + m\angle C = 180°$	5. Substitution Postulate

3. If two angles of a triangle are congruent, the sides opposite these angles are congruent.

Given: $\triangle ABC, \angle A \cong \angle C$

To Prove: $\overline{AB} \cong \overline{BC}$

Plan: When the bisector of $\angle B$ is drawn, the sides to be proved congruent become corresponding sides of congruent triangles.

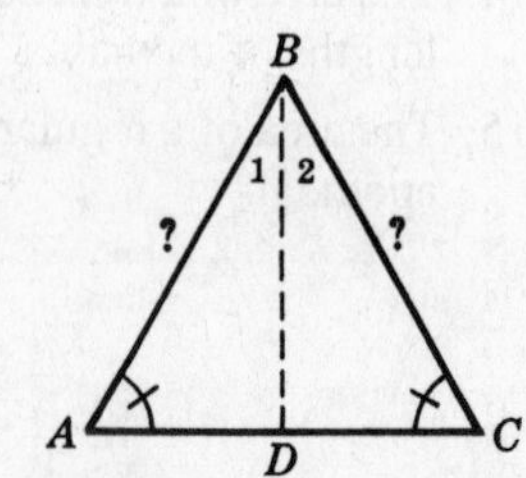

PROOF:

Statements	Reasons
1. Draw $\overline{BD}$ bisecting $\angle B$.	1. An angle may be bisected.
2. $\angle 1 \cong \angle 2$	2. To bisect is to divide into two congruent parts.
3. $\angle A \cong \angle C$	3. Given
4. $\overline{BD} \cong \overline{BD}$	4. Reflexive property
5. $\triangle BDA \cong \triangle BDC$	5. SAA
6. $\overline{AB} \cong \overline{BC}$	6. Corresponding parts of congruent triangles are congruent.

4. Two right triangles are congruent if the hypotenuse and a leg of one are congruent to the corresponding parts of the other.

Given: Right $\triangle ABC$ with right angle at C
Right $\triangle DEF$ with right angle at F
$\overline{AB} \cong \overline{DE}$, $\overline{BC} \cong \overline{EF}$

To Prove: $\triangle ABC \cong \triangle DEF$

Plan: Move the two given triangles together so that $\overline{BC}$ coincides with $\overline{EF}$, forming an isosceles triangle. The given triangles are proved congruent by using Theorem 1 and SAA.

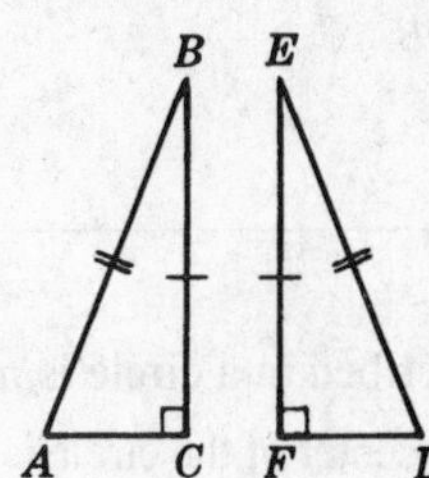

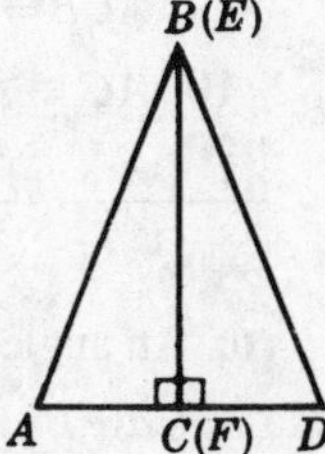

PROOF:

Statements	Reasons
1. $\overline{BC} \cong \overline{EF}$	1. Given
2. Move triangles ABC and DEF together so that $\overline{BC}$ coincides with $\overline{EF}$, and A and D are on opposite sides of $\overline{BC}$.	2. A geometric figure may be moved without changing its size or shape. Equal lines may be made to coincide.
3. $\angle C$ and $\angle F$ are right angles.	3. Given
4. $\angle ACD$ is a straight angle.	4. The whole equals the sum of its parts.
5. $\overline{AD}$ is a straight line segment.	5. The sides of a straight angle lie in a straight line.
6. $\overline{AB} \cong \overline{DE}$	6. Given
7. $\angle A \cong \angle D$	7. If two sides of a triangle are congruent, the angles opposite these sides are congruent.
8. $\triangle ABC \cong \triangle DEF$	8. SAA.

5. A diameter perpendicular to a chord bisects the chord and its arcs.

Given: Circle O, diameter $\overline{AB} \perp \overline{CD}$

To Prove: $\overline{CE} \cong \overline{ED}$, $\overset{\frown}{BC} \cong \overset{\frown}{BD}$, $\overset{\frown}{AC} \cong \overset{\frown}{AD}$

Plan: Congruent triangles are formed when radii are drawn to C and D, proving $\overline{CE} \cong \overline{ED}$. Equal central angles are used to prove $\overset{\frown}{BC} \cong \overset{\frown}{BD}$; then the Subtraction Postulate is used to prove $\overset{\frown}{AC} \cong \overset{\frown}{AD}$.

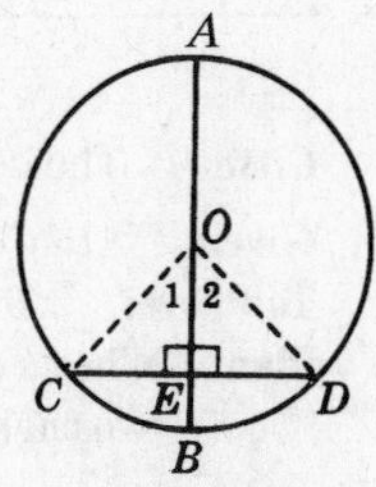

PROOF:

Statements	Reasons
1. Draw $\overline{OC}$ and $\overline{OD}$.	1. A straight line may be drawn between two points.
2. $\overline{OC} \cong \overline{OD}$	2. Radii of a circle are congruent.
3. $\overline{AB} \perp \overline{CD}$	3. Given
4. $\angle OEC$ and $\angle OED$ are right angles.	4. Perpendiculars form right angles.
5. $\overline{OE} \cong \overline{OE}$	5. Reflexive property
6. $\triangle OEC \cong \triangle OED$	6. hy-leg
7. $\overline{CE} \cong \overline{ED}$, $\angle 1 \cong \angle 2$	7. Corresponding parts of congruent triangles are congruent.
8. $\widehat{CB} \cong \widehat{BD}$	8. In a circle, congruent central angles have congruent arcs.
9. $\widehat{ACB} \cong \widehat{ADB}$	9. A diameter bisects a circle.
10. $\widehat{AC} \cong \widehat{AD}$	10. In a circle, congruent arcs are equal arcs; Subtraction Postulate

6. An angle inscribed in a circle is measured by one-half its intercepted arc.

Case I: The center of the circle is on one side of the angle.

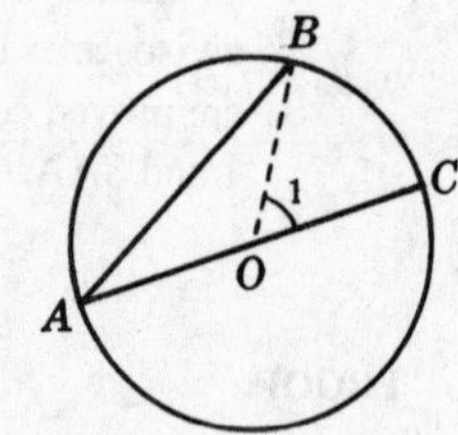

Given: $\angle A$ is inscribed in circle O. O is on side AC.

To Prove: $A\angle \overset{\circ}{=} \frac{1}{2}\widehat{BC}$

Plan: When radius $\overline{OB}$ is drawn, isosceles $\triangle AOB$ is formed. $\angle A$ is proved to be equal in measure to one-half central $\angle 1$, which is measured by $\widehat{BC}$.

PROOF:

Statements	Reasons
1. Draw $\overline{OB}$.	1. A straight line can be drawn between two points.
2. $\overline{AO} \cong \overline{OB}$	2. Radii of a circle are congruent.
3. $\angle A \cong \angle B$	3. If two sides of a triangle are congruent, the angles opposite these sides are congruent.
4. $m\angle A + m\angle B = m\angle 1$	4. In a triangle the measure of an exterior angle equals the sum of the measures of the two adjacent interior angles.
5. $m\angle A + m\angle A = 2m\angle A = m\angle 1$	5. Substitution Postulate
6. $m\angle A \overset{\circ}{=} \frac{1}{2}m\angle 1$	6. Halves of equals are equal.
7. $\angle 1 \overset{\circ}{=} \widehat{BC}$	7. A central angle is measured by its intercepted arc.
8. $\angle A \overset{\circ}{=} \frac{1}{2}\widehat{BC}$	8. Substitution Postulate

Case II: The center is inside the angle.

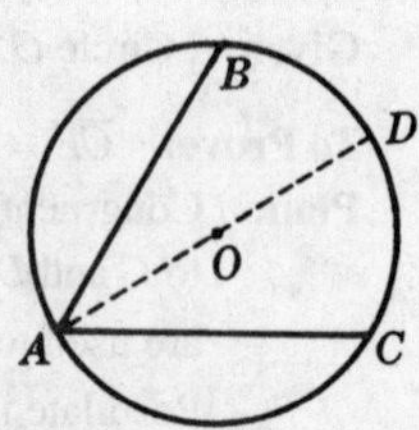

Given: $\angle BAC$ is inscribed in circle O. O is inside $\angle BAC$.

To Prove: $\angle BAC \overset{\circ}{=} \frac{1}{2}\widehat{BC}$

Plan: When a diameter is drawn, $\angle BAC$ is divided into two angles which can be measured by applying Case I.

PROOF:

Statements	Reasons
1. Draw diameter $\overline{AD}$.	1. A straight line may be drawn between two points.
2. $\angle BAD \doteq \frac{1}{2}\overset{\frown}{BD}$, $\angle DAC \doteq \frac{1}{2}\overset{\frown}{DC}$	2. An inscribed angle is measured by one-half its intercepted arc if the center of the circle is on one side.
3. $\angle BAC \doteq \frac{1}{2}\overset{\frown}{BD} + \frac{1}{2}\overset{\frown}{DC}$ or $\angle BAC \doteq \frac{1}{2}(\overset{\frown}{BD} + \frac{1}{2}\overset{\frown}{DC})$	3. If equals are added to equals, the sums are equal.
4. $\angle BAC \doteq \frac{1}{2}\overset{\frown}{BC}$	4. Substitution Postulate

Case III: The center is outside the angle.

Given: $\angle BAC$ is inscribed in circle O. O is outside $\angle BAC$.

To Prove: $\angle BAC \doteq \frac{1}{2}\overset{\frown}{BC}$

Plan: When a diameter is drawn, $\angle BAC$ becomes the difference of two angles which can be measured by applying Case I.

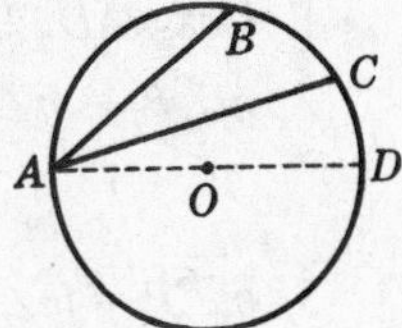

PROOF:

Statements	Reasons
1. Draw diameter $\overline{AD}$.	1. A straight line may be drawn between two points.
2. $\angle BAD \doteq \frac{1}{2}\overset{\frown}{BD}$, $\angle CAD \doteq \frac{1}{2}\overset{\frown}{CD}$	2. An inscribed angle is measured by one-half its intercepted arc if the center of the circle is on one side.
3. $\angle BAC \doteq \frac{1}{2}\overset{\frown}{BD} - \frac{1}{2}\overset{\frown}{CD}$ or $\angle BAC \doteq \frac{1}{2}(\overset{\frown}{BD} - \overset{\frown}{CD})$	3. If equals are subtracted from equals, the differences are equal.
4. $\angle BAC \doteq \frac{1}{2}\overset{\frown}{BC}$	4. Substitution Postulate

7. An angle formed by two chords intersecting inside a circle is measured by one-half the sum of the intercepted arcs.

Given: $\angle 1$ formed by chords $\overline{AB}$ and $\overline{CD}$ intersecting at point E inside circle O

To Prove: $\angle 1 \doteq \frac{1}{2}(\overset{\frown}{AC} + \overset{\frown}{BD})$

Plan: When chord $\overline{AD}$ is drawn, $\angle 1$ becomes an exterior angle of a triangle whose nonadjacent interior angles are inscribed angles measured by $\frac{1}{2}\overset{\frown}{AC}$ and $\frac{1}{2}\overset{\frown}{BD}$.

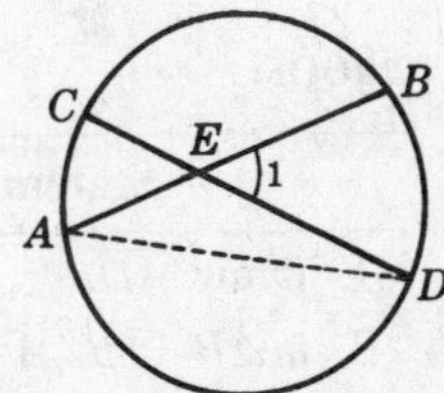

PROOF:

Statements	Reasons
1. Draw $\overline{AD}$.	1. A straight line may be drawn between two points.
2. $m\angle 1 = m\angle A + m\angle D$	2. The measure of an exterior angle of a triangle equals the sum of the measures of the nonadjacent interior angles.
3. $\angle A \doteq \frac{1}{2}\overset{\frown}{BD}$, $\angle D \doteq \frac{1}{2}\overset{\frown}{AC}$	3. An angle inscribed in a circle is measured by one-half its intercepted arc.
4. $\angle 1 \doteq \frac{1}{2}\overset{\frown}{BD} + \frac{1}{2}\overset{\frown}{AC} \doteq \frac{1}{2}(\overset{\frown}{BD} + \overset{\frown}{AC})$	4. Substitution Postulate

8a. An angle formed by two secants intersecting outside a circle is measured by one-half the difference of its intercepted arcs.

Given: $\angle P$ formed by secants $\overline{PBA}$ and $\overline{PDC}$ intersecting at P, a point outside circle O.

To Prove: $\angle P \overset{\circ}{=} \frac{1}{2}(\overset{\frown}{AC} - \overset{\frown}{BD})$

Plan: When $\overline{AD}$ is drawn, $\angle 1$ becomes an exterior angle of $\triangle ADP$, of which $\angle P$ is a nonadjacent interior angle.

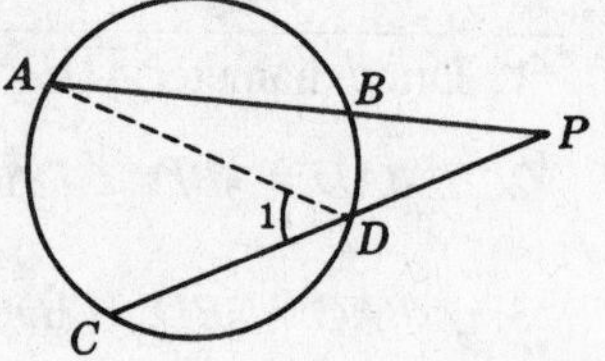

PROOF:

Statements	Reasons
1. Draw $\overline{AD}$.	1. A straight line may be drawn between two points.
2. $m\angle P + m\angle A = m\angle 1$	2. The measure of an exterior angle of a triangle equals the sum of the measures of the nonadjacent interior angles.
3. $m\angle P = m\angle 1 - m\angle A$	3. Subtraction Postulate
4. $\angle 1 \overset{\circ}{=} \frac{1}{2}\overset{\frown}{AC}$, $\angle A \overset{\circ}{=} \frac{1}{2}\overset{\frown}{BD}$	4. An angle inscribed in a circle is measured by one-half its intercepted arc.
5. $\angle P \overset{\circ}{=} \frac{1}{2}\overset{\frown}{AC}, - \frac{1}{2}\overset{\frown}{BD}$ or $\angle P \overset{\circ}{=} \frac{1}{2}(\overset{\frown}{AC} - \overset{\frown}{BD})$	5. Substitution Postulate

8b. An angle formed by a secant and a tangent intersecting outside a circle is measured by one-half the difference of its intercepted arcs.

Given: $\angle P$ formed by secant $\overline{PBA}$ and tangent $\overline{PDC}$ intersecting P, a point outside circle O.

To Prove: $\angle P \overset{\circ}{=} \frac{1}{2}(\overset{\frown}{AD} - \overset{\frown}{BD})$

Plan: When chord $\overline{AD}$ is drawn, $\angle 1$ becomes an exterior angle of $\triangle ADP$, of which $\angle P$ and $\angle A$ are nonadjacent interior angles.

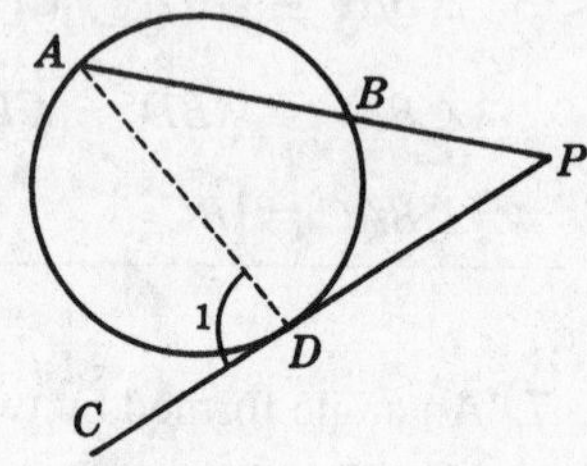

PROOF:

Statements	Reasons
1. Draw $\overline{AD}$.	1. A straight line may be drawn between two points.
2. $m\angle P + m\angle A = m\angle 1$	2. The measure of an exterior angle of a triangle equals the sum of the measures of the nonadjacent interior angles.
3. $m\angle P = m\angle 1 - m\angle A$	3. Subtraction Postulate
4. $\angle 1 \overset{\circ}{=} \frac{1}{2}\overset{\frown}{AD}$	4. An angle formed by a tangent and a chord is measured by one-half its intercepted arc.
5. $\angle A \overset{\circ}{=} \frac{1}{2}\overset{\frown}{BD}$	5. An inscribed angle is measured by one-half its intercepted arc.
6. $\angle P \overset{\circ}{=} \frac{1}{2}\overset{\frown}{AD} - \frac{1}{2}\overset{\frown}{BD}$ or $\angle P \overset{\circ}{=} \frac{1}{2}(\overset{\frown}{AD} - \overset{\frown}{BD})$	6. Substitution Postulate

8c. An angle formed by two tangents intersecting outside a circle is measured by one-half the difference of its intercepted arcs.

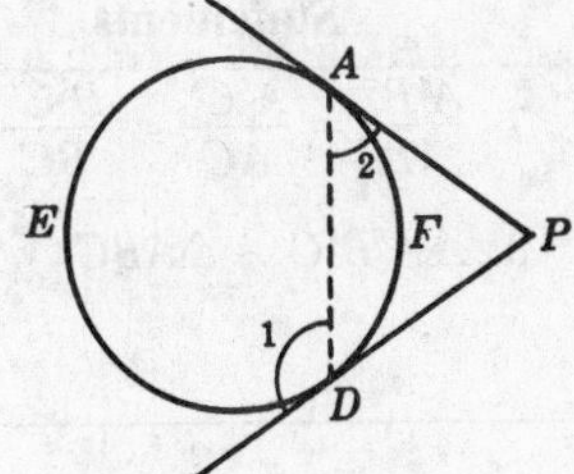

Given: $\angle P$ formed by tangents $\overline{PA}$ and $\overline{PD}$ intersecting at P, a point outside circle O.

To Prove: $\angle P \overset{\circ}{=} \frac{1}{2}(\overset{\frown}{AED} - \overset{\frown}{AFD})$

Plan: When chord $\overline{AD}$ is drawn, $\angle 1$ becomes an exterior angle of $\triangle ADP$, of which $\angle P$ and $\angle 2$ are nonadjacent interior angles.

PROOF:

Statements	Reasons
1. Draw $\overline{AD}$.	1. A straight line may be drawn between two points.
2. $m\angle P + m\angle 2 = m\angle 1$	2. The measure of an exterior angle of a triangle equals the sum of the measures of the nonadjacent interior angles.
3. $m\angle P = m\angle 1 - m\angle 2$	3. Subtraction Postulate.
4. $\angle 1 \overset{\circ}{=} \frac{1}{2}\overset{\frown}{AED}$, $\angle 2 \overset{\circ}{=} \frac{1}{2}\overset{\frown}{AFD}$	4. An angle formed by a tangent and a chord is measured by one-half its intercepted arc.
5. $\angle P \overset{\circ}{=} \frac{1}{2}\overset{\frown}{AED} - \frac{1}{2}\overset{\frown}{AFD}$ or $\angle P \overset{\circ}{=} \frac{1}{2}(\overset{\frown}{AED} - \overset{\frown}{AFD})$	5. Substitution Postulate

9. If three angles of one triangle are congruent to three angles of another triangle, the triangles are similar.

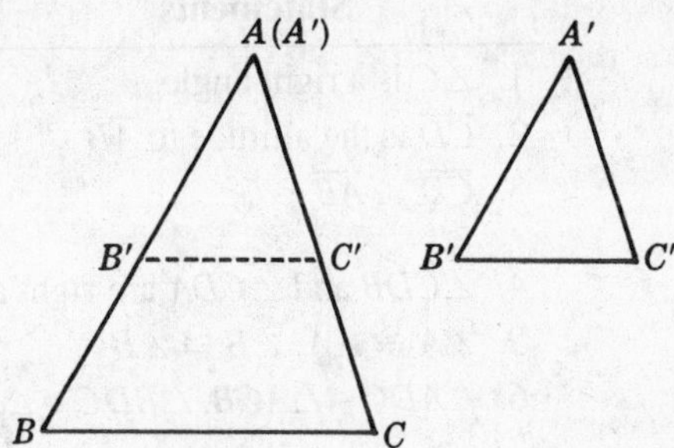

Given: $\triangle ABC$ and $\triangle A'B'C'$ $\angle A \cong \angle A'$, $\angle B \cong \angle B'$, $\angle C \cong \angle C'$

To Prove: $\triangle ABC \sim \triangle A'B'C'$

Plan: To prove the triangles similar, it must be shown that corresponding sides are in proportion. This is done by placing the triangles so that a pair of congruent angles coincide, and then repeating this so that another pair of congruent angles coincide.

PROOF:

Statements	Reasons
1. $\angle A \cong \angle A'$	1. Given
2. Place $\triangle A'B'C'$ on $\triangle ABC$ so that $\angle A'$ coincides with $\angle A$.	2. A geometric figure may be moved without change of its size or shape. Equal angles may be made to coincide.
3. $\angle B \cong \angle B'$	3. Given
4. $\overline{B'C'} \parallel \overline{BC}$	4. Two lines are parallel if their corresponding angles are congruent.
5. $\frac{A'B'}{AB} = \frac{A'C'}{AC}$	5. A line parallel to one side of a triangle divides the other two sides proportionately.
6. In like manner, by placing $\triangle A'B'C'$ on $\triangle ABC$ so that $\angle B'$ coincides with $\angle B$, show that $\frac{A'B'}{AB} = \frac{B'C'}{BC}$	6. Reasons 1 to 5

(*contd.*)

PROOF:

Statements	Reasons
7. $\frac{A'B'}{AB} = \frac{A'C'}{AC} = \frac{B'C'}{BC}$	7. Things (ratios) equal to the same thing are equal to each other.
8. $\triangle A'B'C' \sim \triangle ABC$	8. Two polygons are similar if their corresponding angles are congruent and their corresponding sides are in proportion.

10. If the altitude is drawn to the hypotenuse of a right triangle, then (a) the two triangles thus formed are similar to the given triangle and to each other, and (b) each leg of the given triangle is the mean proportional between the hypotenuse and the projection of that leg upon the hypotenuse.

Given: $\triangle ABC$ with a right angle at C, altitude $\overline{CD}$ to hypotenuse $\overline{AB}$

To Prove: (a) $\triangle ADC \sim \triangle CDB \sim \triangle ACB$
(b) $c{:}a = a{:}p$, $c{:}b = b{:}q$

Plan: The triangles are similar since they have a right angle and a pair of congruent acute angles. The proportions follow from the similar triangles.

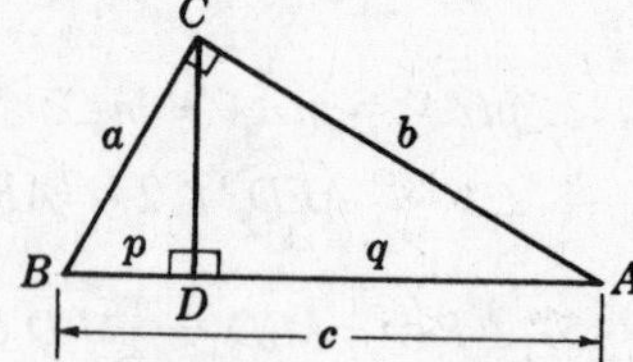

PROOF:

Statements	Reasons
1. $\angle C$ is a right angle.	1. Given
2. $\overline{CD}$ is the altitude to $\overline{AB}$	2. Given
3. $\overline{CD} \perp \overline{AB}$	3. An altitude to a side of a triangle is perpendicular to that side.
4. $\angle CDB$ and $\angle CDA$ are right angles.	4. Perpendiculars form right angles.
5. $\angle A \cong \angle A$, $\angle B \cong \angle B$	5. Reflexive property
6. $\triangle ADC \sim \triangle ACB$, $\triangle BDC \sim \triangle BCA$	6. Right triangles are similar if an acute angle of one is congruent to an acute angle of the other.
7. $\triangle ADC \sim \triangle CDB$	7. Triangles similar to the same triangle are similar to each other.
8. $c{:}a = a{:}p$, $c{:}b = b{:}q$	8. Corresponding sides of similar triangles are in proportion.

11. The square of the length of the hypotenuse of a right triangle equals the sum of the squares of the lengths of the other two sides.

Given: Right $\triangle ABC$, with a right angle at C. Legs have lengths a and b, and hypotenuse has length c.

To Prove: $c^2 = a^2 + b^2$

Plan: Draw $\overline{CD} \perp \overline{AB}$ and apply Theorem 10.

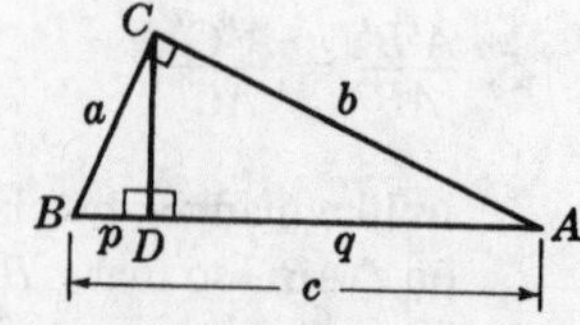

PROOF:

Statements	Reasons
1. Draw $\overline{CD} \perp \overline{AB}$.	1. Through an external point, a line may be drawn perpendicular to a given line.
2. $\frac{c}{a} = \frac{a}{p}, \frac{c}{b} = \frac{b}{q}$	2. If the altitude is drawn to the hypotenuse of a right triangle, either leg is the mean proportional between the hypotenuse and the projection of that leg upon the hypotenuse.
3. $a^2 = cp$, $b^2 = cq$	3. In a proportion, the product of the means equals the product of the extremes.
4. $a^2 + b^2 = cp + cq = c(p + q)$	4. If equals are added to equals, the sums are equal.
5. $c = p + q$	5. The whole equals the sum of its parts.
6. $a^2 + b^2 = c(c) = c^2$	6. Substitution Postulate

12. The area of a parallelogram equals the product of the length of one side and the length of the altitude to that side.

Given: $\square ABCD$, length of base $\overline{AD} = b$, length of altitude $\overline{BE} = h$

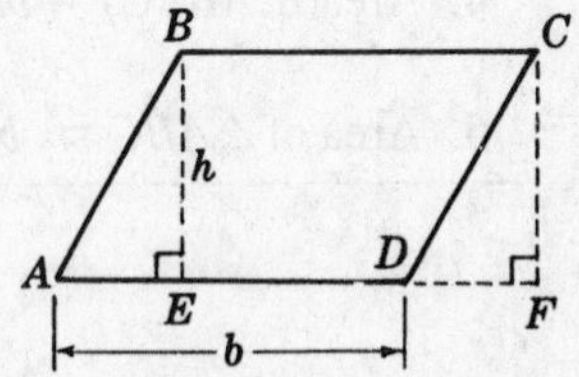

To Prove: Area of $ABCD = bh$

Plan: When a perpendicular is dropped to the base, extended, a rectangle is formed having the same base and altitude as the parallelogram. By adding congruent triangles to a common area, the rectangle and parallelogram are proved equal in area.

PROOF:

Statements	Reasons
1. Draw $\overline{CF} \perp \overrightarrow{AD}$ (extended).	1. Through an external point, a line may be drawn perpendicular to a given line.
2. $\overline{CF} \parallel \overline{BE}$	2. Segments perpendicular to the same line are parallel.
3. $\overline{BC} \parallel \overline{AD}$	3. Opposite sides of a parallelogram are parallel.
4. $\angle CFD$ and $\angle BEA$ are right angles.	4. Perpendiculars form right angles.
5. $BCFE$ is a rectangle.	5. A parallelogram having a right angle is a rectangle.
6. $\overline{AB} \cong \overline{CD}$, $\overline{CF} \cong \overline{BE}$	6. Opposite sides of a parallelogram are equal.
7. $\triangle ABE \cong \triangle DCF$	7. Hy-leg
8. Area(quadrilateral $BCDE$) = area(quadrilateral $BCDE$)	8. Reflexive property
9. Area ($\triangle ABE$) + area (quadrilateral $BCDE$) = area($\triangle DCF$) + area (quadrilateral $BCDE$) or area ($\square ABCD$) = area(rectangle $BCFE$)	9. If equals are added to equals, the sums are equal.
10. Area of rectangle $BCFE = bh$	10. The area of a rectangle equals the product of the lengths of its base and altitude.
11. Area of $\square$ $ABCD = bh$	11. Substitution Postulate

13. The area of a triangle is equal to one-half the product of the length of one side and the length of the altitude to that side.

Given: $\triangle ABC$, length of base $\overline{AC} = b$, length of altitude $\overline{BD} = h$

To Prove: Area of $\triangle ABC = \frac{1}{2}bh$

Plan: Drawing $\overline{BE} \parallel \overline{AC}$ and $\overline{EC} \parallel \overline{AB}$ forms a parallelogram having the same base and altitude as the triangle. Then the area of the triangle is half the area of the parallelogram.

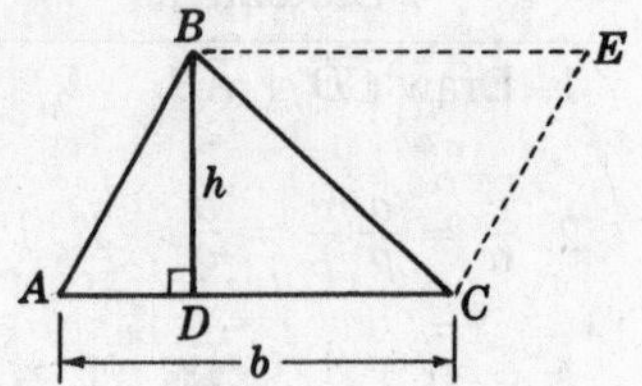

PROOF:

Statements	Reasons
1. Draw $\overline{BE} \parallel \overline{AC}$, $\overline{CE} \parallel \overline{AB}$.	1. Through an external point, a line may be drawn parallel to a given line.
2. $ABEC$ is a parallelogram with base b and altitude h.	2. A quadrilateral is a parallelogram if its opposite sides are parallel.
3. Area($\triangle ABC$) $= \frac{1}{2}$area ($\square ABEC$)	3. A diagonal divides a parallelogram into two congruent triangles.
4. Area($\square ABEC$) $= bh$	4. The area of a parallelogram equals the product of the lengths of its base and altitude.
5. Area of $\triangle ABC = \frac{1}{2}\, bh$	5. Substitution Postulate

14. The area of a trapezoid is equal to one-half the product of the length of the altitude and the sum of the lengths of the bases.

Given: Trapezoid $ABCD$, altitude $\overline{BE}$ with length h, base $\overline{AD}$ with length b, base $\overline{BC}$ with length b'.

To Prove: Area of $ABCD = \frac{1}{2}h(b + b')$

Plan: When a diagonal is drawn, the trapezoid is divided into two triangles having common altitude h and bases b and b'.

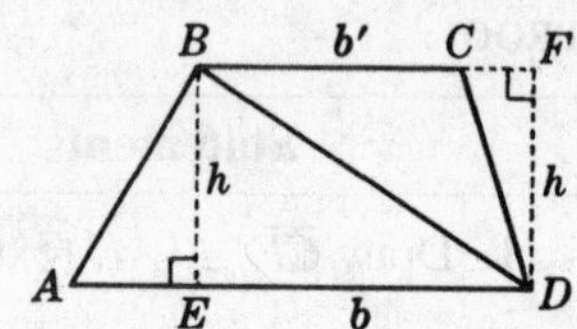

PROOF:

Statements	Reasons
1. Draw $\overline{BD}$.	1. A straight line may be drawn between two points.
2. Draw $\overline{DF} \perp \overrightarrow{BC}$ (extended).	2. Through an external point, a line may be drawn perpendicular to a given line.
3. $DF = BE = h$	3. Parallel lines are everywhere equidistant.
4. Area($\triangle ABD$) $= \frac{1}{2}bh$, Area($\triangle BCD$) $= \frac{1}{2}b'h$	4. The area of a triangle equals one-half the product of the lengths of its base and altitude.
5. Area of $ABCD = \frac{1}{2}bh + \frac{1}{2}b'h$ $= \frac{1}{2}h(b + b')$	5. If equals are added to equals, the sums are equal.

15. The area of a regular polygon is equal to one-half the product of its perimeter and apothem.

Given: Regular polygon *ABCDE* . . . having center *O*, apothem of length *r*, perimeter *p*

To Prove: Area of *ABCDE* . . . $= \frac{1}{2}rp$

Plan: By joining each vertex to the center, congruent triangles are obtained, the sum of whose areas equals the area of the regular polygon.

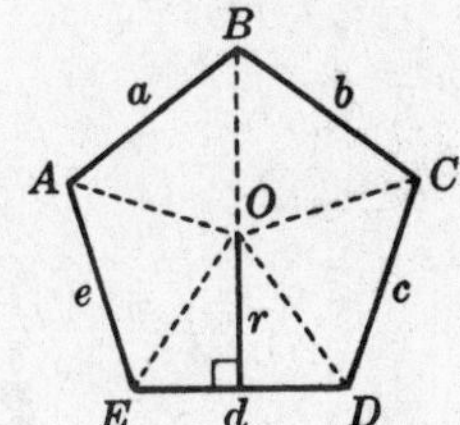

PROOF:

Statements	Reasons
1. Draw $\overline{OA}, \overline{OB}, \overline{OC}, \overline{OD}, \overline{OE}, \ldots$	1. A straight line segment may be drawn between two points.
2. r is the altitude of each triangle formed.	2. Apothems of a regular polygon are congruent.
3. Area of $\triangle AOB = \frac{1}{2}ar$ $\triangle BOC = \frac{1}{2}br$ $\triangle COD = \frac{1}{2}cr$	3. The area of a triangle equals one-half the product of the length of its base and altitude.
4. Area of regular polygon *ABCDE* . . . $= \frac{1}{2}ar + \frac{1}{2}br + \frac{1}{2}cr + \ldots$ $= \frac{1}{2}r(a + b + c + \ldots)$	4. If equals are added to equals, the sums are equal.
5. $p = a + b + c + \ldots$	5. The whole equals the sum of its parts.
6. Area of *ABCDE* . . . $= \frac{1}{2}rp$	6. Substitution Postulate

Extending Plane Geometry into Solid Geometry

17.1 Solids

A *solid* is an enclosed portion of space bounded by plane and curved surfaces.

Thus, the *pyramid*, the *cube*, the *cone*, the *cylinder*, and the *sphere* are solids. A solid has three dimensions: length, width, and thickness.

Practical illustrations of solids include a box, a brick, a block, and a ball. These are not, however, the pure or ideal solids that are the concern of solid geometry. In solid geometry we study the geometric properties of "perfect" solids, such as their shape, their size, the relationships of their parts, and the relationships along solids; physical properties such as their color, weight, or smoothness are disregarded.

17.1A Kinds of Solids

Polyhedra

A *polyhedron* is a solid bounded by plane (flat) surfaces only. Thus, the pyramid and cube are polyhedrons. The cone, cylinder, and sphere are not polyhedra, since each has a curved surface.

The bounding surfaces of the polyhedron are its *faces*; the lines of intersection of the faces are its *edges*, and the points of intersection of its edges are its *vertices*. A diagonal of a polyhedron joins two vertices not in the same face.

Thus, the polyhedron shown in Fig. 17-1 has six faces. Two of them are triangles (*ABC* and *CFG*), and the other four are quadrilaterals. Note $\overline{AF}$ is a diagonal of the polyhedron. The shaded polygon *HJKL* is a *section of the polyhedron* formed by the intersection of the solid and a plane passing through it.

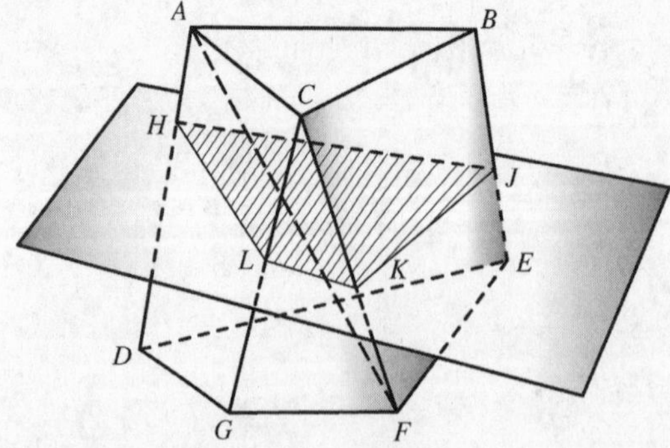

Fig. 17-1

The angle between any two intersecting faces is a *dihedral angle*. The angle between the covers of an open book is a dihedral angle. As the book is opened wider, the dihedral angle grows from one that is acute to one that is right, obtuse, and then straight. The dihedral angle can be measured by measuring the *plane angle* between two lines, one in each face, and in a plane perpendicular to the intersection between the faces.

Prisms

A *prism* (Fig. 17-2) is a polyhedron two of whose faces are parallel polygons, and whose remaining faces are parallelograms. The *bases* of a prism are its parallel polygons. These may have any number of sides. The *lateral* (sides) *faces* are the parallelograms. The distance between the two bases is h; it is measured along a line at right angles to both bases.

A *right prism* is a prism whose lateral faces are rectangles. The distance h is the height of any of the lateral faces.

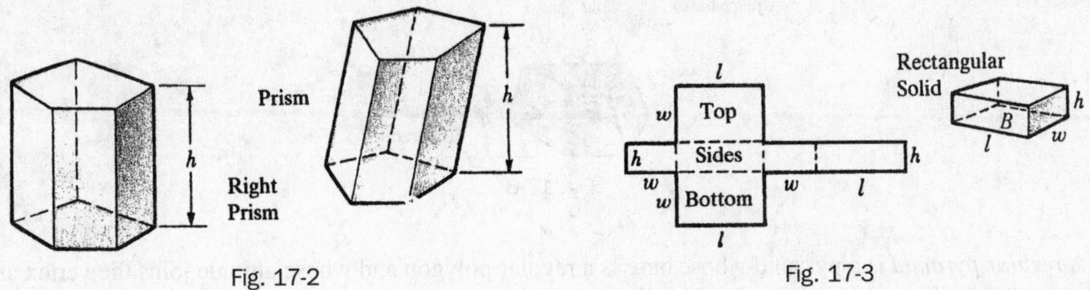

Fig. 17-2

Fig. 17-3

A *rectangular solid* (*box*) is a prism bounded by six rectangles. The solid can be formed from a pattern of six rectangles, as shown in Fig. 17-3, folded along the dashed lines. The length l, width w, and height h are its dimensions.

A *cube* is a rectangular solid bounded by six squares. The cube can be formed from a pattern of six squares, as shown in Fig. 17-4, folded along the dashed lines. Each equal dimension is represented by e in the diagram.

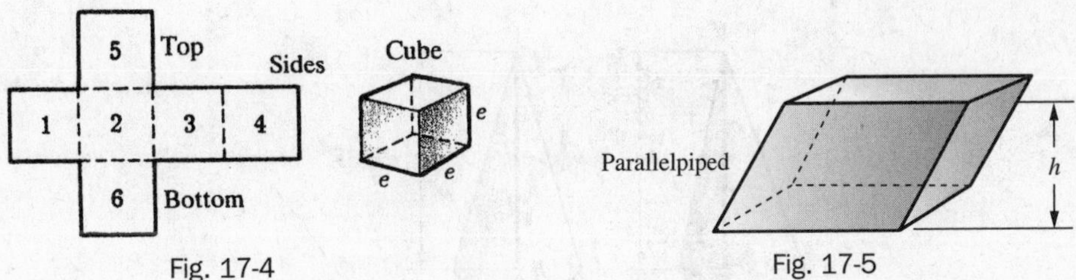

Fig. 17-4

Fig. 17-5

A *cubic unit* is a cube whose edge measures 1 unit. Thus, a cubic inch is a cube whose edge is 1 in long.

A *parallelepiped* is a prism bounded by six parallelograms (Fig. 17-5). Hence, the rectangular solid and cube are special parallelepipeds.

The table that follows shows some relationships among polygons in plane geometry and the corresponding relationships among polyhedra in solid geometry.

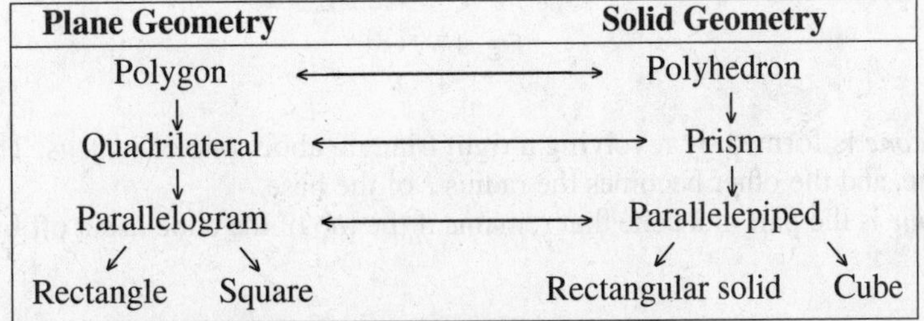

Plane Geometry		Solid Geometry
Polygon	⟷	Polyhedron
↓		↓
Quadrilateral	⟷	Prism
↓		↓
Parallelogram	⟷	Parallelepiped
↙ ↘		↙ ↘
Rectangle Square		Rectangular solid Cube

Pyramids

A *pyramid* is a polyhedron whose base is a polygon and whose other faces meet at a point, its *vertex*. The base (B in Fig. 17-6) may have any number of sides. However, the other faces must be triangles. The distance from the vertex to the base is equal to the measure of the altitude or height h, a line from the vertex at a right angle to the base.

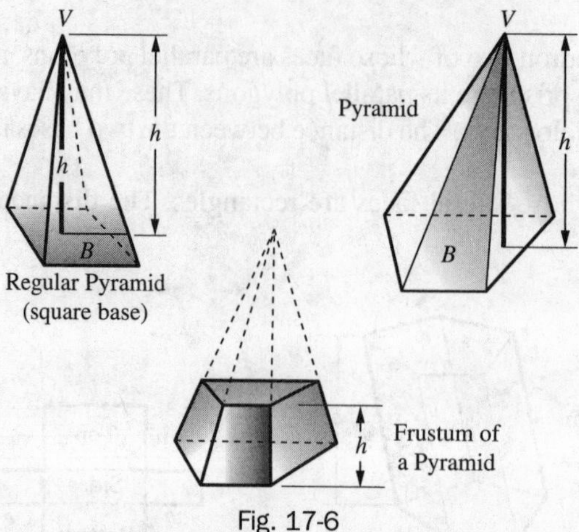

Fig. 17-6

A *regular pyramid* is a pyramid whose base is a regular polygon and whose altitude joins the vertex and the center of the base.

A *frustum of a pyramid* is the part of a pyramid that remains if the top of the pyramid is cut off by a plane parallel to the base. Note in Fig. 17-6 that its lateral faces are trapezoids.

Cones

A *circular cone* (Fig. 17-7) is a solid whose base is a circle and whose lateral surface comes to a point. (A circular cone is usually referred to simply as a cone.)

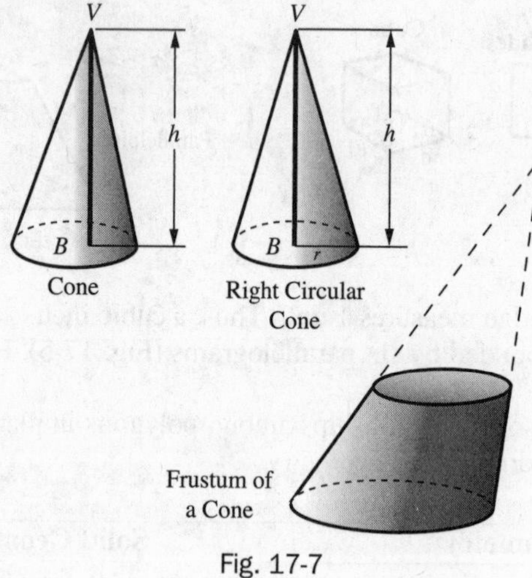

Fig. 17-7

A *right circular cone* is formed by revolving a right triangle about one of its legs. This leg becomes the altitude h of the cone, and the other becomes the radius r of the base.

A *frustum of a cone* is the part of a cone that remains if the top of the cone is cut off by a plane parallel to the base.

Cylinders

A *circular cylinder* (Fig. 17-8) is a solid whose bases are parallel circles and whose cross-sections parallel to the bases are also circles. (A circular cylinder is usually referred to simply as a cylinder.)

A *right circular cylinder* is a circular cylinder such that the line joining the centers of the two bases is perpendicular to the radii of these bases. The line joining the centers is the height h of the cylinder, and the radius of the bases is the radius r of the cylinder.

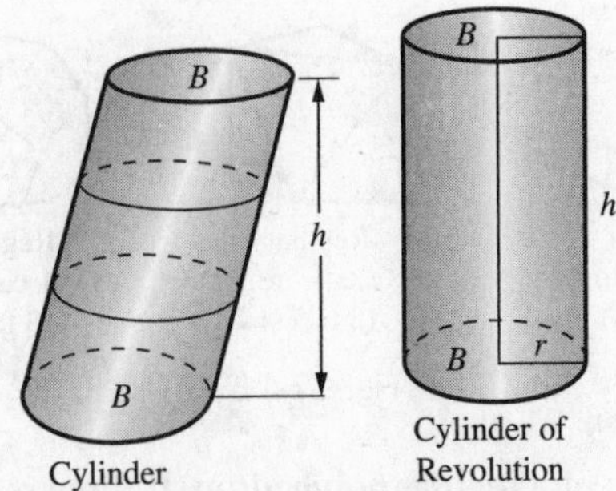

Fig. 17-8

Spheres

A *sphere* is a solid such that every point on its surface is at an equal distance from the same point, its center.

A sphere is formed by revolving a semicircle about its diameter as an *axis*. The outer end of the radius perpendicular to the axis generates a *great circle*, while the outer ends of other chords perpendicular to the diameter generate *small circles*.

Thus, sphere O in Fig. 17-9 is formed by the rotation of semicircle $\overparen{ACB}$ about $\overline{AB}$ as an axis. In the process, point C generates a great circle while points E and G generate small circles.

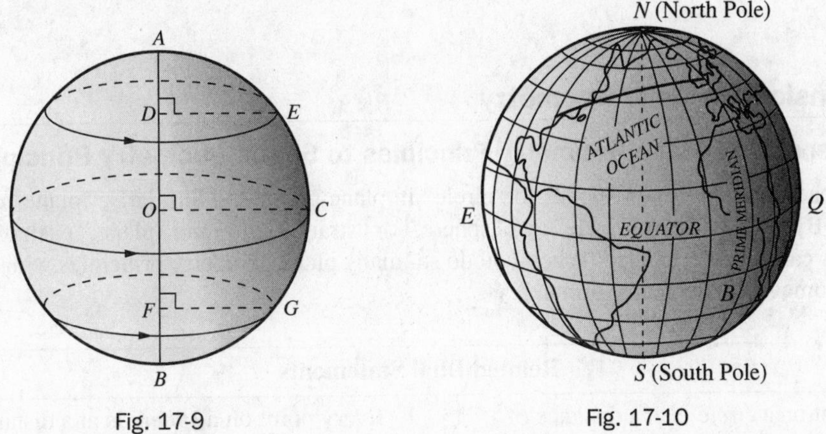

Fig. 17-9 Fig. 17-10

The way in which points on the earth's surface are located is better understood if we think of the earth as a sphere formed by rotating semicircle NQS in Fig. 17-10, which runs through Greenwich, England (near London), about NS as an axis. Point O, halfway between N and S, generates the *equator, EQ*. Points A and B generate *parallels of latitude*, which are small circles on the earth's surface parallel to the equator. Each position of the rotating semicircle is a semi-meridian, or *longitude*. (A *meridian* is a great circle passing through the North and South Poles.) The meridian through Greenwich is called the Prime Meridian.

Using the intersection of the equator and the Prime Meridian as the origin, we would locate New York City at $40°48\frac{1}{2}'$ North Latitude and $73°57\frac{1}{2}'$ West Longitude. The heavy arc shown on the globe is an arc of a great circle through New York City and London. Such an arc is the shortest distance between two points on the earth's surface. We could find this line by stretching a rubber band tightly between New York City and London on a globe.

17.1B Regular Polyhedra

Regular polyhedra are solids having faces that are regular polygons, with the same number of faces meeting at each vertex. There are only five such solids, shown in Fig. 17-11. Note that their faces are equilateral triangles, squares, or regular pentagons.

Regular Tetrahedron (4 faces)

Regular Hexahedron (6 faces)

Regular Octahedron (8 faces)

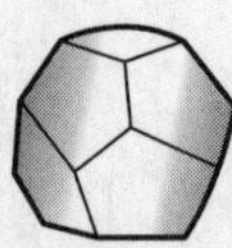
Regular Dodecahedron (12 faces)

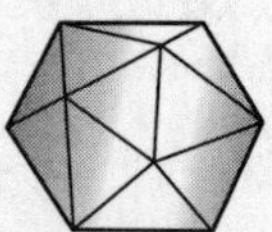
Regular Icosahedron (20 faces)

Fig. 17-11

Regular hexagons cannot be faces of a regular polyhedron. If three regular hexagons had a common vertex (Fig. 17-12), the sum of the measures of the three interior angles at that vertex would be 3(120°) or 360°. As a result, the three regular hexagons would lie in the same plane, so they could not form a solid.

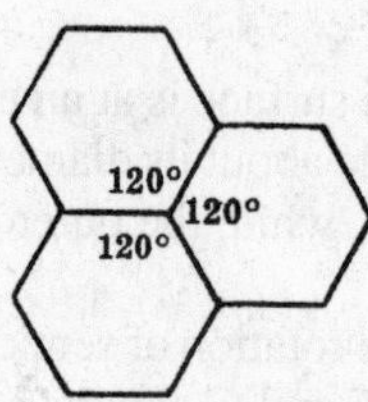

Fig. 17-12

17.2 Extensions to Solid Geometry

17.2A Extension of Plane Geometry Principles to Space Geometry Principles

"Sphere" in space geometry corresponds to "circle" in plane geometry. Similarly, "plane" corresponds to "straight line." By interchanging "circle" and "sphere," or "straight line" and "plane," each of the following *dual statements* can be interchanged. When you do so, many plane geometry principles with which you are acquainted become space geometry principles.

Related Dual Statements	
1. Every point on a *circle* is at a distance of one radius from its center.	1. Every point on a *sphere* is at a distance of one radius from its center.
2. Two intersecting straight *lines* determine a *plane*. 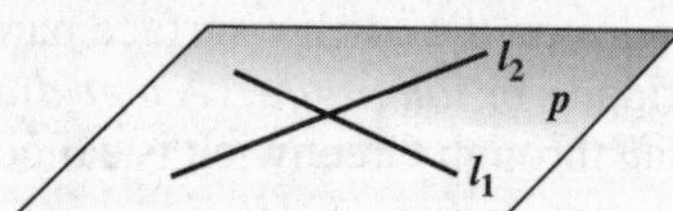	2. Two intersecting *planes* determine a straight *line*. 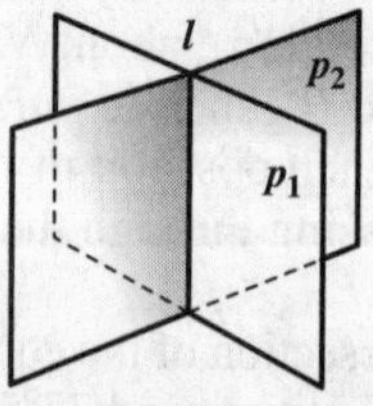
3. Two *lines* perpendicular to the same *plane* are parallel. 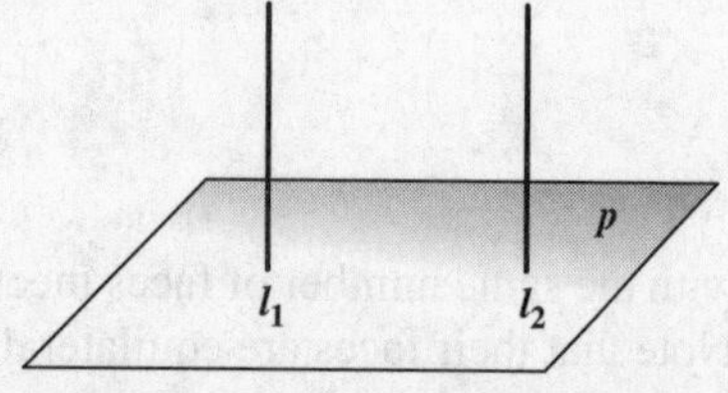	3. Two *planes* perpendicular to the same *line* are parallel. 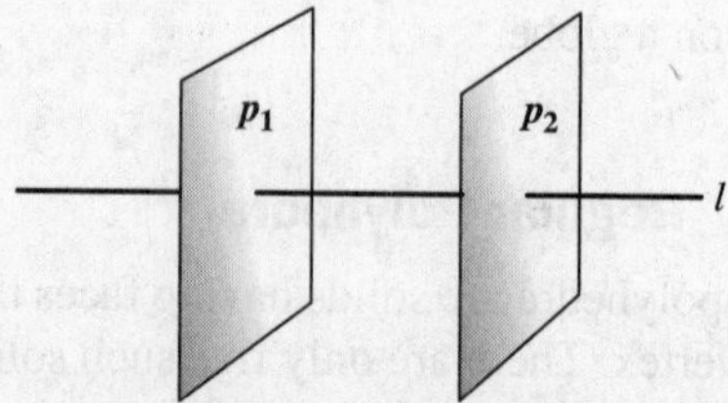

Be sure, in obtaining dual statements, that there is a complete interchange of terms. If the interchange is incomplete, as in the following pair of statements, there is no duality.

4. The locus of a point at a given distance from a given *line* is a pair of lines parallel to the given line and at the given distance from it. 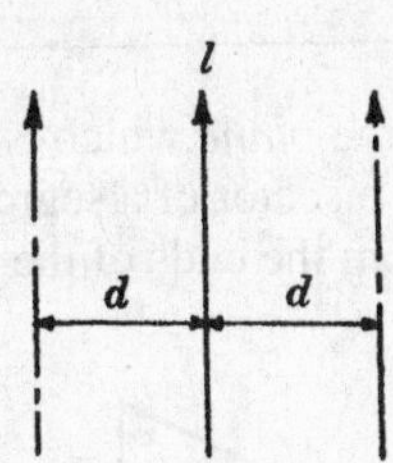	4. The locus of a point at a given distance from a given *line* is a circular cylindrical surface having the given line as axis and the given distance as radius. 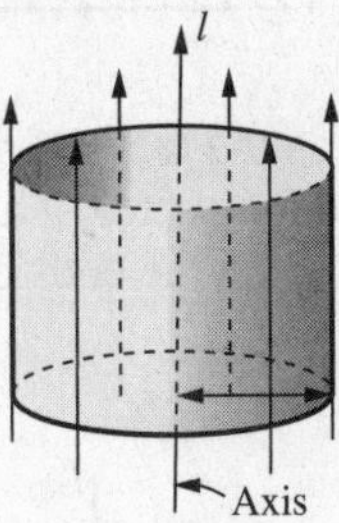

Unlike a cylinder, a cylindrical surface is not limited in extent; nor does it have bases. Similarly, a conical surface is unlimited in the extent and has no base.

Extension of Distance Principles

Here are some dual statements involving distance—in a plane and in space.

Distance in a Plane	Distance in Space
1. The distance from a point to a *line* is the length of the perpendicular from the point to the *line*.	1. The distance from a point to a *plane* is the length of the perpendicular from the point to the *plane*.
2. The distance between two parallel *lines* is the length of a perpendicular between them.	2. The distance between two parallel *planes* is the length of a perpendicular between them.
3. The distance from a point to a *circle* is the external segment of the secant from the point through the center of the *circle*.	3. The distance from a point to a *sphere* is the external segment of the secant from the point through the center of the *sphere*.

Distance in a Plane	Distance in Space
4. Parallel *lines* are everywhere equidistant. 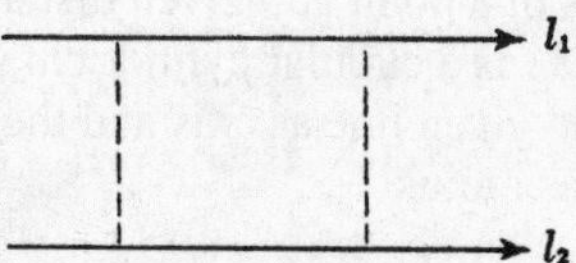 	4. Parallel *planes* are everywhere equidistant. 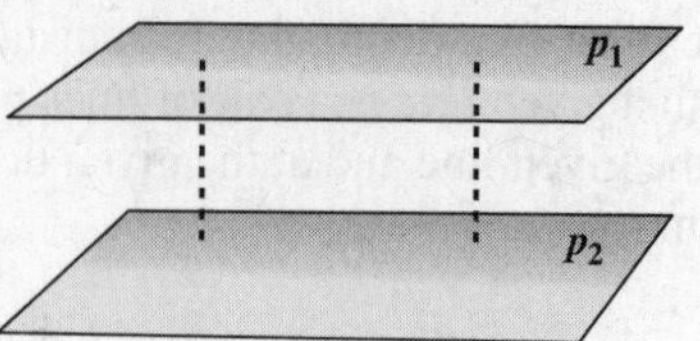
5. Any point on the *line* which is the perpendicular bisector of a segment is equidistant from the ends of the segment. 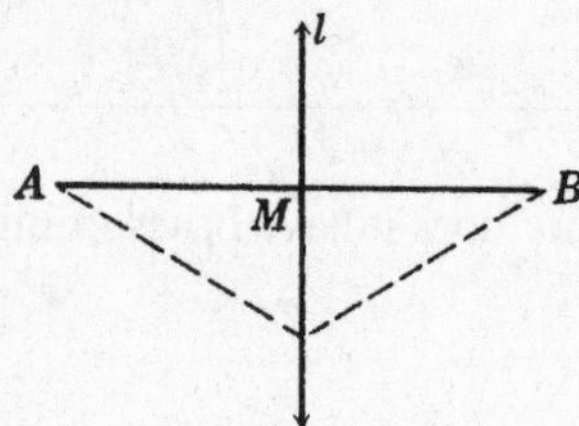 	5. Any point on the *plane* which is the perpendicular bisector of a segment is equidistant from the ends of the segment. 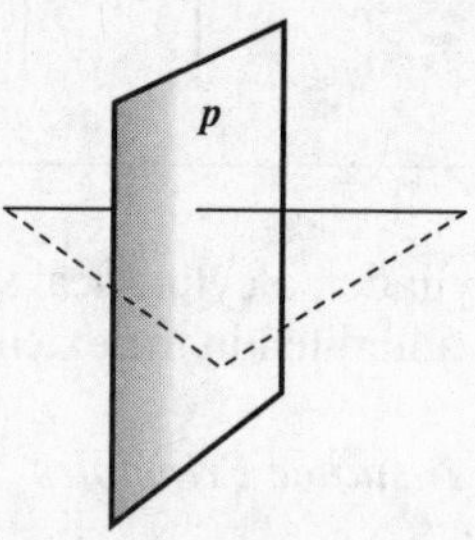
6. Any point on the *line* which is the bisector of the angle between two *lines* is equidistant from the sides of the angle. 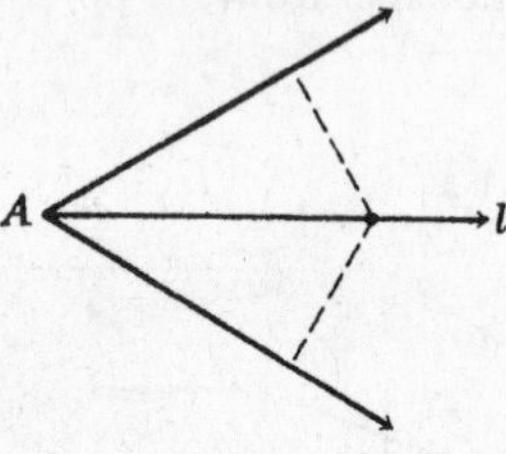 	6. Any point on the *plane* which is the bisector of the dihedral angle between two *planes* is equidistant from the sides of the angle. 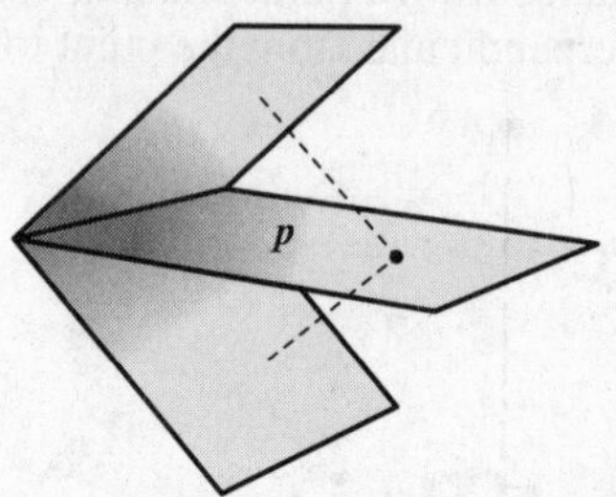

Extension of Locus Principles

The following dual statements involve the locus of points—in a plane and in space.

Locus in a Plane	Locus in Space
1. The locus of points at a given distance from a given point is a *circle* having the given point as center and the given distance as radius.	1. The locus of points at a given distance from a given point is a *sphere* having the given point as center and the given distance as radius.

Locus in a Plane	Locus in Space
2. The locus of points at a given distance from a given *line* is a *pair of lines* parallel to the given *line* and at the given distance from it. 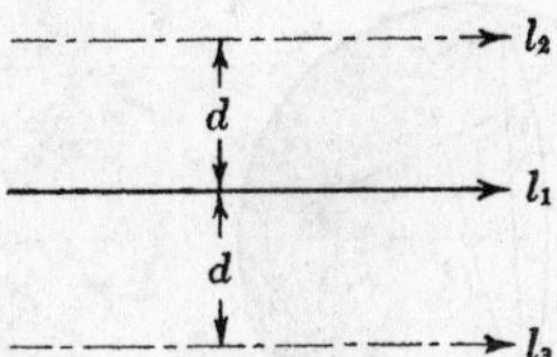	2. The locus of points at a given distance from a given *plane* is a pair of *planes* parallel to the given *plane* and at the given distance from it. 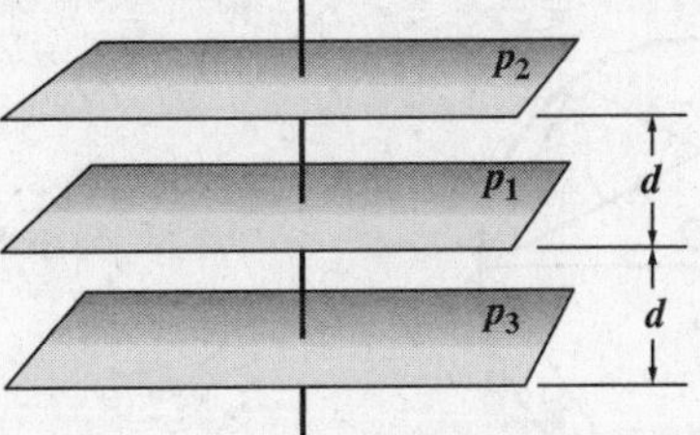
3. The locus of points equidistant from two points is the *line* that is the perpendicular bisector of the segment joining the two points. 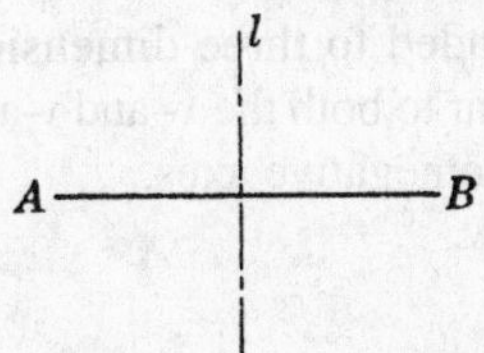	3. The locus of points equidistant from two points is the *plane* that is the perpendicular bisector of the segment joining the two points. 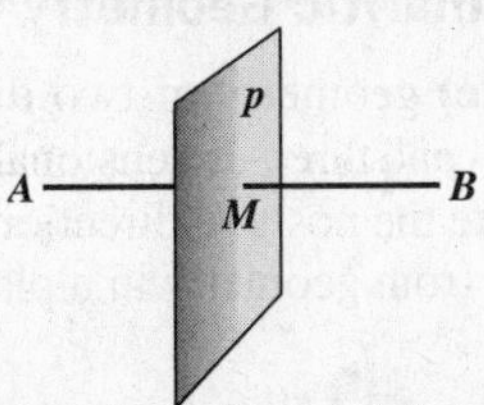
4. The locus of points equidistant from the *lines* that are the sides of an angle is the *line* which is the bisector of the angle between them. 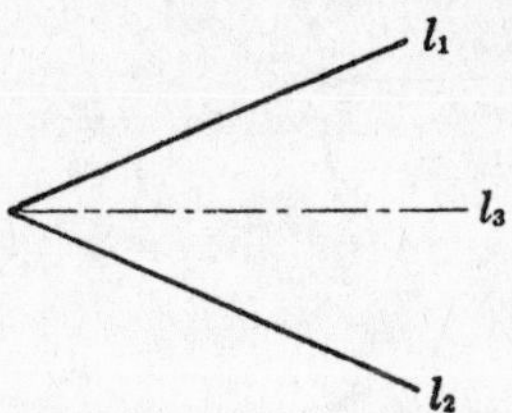	4. The locus of points equidistant from the *planes* that are the sides of a dihedral angle is the *plane* which is the bisector of the angle between them. 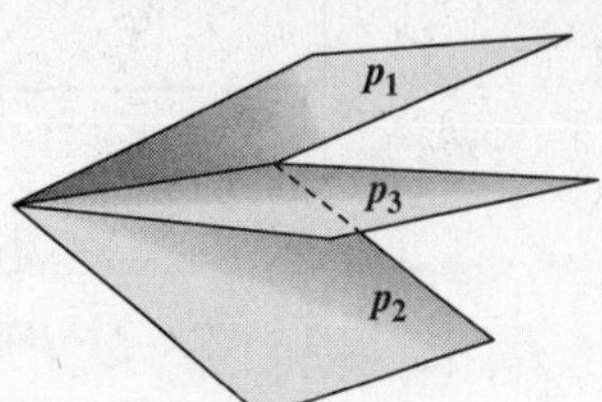
5. The locus of points equidistant from two parallel *lines* is the *line* parallel to them and midway between them. 	5. The locus of points equidistant from two parallel *planes* is the *plane* parallel to them and midway between them. 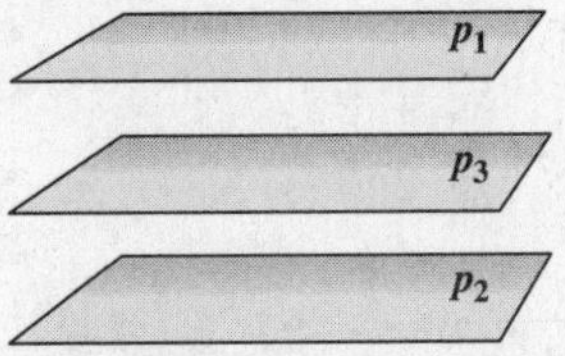
6. The locus of points equidistant from two concentric *circles* is the *circle* midway between the given *circles* and concentric to them. 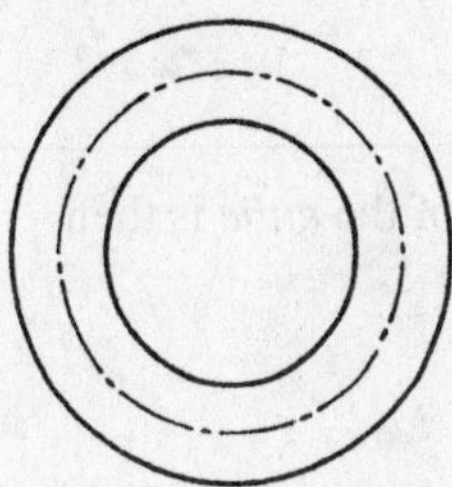	6. The locus of points equidistant from two concentric *spheres* is the *sphere* midway between the given *spheres* and concentric to them. 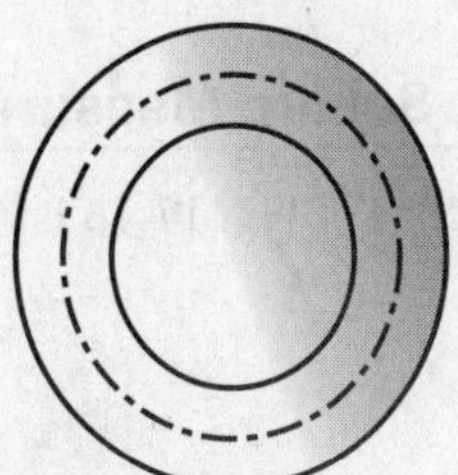

Locus in a Plane	Locus in Space
7. The locus of the vertex of a right triangle having a given hypotenuse is the *circle* having the hypotenuse as its diameter.	7. The locus of the vertex of a right triangle having a given hypotenuse is the *sphere* having the hypotenuse as its diameter.
O	O

17.2B Extension of Analytic Geometry to Three-Dimensional Space

The analytic (or coordinate) geometry of two dimensions can easily be extended to three dimensions. Figure 17-13 shows the two- and three-dimensional axes; the z-axis is perpendicular to both the x- and y-axes. In the figure, arrows indicate the positive directions, and dashed lines indicate the negative axes.

Four specific extensions from geometry in a plane to geometry in space follow.

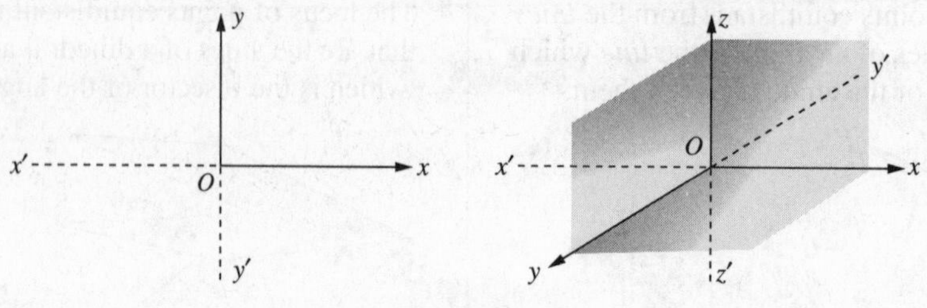

Fig. 17-13

Coordinate Geometry in a Plane	Coordinate Geometry in Space
1. Coordinates: $P_1\ (x_1, y_1),\ P_2(x_2, y_2)$	1. Coordinates: $P_1\ (x_1, y_1, z_1),\ P_2\ (x_2, y_2, z_2)$
2. Midpoint of $\overline{P_1P_2}$: $x_M = \dfrac{x_1 + x_2}{2},\ y_M = \dfrac{y_1 + y_2}{2}$	2. Midpoint of $\overline{P_1P_2}$: $x_M = \dfrac{x_1 + x_2}{2},\ y_M = \dfrac{y_1 + y_2}{2},\ z_M = \dfrac{z_1 + z_2}{2}$
3. Distance P_1P_2: $d = \sqrt{(x_2 - x_1)^2 + (y_2 - y_1)^2}$	3. Distance P_1P_2: $d = \sqrt{(x_2 - x_1)^2 + (y_2 - y_1)^2 + (z_2 - z_1)^2}$
4. Equation of a circle having the origin as center and radius r: $x^2 + y^2 = r^2$	4. Equation of a sphere having the origin as center and radius r: $x^2 + y^2 + z^2 = r^2$

17.3 Areas of Solids: Square Measure

The area of each face of the cube in Fig. 17-14 is $A = e^2$. The total surface area S of the *cube* is then

$$S = 6e^2$$

The areas of the six rectangles that make up the rectangular solid in Fig 17-15 are

$$A = lw \quad \text{for top and bottom faces}$$
$$A = lh \quad \text{for front and back faces}$$
$$A = wh \quad \text{for left and right faces}$$

The total surface area S of the *rectangular solid* is then

$$S = 2lw + 2lh + 2wh$$

The total surface area S of the *sphere* in Fig. 17-16 is

$$S = 4\,\pi r^2$$

The total surface area S of the *right circular cylinder* in Fig. 17-17 is

$$S = 2\pi r^2 + 2\pi rh = 2\pi r(r + h)$$

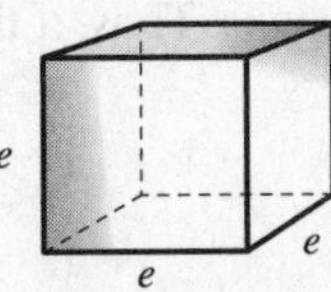

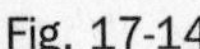
Fig. 17-14

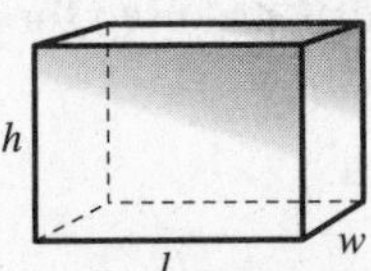

Fig. 17-15

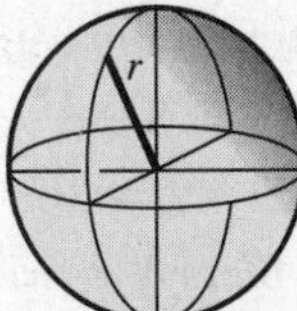

Fig. 17-16

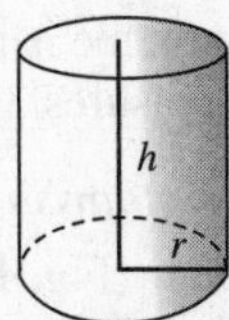

Fig. 17-17

SOLVED PROBLEM

17.1 Finding total surface areas of solids

Find, to the nearest integer, the total surface area of

(a) A cube with an edge of 5 m (Fig. 17-18)

(b) A rectangular solid with dimensions of 10 ft, 7 ft, and $4\frac{1}{2}$ft (Fig. 17-19)

(c) A sphere with a radius of 1.1 cm (Fig. 17-20)

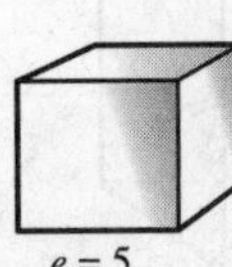

Fig. 17-18

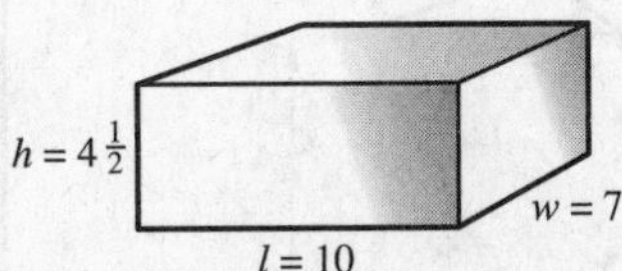

Fig. 17-19

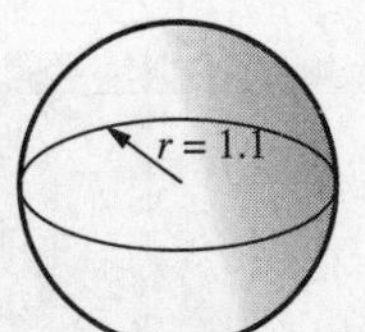

Fig. 17-20

Solutions

(a) $s = 6e^2 = 6(5^2) = 150\text{ m}^2$

(b) $s = 2lw + 2lh + 2wh = 2(10)(7) + 2(10)(4\frac{1}{2}) + 2(7)(4\frac{1}{2}) = 293\text{ ft}^2$

(c) $s = 4\pi r^2 = 4(3.14)(1.1^2) = 15.1976\text{ cm}^2$

17.4 Volumes of Solids: Cubic Measure

A *cubic unit* is a cube whose edge is 1 unit long. Thus, a cubic inch is a cube whose side is 1 in long (Fig. 17-21).

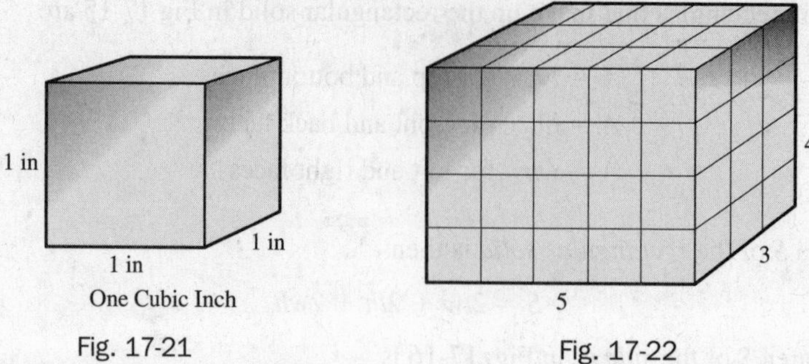

Fig. 17-21 Fig. 17-22

The *volume of a solid* is the number of cubic units that it contains. Thus, a box 5 units long, 3 units wide, and 4 units high has a volume of 60 cubic units; that is, it has a capacity or space large enough to contain 60 cubes, 1 unit on a side. (See Fig. 17-22.)

Here are some formulas for the volumes of solids. In these formulas, V is the volume of the solid, B is the area of a base, and h is the distance between the bases or between the vertex and a base. In volume formulas, the volume is in *cubic units*, the *unit* being the same as that used for the dimensions. Thus, if the edge of a cube measures 3 meters, its volume is 27 cubic meters.

1. *Rectangular solid* (Fig. 17-23): $V = lwh$
2. *Cylinder* (Fig. 17-24): $V = Bh$ or $V = \pi r^2 h$

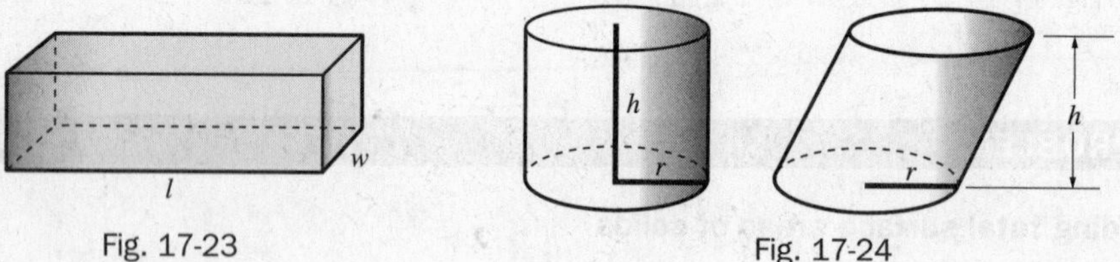

Fig. 17-23 Fig. 17-24

3. *Prism* (Fig. 17-25): $V = Bh$
4. *Cube* (Fig. 17-26): $V = e^3$

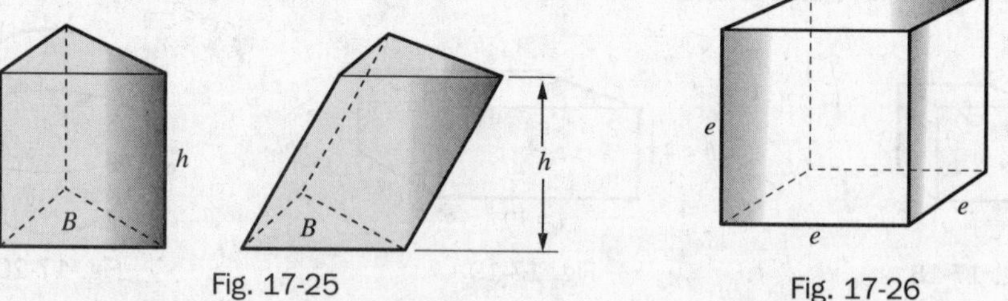

Fig. 17-25 Fig. 17-26

5. *Pyramid* (Fig. 17-27): $V = \frac{1}{3}Bh$
6. *Cone* (Fig. 17-28): $V = \frac{1}{3}Bh$ or $V = \frac{1}{3}\pi r^2 h$
7. *Sphere* (Fig. 17-29): $V = \frac{4}{3}\pi r^3$

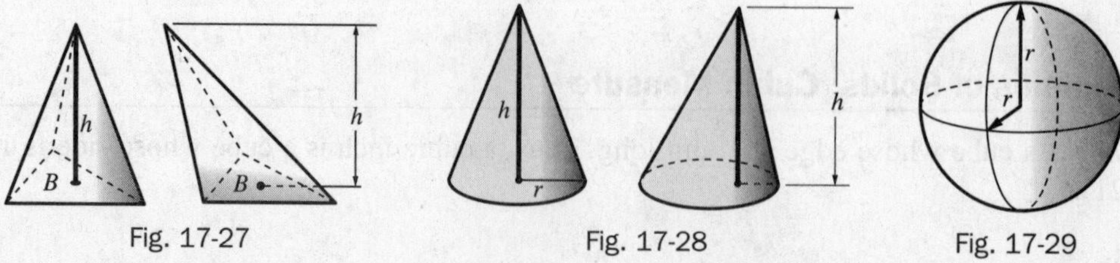

Fig. 17-27 Fig. 17-28 Fig. 17-29

SOLVED PROBLEMS

17.2 Relations among cubic units

Find the volume V of

(a) A cubic foot in cubic inches

(b) A cubic yard in cubic feet

(c) A liter (cubic decimeter) in cubic centimeters

Solutions

(a) $V = e^3$ for a cube. Since 1 ft = 12 in, $V = 12^3 = 1728$
So 1 $\text{ft}^3 = 1728\ \text{in}^3$.

(b) $V = e^3$ for a cube. Since 1 yd = 3 ft, $V = 3^3 = 27$
So 1 $\text{yd}^3 = 27\text{ft}^3$.

(c) $V = e^3$ again. Since 1 $\text{dm}^3 = 10\ \text{cm}^3$, $V = 10^3 = 1000$
So 1 liter = 1000 cm^3.

17.3 Finding volumes or cubes

Find the volume V of a cube, in cubic feet, if one edge is (a) 4 in, (b) 4 ft, (c) 4 yd.

Solutions

To find the volume in cubic feet, we must express the side in feet.

(a) $V = e^3$ and, since 4 in $= \frac{1}{3}$ft, $V = \left(\frac{1}{3}\right)^3 = \frac{1}{27}\ \text{ft}^3$

(b) $V = e^3 = 4^3 = 64\ \text{ft}^3$

(c) $V = e^3$, and since 4 yd = 12 ft, $V = 12^3 = 1728\ \text{ft}^3$

17.4 Finding the volumes of a rectangular solid, prism, and pyramid

Find the volume of

(a) A rectangular solid having a length of 6 in, a width of 4 in, and a height of 1 ft

(b) A prism having a height of 15 yd and a triangular base of 120 ft^2

(c) A pyramid having a height of 8 cm and a square base whose side is $4\frac{1}{2}$ cm

Solutions

(a) $V = lwh = 6(4)(12) = 288\ \text{in}^3$

(b) $V = Bh = 120(45) = 5400\ \text{ft}^3 = 200\ \text{yd}^3$

(c) $V = \frac{1}{3}Bh = \frac{1}{3}\left(\frac{9}{2}\right)^2(8) = 54\ \text{cm}^3$

17.5 Finding the volumes of a sphere, cylinder, and cone

Find the volume of

(a) A sphere with a radius of 10 in

(b) A cylinder with a height of 4 yd and a base whose radius is 2 ft

(c) A cone with a height of 2 ft and a base whose radius is 2 yd

Solutions

In these calculations, we shall let $\pi = 3.14$

(a) $V = \frac{4}{3}\pi r^3 = \frac{4}{3}(3.14)10^3 = 4186\frac{2}{3}\ \text{in}^3$

(b) $V = \pi r^2 h = (3.14)(2^2)12 = 150.72\ \text{ft}^3$

(c) $V = \frac{1}{3}\pi r^2 h = \frac{1}{3}(3.14)(6^2)(2) = 75.36\ \text{ft}^3$

17.6 Deriving formulas from *V = Bh*

From $V = Bh$, the volume formula for a prism or cylinder, derive the volume formulas for the solids in Fig. 17-30.

Solutions

(a) Since $B = lw$, $V = Bh = lwh$

(b) Since $B = e^2$ and $h = e$, $V = Bh = (e^2)\,e = e^3$

(c) Since $B = \pi r^2$, $V = Bh = \pi r^2 h$

(d) Since $B = \frac{h'}{2}(b + b')$, $V = Bh = \frac{h'}{2}(b + b')h = \frac{hh'}{2}(b + b')$

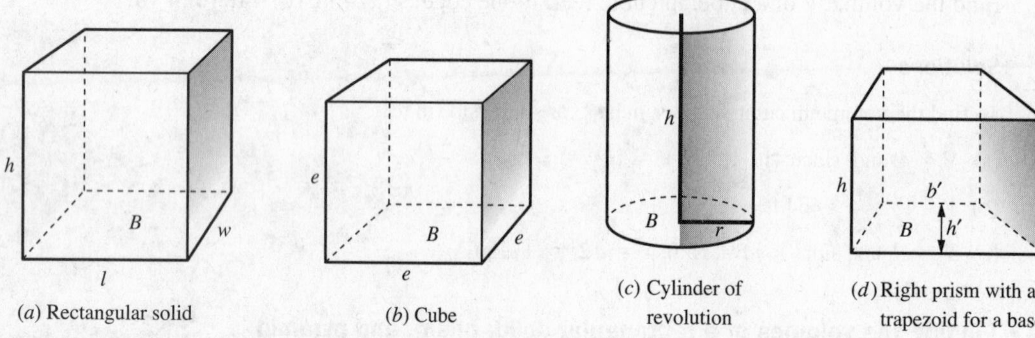

Fig. 17-30

17.7 Formulas for command volumes

State the formula for the volume of each solid in Fig. 17-31.

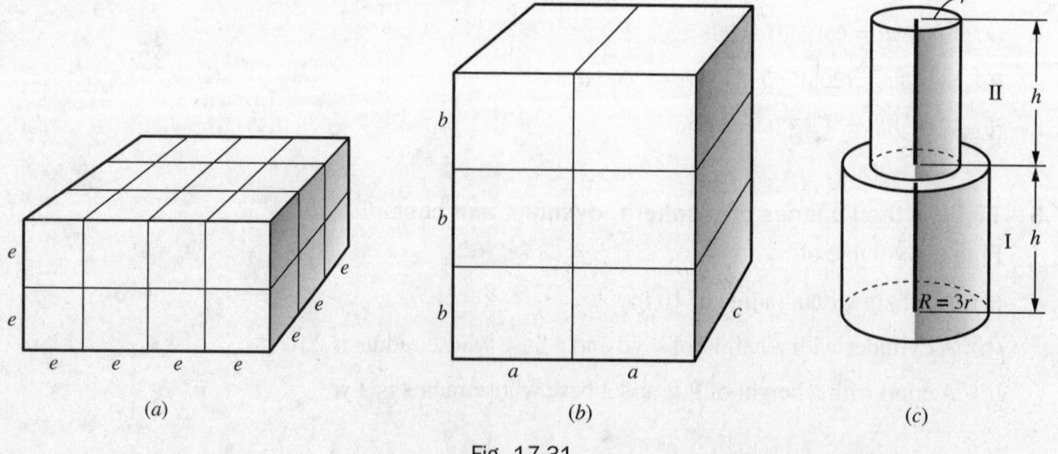

Fig. 17-31

Solutions

(a) $V = lwh$ for this solid. Now $l = 4e$, $w = 3e$, and $h = 2e$. Hence, $V = (4e)(3e)(2e) = 24e^3$

(b) $V = lwh$ again. Here $l = 2a$, $w = c$, and $h = 3b$. Hence, $V = (2a)(c)(3b) = 6abc$

(c) Here $V = V_{\text{cyl. I}} + V_{\text{cyl. II}} = \pi R^2h + \pi r^2h$. But $R = 3r$, so $V = \pi(3r)^2h + \pi r^2h = 10\pi r^2h$

SUPPLEMENTARY PROBLEMS

17.1. Find, to the nearest integer (using $\pi = 3.14$), the total area of (17.1)

(a) A cube with an edge of 7 yd

(b) A rectangular solid with dimensions of 8 ft, $6\frac{1}{2}$ft, and 14 ft

(c) A sphere with radius of 30 m

(d) A cylinder of revolution with a radius of 10 yd and a height of $4\frac{1}{2}$ yd. [*Hint*: Use $T = 2\pi r(r + h)$.]

17.2. Find the volume of (17.2)

(a) A cubic yard in cubic inches

(b) A cubic meter in cubic centimeters (1 m = 100 cm)

17.3. Find, to the nearest cubic inch, the volume of a cube whose edge is (a) 3 in; (b) $4\frac{1}{2}$ in; (c) 7.5 in; (d) 0.3 ft; (e) 1 ft 2 in. (17.3)

17.4. Find, to the nearest integer, the volume of (17.4)

(a) A rectangular solid of length 3 in, width $8\frac{1}{2}$ in, and height 8 in

(b) A prism having a height of 2 ft and a square base whose side is 3 yd

(c) A pyramid having a height of 2 yd and a base whose area is 6.4 ft^2

17.5. Find, to the nearest integer, the volume of (17.5)

(a) A sphere with radius of 6 m

(b) A cylinder having a height of 10 ft and a base whose radius is 2 yd

(c) A cone having a height of 3 yd and a base whose radius is 1.4 ft

17.6. From $V = \frac{1}{3}Bh$, the volume formula for a pyramid or cone, derive volume formulas for each of the solids in Fig. 17-32. (17.6)

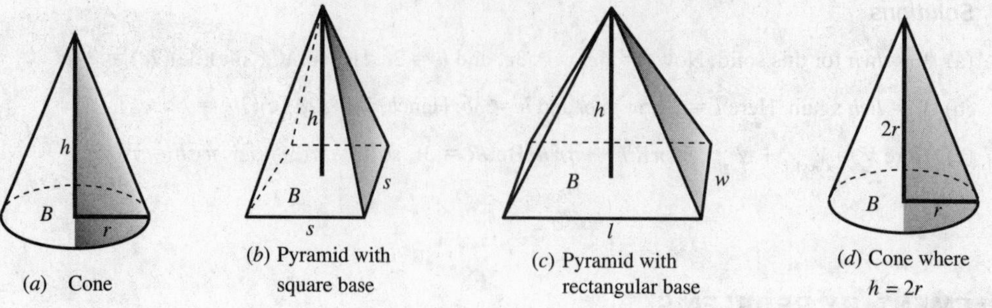

(*a*) Cone (*b*) Pyramid with square base (*c*) Pyramid with rectangular base (*d*) Cone where $h = 2r$

Fig. 17-32

17.7. Find a formula for the volume of each solid in Fig. 17-33. (17.7)

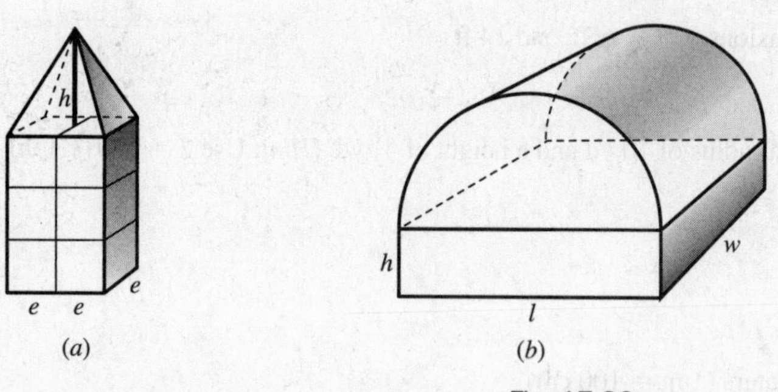

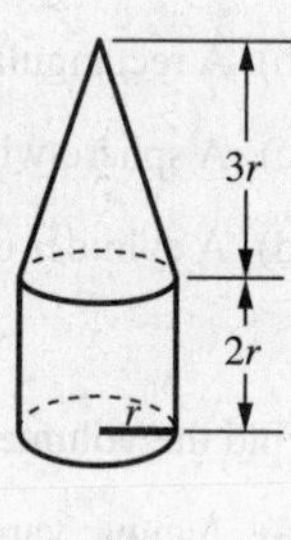

(*a*) (*b*) (*c*)

Fig. 17-33

Transformations

18.1 Introduction to Transformations

Two figures are congruent if one can be moved so that it exactly overlaps the other. A figure cut out of paper can be turned, slid, and flipped over to see if it matches up with another figure. If the figure is put on a graph, then these movements will change the coordinates of the points. A *transformation* is a way to describe such a change of coordinates.

18.2 Transformation Notation

A transformation begins with a general description of a point, such as $P(x, y)$ which represents a point P with coordinates x and y. Following this is an arrow $\mapsto$ and then a description of the point's *image*, the place where it ends up after the move. Usually the image of P is called P', the image of A is called A', and so on.

For example, the transformation $P(x, y) \mapsto P'(x - 5, 4 - y)$ means that the point $A(2, 1)$ is moved to $A'(2 - 5, 4 - 1) = A'(-3, 3)$, the point $B(3, 5)$ is moved to $B'(3 - 5, 4 - 5) = B'(-2, -1)$, and the point $C(6, 1)$ is moved to $C'(6 - 5, 4 - 1) = C'(1, 3)$. This transformation flips the triangle $\triangle ABC$ over and slides it to the left, as shown in Fig. 18-1.

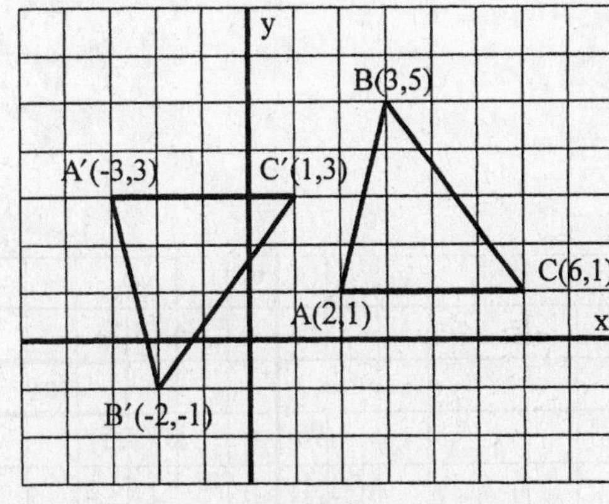

Fig. 18-1

SOLVED PROBLEMS

18.1 Using transformation notation

Name the image of the points $A(3, 1)$, $B(3, 4)$, and $C(5, 1)$ under the following transformations:

(a) $P(x, y) \mapsto P'(x + 2, y - 1)$

(b) $Q(x, y) \mapsto Q'(x + 5, y)$

(c) $R(x, y) \mapsto R'(5x, 5y)$

(d) $S(x, y) \mapsto S'(-y, x)$

(e) $T(x, y) \mapsto T'(y, 5 - x)$

Solutions

(a) $A'(3 + 2, 1 - 1) = A'(5, 0)$, $B'(3 + 2, 4 - 1) = B'(5, 3)$, and $C'(5 + 2, 1 - 1) = C'(7, 0)$

(b) $A'(3 + 5, 1) = A'(8, 1)$, $B'(3 + 5, 4) = B'(8, 4)$, and $C'\ (5 + 5, 1) = C'\ (10, 1)$

(c) $A'(5 \cdot 3, 5 \cdot 1) = A'(15, 5)$, $B'(5 \cdot 3, 5 \cdot 4) = B'(15, 20)$, and $C'(5 \cdot 5, 5 \cdot 1) = C'(25, 5)$

(d) $A'(-1, 3)$, $B'(-4, 3)$, and $C'(-1, 5)$

(e) $A'(1, 5 - 3) = A'(1, 2)$, $B'(4, 5 - 3) = B'(4, 2)$, and $C'(1, 5 - 5) = C'(1, 0)$

18.3 Translations

A transformation that slides figures without flipping or rotating them is called a *translation*. The translation that slides everything to the right a units and up b units is $P(x, y) \mapsto P'(x + a, y + b)$.

SOLVED PROBLEMS

18.2 Performing a translation

Let rectangle $ABCD$ be formed by $A(-1, 4)$, $B(-1, 3)$, $C(3, 3)$, and $D(3, 4)$. Graph rectangle $ABCD$ and its image under the following translations:

(a) $P(x, y) \mapsto P'(x + 4, y + 3)$

(b) $P(x, y) \mapsto P''(x + 2, y - 5)$

(c) $P(x, y) \mapsto P'''(x - 6, y - 2)$

Solutions

See Fig. 18-2.

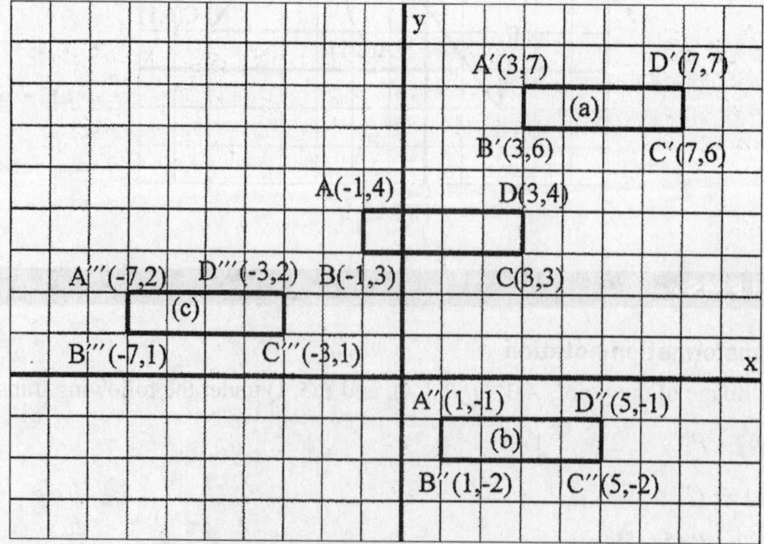

Fig. 18-2

(a) $A'(-1 + 4, 4 + 3) = A'(3, 7), B'(-1 + 4, 3 + 3) = B'(3, 6), C'(3 + 4, 3 + 3) = C'(7, 6)$, and $D'(3 + 4, 4 + 3) = D'(7, 7)$

(b) $A''(-1 + 2, 4 - 5) = A''(1, -1), B''(-1 + 2, 3 - 5) = B''(1, -2)$, $C''(3 + 2, 3 - 5) = C''(5, -2)$, and $D''(3 + 2, 4 - 5) = D''(5, -1)$

(c) $A'''(-1 - 6, 4 - 2) = A'''(-7, 2), B'''(-1 - 6, 3 - 2) = B'''(-7, 1)$, $C'''(3 - 6, 3 - 2) = C'''(-3, 1)$, and $D'''(3 - 6, 4 - 2) = D'''(-3, 2)$

18.3 Recognizing a translation

Name the translation that takes $\triangle ABC$ to (a) $\triangle A'B'C'$, (b) $\triangle A''B''C''$, and (c) $\triangle A'''B'''C'''$ as illustrated in Fig. 18-3.

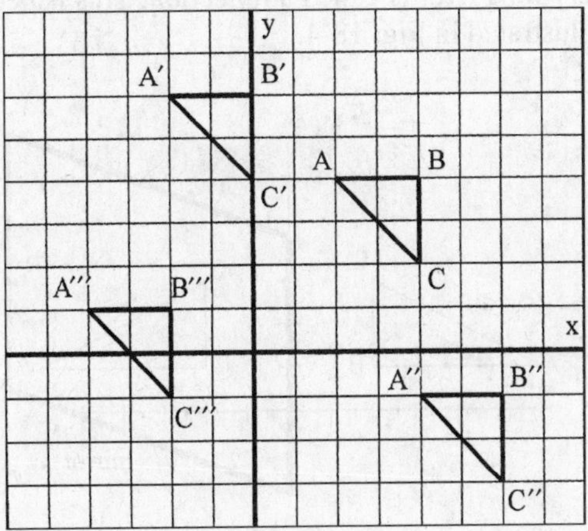

Fig. 18-3

Solutions

(a) $P(x, y) \mapsto P'(x - 4, y + 2)$

(b) $P(x, y) \mapsto P''(x + 2, y - 5)$

(c) $P(x, y) \mapsto P'''(x - 6, y - 3)$

18.4 Naming translations

Name the translation that moves everything:

(a) Up 6 spaces

(b) Down 1 space

(c) To the right 2 spaces

(d) To the left 10 spaces

(e) Up 5 spaces and to the right 3 spaces

(f) Down 7 spaces and to the right 4 spaces

(g) 6 spaces to the left and 4 spaces up

Solutions

(a) $P(x, y) \mapsto P'(x, y + 6)$

(b) $P(x, y) \mapsto P'(x, y - 1)$

(c) $P(x, y) \mapsto P'(x + 2, y)$

(d) $P(x, y) \mapsto P'(x - 10, y)$

(e) $P(x, y) \mapsto P'(x + 3, y + 5)$

(f) $P(x, y) \mapsto P'(x + 4, y - 7)$

(g) $P(x, y) \mapsto P'(x - 6, y + 4)$

18.4 Reflections

A transformation that flips everything over is called a *reflection*. This is because the image of an object in a mirror looks flipped over, as illustrated in Fig. 18-4.

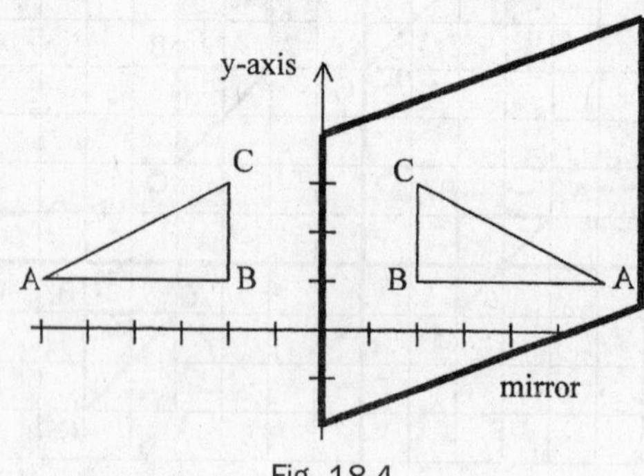

Fig. 18-4

The reflection in Fig. 18-4 is a reflection *across the y-axis* because the edge of the mirror is pressed against the y-axis. The line where the mirror meets the plane is called the *axis of symmetry*.

The reflection across the vertical line $x = a$ is given by $P(x, y) \mapsto P'(2a - x, y)$.

The reflection across the horizontal line $y = a$ is given by $P(x, y) \mapsto P'(x, 2a - y)$.

SOLVED PROBLEMS

18.5 Performing reflections

Let triangle ABC be formed by $A(-1, 1)$, $B(0, 3)$, and $C(3, 1)$. Graph $\triangle ABC$ and its image under:

(a) Reflection across the x axis ($y = 0$), $P(x, y) \mapsto P'(x, -y)$

(b) Reflection across the line $x = 4$, $P(x, y) \mapsto P''(8 - x, y)$

(c) Reflection across the line $y = 5$, $P(x, y) \mapsto P'''(x, 10 - y)$

Solutions

See Fig. 18-5.

(a) $A'(-1, -1)$, $B'(0, -3)$, and $C'(3, -1)$

(b) $A''(8 - (-1), 1) = A''(9, 1)$, $B''(8 - 0, 3) = B''(8, 3)$, and $C''(8 - 3, 1) = C''(5, 1)$

(c) $A'''(-1, 10 - 1) = A'''(-1, 9)$, $B'''(0, 10 - 3) = B'''(0, 7)$, and $C'''(3, 10 - 1) = C'''(3, 9)$

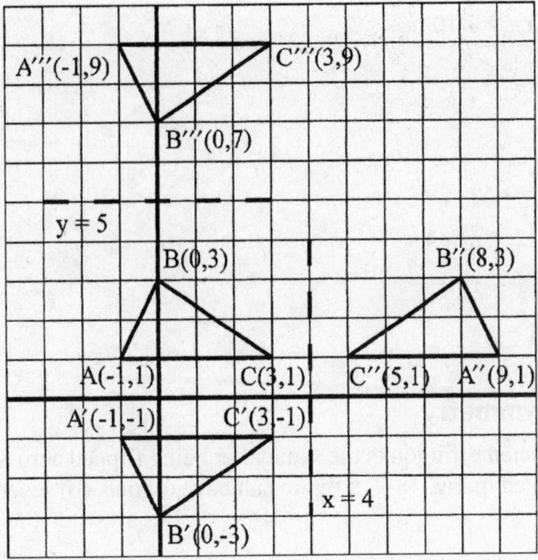

Fig. 18-5

18.6 Recognizing reflections

Name the reflection that takes $\triangle ABC$ to (a) $\triangle A'B'C'$, (b) $\triangle A''B''C''$, and (c) $\triangle A'''B'''C'''$ as illustrated in Fig. 18-6.

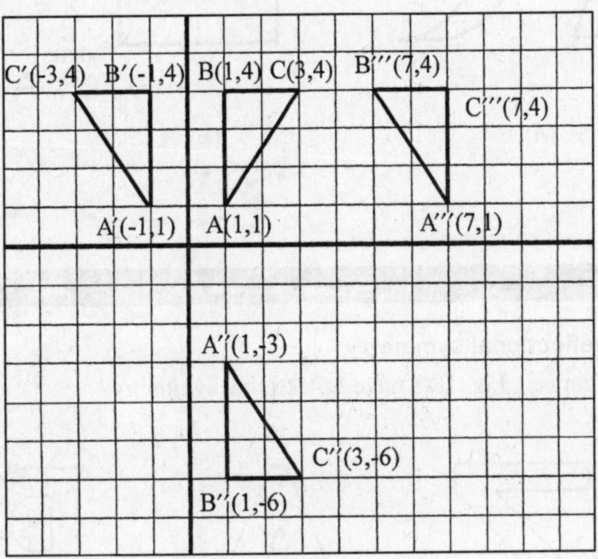

Fig. 18-6

Solutions

(a) Reflection across the y axis, $P(x, y) \mapsto P'(-x, y)$

(b) Reflection across the line $y = -1$, $P(x, y) \mapsto P''(x, -2 - y)$

(c) Reflection across the line $x = 4$, $P(x, y) \mapsto P'''(8 - x, y)$

18.7 Naming reflections

Name the transformation that

(a) Reflects across $x = 2$

(b) Reflects across $y = 6$

(c) Reflects across $x = -10$

(d) Reflects across $y = \frac{1}{2}$

Solutions

(a) $P(x, y) \mapsto P'(4 - x, y)$

(b) $P(x, y) \mapsto P'(x, 12 - y)$

(c) $P(x, y) \mapsto P'(-20 - x, y)$

(d) $P(x, y) \mapsto P'(x, 1 - y)$

18.4A Reflectional Symmetry

A figure has *reflectional symmetry* if it looks the same after being flipped across an axis of symmetry that runs through its center. As illustrated in Fig. 18-7, a figure can have (a) one, (b) several, or (c) no axes of symmetry.

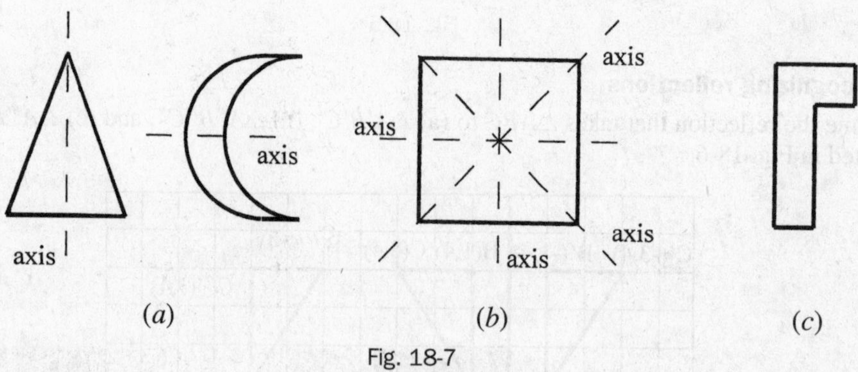

Fig. 18-7

SOLVED PROBLEMS

18.8 Recognizing reflectional symmetry

Which of the figures in Fig. 18-8 have reflectional symmetry?

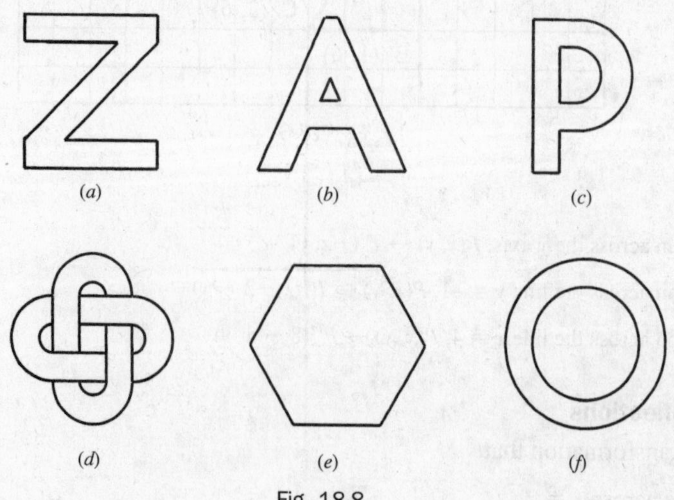

Fig. 18-8

Solutions

Only (b), (e), and (f) have reflectional symmetry. Note that when (d) is flipped, it will look like Fig. 18-9, which is different from the original in that the upper-left-hand crossing is horizontal instead of vertical.

Fig. 18-9

18.5 Rotations

If a pin were pushed through the origin on a graph and the paper were to be turned, the result would be a *rotation about the origin*. A rotation is described by the number of degrees by which the paper is turned.

The 90° clockwise rotation (or 270° counter-clockwise) about the origin is given by $P(x, y) \mapsto P'(y, -x)$.

The 180° rotation about the origin is given by $P(x, y) \mapsto P'(-x, -y)$.

The 270° clockwise (or 90° counter-clockwise) rotation about the origin is given by $P(x, y) \mapsto P'(-y, x)$.

In general, the clockwise rotation about the origin of θ° is given by $P(x, y) \mapsto P'(x \cos\theta + y \sin\theta, y \cos\theta - x \sin\theta)$

SOLVED PROBLEMS

18.9 Performing rotations

Let triangle ABC be given by $A(2, 1)$, $B(3, 1)$, and $C(3, 4)$. Graph the image of $\triangle ABC$ as rotated about the origin by (a) 90° clockwise, (b) 180°, and (c) 270° clockwise.

Solutions

See Fig. 18-10.

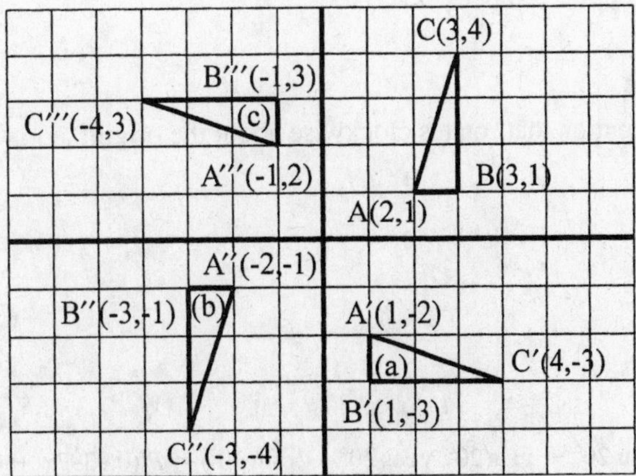

Fig. 18-10

(a) $A'(1,-2)$, $B'(1,-3)$, and $C'(4,-3)$

(b) $A''(-2,-1)$, $B''(-3,-1)$, $C''(-3,-4)$

(c) $A'''(-1, 2)$, $B'''(-1, 3)$, $C'''(-4, 3)$

18.10 Recognizing rotations

Name the rotation that takes $\triangle ABC$ to (a) $\triangle A'B'C'$, (b) $\triangle A''B''C''$, and (c) $\triangle A'''B'''C'''$ as illustrated in Fig. 18-11.

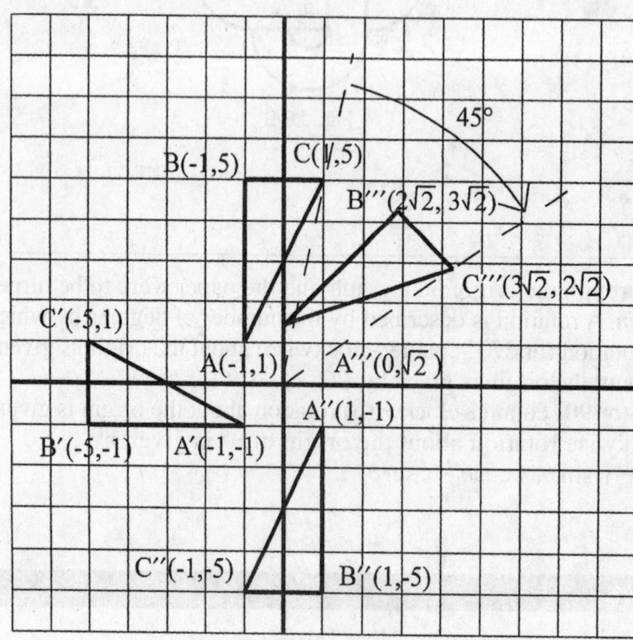

Fig. 18-11

Solutions

(a) 270° clockwise or 90° counter-clockwise about the origin, $P(x, y) \mapsto P'(-y, x)$

(b) 180° about the origin (either clockwise or counter-clockwise), $P(x, y) \mapsto P''(-x,-y)$

(c) 45° clockwise, $P(x, y) \mapsto P'''(x\cos 45° + y\sin 45°, y\cos 45° - x\sin 45°)$

$= P'''\left(\frac{\sqrt{2}}{2}x + \frac{\sqrt{2}}{2}y, \frac{\sqrt{2}}{2}y - \frac{\sqrt{2}}{2}x\right)$

18.11 Naming rotations

Name the transformation that rotates clockwise about the origin:

(a) 20°

(b) 30°

(c) 60°

(d) 75°

Solutions

(a) $P(x, y) \mapsto P'(x\cos 20° + y\sin 20°, y\cos 20° - x\sin 20°) = P'(0.9397x + 0.3420y, 0.9397y - 0.3420x)$

(b) $P(x, y) \mapsto P'(x\cos 30° + y\sin 30°, y\cos 30° - x\sin 30°) = P'\left(\frac{\sqrt{3}}{2}x + \frac{1}{2}y, \frac{\sqrt{3}}{2}y - \frac{1}{2}x\right)$

$= P'(0.866x + 0.5y, 0.866y - 0.5x)$

(c) $P(x, y) \mapsto P'(x\cos 60° + y\sin 60°, y\cos 60° - x\sin 60°) = P'\left(\frac{1}{2}x + \frac{\sqrt{3}}{2}y, \frac{1}{2}y - \frac{\sqrt{3}}{2}x\right)$

$= P'(0.5x + 0.866y, 0.5y - 0.866)$

(d) $P(x, y) \mapsto P'(x\cos 75° + y\sin 75°, y\cos 75° - x\sin 75°) = P'(0.2588x + 0.9659y, 0.2588y - 0.9659x)$

18.5A Rotational Symmetry

A figure has *rotational symmetry* if it can be rotated around its center by fewer than 360° and look the same as it did originally. In Fig. 18-12, there is (a) a figure that looks the same under a 72° rotation, (b) a figure that looks the same under a 120° rotation, (c) a figure that looks the same under a 180°, and (d) a figure without rotational symmetry.

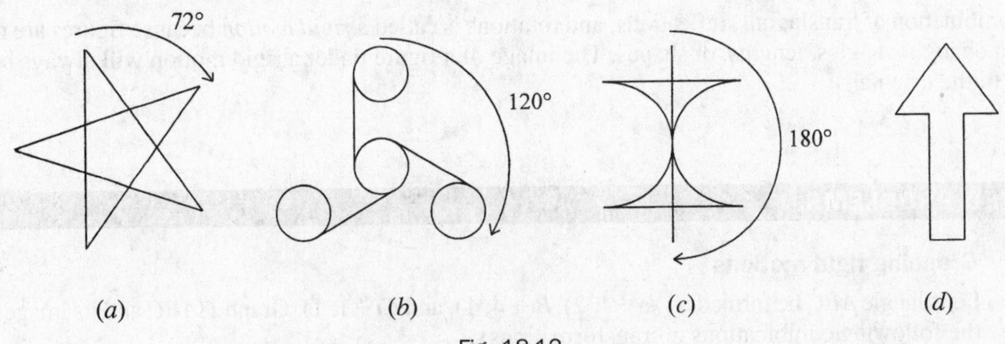

Fig. 18-12

SOLVED PROBLEMS

18.12 Recognizing rotational symmetry

For each figure in Fig. 18-13, give the smallest angle by which the figure could be rotated around its center and still look the same.

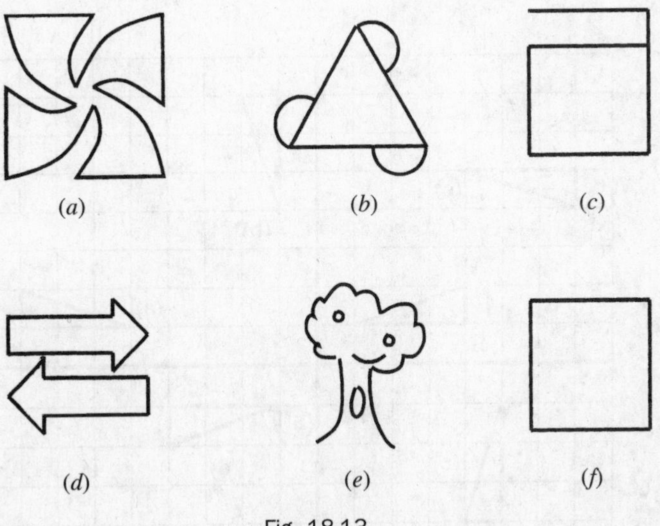

Fig. 18-13

Solutions

(a) 90°

(b) 120°

(c) 360° (no rotational symmetry)

(d) 180°

(e) 360° (no rotational symmetry)

(f) 90°

18.6 Rigid Motions

Any combination of translations, reflections, and rotations is called a *rigid motion* because figures are moved without changing angles, lengths, or shapes. The image of a figure under a rigid motion will always be congruent to the original.

SOLVED PROBLEMS

18.13 Graphing rigid motions

Let triangle ABC be formed by $A(-4, 2)$, $B(-4, 1)$, and $C(-1, 1)$. Graph $\triangle ABC$ and its image under the following combinations of transformations:

(a) Reflect across the y axis and then move to the right 4 spaces.

(b) Rotate 90° clockwise around the origin then move up 3 spaces.

(c) Reflect across $y = 2$ then move up 2 spaces and to the left 3 spaces.

(d) Reflect across the x axis and then reflect across the y axis.

(e) Rotate 90° counter-clockwise around the origin and then reflect across $x = -3$.

Solutions

See Fig. 18-14.

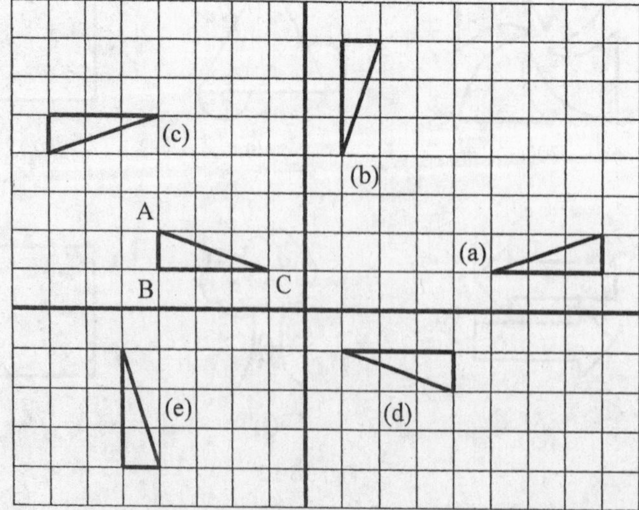

Fig. 18-14

18.14 Combining transformations

Name the single transformation that does the same thing as the combination of:

(a) $P(x, y) \mapsto P'(x + 7, y - 2)$ and then $Q'(x, y) \mapsto Q''(x + 3, y + 5)$

(b) $P(x, y) \mapsto P'(-x, y)$ and then $Q'(x, y) \mapsto Q''(x + 3, y + 2)$

(c) $P(x, y) \mapsto P'(-x, -y)$ and then $Q'(x, y) \mapsto Q''(4 - x, y)$

(d) $P(x, y) \mapsto P'(y, -x)$ and then $Q'(x, y) \mapsto Q''(x - 5, y + 1)$

Solutions

(a) $R(x, y) \mapsto R''((x + 7) + 3, (y - 2) + 5) = R''(x + 10, y + 3)$

(b) $R(x, y) \mapsto R''(-x + 3, y + 2)$

(c) $R(x, y) \mapsto R''(4 - (-x), -y) = R''(4 + x, -y)$

(d) $R(x, y) \mapsto R''(y - 5, -x + 1)$

18.15 Recognizing rigid motions

Name the transformation that takes $\triangle ABC$ to (a) $\triangle A'B'C'$, (b) $\triangle A''B''C''$, and (c) $\triangle A'''B'''C'''$ as illustrated in Fig. 18-15.

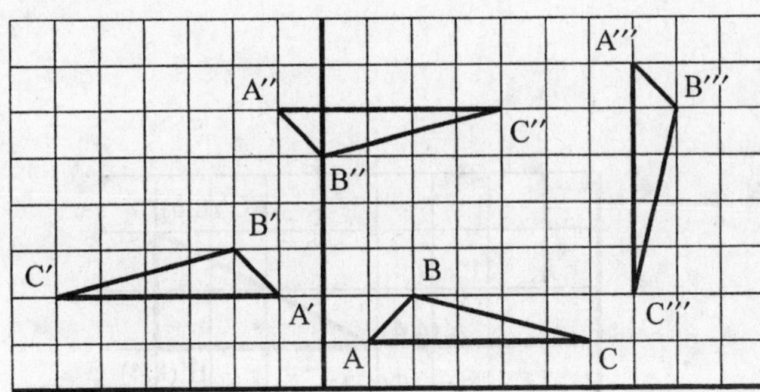

Fig. 18-15

Solutions

(a) The triangle has been reflected across the y axis and then moved up 1 space, so $P(x, y) \mapsto P'(-x, y + 1)$.

(b) The triangle has been reflected across the line $y = 3$, then moved up 1 space and to the left 2 spaces, so $P(x, y) \mapsto P'(x - 2, 7 - y)$.

(c) The triangle has been rotated clockwise around the origin 90° and then moved up 8 spaces and to the right 6 spaces, so $P(x, y) \mapsto P'(y + 6, -x + 8)$.

18.16 Naming rigid motions

Name the transformation that

(a) Rotates everything around the origin 180°, then moves everything up 3 spaces

(b) Reflects across $x = 4$, then slides everything down 2 spaces

(c) Rotates everything 90° clockwise around the origin, then reflects across the y axis

(d) Rotates around the origin 90° counter-clockwise, then slides to the left 5 spaces

Solutions

(a) $P(x, y) \mapsto P'(-x, -y + 3)$

(b) $P(x, y) \mapsto P'(8 - x, y - 2)$

(c) $P(x, y) \mapsto P'(-y, -x)$

(d) $P(x, y) \mapsto P'(-y - 5, x)$

18.7 Dihilations

A *dihilation* (also called a *scaling* or an *enlargement*) is not a rigid motion because it multiplies all lengths by a single *scale factor*. The image of a figure under a dihilation will always be similar to the original.

The dihilation that enlarges everything by a scale factor of k is $P(x, y) \mapsto P'(kx, ky)$.

SOLVED PROBLEMS

18.17 Performing dihilations

Let triangle ABC be formed by $A(2, 2)$, $B(4, 2)$, and $C(4, 3)$. Graph the image of $\triangle ABC$ under (a) magnification by 2 and (b) scaling by $\frac{1}{2}$.

Solutions

See Fig. 18-16.

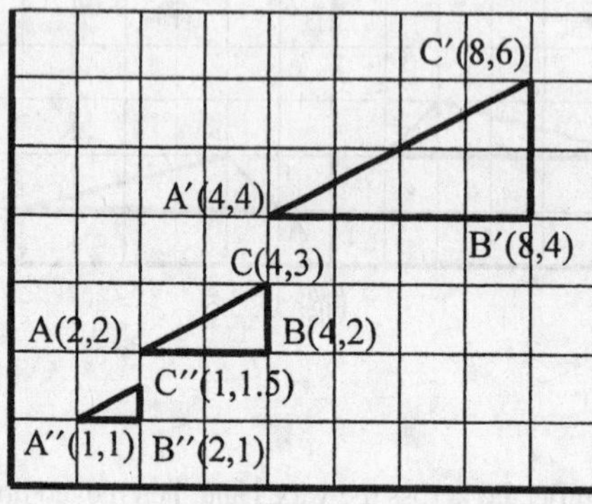

Fig. 18-16

(a) $A'(4, 4)$, $B'(8, 4)$, and $C'(8, 6)$

(b) $A''(1, 1)$, $B''(2, 1)$, and $C''(2, 1.5)$

18.18 Naming dihilations

Name the transformation that

(a) Scales everything 5 times larger

(b) Shrinks every length to half size

(c) Triples all linear dimensions

(d) Depicts everything at $\frac{1}{10}$ scale

(e) Dihilates by a scale factor of 12

Solutions

(a) $P(x, y) \mapsto P'(5x, 5y)$

(b) $P(x, y) \mapsto P'(\frac{1}{2}x, \frac{1}{2}y)$

(c) $P(x, y) \mapsto P'(3x, 3y)$

(d) $P(x, y) \mapsto P'(\frac{1}{10}x, \frac{1}{10}y)$

(e) $P(x, y) \mapsto P'(12x, 12y)$

SUPPLEMENTARY PROBLEMS

18.1. Name the image of points $A(6, 2)$, $B(-1, 4)$, and $C(2, 7)$ under the transformation

(a) $P(x, y) \mapsto P'(x + 3, y)$

(b) $P(x, y) \mapsto P'(2 - x, y)$

(c) $P(x, y) \mapsto P'(-x + 1, -y + 3)$

(d) $P(x, y) \mapsto P'(-y, x + 8)$

(e) $P(x, y) \mapsto P'(2x, 2y)$

(f) $P(x, y) \mapsto P'(4 - 3x, 3y)$

(g) $P(x, y) \mapsto P'(2 - y, 5 - x)$

18.2. Let triangle ABC be defined by $A(1, -1)$, $B(2, 2)$, and $C(3, -1)$. Graph the image of $\triangle ABC$ under

(a) $P(x, y) \mapsto P'(x + 5, y)$

(b) $P(x, y) \mapsto P'(x, y - 4)$

(c) $P(x, y) \mapsto P'(x - 3, y + 2)$

18.3. Name the translation that takes $\triangle ABC$ to (a) $\triangle A'B'C'$, (b) $\triangle A''B''C''$, and (c) $\triangle A'''B'''C'''$ as illustrated in Fig. 18-17.

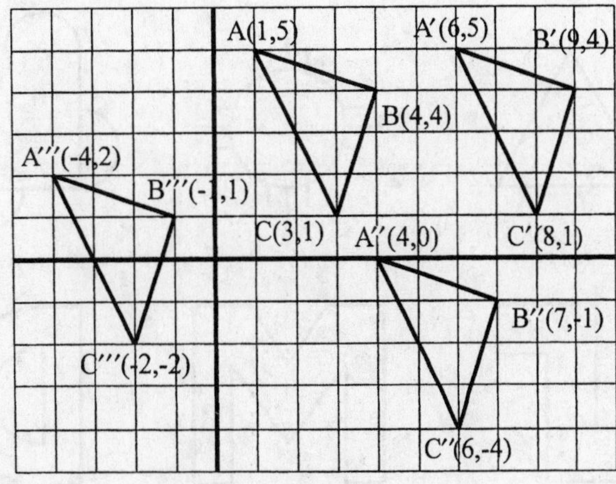

Fig. 18-17

18.4. Name the translation that moves everything

(a) Down 5 spaces

(b) To the right 6 spaces

(c) Up 3 spaces and 7 spaces to the left

(d) Down 2 spaces and 8 spaces to the right

(e) Up 4 spaces and to the left 1 space

18.5. Let trapezoid $ABCD$ be formed by $A(1, 3)$, $B(5, 3)$, $C(4, 1)$, and $D(2, 1)$. Graph trapezoid $ABCD$ and its image under (a) reflection across the y axis $P(x, y) \mapsto P'(-x, y)$, (b) reflection across the line $y = -1$, $P(x, y) \mapsto P''(x, -2 - y)$, and (c) reflection across the line $x = 8$, $P(x, y) \mapsto P'''(16 - x, y)$.

18.6. Name the reflection that takes $\triangle ABC$ to (a) $\triangle A'B'C'$, (b) $\triangle A''B''C''$, and (c) $\triangle A'''B'''C'''$ as illustrated in Fig. 18-18.

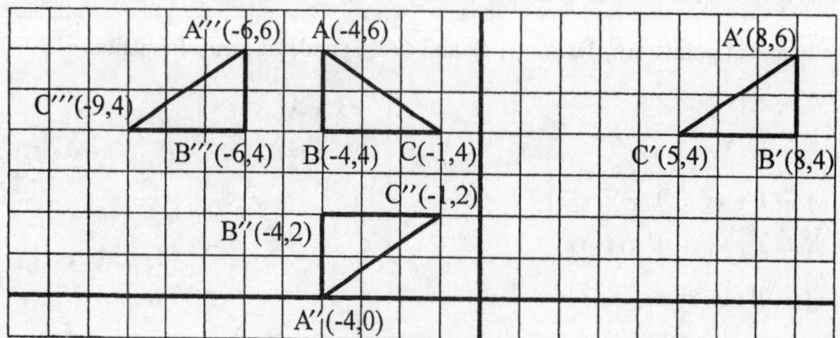

Fig. 18-18

18.7. Name the transformation that

(a) Reflects across $y = 5$

(b) Reflects across $x = -2$

(c) Reflects across $y = -1$

(d) Reflects across $x = \frac{5}{2}$

18.8. Which of the figures in Fig. 18-19 has reflectional symmetry?

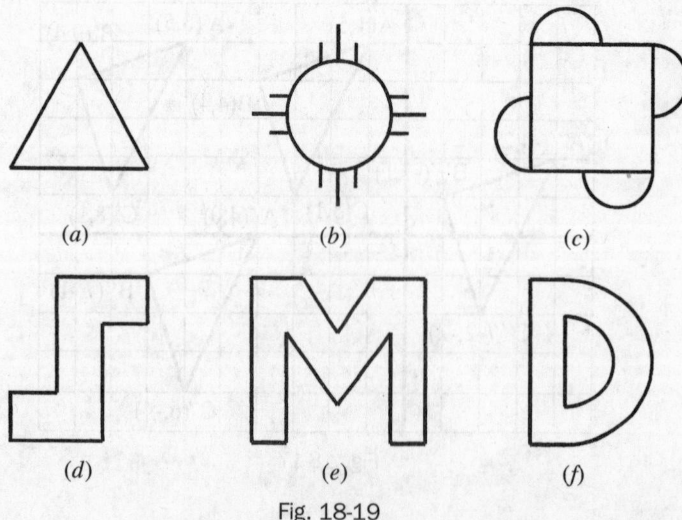

Fig. 18-19

18.9. Let parallelogram $ABCD$ be defined by $A(1, 2)$, $B(4, 2)$, $C(5, 1)$, and $D(2, 1)$. Graph parallelogram $ABCD$ and its image under (a) a 90° clockwise rotation about the origin, (b) a 180° rotation about the origin, and (c) a 270° clockwise rotation about the origin.

18.10. Name the rotation that takes $\triangle ABC$ to (a) $\triangle A'B'C'$, (b) $\triangle A''B''C''$, and (c) $\triangle A'''B'''C'''$ as illustrated in Fig. 18-20.

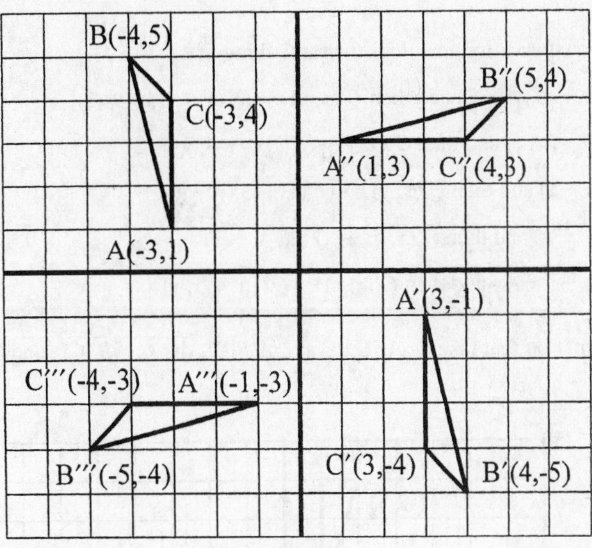

Fig. 18-20

18.11. Name the transformation that rotates clockwise about the origin:

(a) 40°

(b) 50°

(c) 80°

18.12. For each figure in Fig. 18-21, give the smallest angle by which the figure could be rotated around its center and still look the same.

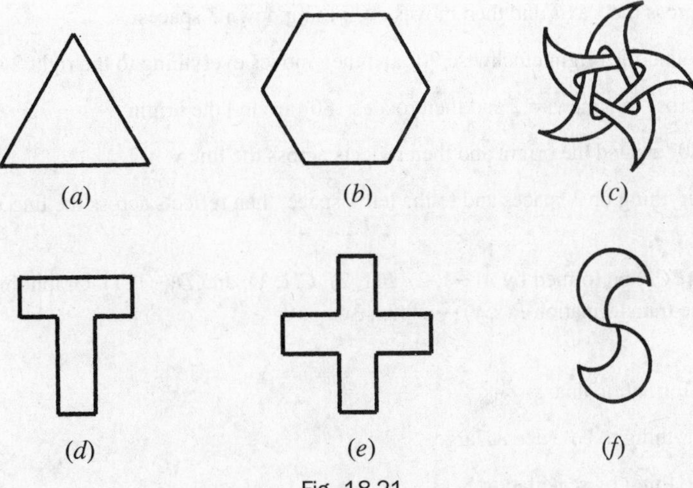

Fig. 18-21

18.13. Let triangle ABC be defined by $A(2, 1)$, $B(3, 2)$, and $C(3, -1)$. Graph $\triangle ABC$ and its image under the following combinations of transformations:

(a) Reflect across the line $y = 3$ and then move to the right 2 spaces.

(b) Rotate about the origin 90° clockwise and then move to the right 1 space and down 3 spaces.

(c) Rotate about the origin 270° clockwise and then reflect across the x axis.

(d) Reflect across the line $x = -1$ and then move up 2 spaces.

18.14. Name the single transformation that does the same thing as

(a) $P(x, y) \mapsto P'(x + 5, y - 3)$ and then $Q'(x, y) \mapsto Q''(x + 1, y + 2)$

(b) $P(x, y) \mapsto P'(5 - x, y)$ and then $Q'(x, y) \mapsto Q''(x - 4, y + 2)$

(c) $P(x, y) \mapsto P'(y, -x)$ and then $Q'(x, y) \mapsto Q''(x + 3, y - 6)$

(d) $P(x, y) \mapsto P'(-y, x)$ and then $Q'(x, y) \mapsto Q''(x, 4 - y)$

(e) $P(x, y) \mapsto P'(x, -3 - y)$ and then $Q'(x, y) \mapsto Q''(6 - x, y)$

18.15. Name the transformation that takes $\triangle ABC$ to (a) $\triangle A'B'C'$, (b) $\triangle A''B''C''$, and (c) $\triangle A'''B'''C'''$ as illustrated in Fig. 18-22.

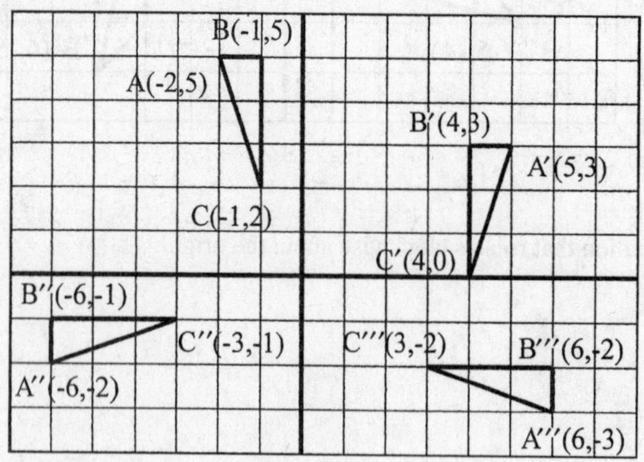

Fig. 18-22

18.16. Name the transformation that

(a) Reflects across the x axis and then moves everything down 3 spaces

(b) Rotates around the origin clockwise 90° and then moves everything to the right 2 spaces

(c) Reflects across the line $y = 2$ and then rotates 180° around the origin

(d) Rotates 180° around the origin and then reflects across the line $y = 2$

(e) Moves everything up 3 spaces and to the left 1 space, then reflects across the line $x = -4$

18.17. Let rectangle $ABCD$ be formed by $A(-1, 2)$, $B(1, 2)$, $C(1, 1)$, and $D(-1, 1)$. Graph this rectangle and also its image under the transformation $P(x, y) \mapsto P'(3x, 3y)$.

18.18. Name the transformation that

(a) Scales everything to be twice as large

(b) Scales everything by scale factor 8

(c) Dihilates everything by a scale factor of $\frac{1}{3}$

Conic Sections

This chapter explains the generation and properties of three important classes of shapes. It begins with an overview and then continues with individual sections on ellipses, parabolas, and hyperbolas.

19.1 The Standard Conic Sections

The *conic sections* are shapes that appear when a double cone is intersected by a flat plane. The *standard conic sections* are circles, ellipses, parabolas, and hyperbolas, as shown in Fig. 19-1. Note that a parabola only occurs if the intersecting plane is parallel to the side of the cone. Also note that only the hyperbola intersects both the top and bottom halves of the cone.

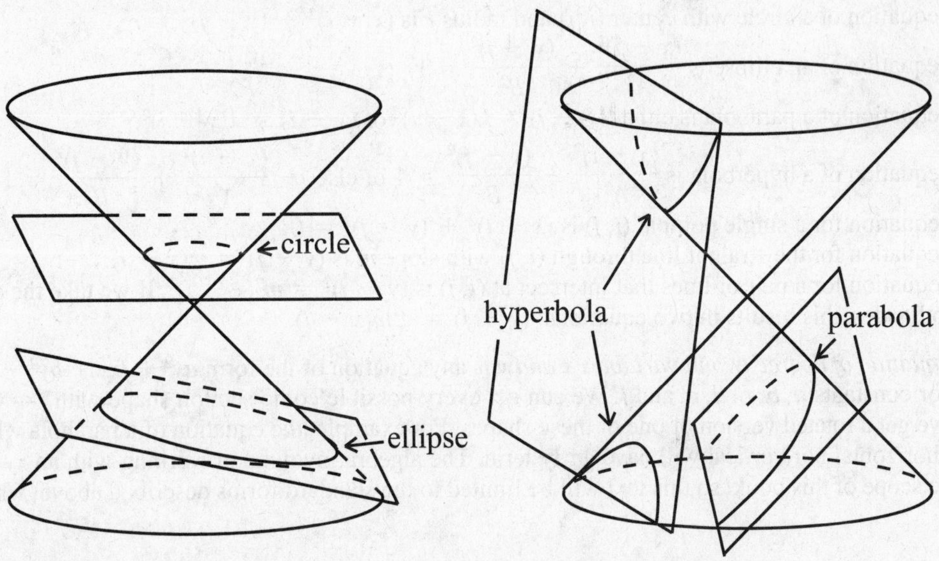

Fig. 19-1

19.1A Degenerate Conics

If the intersecting plane goes through the vertex of the cone, we get *degenerate conics*: straight lines, individual points, and pairs of lines, as shown in Fig. 19-2. Note that we only get a single line if the intersecting plane is parallel to the side of the cone.

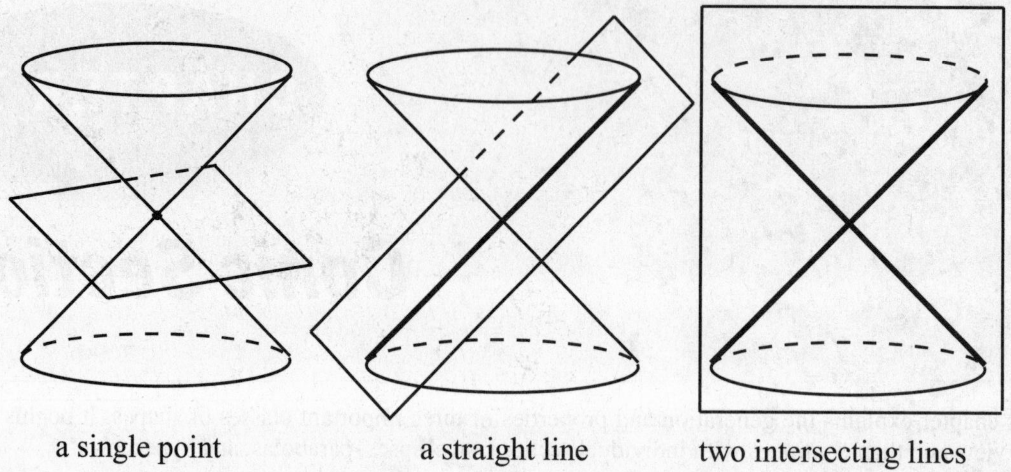

Fig. 19-2

19.1B Conics as Loci of Equations

Any standard or degenerate conic section shape occurs as the locus of points satisfying an equation of the form $ax^2 + by^2 + cx + dy + e = 0$ for some constants a, b, c, d, and e. We can use the algebraic technique known as "completing the square" to rearrange such an equation into one of the following standard forms:

The equation of a circle with center (i, j) and radius r is $(x - i)^2 + (y - j)^2 = r^2$.
The equation of an ellipse is $\dfrac{(x - i)^2}{A^2} + \dfrac{(y - j)^2}{B^2} = 1$.
The equation of a parabola is either $(y - j) = k(x - i)^2$ or $(x - i) = k(y - j)^2$.
The equation of a hyperbola is $\dfrac{(x - i)^2}{A^2} - \dfrac{(y - j)^2}{B^2} = 1$ or else $-\dfrac{(x - i)^2}{A^2} + \dfrac{(y - j)^2}{B^2} = 1$.
The equation for a single point at (i, j) is $(x - i)^2 + (y - j)^2 = 0$.
The equation for the straight line through (i, j) with slope m is $(y - j) = m(x - i)$.
The equation for a pair of lines that intersect at (i, j) is $(y - j)^2 = m^2(x - i)^2$. If we take the square root of both sides, this results in two equations: $(y - j) = \pm m(x - i)$.

An *equation of degree two in variables x and y* is any equation of the form $ax^2 + fxy + by^2 + cx + dy + e = 0$ for constants a, b, c, d, e, and f. We can get every possible conic section shape with $f = 0$. For each $f \neq 0$, we get a rotated version of one of these shapes. For example, the equation of a parabola whose axis is neither horizontal nor vertical will have an xy term. The algebra involved in working with an xy term is beyond the scope of this book, so this text will be limited to the standard forms described above, where $f = 0$.

19.1C Conics from Focus Points and Directrix Lines

A third way to construct a conic section (but not a degenerate or a circle) is with a point F called a *focus*, a line l called a *directrix*, and a number $e > 0$ called the *eccentricity*. The locus of all points P for which $PF = e \times Pl$ will be an ellipse if $0 < e < 1$, a parabola if $e = 1$, and a hyperbola if $e > 1$. That is, a parabola consists of all points P whose distance PF to the focus is exactly the same as its distance Pl to the directrix. An ellipse consists of all points that are closer to F than to l (by exactly the factor e), and a hyperbola consists of all points that are further from F than from l (further by a multiple of $e > 1$). See Fig. 19-3.

Note that when $e > 1$, there are points on both sides of the directrix that are the right amount further from the focus than the directrix, and thus this method generates both pieces of the hyperbola. Because of their reflectional symmetries, hyperbolas and ellipses have two sets of foci and directrices. Parabolas, by contrast, have only one focus and one directrix.

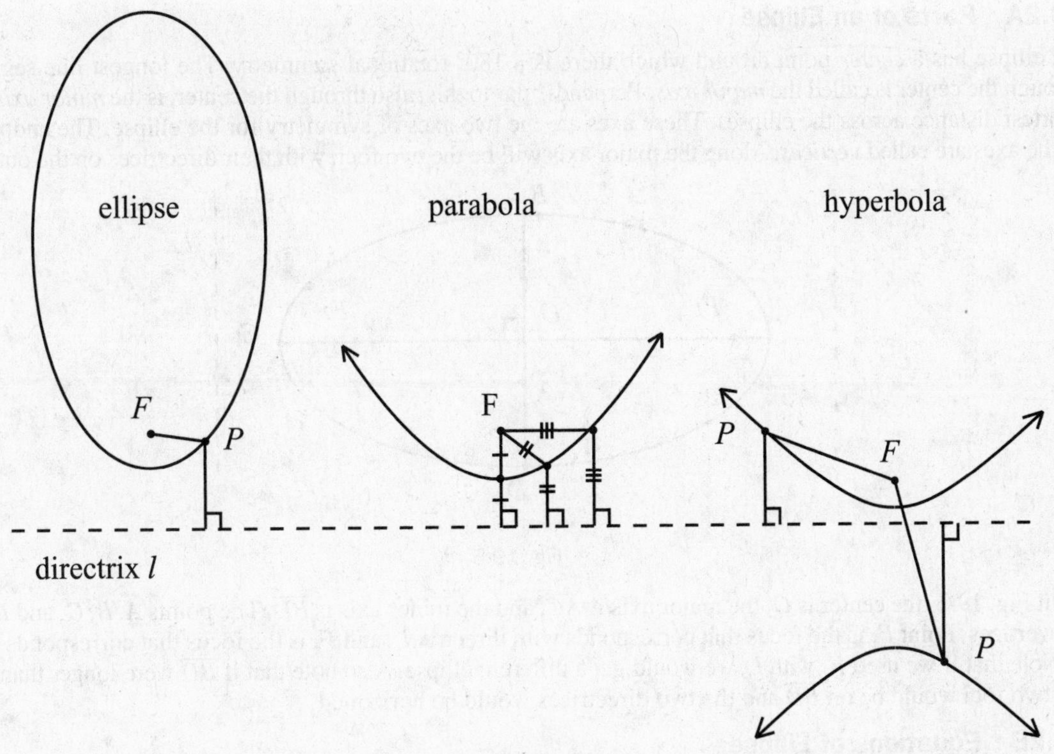

Fig. 19-3

19.1D Reflection Properties

To illustrate this concept, we will imagine ellipses, parabolas, and hyperbolas made of a reflective material and trace the paths of light traveling to and from the focus points.

For ellipses, light traveling toward one focus will reflect away from the other focus. Light traveling from one focus will reflect toward the other.

For parabolas, light traveling from the focus will reflect so as to be parallel to the axis of reflection. Light parallel to this axis will reflect toward the focus on one side of the parabola and away from it on the other.

For hyperbolas, light traveling toward one focus will reflect toward the other, and light traveling away from a focus will reflect away from the other.

These properties are illustrated in the sections that follow, along with real-life applications. The proofs require calculus, so they are omitted.

19.2 Ellipses

An *ellipse* looks like a stretched circle. For example, if we dihilate a unit circle by a factor of 2 in just the x direction using the transformation $T(x, y) \mapsto T'(2x, y)$, we get an ellipse, as shown in Fig. 19-4.

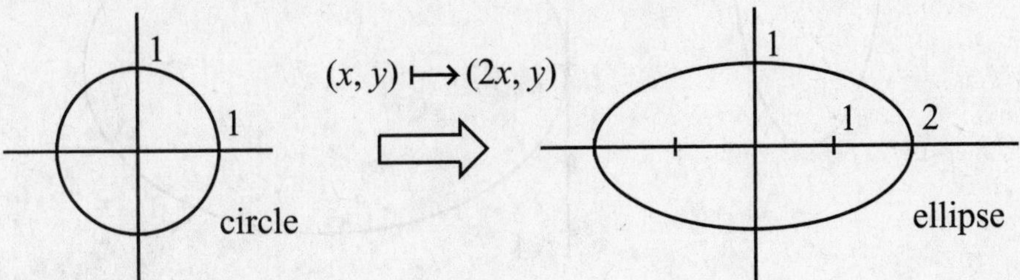

Fig. 19-4

19.2A Parts of an Ellipse

An ellipse has a *center* point around which there is a 180° rotational symmetry. The longest line segment through the center is called the *major axis*. Perpendicular to this, also through the center, is the *minor axis* (the shortest distance across the ellipse). These axes are the two axes of symmetry for the ellipse. The endpoints of the axes are called *vertices*. Along the major axis will be the two foci, with their directrices on the outside.

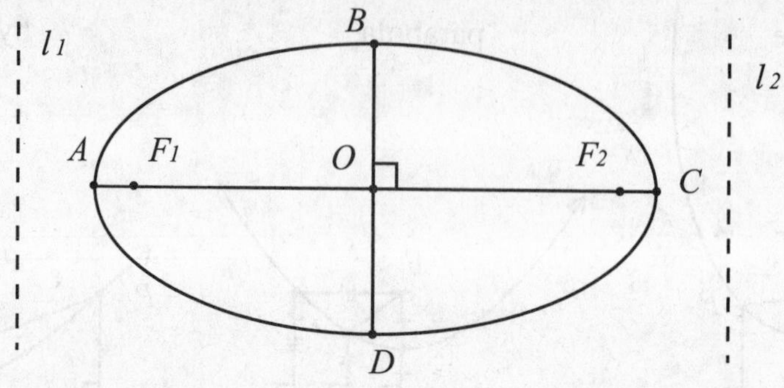

Fig. 19-5

In Fig. 19-5, the center is O, the major axis is $\overline{AC}$, and the minor axis is $\overline{BD}$. The points A, B, C, and D are the vertices. Point F_1 is the focus that corresponds with directrix l_1, and F_2 is the focus that corresponds with l_2. Note that if we used F_1 with l_2, we would get a different ellipse. Also note that if $\overline{BD}$ were longer than $\overline{AC}$, the two foci would be on $\overline{BD}$ and the two directrices would be horizontal.

19.2B Equations of Ellipses

To find the equation of an ellipse, let (i, j) be the coordinates of the center, let A be half the length of the horizontal axis, and let B be half the length of the vertical axis. The ellipse is the locus of the equation $\frac{(x-i)^2}{A^2} + \frac{(y-j)^2}{B^2} = 1$. Furthermore, the two foci are found by moving a distance of $\sqrt{|A^2 - B^2|}$ away from the center in both directions along the major axis.

SOLVED PROBLEMS: FINDING EQUATIONS OF ELLIPSES

19.1 Find the equations and foci of the ellipses in Fig. 19-6.

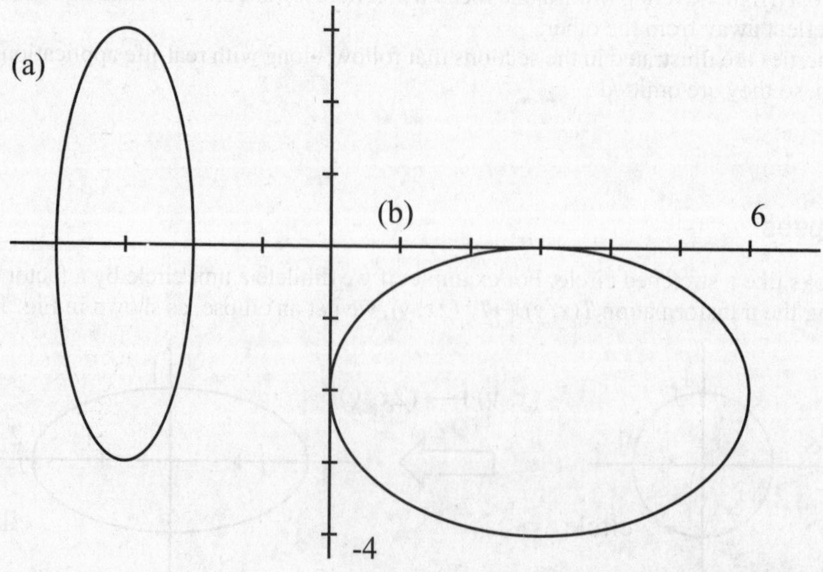

Fig. 19-6

Solutions

(a) The center of this ellipse is $(-3, 0)$, so $i = -3$ and $j = 0$. The horizontal axis is 2 units long, so $A = 1$. The vertical axis is 6 units tall, so $B = 3$. Here, the major axis is the vertical one. The equation is $\frac{(x-(-3))^2}{1^2} + \frac{(y-0)^2}{3^2} = 1$ or $(x+3)^2 + \frac{y^2}{3^2} = 1$. To find the foci, we travel $C = \sqrt{|1^2 - 3^2|} = \sqrt{8} \approx 2.83$ units up and down the major axis from the center to get $(-3, 2.83)$ and $(-3, -2.83)$.

(b) The center of this ellipse is the point $(3, -2)$, so $i = 3$ and $j = -2$. The horizontal axis has length 6 and the vertical axis has length 4, so $A = 3$ and $B = 2$. Thus the equation of this ellipse is $\frac{(x-3)^2}{3^2} + \frac{(y+2)^2}{2^2} = 1$. The major axis is horizontal, so we find the two foci by traveling a distance of $\sqrt{3^2 - 2^2} = \sqrt{5} \approx 2.24$ to the left and right of the center, obtaining $(5.24, -2)$ and $(0.76, 2)$.

SOLVED PROBLEMS: SKETCHING ELLIPSES

19.2 Sketch the graph of each ellipse and plot their foci.

(a) $\frac{(x+1)^2}{3^2} + \frac{(y-1)^2}{4^2} = 1$

(b) $9x^2 + 25(y-2)^2 = 225$

Solutions

(a) Here, the center is $O = (-1, 1)$, the horizontal axis is $2(3) = 6$ units long, and the vertical axis is $2(4) = 8$ units long. After plotting the vertices 3 units to the left and right of the center, and 4 units above and below, we can then sketch a rounded shape whose curves turn around at these points. The foci are $\sqrt{|3^2 - 4^2|} = \sqrt{|-7|} = \sqrt{7} \approx 2.65$ units from the center. Because the major axis is vertical here, the foci are above and below the center at $F_1 = (-1, 3.65)$ and $F_2 = (-1, -1.65)$. This is shown in Fig. 19-7(a).

(b) To get this equation into standard form, we must divide both sides by 225 and then represent the denominators as squares: $\frac{x^2}{5^2} + \frac{(y-2)^2}{3^2} = 1$. This ellipse has its center at $(0, 2)$ and vertices at $(0 \pm 5, 2)$ and $(0, 2 \pm 3)$. That is, the vertices are $(5, 2)$, $(-5, 2)$, $(0, -1)$, and $(0, 5)$. The two foci are found by moving $C = \sqrt{5^2 - 3^2} = \sqrt{16} = 4$ units from the center along the horizontal major axis; thus, $F_1 = (-4, 2)$ and $F_2 = (4, 2)$, as shown in Fig. 19-7(b).

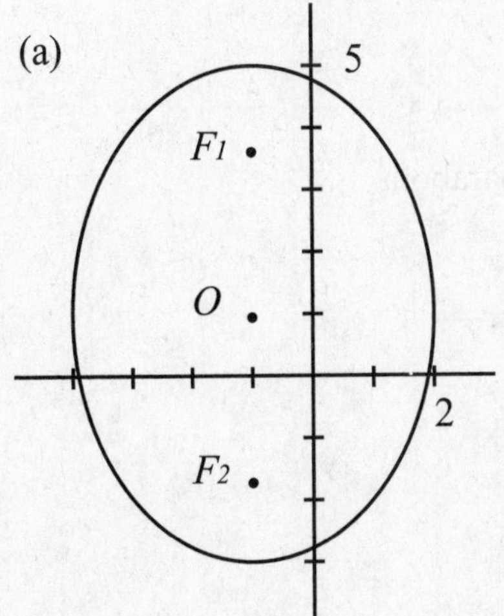

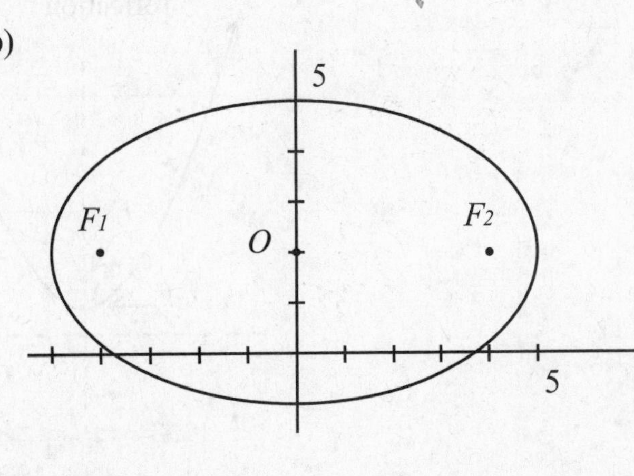

Fig. 19-7

19.2C Elliptical Reflections

Now suppose that we have an ellipse made of a reflective material. Light traveling *toward* a focus from the outside of the ellipse will bounce directly away from the second focus. In Fig. 19-8(a), for example, the ray of light aimed directly at focus F_1 reflects straight away from focus F_2.

Similarly, light traveling *from* one focus will always reflect directly toward the other focus. This is shown in Fig. 19-8(b). Furthermore, when light travels from one focus to the other inside an ellipse like this, the sum of the distances $F_1P + F_2P$ will always be the length of the major axis.

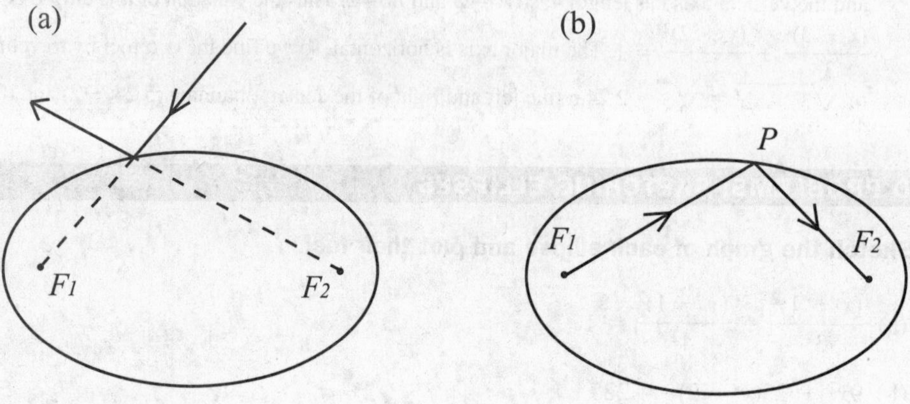

Fig. 19-8

Now suppose that two people stand at the foci of a large, elliptical room. A whisper from one person will travel out in all directions, then bounce off the walls directly toward the second person. Each bit of whisper will travel the same distance, so the sounds will all arrive at the same time and be audible.

19.3 Parabolas

A *parabola* is an infinite curve with a single axis of reflection. On this axis, there is the *focus* point F, a single *vertex* point V, and the point of intersection with the directrix. See Fig. 19-9.

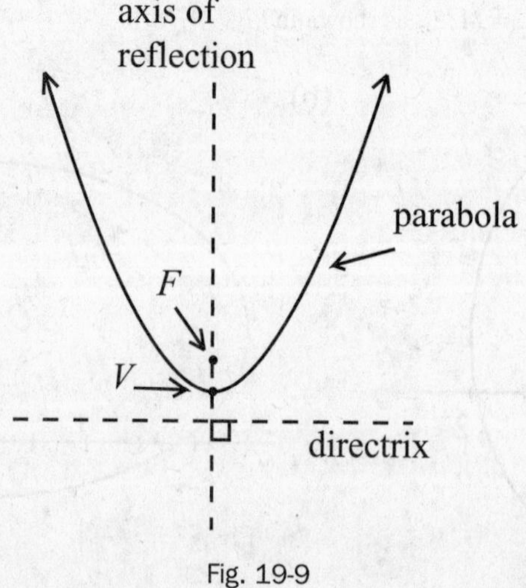

Fig. 19-9

19.3A Equations of Parabolas

The standard equation for a parabola whose vertex is at (i, j) is either

$(y - j) = k(x - i)^2$ (opening up if $k > 0$ and down if $k < 0$) or else
$(x - i) = k(y - j)^2$ (opening to the right if $k > 0$, and to the left if $k < 0$).

If we step 1 unit perpendicular to the axis, a point of the parabola will be $|k|$ units in the direction of the axis. This can be used to sketch parabolas and identify their equations.

The focus of a parabola is found by traveling $\frac{1}{4k}$ units along the axis from the vertex.

SOLVED PROBLEMS: FINDING EQUATIONS OF PARABOLAS

19.3 Find the equation and focus for each of the parabolas in Fig. 19-10.

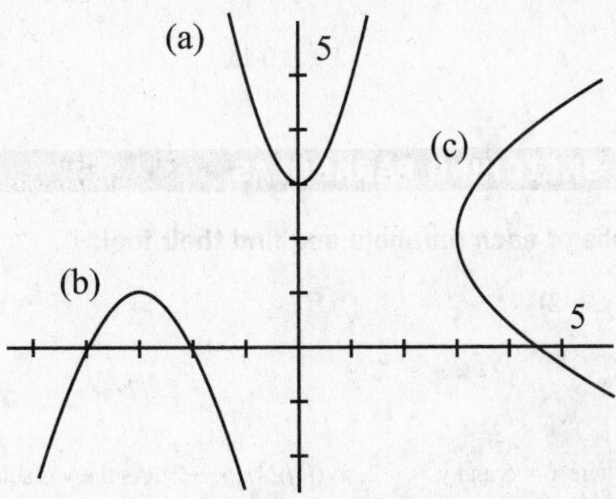

Fig. 19-10

Solutions

(a) The vertex is $(i, j) = (0, 3)$ and the parabola opens upward, so the equation has the form $(y - 3) = k(x - 0)^2$ for a positive k. If we take a step to the right from the vertex at $x = 0$, we have to move up $k = 2$ units to the point $(1, 5)$ on the parabola. Thus, the equation is $y - 3 = 2x^2$, although many would write this as $y = 2x^2 + 3$. Since $k = 2$, the focus is $\frac{1}{4k} = \frac{1}{8}$ of a unit up from the vertex, at $(0, 3.125)$.

(b) The vertex is $(i, j) = (-3, 1)$ and the parabola opens downward, so we have $(y - 1) = k(x - (-3))^2$ with a negative k. If we move 1 unit to the right of the vertex, the next point is down 1 unit at $(-2, 0)$. Thus $k = -1$ and the equation is $y - 1 = -(x + 3)^2$. To find the focus, we compute $\frac{1}{4k} = -\frac{1}{4}$ and go 1/4 of a unit below the vertex to $(-3, 0.75)$.

(c) Here, the vertex is $(i, j) = (3, 2)$ and the parabola opens to the right, so our equation has the form $(x - 3) = k(y - 2)^2$ for a positive k. If we step down 1 unit from the vertex, it appears as though the next point is $k = 1/2$ of a unit over at $(3.5, 1)$. Thus, the equation is $(x - 3) = \frac{1}{2}(y - 2)^2$. Precision is difficult when reading points on a graph, so it is possible that k is really 0.49 or 0.52 or something else close to 0.5. Assuming $k = 1/2$, the focus is $\frac{1}{4k} = \frac{1}{2}$ a unit to the right of the vertex, at $(3.5, 2)$.

The manner in which we obtain the values of k is shown in Fig. 19-11.

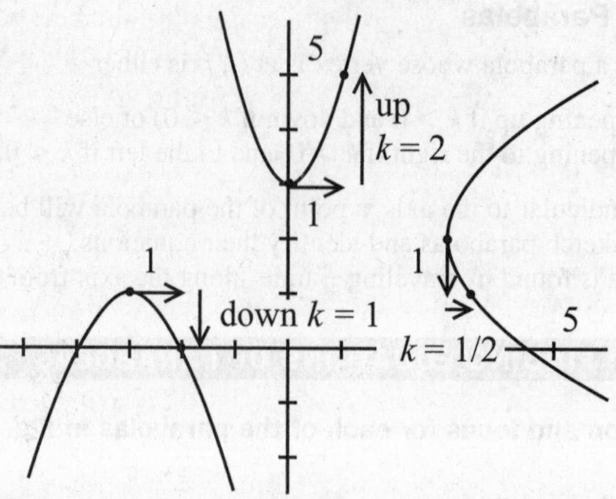

Fig. 19-11

SOLVED PROBLEMS: SKETCHING PARABOLAS

19.4 Sketch the graphs of each parabola and find their foci:

(a) $x - 5 = -2(y + 2)^2$

(b) $y - 1 = \frac{1}{2}(x + 3)^2$

Solutions

(a) The vertex is where $x = 5$ and $y = -2$, so $(i, j) = (5, -2)$. As the y is squared, this parabola has a horizontal axis of symmetry. Because $k = -2$, it opens to the left, and when we move 1 unit up or down from the vertex, we must move 2 units to the left. The focus is $\frac{1}{4k} = -\frac{1}{8}$ units from the vertex along the horizontal axis, at $(4.875, -2)$. This is sketched in Fig. 19-12(a).

(b) Here, the vertex is $(i, j) = (-3, 1)$ and the parabola opens upward. If we move 1 unit to the left or right, we move up by $k = 1/2$. The focus is found by going $\frac{1}{4k} = \frac{1}{2}$ of a unit along the axis from the vertex at $(-3, 1.5)$, as shown in Fig. 19-12(b).

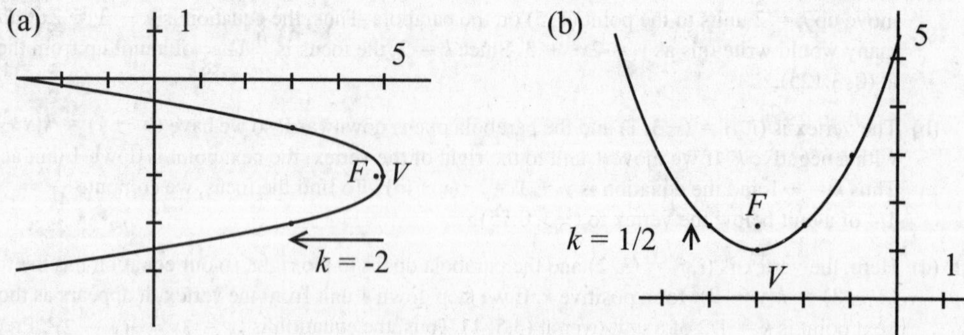

Fig. 19-12

19.3B Parabolic Reflections

The reflective property of parabolas is shown in Fig. 19-13. Light beams coming from the focus will all reflect to beams parallel to the axis, as shown in (a). If we reverse the light, light parallel to the axis will reflect directly to the focus, as shown in (b). If light beams in the direction of the axis strike the convex side of the parabola, they will each scatter directly away from the focus, as shown in (c). If this is reversed, all beams headed toward the focus will reflect into parallel beams, as shown in (d).

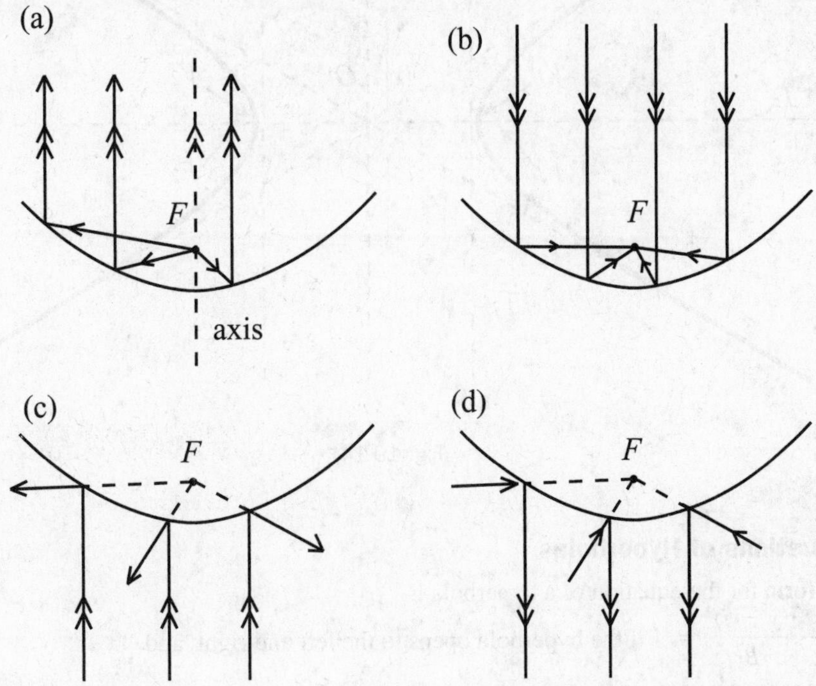

Fig. 19-13

The property in (a) is used when a concave parabolic mirror is placed behind the bulb of a flashlight. If the light source is at the focus, the light will shine out in a straight, narrow beam.

The property in (b) is used when a satellite dish is shaped like a parabola. If the dish is pointed directly at the satellite, then the incoming transmission waves will be basically parallel and reflect toward a single focal point. A small receiver suspended at this point will thus receive a greatly magnified signal.

The property in (d) is used when a person installs a convex parabolic mirror across the street from a driveway. Everything that the focal point can "see" (have a beam of light travel straight to it) will appear on the surface of the mirror to someone in the direction of the axis. Similarly, a parabolic mirror in the corner of a large room will allow someone (in the direction of the axis) to see everything in the room at once. If that person were to shine a light at the mirror, everyone in the room would see the spot of light as if it were at the mirror's focal point, because of (c).

19.4 Hyperbolas

A *hyperbola* consists of two disconnected curves, shown with solid lines in Fig. 19-14. There is 180° rotational symmetry around the *center* point O, and two axes of symmetry. The *transverse axis* T runs through the two foci (F_1 and F_2) and the two vertices (V_1 and V_2). The *conjugate axis* C is perpendicular to the transverse axis at the center point and contains no points of the hyperbola. The two *directrix* lines l_1 and l_2 are parallel to the conjugate axis. Finally, as the points of the hyperbola get further from the center, they get closer to two *slant asymptotes* S_1 and S_2.

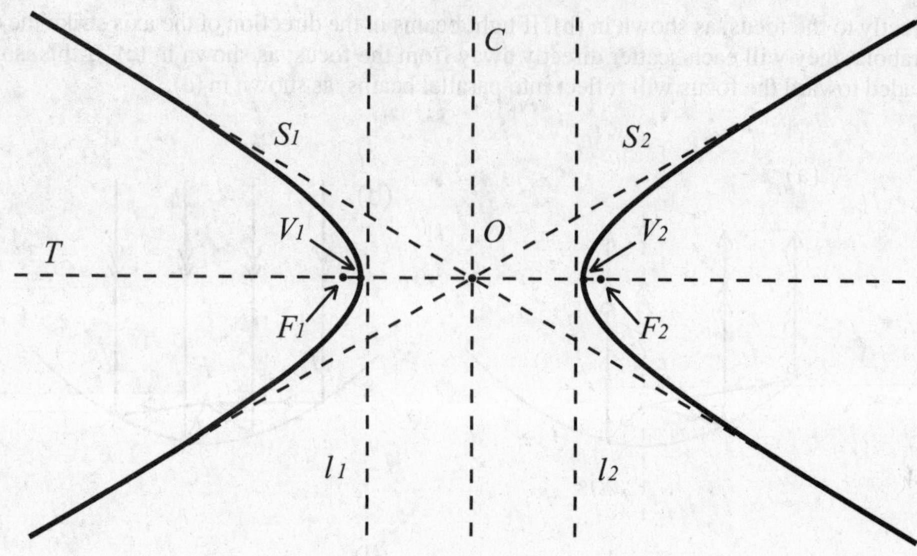

Fig. 19-14

19.4A Equations of Hyperbolas

The standard form for the equation of a hyperbola is

(a) $\dfrac{(x-i)^2}{A^2} - \dfrac{(y-j)^2}{B^2} = 1$ if the hyperbola opens to the left and right, and

(b) $-\dfrac{(x-i)^2}{A^2} + \dfrac{(y-j)^2}{B^2} = 1$ if the hyperbola opens up and down.

In both cases, the center is (i, j). If we draw a rectangle that is $2A$ units wide, $2B$ units tall, and centered at (i, j), then the diagonals of the rectangle will extend out to be the slant asymptotes.

In (a), the conjugate axis is $x = i$ (because if $x = i$, the equation $-\dfrac{(y-j)^2}{B^2} = 1$ can't be solved) and the transverse axis is $y = j$. In (b), this is reversed: the conjugate axis is $y = j$ and the transverse axis is $x = i$.

The two vertices are where the transverse axis crosses the rectangle we have drawn. The two foci are the points that are $\sqrt{A^2 + B^2}$ units from the center along the transverse axis.

SOLVED PROBLEMS: SKETCHING HYPERBOLAS

19.5 Sketch the graphs and find the foci of:

(a) $\dfrac{(x-3)^2}{4^2} - \dfrac{(y+1)^2}{2^2} = 1$

(b) $-\dfrac{x^2}{3^2} + \dfrac{(y-2)^2}{2^2} = 1.$

Solutions

(a) We have center $(3, -1)$, conjugate axis $x = 3$, and transverse axis $y = -1$. As $A = 4$ and $B = 2$, we draw an 8 by 4 rectangle around the center and extend out its diagonals for the slant asymptotes. Since the transverse axis is horizontal, the midpoints of the two sides of the rectangle are the vertices: $(-1, -1)$ and $(7, 1)$. By drawing curving lines from these vertices out, away from the center, and approaching the slant asymptotes, we get a decent approximation of the curve, as shown in Fig. 19-15(a). To obtain the foci, we compute $\sqrt{A^2 + B^2} = \sqrt{4^2 + 2^2} = \sqrt{20} \approx 4.47$ and travel this distance from the center along the transverse axis to just beyond the curve to $F_1 = (-1.47, -1)$ and $F_2 = (7.47, -1)$.

(b) The center is $(0, 2)$ with $A = 3$ and $B = 2$. We draw a rectangle that is 6 units wide and 4 units tall around the center and extend out its diagonals for the slant asymptotes. Of the two axes, if we set $x = 0$, we get $\frac{(y-2)^2}{2^2} = 1$, whose two solutions $y = 4$ and $y = 0$ are the vertices. If we set $y = 2$, however, we get $-\frac{x^2}{3^2} = 1$, which is impossible; thus $y = 2$ is the conjugate axis. As before, we plot the two vertices and draw curves that approach the asymptotes. The foci are then $\sqrt{3^2 + 2^2} = \sqrt{13} \approx 3.6$ units along the transverse axis from the center, at $F_1 = (0, -1.6)$ and $F_2 = (0, 5.6)$, as shown in Fig. 19-15(b).

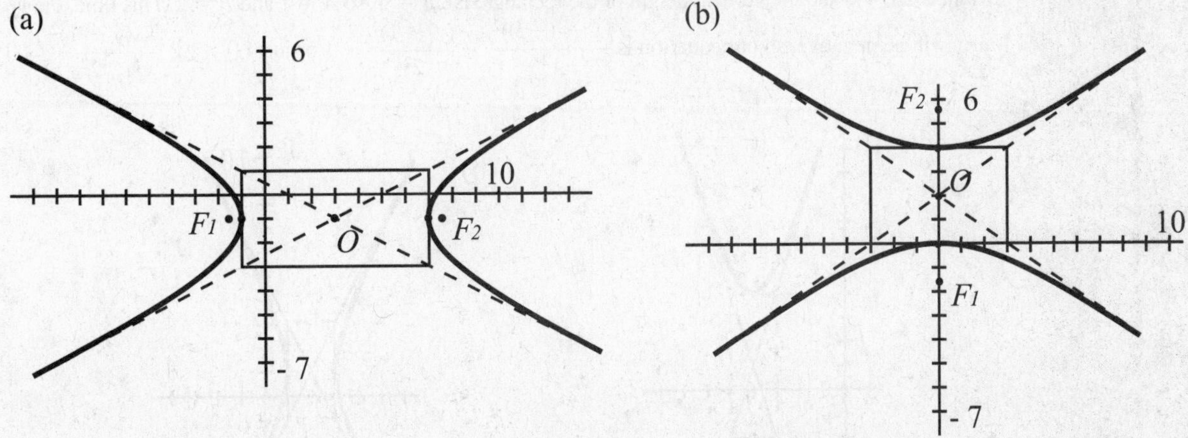

Fig. 19-15

SOLVED PROBLEMS: FINDING EQUATIONS OF HYPERBOLAS

19.6 Estimate the equations of the hyperbolas in Fig. 19-16.

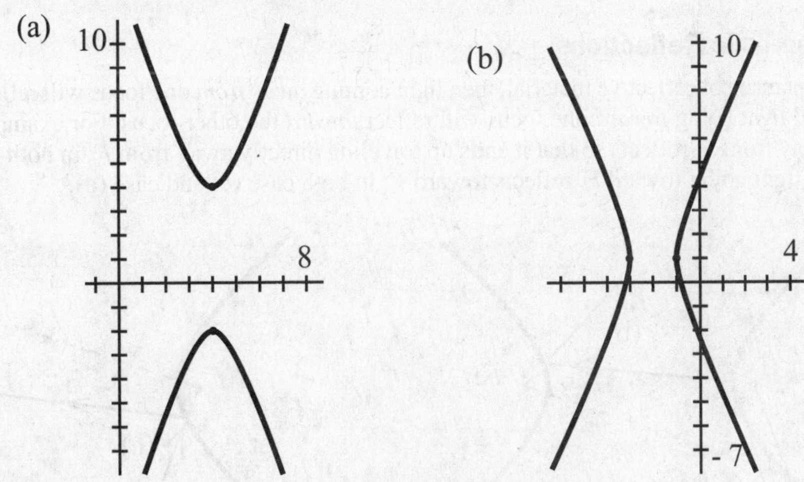

Fig. 19-16

Solutions

(a) It is clear that the two vertices are at $(4, 4)$ and $(4, -2)$, and thus the center is the midpoint of these, $(4, 1) = (i, j)$. Furthermore, since the vertices are each 3 vertical units away from the center, this means that $B = 3$ (it would have been A if the transverse axis were horizontal). Next, we sketch in slant asymptotes through the center and connect them with horizontal lines through the vertices, as shown in Fig. 19-17(a).

We then complete the rectangle, as shown with broken vertical lines, and see that it measures 2 horizontally and 6 vertically. This means that $A = 1, B = 3$, and the equation must be $-\frac{(x-4)^2}{1^2} + \frac{(y-1)^2}{3^2} = 1$ (we know the x^2 term must be negative, because the conjugate axis is horizontal).

(b) Here, the vertices are $(-3, 1)$ and $(-1, 1)$, so the center is $(i, j) = (-2, 1)$. To sketch the slant asymptotes, we go an equal distance up and down the hyperbola from the center, say 9 units, and trace lines through the center. If we draw vertical lines through the vertices until they meet these slant lines and then connect them with horizontal, broken lines, we get the rectangle in Fig. 19-17(b). Clearly, the horizontal width is $2A = 2$ and the vertical height of the rectangle is $2B = 4$, so $A = 1$ and $B = 2$. This time, the y^2 term will be negative, so our equation is $\frac{(x-(-2))^2}{1^2} - \frac{(y-1)^2}{2^2} = 1$ or just $(x+2)^2 - \frac{(y-1)^2}{2^2} = 1$.

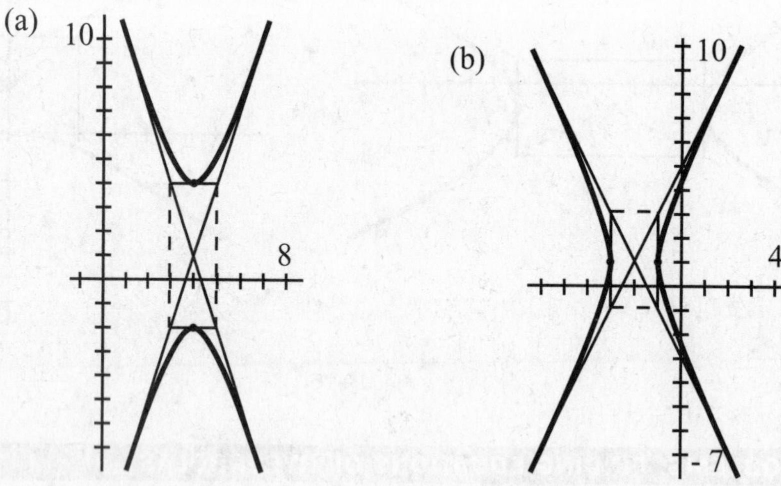

Fig. 19-17

19.4B Hyperbolic Reflections

If a hyperbola is made of reflective material, then light coming *away from* one focus will reflect *away from* the other focus and light going *toward* one focus will reflect *toward* the other focus. For example, in Fig. 19-18, light going away from F_1 reflects so that it ends up traveling directly away from F_2 in both situations (a) and (b). Similarly, light going toward F_1 reflects toward F_2 in both case (c) and case (d).

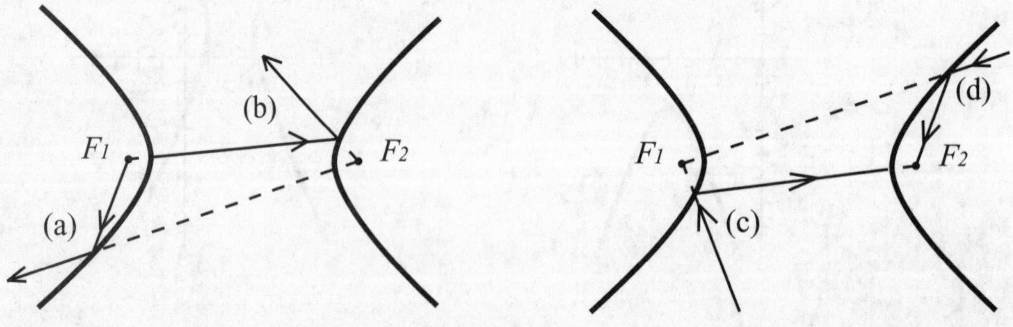

Fig. 19-18

If we have a mirror shaped like half of a hyperbola and take a picture of it from the outside focus (Fig. 19-19), the result will be a panoramic photo. Every illuminated object within sight of the mirror will send some light toward the inner focus F_1, and all of this light will be reflected to the camera. The photograph of the mirror will thus capture all the light rays going toward F_1, and thus look exactly like the view from that location: a panorama.

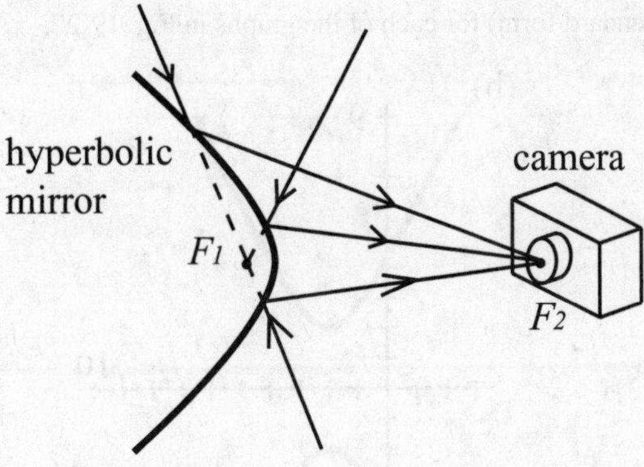

Fig. 19-19

SUPPLEMENTARY PROBLEMS

19.1. For each of the following ellipses, find the center, both axes, all vertices, and both foci.

(a) $\dfrac{(x-5)^2}{10^2}+\dfrac{(y+3)^2}{2^2}=1$

(b) $\dfrac{(x+2)^2}{3^2}+y^2=1$

(c) $\dfrac{(x-1)^2}{6^2}+\dfrac{(y-4)^2}{7^2}=1$

19.2. For each of the following parabolas, find the vertex, the axis of symmetry, and the focus:.

(a) $(x-4)=3(y+2)^2$

(b) $y=-0.1x^2-2$

(c) $(x+2)=(y+1)^2$

19.3. For each of the following hyperbolas, find the center, axes, vertices, and foci.

(a) $-\dfrac{(x+7)^2}{3^2}+\dfrac{(y+5)^2}{2^2}=1$

(b) $-\dfrac{(x-3)^2}{4^2}+\dfrac{y^2}{25}=1$

(c) $\dfrac{(x+2)^2}{8^2}-(y-1)^2=1$

19.4. Sketch the graph of each of the following equations, including focus point(s):

(a) $(y-1)+(x+1)^2=0$

(b) $\dfrac{(x+2)^2}{2^2}+\dfrac{(y-1)^2}{4^2}=1$

(c) $(x+3)-\dfrac{(y+1)^2}{2}=0$

(d) $\dfrac{(x+1)^2}{4^2}-y^2=1$

(e) $\dfrac{(x-3)^2}{4^2}+(y+2)^2=1$

(f) $-\frac{(x-2)^2}{2^2}+\frac{(y+2)^2}{3^2}=1$

19.5. Find the equation (in standard form) for each of the graphs in Fig. 19-20.

(a)

(b)

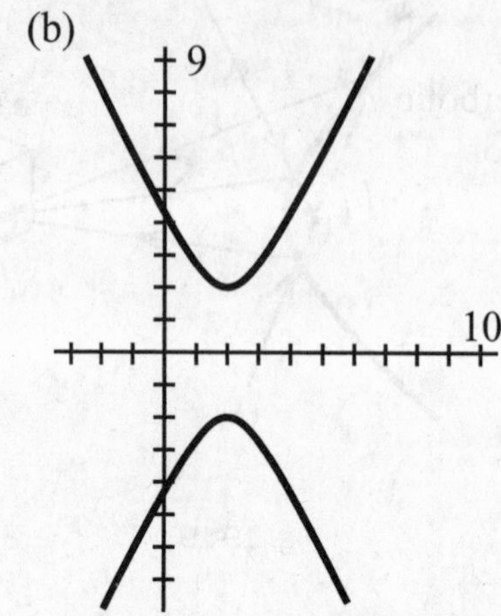

(c)

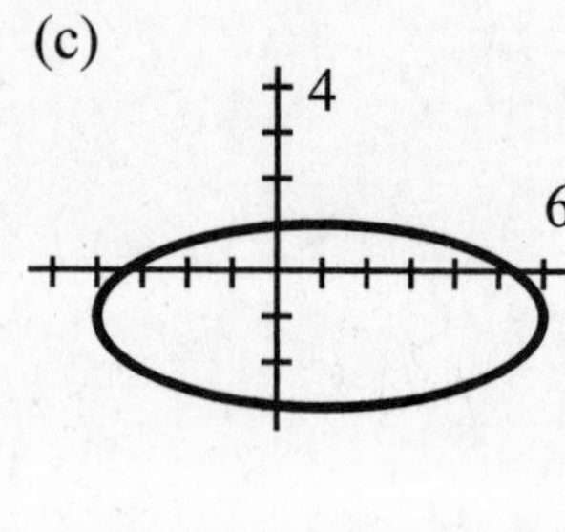

(d)

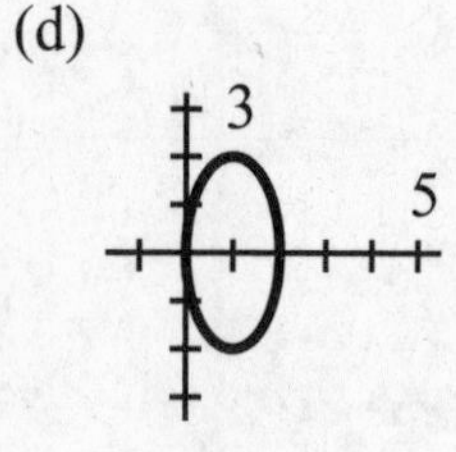

(e)

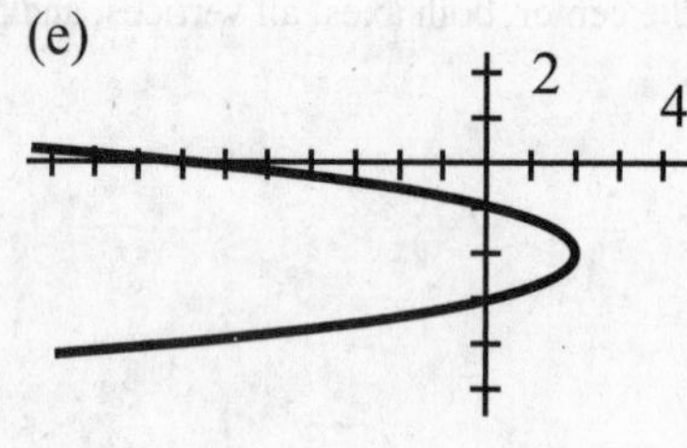

(f)

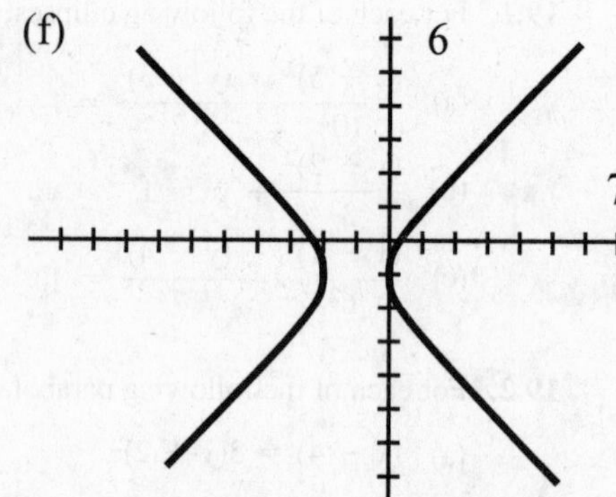

Fig. 19-20

19.6. Suppose that the curve in Fig. 19-21 is a reflective parabola with focus F. Suppose further that the rays at (a) and (d) are parallel to the axis of symmetry, and that (b) is traveling toward F. Illustrate the directions in which each of these rays will bounce.

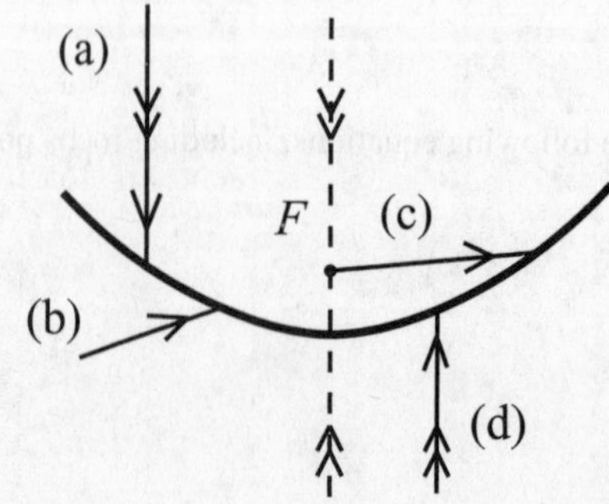

Fig. 19-21

19.7. Suppose that the curve in Fig. 19-22 is a reflective ellipse with foci F_1 and F_2, and that the rays in (c) and (d) are traveling toward foci F_2 and F_1, respectively. Illustrate the directions in which each of these rays will bounce.

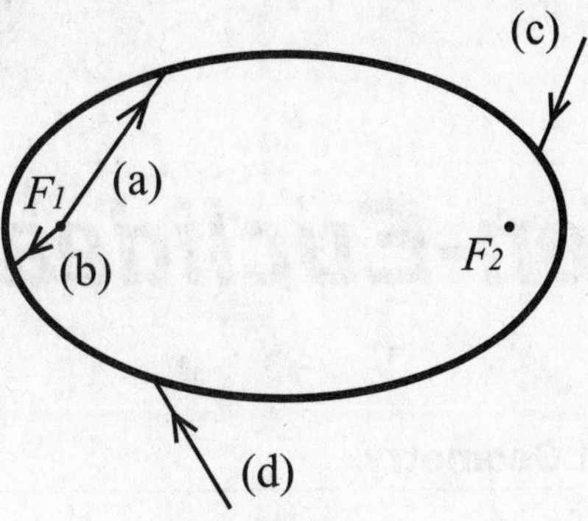

Fig. 19-22

19.8. Suppose that the curve in Fig. 19-23 is a reflective hyperbola with foci F_1 and F_2. Suppose further that each ray is either traveling directly toward or directly away from one of these foci. Illustrate the directions in which each of these rays will bounce.

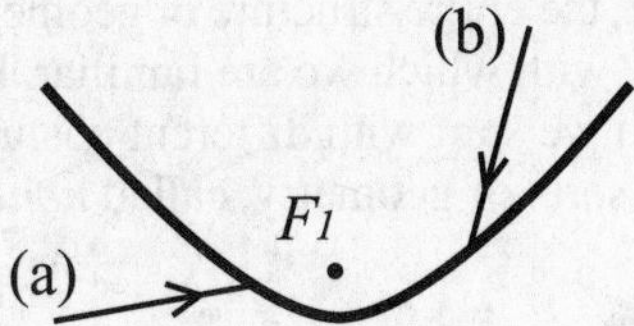

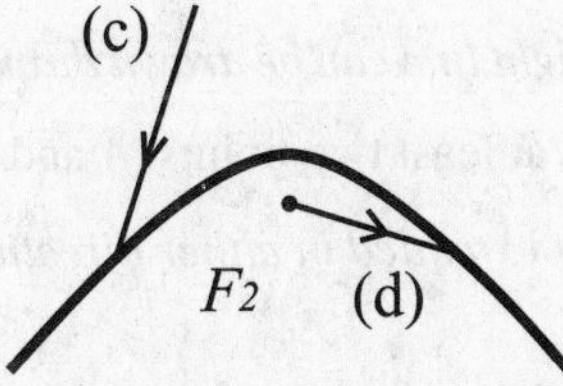

Fig. 19-23

Non-Euclidean Geometry

20.1 The Foundations of Geometry

For most of the years since Euclid wrote the *Elements* in 325 B.C., people felt that only one sort of geometry was possible. Planes looked like infinitely large, flat sheets of paper, lines went on forever as straight as the mind could imagine, and a grid of parallel lines could be drawn to make a plane look like graph paper. However, the foundations of this geometry were unfortunately vague.

As discussed in Chapter 1, the concepts of *point*, *line*, and *plane* were not given formal definitions. The individual points, infinite straight lines, and flat planes discussed throughout this book all fit the properties and descriptions of points, lines, and planes, but other objects fit these general descriptions as well. When these basic objects are different, the resulting geometry is different as well.

Similarly, as discussed in Chapter 2, the entire structure of geometric proof rests upon unproved postulates. These postulates lead to the geometry with which we are familiar. However, why should we believe one set of postulates and not a different set? If we start with different postulates, then our theorems will be different as well. Our choices lead to different sorts of geometry, called *non-euclidean geometries*.

20.2 The Postulates of Euclidean Geometry

Euclid's *Elements* began with only five postulates. We shall go through them with an eye for alternatives.

POSTULATE 1: *One and only one straight line can be drawn through any two points*. (Chapter 2, Postulate 11)

We further postulate that there exist at least two points, A and B, so there exists a straight line.

POSTULATE 2: *A line segment can be extended in either direction indefinitely.* (Chapter 1, description of a line)

Traditionally, we suppose that the line through A and B looks like Fig. 20-1(a), but our postulates do not rule out the possibility of Fig. 20-1(b). In Fig. 20-1(b), the line reaches no end in either direction, and thus in some sense can be extended indefinitely. Remember that the original meaning of "geometry" comes from "earth" and "measure." The straightest line that could be drawn on the earth would wrap around to form a great circle, as discussed in Chapter 17.

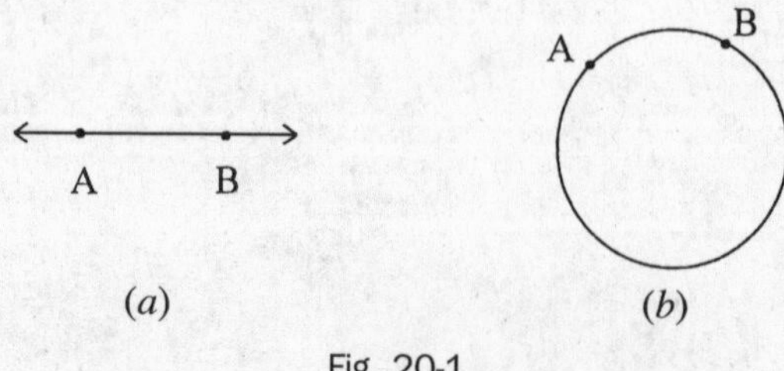

Fig. 20-1

POSTULATE 3: *A circle can be drawn with any center and radius.* (Chapter 2, Postulate 14)

When we believe this postulate, we believe that we have a plane with no end in any direction on which we could find circles with ever-larger radii. It could be that this leads to circles like ripples on the surface of a still pond, as shown in Fig. 20-2(a). However, if our plane were more like the surface of the Earth, our circles might look more like the circles of latitude, shown in Fig. 20-2(b).

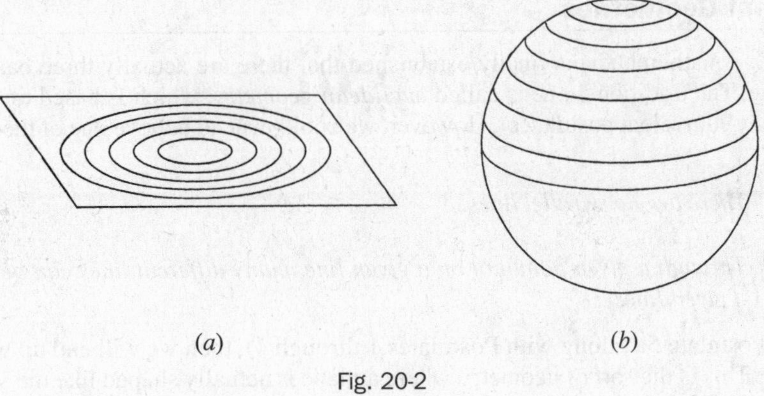

(*a*) (*b*)

Fig. 20-2

POSTULATE 4: *All right angles have the same measure.*

This postulate enables us to measure angles in degrees. A plane would have to be somehow lumpy or uneven if some right angles could be bigger than others.

POSTULATE 5: *Through a given point not on a given line, one and only one line can be drawn parallel to a given line.* (Chapter 4, Parallel-Line Postulate)

This postulate is the one which establishes that our plane cannot look like a giant sphere. On a flat plane, given point P not on line $\overleftrightarrow{AB}$, only a point C which makes $\angle APC \cong \angle PAB$ will make $\overleftrightarrow{PC} \parallel \overleftrightarrow{AB}$, as shown in Fig. 20-3(a). On a giant sphere, straight lines are great circles which divide the sphere into two equal-sized pieces. Any two such lines will always meet at two spots at opposite points of the sphere (called *antipodal points*), such as X and Y, as shown in Fig. 20-3(b). Because any two straight lines meet, it is impossible for there to be parallel lines on a sphere.

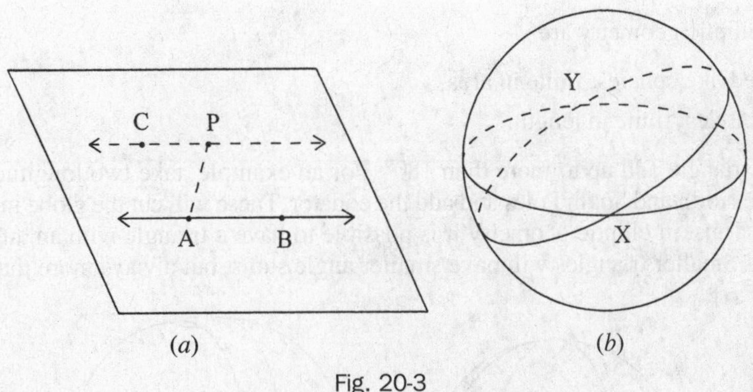

(*a*) (*b*)

Fig. 20-3

20.3 The Fifth Postulate Problem

For about 2000 years, certain mathematicians tried to use the first four postulates to prove the fifth. This challenge was called the *fifth postulate problem*. The first four seem to come straight from the basic tools of geometry: the straight edge (connecting points and extending straight lines), the compass (drawing circles), and the square (measuring right angles). The fifth, it seemed, was something that ought to be provable. However, no one was able to succeed in proving the fifth postulate without introducing new, equivalent postulates that required belief without proof. (Actually, the fifth postulate stated previously is a simpler alternative to the one Euclid actually used.) If a postulate about parallel lines must be accepted without proof, why must it be the traditional fifth postulate of euclidean geometry?

20.4 Different Geometries

In the 19th century, mathematicians finally established that there are actually three basic types of planar geometry, not one. The traditional one is called *euclidean geometry*, which is based on the fifth postulate (or one of the many equivalent postulates). However, we could instead believe one of the two alternate fifth postulates:

POSTULATE 5a: *There are no parallel lines.*

POSTULATE 5b: *Through a given point not on a given line, many different lines can be drawn parallel to a given line.*

If we believe Postulate 5a (along with Postulates 1 through 4), then we will end up with what is called *elliptic geometry.* This is the sort of geometry where a plane is actually shaped like the surface of a sphere and lines are great circles.

If we believe Postulate 5b, we will end up with an even stranger geometry called *hyperbolic geometry*.

20.4A Euclidean Geometry

Just for comparison, here are several properties of euclidean geometry:

Planes are infinite in area and flat.

Lines are infinite in length.

The angles of triangles always sum to 180°.

There are similar triangles of different sizes.

The circumference of a circle with radius r is $C = 2\pi r$.

The area of a circle with radius r is $A = \pi r^2$.

20.4B Elliptic Geometry

The properties of elliptic geometry are

Planes are shaped like spheres, finite in area.

Lines are great circles, finite in length.

The angles of a triangle add up to more than 180°. For an example, take two longitude circles that meet at right angles at the North and South Poles and add the equator. These will cut the globe into 8 triangles where every angle is 90°. Thus, in elliptic geometry it is possible to have a triangle with an angle sum of 270°, as shown in Fig. 20-4. Smaller triangles will have smaller angle sums, but always more than 180°.

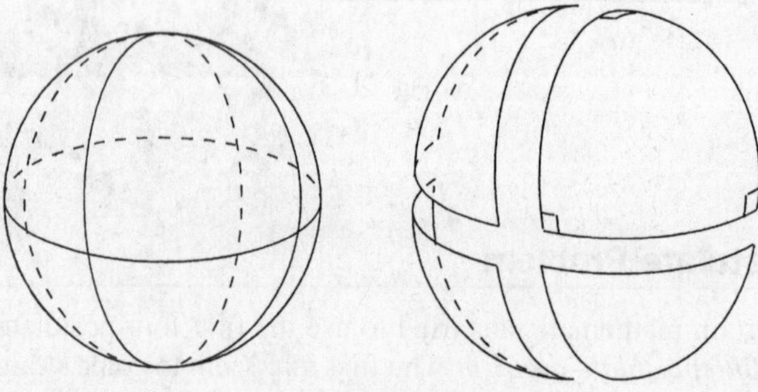

Fig. 20-4

The size of a triangle on a sphere is related to the sum of all its angles. For example, every triangle with an angle sum of 270° will have area equal to one eighth of the whole sphere, just like the triangles in Fig. 20-4. Smaller triangles will have smaller angle sums. Because of this, it is impossible to have two triangles of different sizes that have the same angle measures. Thus, there cannot be similar triangles of different sizes in elliptic space.

A circle in elliptic space is defined as usual: the set of all points a given distance from a point. Because distances are measured along lines which curve around great circles on a sphere, this means that a larger radius does not always lead to a larger circle. Once the radius exceeds one quarter of a great circle's circumference, circles actually begin to shrink, as illustrated in Fig. 20-5. The distance from the North Pole N to point A is one eighth of the circumference of the sphere, from N to B is one quarter of the circumference, and from N to C is three quarters of the circumference. However, the circumference of the circle around N with radius NB (the equator) is greater than the circumference of the latitudinal circle around N with radius NC. A larger radius does not guarantee a larger circumference. In fact, the circumference of a circle with radius r in elliptic space is less than $2\pi r$.

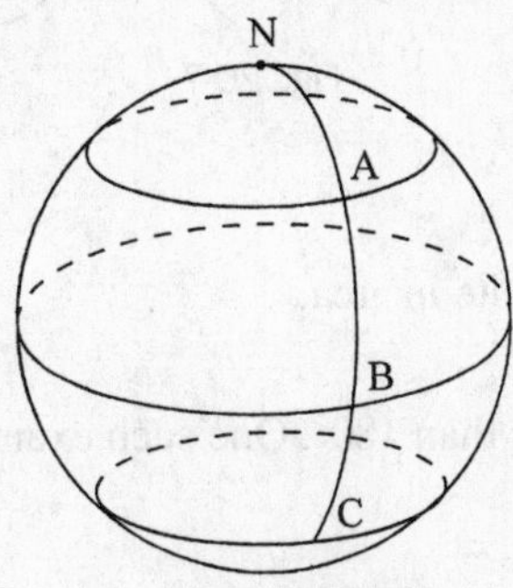

Fig. 20-5

Similarly, the area of a circle with radius r in elliptic space is $A < \pi r^2$. If a circle has a very small radius, then it will be almost flat and have area close to πr^2 and circumference near $2\pi r$ (though slightly less in both cases). The larger a circle's radius becomes, the further it will be from having the area and circumference of a circle in Euclidean space with the same radius.

Technically, elliptic space violates the first postulate. Between the North and South Poles, any of the longitudinal great circles counts as a straight line, thus, there are many straight lines between two points. A trick to overcome this is to call the North and South Poles together as one single point. If each pair of antipodal spots on a globe is viewed as a single point, then the resulting geometry is called *projective space*, which has all of the above properties in common with elliptic space.

20.4C Hyperbolic Space

Euclidean geometry is based on flat planes. A flat object is said to have *zero curvature*. Elliptic geometry is based on planes that are *positively curved*, like the surface of a sphere. Hyperbolic geometry is based on planes that are *negatively curved* and frilly like certain kinds of seaweed. The three kinds of curvature are illustrated in Fig. 20-6.

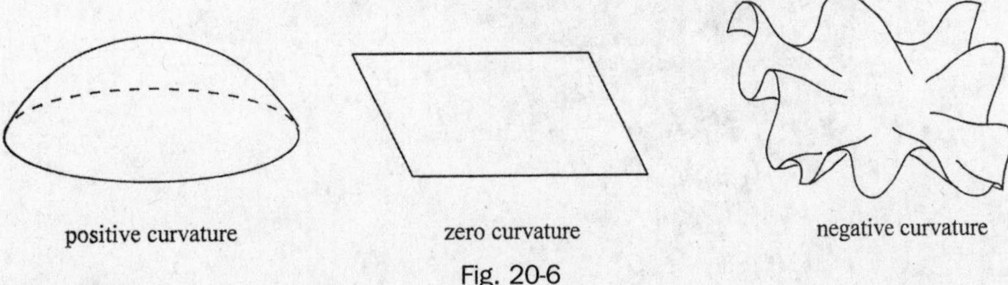

Fig. 20-6

Just as with elliptic space, any tiny piece of hyperbolic space is almost flat and will have properties close to those of euclidean space. As pieces get larger, however, they contain more of the warps, bends, and frills that are symptomatic of this space. The word "hyperbolic," by the way, has roots meaning "excessive" and "exaggerated."

The reason why hyperbolic space is the one that satisfies Postulate 5b can be seen in Fig. 20-7. Because of all the frilly waves of space through which a line can travel, there is more than one line through point P that will never cross $\overleftrightarrow{AB}$. Note that these do not have the usual property of parallel lines; they are not everywhere equidistant.

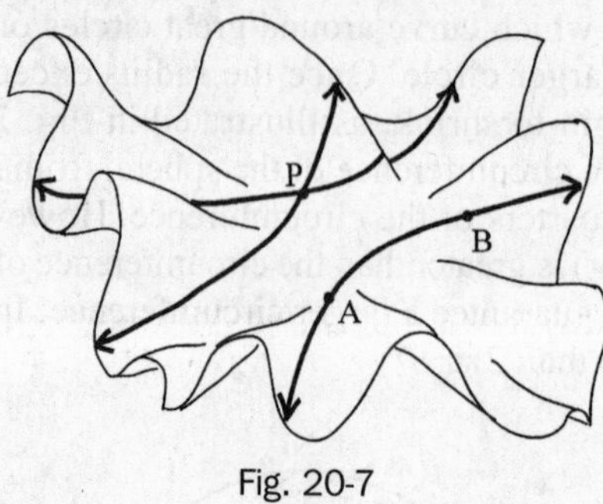

Fig. 20-7

In hyperbolic space

Planes are negatively curved and infinite in area.

Lines are infinite in length.

The angles of a triangle add up to less than 180°. One such example is illustrated in Fig. 20-8.

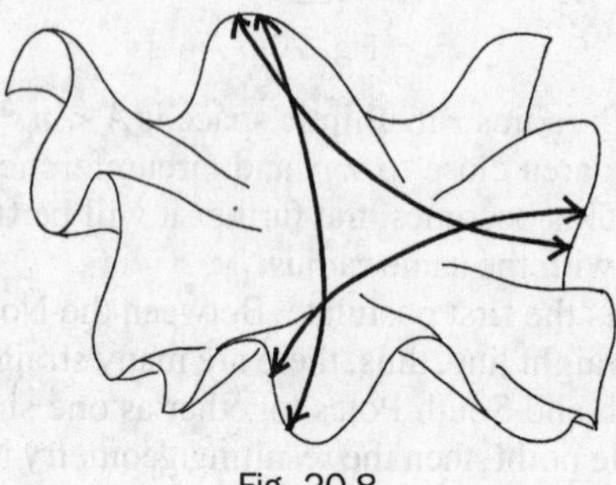

Fig. 20-8

Just as with elliptic space, the larger a triangle is, the further its angle sum will be from 180°. A tiny triangle will have angles that add up to just less than 180°. The largest triangles in hyperbolic space have angle sums of almost zero. Because of this, there are no similar triangles of different sizes in hyperbolic space.

Finally, a circle has more circumference and encompasses more area in hyperbolic space than it would in the euclidean plane. Thus, a circle with radius r in hyperbolic space has circumference $C > 2\pi r$ and area $A > \pi r^2$.

Formulas for Reference

Angle Formulas

1. Complement of $a°$	1. $c = 90° - a°$
2. Supplement of $a°$	2. $s = 180° - a°$
3. Sum of measures of angles of a triangle	3. $S = 180°$
4. Sum of measures of angles of a quadrilateral	4. $S = 360°$
5. Sum of measures of exterior angles of an n-gon	5. $S = 360°$
6. Sum of measures of interior angles of an n-gon	6. $S = 180°(n - 2)$
7. Measure of each interior angle of an equiangular or regular n-gon	7. $S = \frac{180°(n - 2)}{n}$
8. Measure of each exterior angle of an equiangular or regular n-gon	8. $S = \frac{360°}{n}$
9. Measure of central $\angle O$ intercepting an arc of $a°$	9. $m\angle O = a°$
10. Measure of inscribed $\angle A$ intercepting an arc of $a°$	10. $m\angle A = \frac{1}{2}a°$
11. Measure of $\angle A$ formed by a tangent and a chord and intercepting an arc of $a°$	11. $m\angle A = \frac{1}{2}a°$
12. Measure of $\angle A$ formed by two intersecting chords and intercepting arcs of $a°$ and $b°$	12. $m\angle A = \frac{1}{2}(a° + b°)$
13. Measure of $\angle A$ formed by two intersecting tangents, two intersecting secants, or an intersecting tangent and secant and intercepting arcs of $a°$ and $b°$	13. $m\angle A = \frac{1}{2}(a° - b°)$
14. Measure of $\angle A$ inscribed in a semicircle	14. $m\angle A = 90°$
15. Opposite $\angle$s A and B of an inscribed quadrilateral	15. $m\angle A = 180° - m\angle B$

Area Formulas

1. Area of a rectangle	1. $A = bh$	
2. Area of a square	2. $A = s^2$,	$A = \frac{1}{2}d^2$
3. Area of a parallelogram	3. $A = bh$,	$A = ab \sin C$
4. Area of a triangle	4. $A = \frac{1}{2}bh$,	$A = \frac{1}{2}ab \sin C$
5. Area of a trapezoid	5. $A = \frac{1}{2}h(b + {}^{n}b)$,	$A = hm$
6. Area of an equilateral triangle	6. $A = \frac{1}{4}s^2\sqrt{3}$,	$A = \frac{1}{3}h^2\sqrt{3}$
7. Area of a rhombus	7. $A = \frac{1}{2}d\,{}^{n}d$	
8. Area of a regular polygon	8. $A = \frac{1}{2}pr$	
9. Area of a circle	9. $A = \pi r^2$,	$A = \frac{1}{4}\pi d^2$
10. Area of a sector	10. $A = \frac{n}{360}(\pi r^2)$	
11. Area of a minor segment	11. $A =$ area of sector $-$ area of triangle	

Circle Intersection Formulas

1.	2.	3.
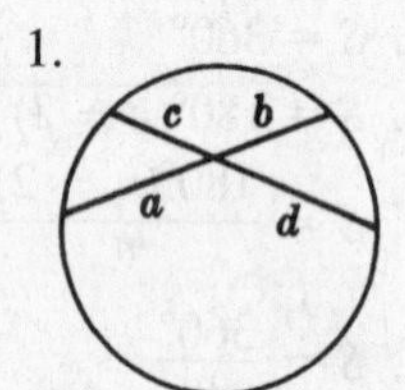 Intersecting Chords $ab = cd$	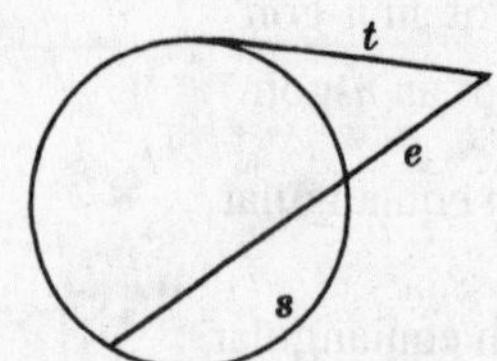 Intersecting Tangent and Secant $\frac{s}{t} = \frac{t}{e}$, $t^2 = se$	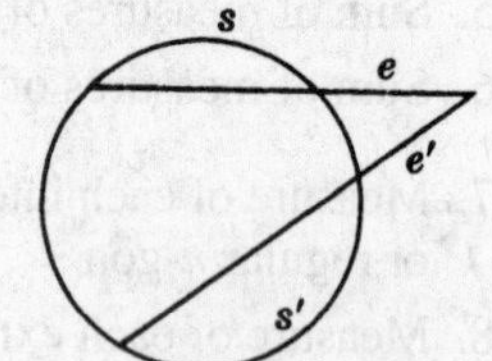 Intersecting Secants $se = s'e'$

Right-Triangle Formulas

1.	Pythagorean Theorem	1. $c^2 = a^2 + b^2$
2.	Leg opposite 30° angle Leg opposite 45° angle Leg opposite 60° angle	2. $b = \frac{1}{2}c$ $b = \frac{1}{2}c\sqrt{2}, b = a$ $a = \frac{1}{2}c\sqrt{3}, a = b\sqrt{3}$
3.	Altitude of equilateral triangle Side of equilateral triangle	3. $h = \frac{1}{2}s\sqrt{3}$ $s = \frac{2}{3}h\sqrt{3}$

Right-Triangle Formulas

4. 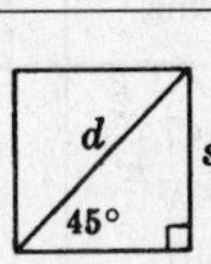	Side of square Diagonal of square	4. $s = \frac{1}{2}d\sqrt{2}$ $d = s\sqrt{2}$
5. 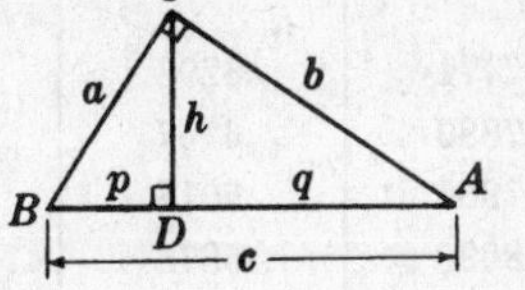	Altitude to hypotenuse Leg of right triangle	5. $\frac{p}{h} = \frac{h}{q}, h^2 = pq, h = \sqrt{pq}$ $\frac{c}{a} = \frac{a}{p}, a^2 = pc, a = \sqrt{pc}$ $\frac{c}{b} = \frac{b}{q}, b^2 = qc, b = \sqrt{qc}$

Coordinate-Geometry Formulas

1. 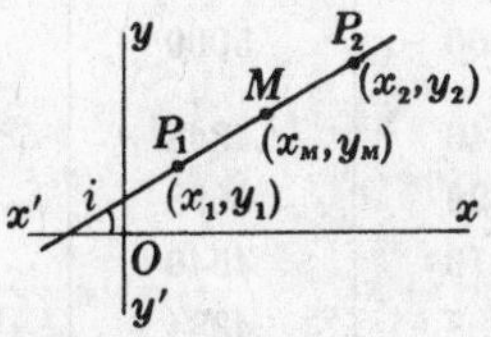	Midpoint M Distance P_1P_2 Slope of $\overleftrightarrow{P_1P_2}$	1. $x_M = \frac{x_1 + x_2}{2}, y_M = \frac{y_1 + y_2}{2}$ $d = \sqrt{(x_2 - x_1)^2 + (y_2 - y_1)^2}$ $m = \frac{y_2 - y_1}{x_2 - x_1}, m = \frac{\Delta y}{\Delta x}, m = \tan i$
2. 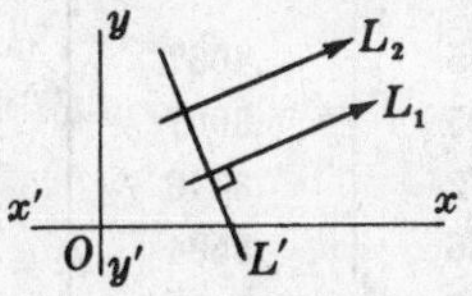	Slopes of parallels, L_1 and L_2 Slopes of perpendiculars, L_1 and L'	2. Same slope, m $mm' = -1$ $m' = -\frac{1}{m}, m = -\frac{1}{m'}$
3. 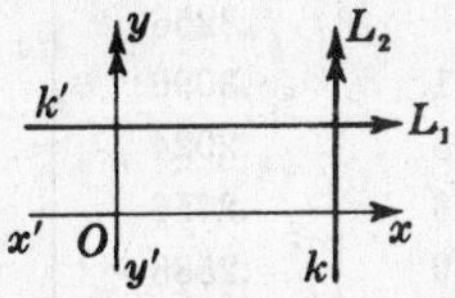	Equation of L_1, parallel to x-axis Equation of L_2, parallel to y-axis	3. $y = k'$ $x = k$
4. 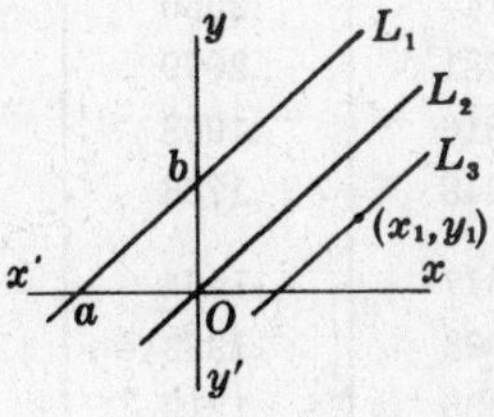	Equation of L_1 with slope m and y-intercept b Equation of L_2 with slope m passing through the origin Equation of L_1 with x-intercept a and y-intercept b Equation of L_3 with slope m and passing through (x_1, y_1)	4. $y = mx + b$ $y = mx$ $\frac{x}{a} + \frac{y}{b} = 1$ $y - y_1 = m(x - x_1)$
5. 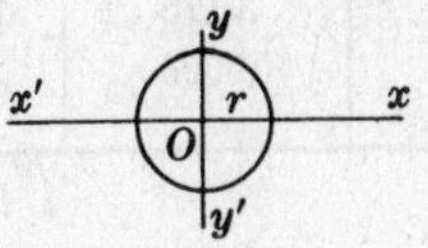	Equation of circle with center at origin and radius r	5. $x^2 + y^2 = r^2$

Table of Trigonometric Functions

Angle Measure	Sine	Cosine	Tangent	Angle Measure	Sine	Cosine	Tangent
1°	.0175	.9998	.0175	46°	.7193	.6947	1.0355
2°	.0349	.9994	.0349	47°	.7314	.6820	1.0724
3°	.0523	.9986	.0524	48°	.7431	.6691	1.1106
4°	.0698	.9976	.0699	49°	.7547	.6561	1.1504
5°	.0872	.9962	.0875	50°	.7660	.6428	1.1918
6°	.1045	.9945	.1051	51°	.7771	.6293	1.2349
7°	.1219	.9925	.1228	52°	.7880	.6157	1.2799
8°	.1392	.9903	.1405	53°	.7986	.6018	1.3270
9°	.1564	.9877	.1584	54°	.8090	.5878	1.3764
10°	.1736	.9848	.1763	55°	.8192	.5736	1.4281
11°	.1908	.9816	.1944	56°	.8290	.5592	1.4826
12°	.2097	.9781	.2126	57°	.8387	.5446	1.5399
13°	.2250	.9744	.2309	58°	.8480	.5299	1.6003
14°	.2419	.9703	.2493	59°	.8572	.5150	1.6643
15°	.2588	.9659	.2679	60°	.8660	.5000	1.7321
16°	.2756	.9613	.2867	61°	.8746	.4848	1.8040
17°	.2924	.9563	.3057	62°	.8829	.4695	1.8807
18°	.3090	.9511	.3249	63°	.8910	.4540	1.9626
19°	.3256	.9455	.3443	64°	.8988	.4384	2.0503
20°	.3420	.9397	.3640	65°	.9063	.4226	2.1445
21°	.3584	.9336	.3839	66°	.9135	.4067	2.2460
22°	.3746	.9272	.4040	67°	.9205	.3907	2.3559
23°	.3907	.9205	.4245	68°	.9272	.3746	2.4751
24°	.4067	.9135	.4452	69°	.9336	.3584	2.6051
25°	.4226	.9063	.4663	70°	.9397	.3420	2.7475
26°	.4384	.8988	.4877	71°	.9455	.3256	2.9042
27°	.4540	.8910	.5095	72°	.9511	.3090	3.0777
28°	.4695	.8829	.5317	73°	.9563	.2924	3.2709
29°	.4848	.8746	.5543	74°	.9613	.2756	3.4874
30°	.5000	.8660	.5774	75°	.9659	.2588	3.7321
31°	.5150	.8572	.6009	76°	.9703	.2419	4.0108
32°	.5299	.8480	.6249	77°	.9744	.2250	4.3315
33°	.5446	.8387	.6494	78°	.9781	.2079	4.7046
34°	.5592	.8290	.6745	79°	.9816	.1908	5.1446
35°	.5736	.8192	.7002	80°	.9848	.1736	5.6713
36°	.5878	.8090	.7265	81°	.9877	.1564	6.3138
37°	.6018	.7986	.7536	82°	.9903	.1392	7.1154
38°	.6157	.7880	.7813	83°	.9925	.1219	8.1443
39°	.6293	.7771	.8098	84°	.9945	.1045	9.5144
40°	.6428	.7660	.8391	85°	.9962	.0872	11.4301
41°	.6561	.7547	.8693	86°	.9976	.0698	14.3007
42°	.6691	.7431	.9004	87°	.9986	.0523	19.0811
43°	.6820	.7314	.9325	88°	.9994	.0349	28.6363
44°	.6947	.7193	.9657	89°	.9998	.0175	57.2900
45°	.7071	.7071	1.0000	90°	1.0000	.0000	

Answers to Supplementary Problems

Chapter 1

1. (a) point; (b) line; (c) plane; (d) plane; (e) line; (f) point
2. (a) $\overline{AE}, \overline{DE}$; (b) $\overline{ED}, \overline{CD}, \overline{BD}, \overline{FD}$; (c) $\overline{AD}, \overline{BE}, \overline{CE}, \overline{EF}$; (d) F
3. (a) $AB = 16$; (b) $AE = 10\frac{1}{2}$
4. (a) 18; (b) 90°; (c) 50°; (d) 130°; (e) 230°
5. (a) $\angle CBE$; (b) $\angle AEB$; (c) $\angle ABE$; (d) $ABC, \angle BCD, \angle BED$; (e) $\angle AED$
6. (a) 130°; (b) 120°; (c) 75°; (d) 132°
7. (a) 75°; (b) 40°; (c) $10\frac{1}{3}°$ or 10°20′; (d) 9°11′
8. (a) 90°; (b) 120°; (c) 135°; (d) 270°; (e) 180°
9. (a) 90°; (b) 60°; (c) 15°; (d) 165°
10. (a) $\overline{AB} \perp \overline{BC}$ and $\overline{AC} \perp \overline{CD}$; (b) 129°; (c) 102°; (d) 51°; (e) 129°
11. (a) $\triangle ABC$, hypotenuse $\overline{AB}$, legs $\overline{AC}$ and $\overline{BC}$
    $\triangle ACD$, hypotenuse $\overline{AC}$, legs $\overline{AD}$ and $\overline{CD}$
    $\triangle BCD$, hypotenuse $\overline{BC}$, legs $\overline{BD}$ and $\overline{CD}$
    (b) $\triangle DAB$ and $\triangle ABC$
    (c) $\triangle AEB$, legs $\overline{AE}$ and $\overline{BE}$, base $\overline{AB}$, vertex angle $\angle AEB$
    $\triangle CED$, legs $\overline{DE}$ and $\overline{CE}$, base $\overline{CD}$, vertex angle $\angle CED$
12. (a) $\overline{AR} \cong \overline{BR}$ and $\angle PRA \cong \angle PRB$; (b) $\angle ABF \cong \angle CBF$; (c) $\angle CGA \cong \angle CGD$; (d) $\overline{AM} \cong \overline{MD}$
13. (a) vert. ∠s; (b) comp. adj. ∠s; (c) adj. ∠s; (d) supp. adj. ∠s; (e) comp. ∠s; (f) vert. ∠s
14. (a) 25°, 65°; (b) 18°, 72°; (c) 60°, 120°; (d) 61°, 119°; (e) 50°, 130°; (f) 56°, 84°; (g) 90°, 90°
15. (a) $a + b = 75°, a - b = 21°, a = 48°$ and $b = 27°$
    (b) $a + b = 90°, a = 3b - 10°, a = 65°$ and $b = 25°$
    (c) $a + b = 180°, a = 4b + 20°, a = 148°$ and $b = 32°$

Chapter 2

1. (a) A is H; (b) P is D; (c) R is S; (d) E is K; (e) A is G; (f) triangles are geometric figures; (g) a rectangle is a quadrilateral

2. (a) $a = c = f$; (b) $g = 15$; (c) $f = a$; (d) $a = f, a = h, c = h$; (e) $b = g, b = e, d = e$

3. (a) 130; (b) 4; (c) yes; (d) $x = 8\frac{1}{2}°$; (e) $y = 15$; (f) $x = 6$; (g) $x = \pm 6$

4. (a) $AC = 12, AE = 11, AF = 15, DF = 9$
   (b) $m\angle ADC = 92°, m\angle BAE = 68°, m\angle FAD = 86°, m\angle BAD = 128°$

5. (a) $AB = DF$; (b) $AB = AC$; (c) $\angle ECA \cong \angle DCB$; (d) $\angle BAD \cong \angle BCD$

6. (a) If equals are divided by equals, the quotients are equal.
   (b) Doubles of equals are equal.
   (c) If equals are multiplied by equals, the products are equal.
   (d) Halves of equals are equal.

7. (a) If equals are divided by equals, the quotients are equal.
   (b) If equals are multiplied by equals, the products are equal.
   (c) Doubles of equals are equal.
   (d) Halves of equals are equal.

8. (a) Their new rates of pay per hour will be the same. (Add. Post.)
   (b) Those stocks have the same value now. (Mult. Post.)
   (c) The classes have the same number of pupils now. (Subt. Post.)
   (d) 100°C = 212°F (Trans. Post.)
   (e) Their parts will be the same length. (Div. Post.)
   (f) He has a total of $10,000 in Banks A, B, and C. (Part. Post.)
   (g) Their values are the same. (Trans. Post.)

9. (a) Vertical angles are congruent.
   (b) All straight angles are congruent.
   (c) Supplements of congruent angles are congruent.
   (d) Perpendiculars form right angles and all right angles are congruent.
   (e) Complements of congruent angles are congruent.

10. In each answer, (H) indicates the hypothesis and (C) indicates the conclusion.
    (a) (H) Stars, (C) twinkle.
    (b) (H) Jet planes, (C) are the speediest.
    (c) (H) Water, (C) boils at 212° Fahrenheit.
    (d) (H) If it is the American flag, (C) its colors are red, white, and blue.
    (e) (H) If you fail to do homework in the subject, (C) you cannot learn geometry.
    (f) (H) If the umpire calls a fourth ball, (C) a batter goes to first base.
    (g) (H) If A is B's brother and C is B's daughter, (C) then A is C's uncle.
    (h) (H) An angle bisector, (C) divides the angle into two equal parts.
    (i) (H) If it is divided into three equal parts, (C) a segment is trisected.
    (j) (H) A pentagon (C) has five sides and five angles.
    (k) (H) Some rectangles (C) are squares.
    (l) (H) If their sides are made longer, (C) angles do not become larger.
    (m) (H) If they are congruent and supplementary, (C) angles are right angles.
    (n) (H) If one of its sides is not a straight line segment, (C) the figure cannot be a polygon.

11. (a) An acute angle is half a right angle. Not necessarily true.
    (b) A triangle having one obtuse angle is an obtuse triangle. True.
    (c) If the batter is out, then the umpire called a third strike. Not necessarily true.
    (d) If you are shorter than I, then I am taller than you. True.
    (e) If our weights are unequal, then I am heavier than you. Not necessarily true.

Chapter 3

1. (a) $\Delta I \cong \Delta II \cong \Delta III$, SAS; (b) $\Delta I \cong \Delta III$, ASA; (c) $\Delta I \cong \Delta II \cong \Delta III$, SSS.
2. (a) ASA; (b) SAS; (c) SSS; (d) SAS; (e) ASA; (f) SAS; (g) SAS; (h) ASA.
3. (a) $\overline{AD} \cong \overline{DC}$; (b) $\angle ABD = \angle DBC$; (c) $\angle 1 \cong \angle 4$; (d) $\overline{BE} \cong \overline{ED}$; (e) $\overline{BD} \cong \overline{AC}$; (f) $\angle BAD \cong \angle CDA$
4. (a) $\angle 1 \cong \angle 3, \angle 2 \cong \angle 4, \overline{BD} \cong \overline{BE}$; (b) $\overline{AB} \cong \overline{AC}, \overline{BD} \cong \overline{DC}, \angle B \cong \angle C$;
   (c) $\angle E \cong \angle C, \angle A \cong \angle F, \angle EDF \cong \angle ABC$
5. (a) $x = 19, y = 8$; (b) $x = 4, y = 12$; (c) $x = 48, y = 12$
8. (a) $\angle b \cong \angle d, \angle E \cong \angle G$; (b) $\angle A \cong \angle 1 \cong \angle 4, \angle 2 \cong \angle C$; (c) $\angle 1 \cong \angle 5, \angle 4 \cong \angle 6, \angle EAD \cong \angle EDA$
9. (a) $\overline{BE} \cong \overline{EC}$; (b) $\overline{AB} \cong \overline{BD} \cong \overline{AD}, \overline{BC} \cong \overline{CD}$; (c) $\overline{BD} \cong \overline{DE}, \overline{EF} \cong \overline{FC}, \overline{AB} \cong \overline{AC}$

Chapter 4

1. (a) $x = 105°, y = 75°$; (b) $x = 60°, y = 40°$; (c) $x = 85°, y = 95°$; (d) $x = 50°, y = 50°$;
   (e) $x = 65°, y = 65°$; (f) $x = 40°, y = 30°$; (g) $x = 60°, y = 120°$; (h) $x = 90°, y = 35°$;
   (i) $x = 30°, y = 40°$; (j) $x = 80°, y = 10°$; (k) $x = 30°, y = 150°$; (l) $x = 85°, y = 95°$
2. (a) $x = 22°, y = 102°$; (b) $x = 40°, y = 100°$; (c) $x = 80°, y = 40°$
3. (a) Each angle measures 105°. (b) Each angle measures 70°. (c) Angles measure 72° and 108°.
7. (a) 25; (b) 9; (c) 20; (d) 8
8. (a) 8; (b) 10; (c) 2; (d) 14
10. (a) P is equidistant from B and C. P is on $\perp$ bisector of $\overline{BC}$.
    Q is equidistant from A and B. Q is on $\perp$ bisector of $\overline{AB}$.
    R is equidistant from A, C, and D. R is on $\perp$ bisectors of $\overline{AD}$ and $\overline{CD}$.
    (b) P is equidistant from $\overline{AB}$ and $\overline{AD}$. P is on bisector of $\angle A$.
    Q is equidistant from $\overline{AB}$ and $\overline{BC}$. Q is on bisector of $\angle B$.
    R is equidistant from $\overline{BC}$, $\overline{CD}$, and $\overline{AD}$. R is on the bisectors of $\angle C$ and $\angle D$.
11. (a) P is equidistant from $\overline{AD}$, $\overline{AB}$, and $\overline{BC}$. Q is equidistant from $\overline{AD}$ and $\overline{AB}$ and equidistant from A and D. R is equidistant from $\overline{AB}$ and $\overline{BC}$ and equidistant from A and D.
    (b) P is equidistant from $\overline{AD}$ and $\overline{CD}$ and equidistant from B and C. Q is equidistant from A, B, and C. R is equidistant from $\overline{AD}$ and $\overline{CD}$ and equidistant from A and B.
12. (a) $x = 50°, y = 110°$; (b) $x = 65°, y = 65°$; (c) $x = 30°, y = 100°$; (d) $x = 51°, y = 112°$;
    (e) $x = 52°, y = 40°$; (f) $x = 120°, y = 90°$
13. (a) $x = 55°, y = 125°$; (b) $x = 80°, y = 90°$; (c) $x = 56°, y = 68°$; (d) $x = 100°, y = 30°$;
    (e) $x = 30°, y = 120°$; (f) $x = 90°, y = 30°$
14. (a) 18°, 54°, 108°; (b) 40°, 50°, 90°; (c) 36°, 36°, 108°; (d) 36°, 72°, 108°, 144°; (e) 50°, 75°;
    (f) 100°, 60°, and 20°
16. (a) Since $x = 45$, each angle measures 60°.
    (b) Since $x = 25, x + 15 = 40$ and $3x - 35 = 40$; that is, two angles each measure 40°.
    (c) If $2x$, $3x$, and $5x$ represent the angles, $x = 18$ and $5x = 90$; that is, one of the angles measures 95°.
    (d) If x and $5x - 10$ represent the unknown angles, $x = 21$ and $5x - 10 = 95$; that is, one of the angles measures 95°.
17. (a) 7 st. ⦞s, 30 st. ⦞s; (b) 1620°, 5400°, 180,000°; (c) 30, 12, 27, 202
18. (a) 20°, 18°, 9°; (b) 160°, 162°, 171°; (c) 3, 9, 20, 180; (d) 3, 12, 36, 72, 360
19. (a) 65°, 90°, 95°, 110°; (b) 140°, 100°, 60°, 60°
20. (a) $\Delta I \cong \Delta III$ by hy-leg; (b) $\Delta I \cong \Delta III$ by SAA.

Chapter 5

1. (a) $x = 15, y = 25$; (b) $x = 20, y = 130$; (c) $x = 20, y = 140$

4. (a) ▱*EFGH*; (b) ▱*ABCD* and *EBFD*; (c) ▱ *GHKJ*, *HILK*, *GILJ*; (d) ▱ *ACHB*, *CEFH*

5. (a) Two sides are congruent and ‖. (b) Opposite sides are congruent.
(c) Opposite angles are congruent. (d) $\overline{AD}$ and $\overline{BC}$ are congruent and parallel ($\overline{AD} \cong \overline{EF} \cong \overline{BC}$).

6. (a) $x = 6, y = 12$; (b) $x = 5, y = 9$; (c) $x = 120, y = 30$; (d) $x = 15, y = 45$

7. (a) $x = 14, y = 6$; (b) $x = 18, y = 4\frac{1}{2}$; (c) $x = 8, y = 5$; (d) $x = 3, y = 9$

10. (a) $x = 5, y = 7$; (b) $x = 10, y = 35$; (c) $x = 2\frac{1}{2}, y = 17\frac{1}{2}$; (d) $x = 8, y = 4$; (e) $x = 25, y = 25$;
(f) $x = 11, y = 118$

13. (a) $x = 6, y = 40$; (b) $x = 3, y = 5\frac{1}{2}$; (c) $x = 8\frac{1}{3}, y = 22$

14. (a) $x = 28, y = 25\frac{1}{2}$; (b) $x = 12$ (since *y* does not join midpoints, Pr. 3 does not apply);
(c) $x = 19,\ y = 23\frac{1}{2}$

15. (a) $m = 19$; (b) $b' = 36$; (c) $b = 73$

16. (a) $x = 11, y = 33$; (b) $x = 32, y = 26$; (c) $x = 12, y = 36$

17. (a) $22\frac{1}{2}$; (b) 70

18. (a) 21; (b) 30; (c) 14; (d) 26

Chapter 6

5. (a) square; (b) isosceles triangle; (c) trapezoid; (d) right triangle

6. (a) 140°; (b) 60°; (c) 90°; (d) $(180 - x)°$; (e) $x°$; (f) $(90 + x)°$

7. (a) 100°; (b) 50°, 80°; (c) 54°, 27°; (d) 45°; (e) 35°; (f) 45°

8. (a) $x = 22$; (b) $y = 6$; (c) $AB + CD = 22$; (d) perimeter $= 44$; (e) $x = 21$; (f) $r = 14$

9. (a) 0; (b) 40; (c) 33; (d) 7

10. (a) tangent externally; (b) tangent internally; (c) the circles are 5 units apart; (d) overlapping

11. (a) concentric; (b) tangent internally; (c) tangent externally; (d) outside each other; (e) the smaller entirely inside the larger; (f) overlapping

13. (a) 40; (b) 90; (c) 170; (d) 180; (e) $2x$; (f) $180 - x$; (g) $2x - 2y$

14. (a) 20; (b) 45; (c) 85; (d) 90; (e) 130; (f) 174; (g) x; (h) $90 - \frac{1}{2}x$; (i) $x - y$

15. (a) 85; (b) 170; (c) c; (d) $2i$; (e) 60; (f) 30

16. (a) 60, 120, 180; (b) 80, 120, 160; (c) 100, 120, 140; (d) 36, 144, 180

17. (a) $m\angle x = 136°$; (b) $m\widehat{y} = 11°$; (c) $m\angle x = 130°$; (d) $m\angle y = 126°$; (e) $m\angle x = 110°$; $m\widehat{y} = 77°$

18. (a) 135°; (b) 90°; (c) $(180 - x)°$; (d) $(90 + x)°$; (e) 100°; (f) 80°; (g) 55°; (h) 72°

19. (a) 85°; (b) $y°$; (c) 110°; (d) 95°; (e) 72°; (f) 50°; (g) 145°; (h) 87°

20. (a) 50; (b) 60

21. (a) $m\widehat{x} = 65°, m\widehat{y} = 65°$; (b) $m\angle x = 90°, m\angle y = 55°$; (c) $m\angle x = 37°, m\angle y = 50°$

22. (a) 19; (b) 45; (c) 69; (d) 90; (e) 125; (f) 167; (g) $\frac{1}{2}x$; (h) $180 - \frac{1}{2}x$; (i) $x + y$

23. (a) 110; (b) 135; (c) 180; (d) 270; (e) $180 - 2x$; (f) $360 - 2x$; (g) $2x - 2y$; (h) $7x$

24. (a) 45°; (b) 60°; (c) 30°; (d) 18°

25. (a) $m\widehat{x} = 120°, m\angle y = 60°$; (b) $m\angle x = 62°, m\angle y = 28°$; (c) $m\angle x = 46°, m\angle y = 58°$

26. (a) 75°; (b) 75°; (c) 115°; (d) 100°; (e) 140°; (f) 230°; (g) 80°; (h) 48°

27. (a) 85°; (b) 103°; (c) 80°; (d) 72°; (e) 90°; (f) 110°; (g) 130°; (h) 110°

28. (a) $m\widehat{x} = 68°, m\angle y = 95°$; (b) $m\angle x = 90°, m\angle y = 120°$; (c) $m\widehat{x} = 34°, m\angle \widehat{y} = 68°$

29. (a) 30°; (b) 37°; (c) 20°; (d) 36°; (e) 120°; (f) 130°; (g) 94°; (h) 25°

30. (a) 45°; (b) 75°; (c) 50°; (d) $36\frac{1}{2}°$; (e) 90°; (f) 140°; (g) 115°; (h) 45°; (i) 80°

31. (a) 20°; (b) 85°; (c) $(180 - x)°$; (d) $(90 + x)°$; (e) 90°; (f) 25°; (g) 42°; (h) 120°; (i) 72°; (j) 110°; (k) 145°; (l) $(180 - y)°$; (m) 240°; (n) $(180 + x)°$; (o) 270°

32. (a) $m\widehat{x} = 43°, m\angle y = 43°$, (b) $m\widehat{x} = 190°, m\angle y = 55°$; (c) $m\widehat{x} = 140°, m\angle y = 40°$

33. (a) 120°; (b) 150°; (c) 180°; (d) 50°; (e) $22\frac{1}{2}$; (f) 45°

34. (a) $m\widehat{x} = 150°, m\widehat{y} = 40°$; (b) $m\widehat{x} = 190°, m\widehat{y} = 70°$; (c) $m\widehat{x} = 252°, m\widehat{y} = 108°$

35. (a) 25°; (b) 39°; (c) 50°; (d) 30°; (e) 40°; (f) 76°; (g) 45°; (h) 95°; (i) 75°; (j) 120°

36. (a) 74°; (b) 90°; (c) 55°; (d) 60°; (e) 40°; (f) 37°; (g) 84°; (h) 110°; (i) 66°; (j) 98°; (k) 75°; (l) 79°

37. (a) $m\angle x = 120°, m\angle y = 60°$; (b) $m\angle x = 45°, m\angle y = 22\frac{1}{2}°$; (c) $m\angle x = 36°, m\angle y = 72°$

38. (a) $m\widehat{x} = 40°, m\angle y = 80°$; (b) $m\widehat{x} = 45°, m\angle y = 67\frac{1}{2}°$; (c) $m\angle x = 78°, m\angle y = 103°$

Chapter 7

1. (a) 4; (b) $\frac{1}{3}$; (c) $\frac{6}{5}$; (d) $\frac{10}{7}$; (e) $\frac{9}{7}$; (f) 2; (g) $\frac{1}{5}$; (h) $\frac{3}{7}$; (i) $\frac{2}{3}$; (j) $\frac{7}{8}$; (k) 2; (l) $\frac{5}{7}$;(m) 20; (n) $\frac{1}{3}$; (o) 3

2. (a) 6; (b) $\frac{14}{5}$; (c) $\frac{1}{7}$; (d) $\frac{3}{2}$; (e) 3; (f) $\frac{7}{2}$; (g) 2; (h) 250; (i) $\frac{1}{20}$; (j) 8; (k) $\frac{5}{3}$; (l) $\frac{9}{2}$

3. (a) 2:3:10; (b) 12:6:1; (c) 5:2:1; (d) 1:4:7; (e) 4:3:1; (f) 8:2:1; (g) 50:5:1; (h) 6:2:1; (i) 8:2:1

4. (a) $\frac{6}{7}$; (b) 12; (c) $\frac{13}{3}$; (d) $\frac{1}{4}$; (e) 6; (f) $\frac{16}{9}$; (g) $\frac{1}{3}$; (h) $\frac{3}{2}$; (i) $\frac{2}{7}$; (j) 11; (k) $\frac{4}{5}$; (l) 60; (m) 3; (n) $\frac{3}{20}$; (o) $\frac{1}{2}$; (p) 14

5. (a) $\frac{1}{3}$; (b) $3c$; (c) $\frac{d}{2}$; (d) $\frac{2r}{D}$; (e) $\frac{b}{a}$; (f) $\frac{4}{S}$; (g) $\frac{S}{6}$; (h) $\frac{3r}{2t}$; (i) 1:4: 10; (j) 3:2:1; (k) x^2: x:1; (l) 6:5:4:1

6. (a) $5x$ and $4x$, sum $= 9x$; (b) $9x$ and x, sum $= 10x$; (c) $2x$, $5x$, and $11x$, sum $= 18x$; (d) x, $2x$, $2x$, $3x$, and $7x$, sum $= 15x$

7. (a) $5x + 4x = 45, x = 5$, 25° and 20°; (b) $5x + 4x = 90, x = 10$, 50° and 40°; (c) $5x + 4x = 180, x = 20$, 100° and 80°; (d) $5x + 4x + x = 180, x = 18$, 90° and 72°

8. (a) $7x + 6x = 91, x = 7$, 49°, 42° and 35°; (b) $7x + 5x = 180, x = 15$, 105°, 90° and 75°; (c) $7x + 3x = 90, x = 9$, 63°, 54° and 45°; (d) $7x + 6x + 5x = 180, x = 10$, 70°, 60° and 50°

9. (a) 16; (b) 16; (c) ± 6; (d) $\pm 2\sqrt{5}$; (e) ± 5; (f) 2; (g) $\frac{bc}{a}$; (h) $\pm 6y$

10. (a) 21; (b) $4\frac{2}{3}$; (c) ± 6; (d) $\pm \sqrt{5}$; (e) 8; (f) ± 4; (g) 3; (h) $\pm \sqrt{ab}$

11. (a) 15; (b) 3; (c) 6; (d) $2\frac{2}{3}$; (e) $3\frac{1}{3}$; (f) 30; (g) 32; (h) $6a$

12. (a) 6; (b) 6; (c) 3; (d) $4b$; (e) $\sqrt{10}$; (f) $\sqrt{27}$ or $3\sqrt{3}$; (g) $\sqrt{pq}$; (h) $a\sqrt{b}$

13. (a) $\frac{c}{b} = \frac{d}{x}$; (b) $\frac{a}{p} = \frac{q}{x}$; (c) $\frac{h}{a} = \frac{a}{x}$; (d) $\frac{3}{7} = \frac{1}{x}$; (e) $\frac{c}{a} = \frac{b}{x}$

14. (a) $\frac{x}{y} = \frac{1}{2}$; (b) $\frac{x}{y} = \frac{3}{4}$; (c) $\frac{x}{y} = \frac{1}{2}$; (d) $\frac{x}{y} = \frac{h}{a}$; (e) $\frac{x}{y} = b$

15. Only (b) is not a proportion since $3(12) \neq 5(7)$; that is, $36 \neq 35$.

16. (a) $\frac{x}{2} = \frac{9}{3}, x = 6$; (b) $\frac{x}{1} = \frac{4}{5}, x = \frac{4}{5}$; (c) $\frac{x}{a} = \frac{b}{2}, x = \frac{ab}{2}$; (d) $\frac{x}{5} = \frac{1}{10}, x = \frac{1}{2}$; (e) $\frac{x}{20} = \frac{5}{4}, x = 25$

17. (a) d; (b) 35; (c) 5 (d) 4

18. (a) 21; (b) $\frac{3}{2}$; (c) 5

19. (a) 16; (b) $6\frac{2}{3}$; (c) 10

20. (a) yes, since $\frac{15}{10} = \frac{18}{12}$; (b) no, since $\frac{10}{13} \neq \frac{7}{9}$; (c) yes, since $\frac{3x}{5x} = \frac{36}{60}$

21. (a) 12; (b) 8; (c) 60

22. (a) 15; (b) 15; (c) $6\frac{1}{2}$

24. (a) 35°; (b) 53°

25. (a) $a = 16$; (b) $b = 15$; (c) $c = 126$

27. (a) $\angle ABE \cong \angle EDC, \angle BAE \cong \angle DCE$ (also vert. $\angle$s at E)
(b) $\angle BAF \cong \angle FEC, \angle B \cong \angle D$ (also $\angle EAD \cong \angle BFA$)
(c) $\angle A \cong \angle EDF, \angle F \cong \angle BCA$
(d) $\angle A \cong \angle A, \angle B \cong \angle C$
(e) $\angle C \cong \angle D, \angle CAB \cong \angle CAD$
(f) $\angle A \cong \angle A, \angle C \cong \angle DBA$

28. (a) $\angle D \cong \angle B, \angle AED \cong \angle FGB$; (b) $\angle ADB \cong \angle ABC, \angle A \cong \angle A$; (c) $\angle ABC \cong \angle AED, \angle BAE \cong \angle EDA$

29. (a) $\angle C \cong \angle F, \frac{14}{20} = \frac{21}{30}$; (b) $\angle A \cong \angle A, \frac{10}{25} = \frac{6}{15}$; (c) $\angle B \cong \angle B, \frac{16}{28} = \frac{20}{35}$

30. (a) $\frac{6}{18} = \frac{8}{24} = \frac{10}{30}$; (b) $\frac{24}{36} = \frac{28}{42} = \frac{30}{45}$; (c) $\frac{12}{18} = \frac{16}{24} = \frac{18}{27}$

32. (a) $q = 20$; (b) $p = 8$; (c) $b = 7$; (d) $a = 12$; (e) $AB = 35$; (f) $d = 2\frac{1}{4}$

33. (a) 8; (b) 6; (c) $26\frac{2}{3}$

34. (a) 42 ft; (b) 66ft

37. (a) 8:5; (b) 3:5; (c) halved (in each case)

38. (a) 15; (b) 60; (c) 25, 35, 40; (d) 4; (e) 6, 3

39. (a) 3:7; (b) 7:2; (c) quadrupled; (d) 7

43. (a) 5; (b) 14; (c) 6; (d) 5; (e) 12; (f) 13; (g) 48; (h) 2

44. 30, 18

45. (a) 8; (b) 6; (c) 12; (d) 5; (e) 7; (f) 12; (g) 30; (h) $7\frac{1}{2}$; (i) 5; (j) 8

46. (a) 8; (b) 13; (c) 21; (d) 6; (e) 9; (f) 14; (g) 3; (h) 8

47. (a) $a = 4, h = \sqrt{12}$ or $2\sqrt{3}$; (b) $c = 9, h = \sqrt{20}$ or $2\sqrt{5}$; (c) $q = 4$ and $b = \sqrt{80}$ or $4\sqrt{5}$; (d) $p = 18, h = \sqrt{108} = 6\sqrt{3}$

48. (a) 25; (b) 39; (c) $\sqrt{41}$; (d) 10; (e) $7\sqrt{2}$

49. (a) $b = 16$; (b) $a = 2\sqrt{7}$; (c) $a = 8$; (d) $b = 2\sqrt{3}$; (e) $b = 5\sqrt{2}$; $b = \sqrt{3}$

50. (a) 9, 12; (b) 10, 24; (c) 80, 150; (d) $2\sqrt{5}, 4\sqrt{5}$

51. (a) 41; (b) $5\sqrt{5}$

52. (a) 12; (b) $10\sqrt{2}$; (c) $5\sqrt{5}$

53. All except (h)

54. (a) yes; (b) no, since $(2x)^2 + (3x)^2 \neq (4x)^2$

55. (a) 8; (b) 6; (c) $\sqrt{19}$; (d) $5\sqrt{3}$

56. (a) 15; (b) $2\sqrt{5}$; (c) 6

57. (a) 16; (b) 30; (c) $4\sqrt{3}$; (d) 10

58. (a) 10; (b) 12; (c) 28; (d) 15

59. (a) 5; (b) 20; (c) 15; (d) 25

60. (a) 12; (b) 24

61. 12

62. 30

63. (a) 10 and $10\sqrt{3}$; (b) $7\sqrt{3}$ and 14; (c) 5 and 10

64. (a) $11\sqrt{3}$; (b) $a\sqrt{3}$; (c) 48; (d) $16\sqrt{3}$

65. (a) 25 and $25\sqrt{3}$; (b) 35 and $35\sqrt{3}$

66. (a) 28, $8\sqrt{3}$; (b) 17, $14\sqrt{3}$

67. (a) $17\sqrt{2}$; (b) $a\sqrt{2}$; (c) $34\sqrt{2}$; (d) 30

68. (a) $20\sqrt{2}$; (b) $40\sqrt{2}$

69. (a) 45, $13\sqrt{2}$; (b) 11, $27\sqrt{2}$; (c) $15\sqrt{2}$, 55

70. $6\sqrt{2}, 5\sqrt{2}$

Chapter 8

1. (a) 0.4226, 0.7431, 0.8572, 0.9998; (b) 0.9659, 0.6157, 0.2756, 0.0349; (c) 0.0699, 0.6745, 1.4281, 19.0811; (d) sine and tangent; (e) cosine; (f) tangent

2. (a) $x = 20°$; (b) $A = 29°$; (c) $B = 71°$; (d) $A' = 21°$; (e) $y = 45°$; (f) $Q = 69°$; (g) $W = 19°$; (h) $B' = 67°$

3. (a) 26°; (b) 47°; (c) 69°; (d) 8°; (e) 40°; (f) 74°; (g) 7°; (h) 27°; (i) 80°; (j) 13° since $\sin x = 0.2200$; (k) 45° since $\sin x = 0.707$; (l) 59° since $\cos x = 0.5200$; (m) 68° since $\cos x = 0.3750$; (n) 30° since $\cos x = 0.866$; (o) 16° since $\tan x = 0.2857$; (p) 10° since $\tan x = 0.1732$

4. (a) $\sin A = \frac{4}{5}, \cos A = \frac{3}{4}, \tan A = \frac{4}{5}$; (b) $\sin A = \frac{3}{5}, \cos A = \frac{4}{5}, \tan A = \frac{3}{4}$;
(c) $\sin A = \dfrac{\sqrt{7}}{4}, \cos A = \frac{3}{4}, \tan A = \dfrac{\sqrt{7}}{3}$

5. (a) $m\angle A = 27°$ since $\cos A = 0.8900$; (b) $m\angle A = 58°$ since $\sin A = 0.8500$; (c) $m\angle A = 52°$ since $\tan A = 1.2800$

6. (a) $m\angle A = 42°$ since $\sin B = 0.6700$; (b) $m\angle B = 74°$ since $\cos B = 0.2800$; (c) $m\angle B = 68°$ since $\tan B = 2.500$; (d) $m\angle B = 30°$ since $\tan B = 0.577$

8. (a) 23°, 67°; (b) 28°, 62°; (c) 16°, 74°; (d) 10°, 80°

9. (a) $x = 188, y = 313$; (b) $x = 174, y = 250$; (c) $x = 123, y = 182$

10. (a) 82 ft; (b) 88 ft

11. 156 ft

12. (a) 2530 ft; (b) 2560 ft

13. (a) 21 in; (b) 79 in

14. 14

15. 16 and 18 in

16. 31 ft

17. 15 yd

18. (a) 1050 ft; (b) 9950 ft

19. 7°

20. 282 ft

21. (a) 81°; (b) 45°

22. (a) 22 ft; (b) 104 ft

23. 754 ft

24. 404 ft

25. (a) 295 ft; (b) 245 ft; (c) 960 ft

26. (a) 234 ft; (b) 343 ft

27. (a) 96 ft; (b) 166 ft

28. 9.1

Chapter 9

1. (a) 99 in²; (b) 3 ft² or 432 in²; (c) 500; (d) 120; (e) $36\sqrt{3}$; (f) $100\sqrt{3}$; (g) 300; (h) 150

2. (a) 48; (b) 432; (c) $25\sqrt{3}$; (d) 240

3. (a) 7 and 4; (b) 12 and 6; (c) 9 and 6; (d) 6 and 2; (e) 10 and 7; (f) 20 and 8

4. (a) 1296 in²; (b) 100 square decimeters (100 dm²)

5. (a) 225; (b) $12\frac{1}{4}$; (c) 3.24; (d) $64a^2$; (e) 121; (f) $6\frac{1}{4}$; (g) $9b^2$; (h) 32; (i) $40\frac{1}{2}$; (j) 64

6. (a) 128; (b) 72; (c) 100; (d) 49; (e) 400

7. (a) 1600; (b) 400; (c) 100

8. (a) 9; (b) 36; (c) $9\sqrt{2}$; (d) $4\frac{1}{2}$; (e) $\frac{9}{2}\sqrt{2}$

9. (a) $2\frac{1}{2}$; (b) 52; (c) 10; (d) $5\sqrt{2}$; (e) 6; (f) 4

10. (a) 16 ft²; (b) 6 ft² or 864 in²; (c) 70; (d) 1.62 m²

11. (a) $3x^2$; (b) $x^2 + 3x$; (c) $x^2 - 25$; (d) $12x^2 + 11x + 2$

12. (a) 36; (b) 15; (c) 16

13. (a) $2\frac{2}{3}$; (b) 20; (c) 9; (d) 3; (e) 15; (f) 12; (g) 8; (h) 7

14. (a) 11 in²; (b) 3 ft² (c) $4x - 28$; (d) $10x^2$; (e) $2x^2 + 18x$; (f) $\frac{1}{2}(x^2 - 16)$; (g) $x^2 - 9$

15. (a) 84; (b) 48; (c) 30; (d) 120; (e) 148; (f) 423; (g) $8\sqrt{3}$; (h) 9

16. (a) 24; (b) 2; (c) 4

17. (a) 8; (b) 10; (c) 8; (d) 18; (e) $9\frac{3}{5}$; (f) $12\frac{1}{2}$; (g) 12; (h) 18

18. (a) $25\sqrt{3}$; (b) $36\sqrt{3}$; (c) $12\sqrt{3}$; (d) $25\sqrt{3}$; (e) $b^2\sqrt{3}$; (f) $4x^2\sqrt{3}$; (g) $3r^2\sqrt{3}$

19. (a) $2\sqrt{3}$; (b) $\frac{49}{2}\sqrt{3}$; (c) $24\sqrt{3}$ (d) $18\sqrt{3}$

20. (a) $24\sqrt{3}$; (b) $54\sqrt{3}$; (c) $150\sqrt{3}$

21. (a) 15; (b) 8; (c) 12; (d) 5

22. (a) 140; (b) 69; (c) 225; (d) $60\sqrt{2}$; (e) 94

23. (a) 150; (b) 204; (c) 39; (d) $64\sqrt{2}$; (e) 160

24. (a) 4; (b) 7; (c) 18 and 9; (d) 9 and 6; (e) 10 and 5

25. (a) 17 and 9; (b) 23 and 13; (c) 17 and 11; (d) 5; (e) 13

26. (a) 36; (b) $38\frac{1}{2}$; (c) $12\sqrt{3}$; (d) $12x^2$; (e) 120; (f) 96; (g) 18; (h) $\frac{49}{2}\sqrt{2}$; (i) $32\sqrt{3}$; $98\sqrt{3}$

27. (a) 737; (b) 14; (c) 77

28. (a) 10; (b) 12 and 9; (c) 20 and 10; (d) 5; (e) $\sqrt{10}$

29. 12

34. (a) 1:49; (b) 49:4; (c) 1:3; (d) 1:25; (e) 81: x^2; (f) 9:x; (g) 1:2

35. (a) 49:100; (b) 4:9; (c) 25:36; (d) 1:9; (e) 9:4; (f) 1:2

36. (a) 10:1; (b) 1:7; (c) 20:9; (d) 5:11; (e) 2:y; (f) $3x$:1; (g) $\sqrt{3}$:2; (h) 1:$\sqrt{2}$; (i) x:$\sqrt{5}$; (j) $\sqrt{x}$:4

37. (a) 6:5; (b) 3:7; (c) $\sqrt{3}$:1; (d) $\sqrt{5}$:2; (e) $\sqrt{3}$:3 or 1:$\sqrt{3}$

38. (a) 100; (b) $12\frac{1}{2}$; (c) 12; (d) 100; (e) 105; (f) 18; (g) $20\sqrt{3}$

39. (a) 12; (b) 63; (c) 48; (d) $2\frac{1}{2}$; (e) 45

Chapter 10

1. (a) 200; (b) 24.5; (c) 112; (d) 13; (e) 9; (f) $3\frac{1}{3}$; (g) 4.5

2. (a) $12\frac{1}{2}$; (b) 23.47; (c) $7\sqrt{3}$; (d) 18.5; (e) $3\sqrt{2}$

3. (a) 24°; (b) 24°; (c) 156°

4. (a) 40°; (b) 9; (c) 140°

5. (a) 15°; (b) 15°; (c) 24

6. (a) 5°; (b) 72°; (c) 175°

7. (a) regular octagon; (b) regular hexagon; (c) equilateral triangle; (d) regular decagon; (e) square; (f) regular dodecagon (12 sides)

9. (a) 9; (b) 30; (c) $6\sqrt{3}$; (d) 6; (e) $13\sqrt{3}$; (f) 6; (g) $20\sqrt{3}$; (h) 60

10. (a) $18\sqrt{2}$; (b) $7\sqrt{2}$; (c) 40; (d) $8\sqrt{2}$; (e) 3.4; (f) 28; (g) $5\sqrt{2}$; (h) $2\sqrt{2}$

11. (a) $30\sqrt{3}$; (b) 14; (c) 27; (d) 18; (e) $8\sqrt{3}$; (f) $4\sqrt{3}$; (g) $48\sqrt{3}$; (h) 42; (i) 6; (j) 10; (k) $\frac{5}{2}\sqrt{3}$; (l) $3\sqrt{3}$

12. (a) 817; (b) 3078

13. (a) $54\sqrt{3}$; (b) $96\sqrt{3}$; (c) $600\sqrt{3}$

14. (a) 576; (b) 324; (c) 100

15. (a) $36\sqrt{3}$; (b) $27\sqrt{3}$; (c) $\frac{16}{3}\sqrt{3}$; (d) $144\sqrt{3}$; (e) $3\sqrt{3}$; (f) $48\sqrt{3}$

16. (a) 10; (b) 10; (c) $5\sqrt{3}$

17. (a) 18; (b) $9\sqrt{3}$; (c) $6\sqrt{3}$; (d) $3\sqrt{3}$

18. (a) 1:8; (b) 4:9; (c) 9:10; (d) 8:11; (e) 3:1; (f) 2:5; (g) $4\sqrt{2}$:3; (h) 5:2

19. (a) 5:2; (b) 1:5; (c) 1:3; (d) 3:4; (e) 5:1

20. (a) 5:1; (b) 4:7; (c) x:2; (d) $\sqrt{2}$:1; (e) $\sqrt{3}$:y; (f) $\sqrt{x}$:$3\sqrt{2}$ or $\sqrt{2x}$:6

21. (a) 1:4; (b) 1:25; (c) 36:1; (d) 9:100; (e) 49:25

22. (a) 12π; (b) 14π; (c) 10π; (d) $2\pi\sqrt{3}$

23. (a) 9π; (b) 25π; (c) 64π; (d) $\frac{1}{4}\pi$, (e) 18π

24. (a) $C = 10\pi, A = 25\pi$; (b) $r = 8, A = 64\pi$; (c) $r = 4, C = 8\pi$

25. (a) 12π; (b) 4π; (c) 7π; (d) 26π; (e) $8\sqrt{3}\pi$ (f) 3π

26. (a) 98π; (b) 18π; (c) 32π; (d) 25π; (e) 72π; (f) 100π

27. (a) (1) $C = 8\pi, A = 16\pi$; (2) $C = 4\sqrt{3}\pi, A = 12\pi$

(b) (1) $C = 16\pi, A = 64\pi$; (2) $C = 8\sqrt{3}\pi, A = 48\pi$

(c) (1) $C = 12\pi, A = 36\pi$; (2) $C = 6\pi, A = 9\pi$

(d) (1) $C = 16\pi, A = 64\pi$; (2) $C = 8\pi, A = 16\pi$

(e) (1) $C = 20\sqrt{2}\pi, A = 200\pi$; (2) $C = 20\pi, A = 100\pi$

(f) (1) $C = 6\sqrt{2}\pi, A = 18\pi$; (2) $C = 6\pi, A = 9\pi$

28. (a) 10 ft; (b) 17 ft; (c) $3\sqrt{5}$ ft or 6.7 ft

29. (a) 2π; (b) 10π; (c) 8; (d) 11π; (e) 6π; (f) 10π

30. (a) 3π; (b) $12\frac{1}{2}$; (c) 5π; (d) 2π; (e) π; (f) 4π

31. (a) 6π; (b) $\pi/6$; (c) $25\pi/6$; (d) 25π; (e) $4\frac{1}{2}$; (f) 13; (g) 24π; (h) $8\pi/3$

32. (a) 6π; (b) 20; (c) 3π; (d) 16π

33. (a) 120°; (b) 240°; (c) 36°; (d) 180°; (e) 135°; (f) $(180/\pi)°$ or 57.3° to nearest tenth

34. (a) 72°; (b) 270°; (c) 40°; (d) 150°; (e) 320°

35. (a) 90°; (b) 270°; (c) 45°; (d) 36°

36. (a) 12; (b) 9; (c) 10; (d) 6; (e) 5; (f) $3\sqrt{2}$

37. (a) 4; (b) 10; (c) 10 cm; (d) 9

38. (a) $6\pi - 9\sqrt{3}$; (b) $24\pi - 36\sqrt{3}$; (c) $\frac{3}{2}\pi - \frac{9}{4}\sqrt{3}$; (d) $\frac{\pi r^2}{6} - \frac{r^2\sqrt{3}}{4}$; (e) $\frac{2\pi r^2}{3} - r^2\sqrt{3}$

39. (a) $4\pi - 8$; (b) $150\pi - 225\sqrt{3}$; (c) $24\pi - 36\sqrt{3}$; (d) $16\pi - 32$; (e) $50\pi - 100$

40. (a) $\frac{64\pi}{3} - 16\sqrt{3}$; (b) $24\pi - 16\sqrt{2}$; (c) $\frac{80\pi}{3} - 16$

41. (a) $\frac{16\pi}{3} - 4\sqrt{3}$; (b) $\frac{8\pi}{3} - 4\sqrt{3}$; (c) $4\pi - 8$

42. (a) $12\pi - 9\sqrt{3}$; (b) $\frac{3}{2}\pi - \frac{9}{4}\sqrt{3}$; (c) $9\pi - 18$

43. (a) $200 - 25\pi/2$; (b) $48 + 26\pi$; (c) $25\sqrt{3} - 25\pi/2$; (d) $100\pi - 96$; (e) $128 - 32\pi$;
(f) $300\pi + 400$; (g) 39π; (h) 100

44. (a) 36π; (b) $36\sqrt{3} + 18\pi$, (c) 14π

Chapter 11

1. The description of each locus is left for the reader.

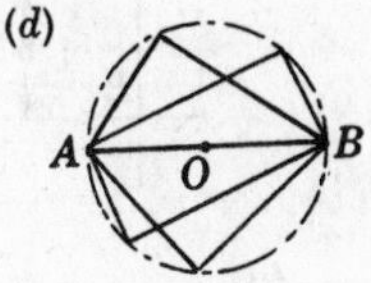

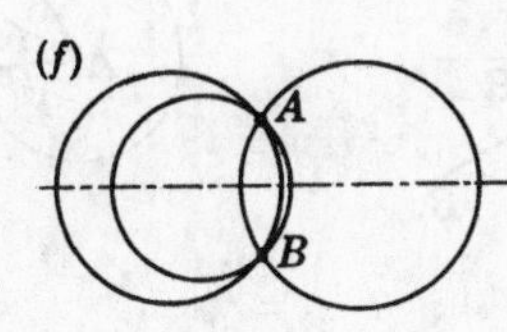

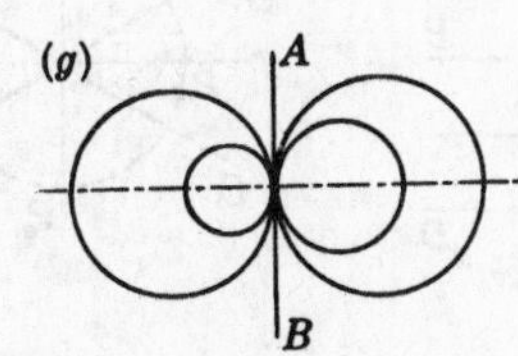

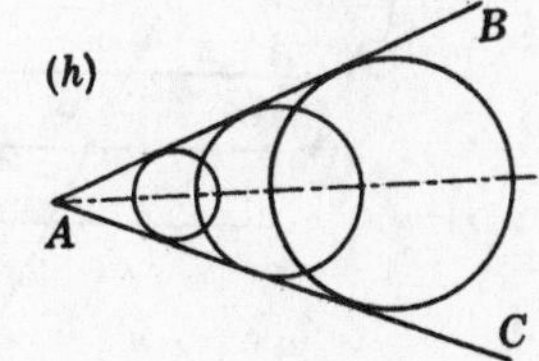

2. The diagrams are left for the reader.
(a) The line parallel to the banks and midway between them
(b) The perpendicular bisector of the segment joining the two floats
(c) The bisector of the angle between the roads
(d) The pair of bisectors of the angles between the roads

3. The diagrams are left for the reader.
(a) A circle having the sun as its center and the fixed distance as its radius
(b) A circle concentric to the coast, outside it, and at the fixed distance from it
(c) A pair of parallel lines on either side of the row and 20 ft from it
(d) A circle having the center of the clock as its center and the length of the clock hand as its radius.

4. (a) $\overline{EF}$; (b) $\overline{GF}$; (c) $\overline{EF}$; (d) $\overline{GH}$; (e) $\overline{EF}$; (f) $\overline{GH}$; (g) $\overline{AB}$; (h) a 90° arc from A to G with B as center

5. (a) $\overline{AC}$; (b) $\overline{BD}$; (c) $\overline{BD}$; (d) $\overline{AC}$; (e) E

6. In each case, the letter refers to the circumference of the circle. (a) A; (b) C; (c) B; (d) A; (e) C; (f) A and C; (g) B

7. The description of each locus is left for the reader.

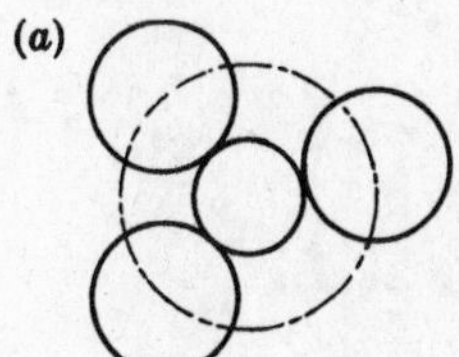

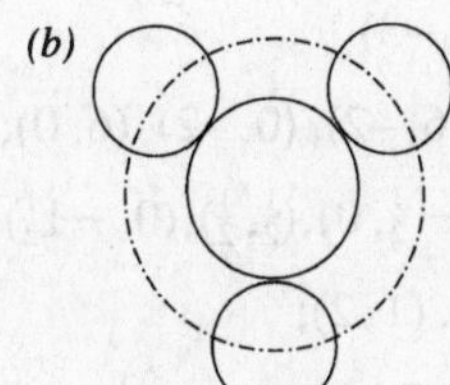

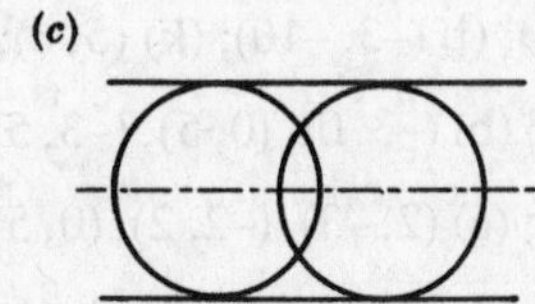

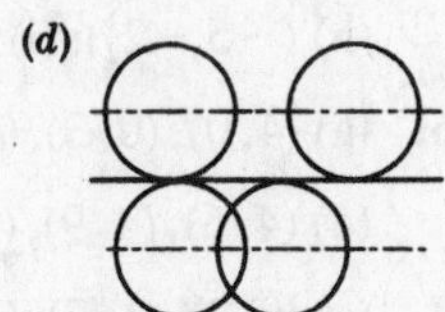

8. (a) $\overline{EF}$; (b) $\overline{GH}$; (c) line parallel to $\overline{AD}$ and $\overline{EF}$ midway between them; (d) $\overline{EF}$; (e) $\overline{BC}$; (f) $\overleftrightarrow{GH}$

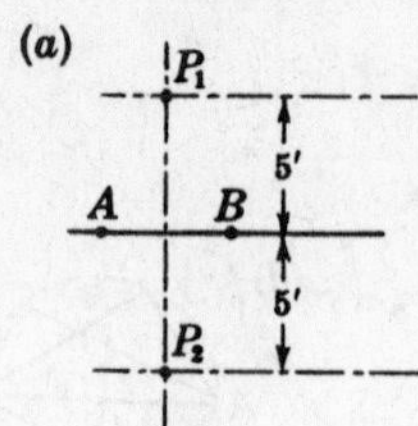

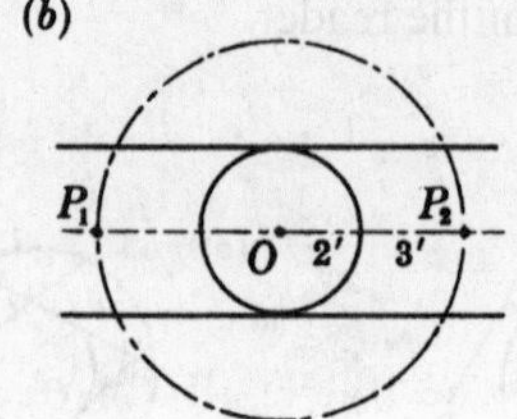

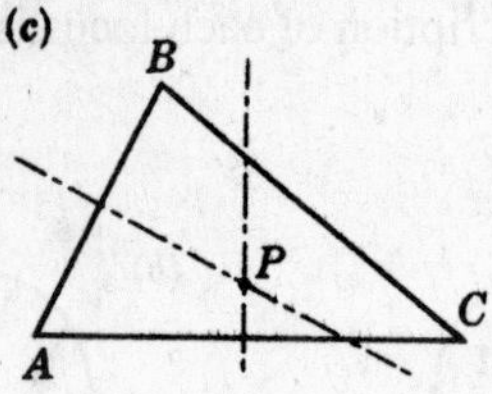

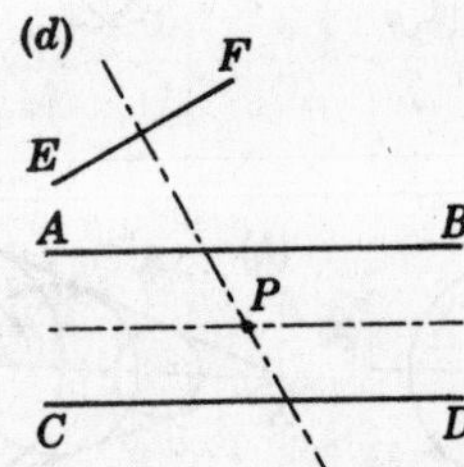

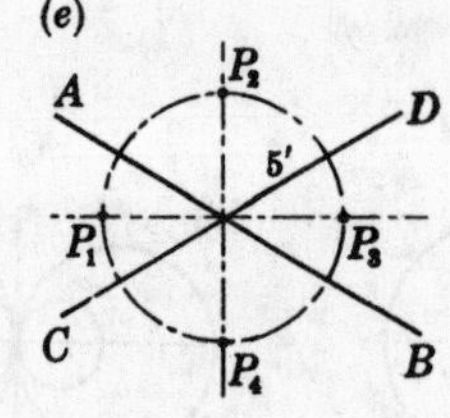

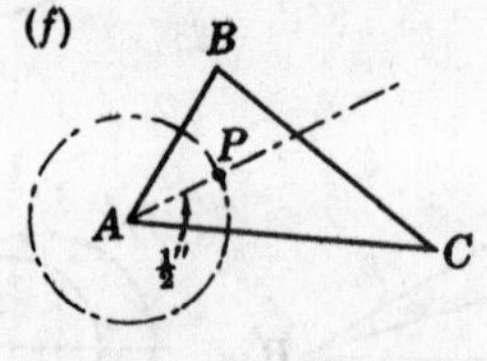

9. The explanation is left for the reader.

10. (a) The intersection of two of the angle bisectors
(b) The intersection of two of the ⊥ bisectors of the sides
(c) The intersection of the ⊥ bisector of $\overline{AB}$ and the bisector of ∠B
(d) The intersection of the bisector of ∠C and a circle with C as center and 5 as radius
(e) The intersections of two circles, one with B as center and 5 as radius and the other with A as center and 10 as radius

11. (a) 1; (b) 1; (c) 4; (d) 2; (e) 2; (f) 1

Chapter 12

1. $A(3, 0)$; $B(4, 3)$; $C(3, 4)$; $D(0, 2)$; $E(-2, 4)$; $F(-4, 2)$; $G(-1, 0)$; $H(-3\frac{1}{2}, -2)$; $I(-2, -3)$; $J(0, -4)$; $K(1\frac{1}{2}, -2\frac{1}{2})$; $L(4, -2\frac{1}{2})$

3. Perimeter of square formed is 20 units; its area is 25 square units.

4. Area of parallelogram = 30 square units.
Area of △BCD = 15 square units.

5. (a) $(4, 3)$; (b) $(2\frac{1}{2}, 3\frac{1}{2})$; (c) $(-4, 6)$; (d) $(7, -5)$; (e) $(-10, -2\frac{1}{2})$; (f) $(0, 10)$; (g) $(4, -1)$; (h) $(-5, -2\frac{1}{2})$; (i) $(5, 5)$; (j) $(-3, -10)$; (k) $(5, 6)$; (l) $(0, -3)$

6. (a) $(4, 0), (0, 3), (4, 3)$; (b) $(-3, 0), (0, 5), (-3, 5)$; (c) $(6, -2), (0, -2), (6, 0)$; (d) $(4, 6), (4, 9), (3, 8)$; (e) $(2, -3), (-2, 2), (0, 5)$; (f) $(-\frac{1}{2}, 0), (\frac{1}{2}, \frac{1}{2}), (0, -1\frac{1}{2})$

7. (a) $(0, 2), (1, 7), (4, 5), (3, 0)$; (b) $(-2, 7), (3, 6), (6, 1), (1, 2)$; (c) $(-1, 2), (3, 3), (3, -4), (-1, -5)$; (d) $(-2, 1), (4, 2\frac{1}{2}), (7, -4), (1, -7\frac{1}{2})$

8. (a) $(2, 6), (4, 3)$; (b) $(1, 0), (0, -2\frac{1}{2})$; (c) common midpoint, $(2, 2)$

9. (a) $(-2, 3)$; (b) $(-3, -6)$; (c) $(-1\frac{1}{2}, -2)$; (d) (a, b); (e) $(2a, 3b)$; (f) $(a, b + c)$

10. (a) $M(4, 8)$; (b) $A(-1, 0)$; (c) $B(6, -3)$

11. (a) $B(2, 3\frac{1}{2})$; (b) $D(3, 3)$; (c) $A(-2, 9)$

12. (a) Prove that $ABCD$ is a parallelogram (since opposite sides are congruent) and has a rt. ∠.
(b) The point $(3, 2\frac{1}{2})$ is the midpoint of each diagonal.
(c) Yes, since the midpoint of each diagonal is their common point.

13. (a) $D(3, 2)$, $(1\frac{1}{2}, 1)$; (b) $E(0, 2)$, $(3, 1)$; (c) no, since the midpoint of each median is not a common point

14. (a) 5; (b) 6; (c) 10; (d) 12; (e) 5.4; (f) 7.5; (g) 9; (h) a

15. (a) 3, 3, 6; (b) 4, 14, 18; (c) 1, 3, 4; (d) a, $2a$, $3a$

16. (a) 13; (b) 5; (c) 15; (d) 5; (e) 10; (f) 15; (g) $3\sqrt{2}$; (h) $5\sqrt{2}$; (i) $\sqrt{10}$; (j) $2\sqrt{5}$; (k) 4; (l) $a\sqrt{2}$

18. (a) $\triangle ABC$; (b) $\triangle DEF$; (c) $\triangle GHJ$; (d) $\triangle KLM$ is *not* a rt. △

19. (a) $5\sqrt{2}$; (b) $\sqrt{5}$; (c) $\sqrt{65}$

21. (a) 10; (b) 5; (c) $5\sqrt{2}$; (d) 13; (e) 4; (f) 3

22. (a) on; (b) on; (c) outside; (d) on; (e) inside; (f) inside; (g) on

23. (a) $\frac{9}{5}$; (b) $\frac{5}{9}$; (c) $\frac{5}{2}$; (d) 3; (e) 2; (f) 1; (g) 5; (h) –2; (i) –3; (j) $\frac{3}{2}$; (k) –1; (l) 1

24. (a) 3; (b) 4; (c) $-\frac{1}{2}$; (d) –7; (e) 5; (f) 0; (g) 3; (h) 5; (i) –4; (j) $-\frac{2}{3}$; (k) –1; (l) –2; (m) 5; (n) 6; (o) –4; (p) –8

25. (a) 72°; (b) 18°; (c) 68°; (d) 22°; (e) 45°; (f) 0°

26. (a) 0.0875; (b) 0.3057; (c) 0.3640; (d) 0.7002; (e) 1; (f) 3.2709; (g) 11.430

27. (a) 0°; (b) 25°; (c) 45°; (d) 55°; (e) 7°; (f) 27°; (g) 37°; (h) 53°; (i) 66°

28. (a) $\overline{BC}$, $\overline{BD}$, $\overline{AD}$, $\overline{AE}$; (b) $\overline{BF}$, $\overline{CF}$, $\overline{DE}$; (c) $\overline{AF}$, $\overline{CD}$; (d) $\overline{AB}$, $\overline{EF}$

29. (a) 0; (b) no slope; (c) 5; (d) –5; (e) 0.5; (f) –0.0005

30. (a) 0; (b) no slope; (c) no slope; (d) 0; (e) 5; (f) –1; (g) 2

31. (a) $\frac{3}{2}$; (b) $\frac{7}{3}$; (c) –1; (d) 6

32. (a) –2; (b) –1; (c) $-\frac{1}{3}$; (d) $-\frac{2}{5}$; (e) –10; (f) 1; (g) $\frac{5}{4}$; (h) $\frac{4}{13}$; (i) no slope; (j) 0

33. (a) 0; (b) –2; (c) 3; (d) –1

34. (a) $-\frac{3}{2}$; (b) $\frac{2}{3}$; (c) $-\frac{3}{2}$

35. (a) $-\frac{1}{2}$; (b) 1; (c) 2; (d) –1

36. (a) $-\frac{3}{2}$; (b) $\frac{1}{4}$; (c) $\frac{5}{6}$

37. (a) and (b)

38. (a) 19; (b) 9; (c) 2

39. (a) $x = -5$; (b) $y = 3\frac{1}{2}$; (c) $y = 3$ and $y = -3$; (d) $y = -5$; (e) $x = 4$ and $x = -4$; (f) $x = 5$ and $x = -1$; (g) $y = 4$; (h) $x = 1$; (i) $x = 9$

40. (a) $x = 6$; (b) $y = 5$; (c) $x = 6$; (d) $x = 5$; (e) $x = 6$; (f) $y = 3$

41. (a) $x = y$; (b) $y = x + 5$; (c) $x = y - 4$; (d) $y - x = 10$; (e) $x + y = 12$; (f) $x - y = 2$ or $y - x = 2$; (g) $x = y$ and $x = -y$; (h) $x + y = 5$

42. (a) line having y-intercept 5, slope 2; (b) line passing through (2, 3), slope 4; (c) line passing through (–2, –3), slope $\frac{5}{4}$; (d) line passing through origin, slope $\frac{1}{2}$; (e) line having y-intercept 7, slope –1; (f) line passing through origin, slope $\frac{1}{3}$

43. (a) $y = 4x$; (b) $y = -2x$; (c) $y = \frac{3}{2}x$ or $2y = 3x$; (d) $y = -\frac{2}{5}x$ or $5y = -2x$; (e) $y = 0$

44. (a) $y = 4x + 5$; (b) $y = -3x + 2$; (c) $y = \frac{1}{3}x - 1$ or $3y = x - 3$; (d) $y = 3x + 8$; (e) $y = -4x - 3$; (f) $y = 2x$ or $y - 2x = 0$

45. (a) $\frac{y-4}{x-1}=2$ or $y=2x+2$; (b) $\frac{y-3}{x+2}=2$ or $y=2x+7$; (c) $\frac{y}{x+4}=2$ or $y=2x+8$; (d) $\frac{y+7}{x}=2$ or $y=2x-7$

46. (a) $y=4x$; (b) $y=\frac{1}{2}x+3$; (c) $\frac{y-2}{x-1}=3$; (d) $\frac{y+2}{x+1}=\frac{1}{3}$; (e) $y=2x$

47. (a) circle with center at origin and radius 7; (b) $x^2+y^2=16$; (c) $x^2+y^2=64$ and $x^2+y^2=4$

48. (a) $x^2+y^2=25$; (b) $x^2+y^2=81$; (c) $x^2+y^2=4$ or $x^2+y^2=144$

49. (a) 3; (b) $\frac{4}{3}$; (c) 2; (d) $\sqrt{3}$

50. (a) $x^2+y^2=16$; (b) $x^2+y^2=121$; (c) $x^2+y^2=\frac{4}{9}$ or $9x^2+9y^2=4$; (d) $x^2+y^2=\frac{9}{4}$ or $4x^2+4y^2=9$; (e) $x^2+y^2=5$; (f) $x^2+y^2=\frac{3}{4}$ or $4x^2+4y^2=3$

51. (a) 10; (b) 10; (c) 20; (d) 20; (e) 7; (f) 25

52. (a) 16; (b) 12; (c) 20; (d) 24

53. (a) 10; (b) 12; (c) 22

54. (a) 5; (b) 13; (c) $7\frac{1}{2}$

55. (a) 6; (b) 10; (c) 1.2

56. (a) 15; (b) 49; (c) 53

57. (a) 30; (b) 49; (c) 88; (d) 24; (e) 16; (f) 18

Chapter 13

1. (a) $<$; (b) $>$; (c) $>$; (d) $>$; (e) $>$; (f) $<$

2. (a) $>$; (b) $>$; (c) $<$; (d) $>$

3. (a) $>$; (b) $<$; (c) $<$; (d) $>$

4. (a) more; (b) less

5. (a) $>$; (b) $>$; (c) $<$; (d) $>$; (e) $<$; (f) $<$

6. (c), (d), and (e)

7. (a) 5 to 7; (b) 6 to 10; (c) 4 to 10; (d) 3 to 9; (e) 2 to 8; (f) 1 to 13

8. (a) $\angle B, \angle A, \angle C$; (b) $\overline{DF}, \overline{EF}, \overline{DE}$; (c) $\angle 3, \angle 2, \angle 1$

9. (a) $m\angle BAC > m\angle ACD$; (b) $AB > BC$

10. (a) $\overline{BC}, \overline{AB}, \overline{AC}$; (b) $\angle BOC, \angle AOB, \angle AOC$; (c) $\overline{AD}, \overline{AB} \cong \overline{CD}, \overline{BC}$; (d) $\overline{OG}, \overline{OH}, \overline{OJ}$

Chapter 14

1. (a) Ornament, jewelry, ring, wedding ring; (b) vehicle, automobile, commercial automobile, taxi; (c) polygon, quadrilateral, parallelogram, rhombus; (d) angle, obtuse angle, obtuse triangle, isosceles obtuse triangle

2. (a) A regular polygon is an equilateral and an equiangular polygon.
(b) An isosceles triangle is a triangle having at least two congruent sides.

(c) A pentagon is a polygon having five sides.
(d) A rectangle is a parallelogram having one right angle.
(e) An inscribed angle is an angle formed by two chords and having its vertex on the circumference of the circle.
(f) A parallelogram is a quadrilateral whose opposite sides are parallel.
(g) An obtuse angle is an angle larger than a right angle and less than a straight angle.

3. (a) $x + 2 \neq 4$; (b) $3y = 15$; (c) she does not love you; (d) his mark was not more than 65; (e) Joe is not heavier than Dick; (f) $a + b = c$

4. (a) A nonsquare does not have congruent diagonals. False (for example, when applied to a rectangle or a regular pentagon).
(b) A non-equiangular triangle is not equilateral. True.
(c) A person who is not a bachelor is a married person. This inverse is false when applied to an unmarried female.
(d) A number that is not zero is a positive number. This inverse is false when applied to negative numbers.

5. (a) Converse true, inverse true, contrapositive true
(b) Converse false, inverse false, contrapositive true
(c) Converse true, inverse true, contrapositive true
(d) Converse false, inverse false, contrapositive true

6. (a) Partial converses: interchange (2) and (3) or (1) and (3)
Partial inverses: negate (1) and (3) or (2) and (3)
(b) Partial converses: interchange (1) and (4) or (2) and (4) or (3) and (4)
Partial inverses: negate (1) and (4) or (2) and (4) or (3) and (4)

7. (a) Necessary and sufficient; (b) necessary but not sufficient; (c) neither necessary nor sufficient; (d) sufficient but not necessary; (e) necessary and sufficient; (f) sufficient but not necessary; (g) necessary but not sufficient

Chapter 17

1. (a) $6(7^2)$ or 294 yd^2; (b) $2(8)(6\frac{1}{2}) + 2(8)(14) + 2(6\frac{1}{2})(14)$ or 510 ft^2;
(c) $4(3.14)30^2$ or 11,304 m^2; (d) $(3.14)(10)(10 + 4\frac{1}{2})$ or 911 yd^2

2. (a) 36^3 or 46,656 in^3; (b) 100^3 or 1,000,000 cm^3

3. (a) 27 in^3; (b) 91 in^3; (c) 422 in^3; (d) 47 in^3; (e) 2744 in^3

4. (a) $3(8\frac{1}{2})(8)$ or 204 in^3; (b) $2(9)(9)$ or 162 ft^3; (c) $\frac{1}{2}(6)(6.4)$ or 13 ft^3

5. (a) 904 m^3; (b) 1130 ft^3; (c) 18 ft^3

6. (a) $V = \frac{1}{3}\pi r^2 h$; (b) $V = \frac{1}{3}s^2 h$; (c) $V = \frac{1}{3}lwh$; (d) $V = \frac{2}{3}\pi r^3$

7. (a) $6e^3 + \dfrac{2e^2h}{3}$; (b) $lwh + \dfrac{\pi l^2 w}{8}$; (c) $3\pi r^3$

Chapter 18

1. (a) $A'(9, 2)$, $B'(2, 4)$, and $C'(5, 7)$; (b) $A'(-4, 2)$, $B'(3, 4)$, and $C'(0, 7)$; (c) $A'(-5, 1)$, $B'(2, -1)$, and $C'(-1, -4)$; (d) $A'(-2, 14)$, $B'(-4, 7)$, and $C'(-7, 10)$; (e) $A'(12, 4)$, $B'(-2, 8)$, and $C'(4, 14)$; (f) $A'(-14, 6)$, $B'(7, 12)$, and $C'(-2, 21)$; (g) $A'(0, -1)$, $B'(-2, 6)$, and $C'(-5, 3)$

2. (a) $A'(6,-1)$, $B'(7, 2)$, and $C'(8,-1)$; (b) $A'(1,-5)$, $B'(2,-2)$, and $C'(3,-5)$;
(c) $A'(-2, 1)$, $B'(-1, 4)$, and $C'(0, 1)$

3. (a) translation to the right 5 spaces, $P(x, y) \mapsto P'(x + 5, y)$; (b) translation down 5 spaces and to the right 3, $P(x, y) \mapsto P'(x + 3, y-5)$; (c) translation down 3 spaces and to the left 5, $P(x, y) \mapsto P'(x-5, y-3)$

4. (a) $P(x, y) \mapsto P'(x, y-5)$; (b) $P(x, y) \mapsto P'(x + 6, y)$; (c) $P(x, y) \mapsto P'(x-7, y + 3)$; (d) $P(x, y) \mapsto P'(x + 8, y-2)$; (e) $P(x, y) \mapsto P'(x-1, y + 4)$

5. (a) $A'(-1, 3)$, $B'(-5, 3)$, $C'(-4, 1)$, and $D'(-2, 1)$; (b) $A''(1,-5)$, $B''(5,-5)$, $C''(4,-3)$, and $D''(2,-3)$; (c) $A'''(15, 3)$, $B'''(11, 3)$, $C'''(12, 1)$, and $D'''(14, 1)$

6. (a) reflection across the line $x = 2$, $P(x, y) \mapsto P'(4-x, y)$; (b) reflection across $y = 3$, $P(x, y) \mapsto P''(x, 6-y)$; (c) reflection across $x = -5$, $P(x, y) \mapsto P'''(-10-x, y)$

7. (a) $P(x, y) \mapsto P'(x, 10-y)$; (b) $P(x, y) \mapsto P'(-4-x, y)$; (c) $P(x, y) \mapsto P'(x,-2-y)$; (d) $P(x, y) \mapsto P'(5-x, y)$

8. (a), (b), (e), and (f)

9. (a) $A'(2, -1)$, $B'(2, -4)$, $C'(1, -5)$, and $D'(1, -2)$; (b) $A''(-1, -2)$, $B''(-4, -2)$, $C''(-5, -1)$, and $D''(-2,-1)$; (c) $A'''(-2, 1)$, $B'''(-2, 4)$, $C'''(-1, 5)$, and $D'''(-1, 2)$

10. (a) 180° rotation about the origin, $P(x, y) \mapsto P'(-x,-y)$; (b) 90° clockwise rotation about the origin, $P(x, y) \mapsto P''(y,-x)$; (c) 270° clockwise rotation about the origin (or 90° counter-clockwise), $P(x, y) \mapsto P'''(-x, y)$

11. (a) $P(x, y) \mapsto P'(x \cos 40° + y \sin 40°, y \cos 40°-x \sin 40°) = P'(0.766x + 0.6428y, 0.766y-0.6428x)$; (b) $P(x, y) \mapsto P'(x \cos 50° + y \sin 50°, y \cos 50°-x \sin 50°) = P'(0.6428x + 0.766y, 0.6428y-0.766x)$; (c) $P(x, y) \mapsto P'(x \cos 80° + y \sin 80°, y \cos 80°-x \sin 80°) = P'(0.1736x + 0.9848y, 0.1736y- 0.9848x)$

12. (a) 120°; (b) 60°; (c) 72°; (d) 360° (no rotational symmetry); (e) 90°; (f) 180°

13. (a) $A'(4, 5)$, $B'(5, 4)$, and $C'(5, 7)$; (b) $A''(2,-5)$, $B''(3,-6)$, and $C''(0,-6)$; (c) $A'''(-1,-2)$, $B'''(-2,-3)$, and $C'''(1,-3)$; (d) $A''''(-4, 3)$, $B''''(-5, 4)$, and $C''''(-5, 1)$

14. (a) $R(x, y) \mapsto R''(x + 6, y-1)$; (b) $R(x, y) \mapsto R''(1-x, y + 2)$; (c) $R(x, y) \mapsto R''(y + 3,-x-6)$; (d) $R(x, y) \mapsto R''(-y, 4-x)$; (e) $R(x, y) \mapsto R''(6-x,-3-y)$

15. (a) reflect across the y axis and then move down 2 and to the right 3 spaces, $P(x, y) \mapsto P'(-x + 3, y-2)$; (b) rotate about the origin counterclockwise 90°, then move to the left 1 space, $P(x, y) \mapsto P''(-y-1, x)$; (c) rotate around the origin 90° clockwise, then reflect across the x axis, then move down and to the right 1 space, $P(x, y) \mapsto P'''(y + 1, x-1)$

16. (a) $P(x, y) \mapsto P'(x,-y-3)$; (b) $P(x, y) \mapsto P'(y + 2,-x)$; (c) $P(x, y) \mapsto P'(-x, y-4)$; (d) $P(x, y) \mapsto P'(-x, 4 + y)$; (e) $P(x, y) \mapsto P'(-7-x, y + 3)$

17. $A'(-3, 6)$, $B'(3, 6)$, $C'(3, 3)$, and $D'(-3, 3)$

18. (a) $P(x, y) \mapsto P'(2x, 2y)$; (b) $P(x, y) \mapsto P'(8x, 8y)$; (c) $P(x, y) \mapsto P'(\frac{1}{3}x, \frac{1}{3}x)$

Chapter 19

1. (a) Center: $(5, -3)$. Major axis: $y = -3$. Minor axis: $x = 5$. Vertices: $(-5, -3)$, $(15, -3)$, $(5, -1)$, and $(5, -5)$. Foci: $(5 \pm \sqrt{96}, -3)$.

(b) Center: $(-2, 0)$. Major axis: $y = 0$ (the x axis). Minor axis: $x = -2$. Vertices: $(-5, 0)$, $(1, 0)$, $(-2, 1)$, and $(-2, -1)$. Foci: $(-2 \pm \sqrt{8}, 0)$.

(c) Center: $(1, 4)$. Major axis: $x = 1$. Minor axis: $y = 4$. Vertices: $(-5, 4)$, $(7, 4)$, $(1, -3)$, and $(1, 11)$. Foci: $(1, 4 \pm \sqrt{13})$.

2. (a) Vertex: $(4, -2)$. Axis: $y = -2$. Focus: $(4 + \frac{1}{12}, -2)$.

(b) Vertex: $(0, -2)$. Axis: $x = 0$ (the y axis). Focus: $(0, -4.5)$.

(c) Vertex: $(-2, -1)$. Axis: $y = -1$. Focus $(-1.75, -1)$.

3. (a) Center: $(-7, -5)$. Transverse axis: $x = -7$. Conjugate axis: $y = -5$. Vertices: $(-7, -3)$ and $(-7, -7)$. Foci: $(-7, -5 \pm \sqrt{13})$.

(b) Center: $(3, 0)$. Transverse axis: $x = 3$. Conjugate axis: $y = 0$ (x axis). Vertices: $(3, \pm 5)$. Foci: $(3, \pm\sqrt{41})$.

(c) Center: $(-2, 1)$. Transverse axis: $y = 1$. Conjugate axis: $x = -2$. Vertices: $(-10, 1)$ and $(6, 1)$. Foci: $(-2 \pm \sqrt{65}, 1)$.

4. See Fig. 19-24.

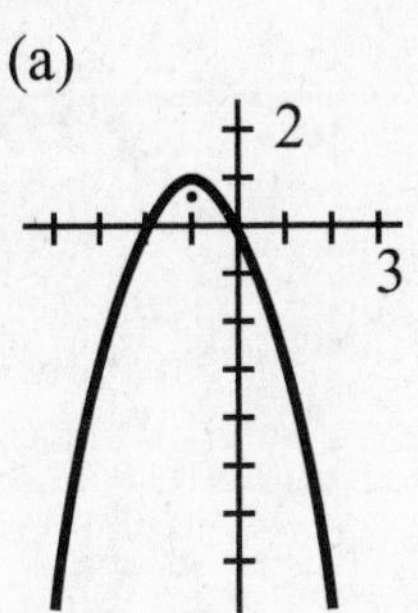

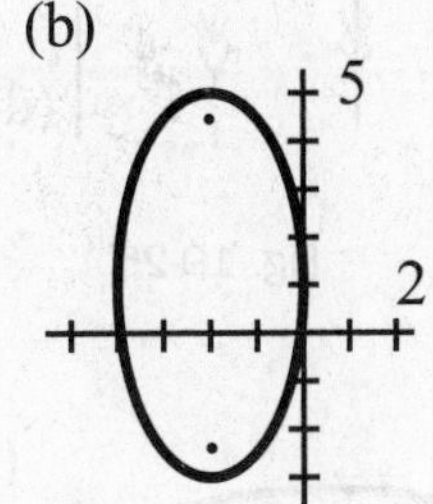

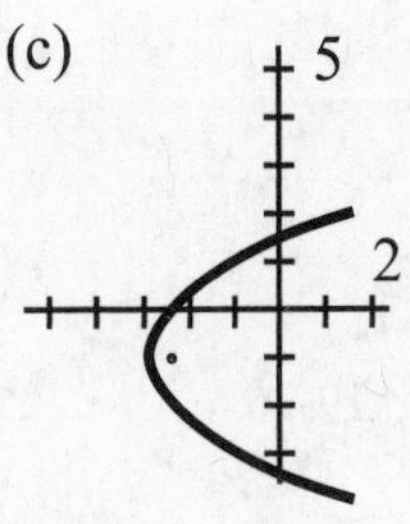

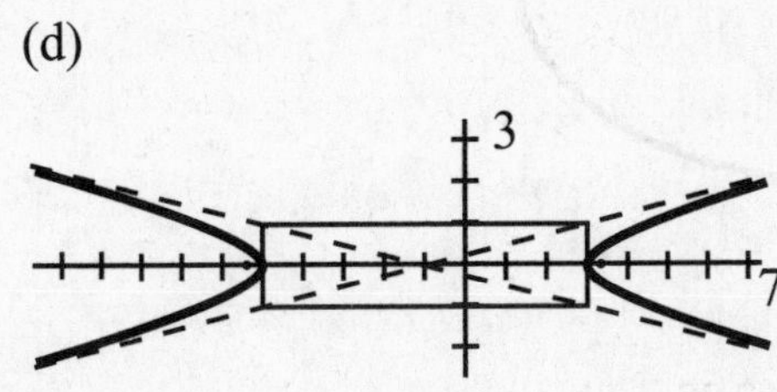

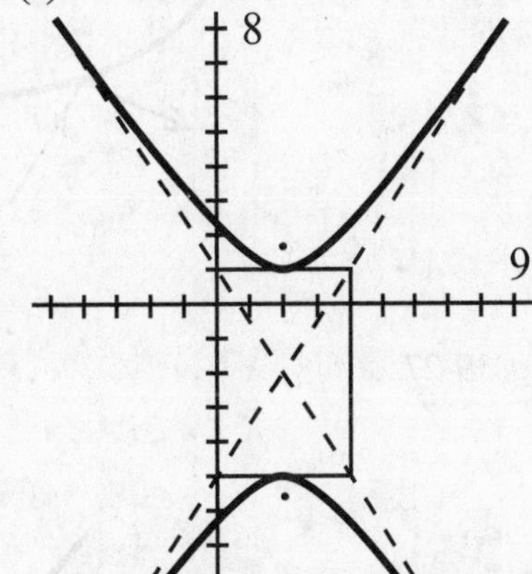

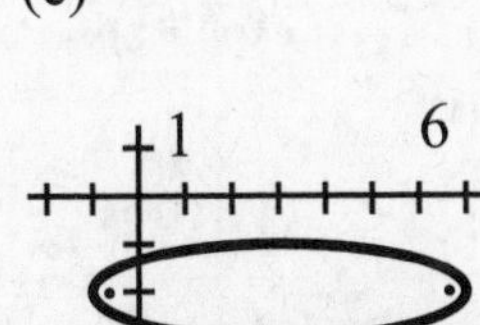

Fig. 19-24

5. (a) $y + 3 = (x - 2)^2$

(b) $-(x - 2)^2 + \dfrac{y^2}{2^2} = 1$

(c) $\dfrac{(x - 1)^2}{5^2} + \dfrac{(y + 1)^2}{2^2} = 1$

(d) $(x - 1)^2 + \dfrac{y^2}{2^2} = 1$

(e) $x - 2 = -2(y + 2)^2$

(f) $(x + 1)^2 - (y + 1)^2 = 1$

6. See Fig. 19-25.

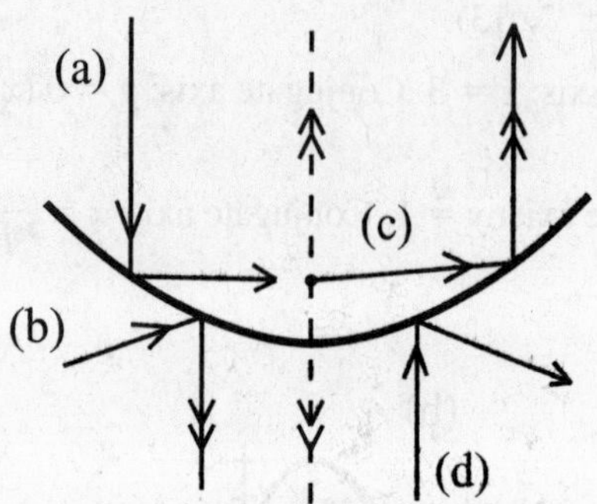

Fig. 19-25

7. See Fig. 19-26.

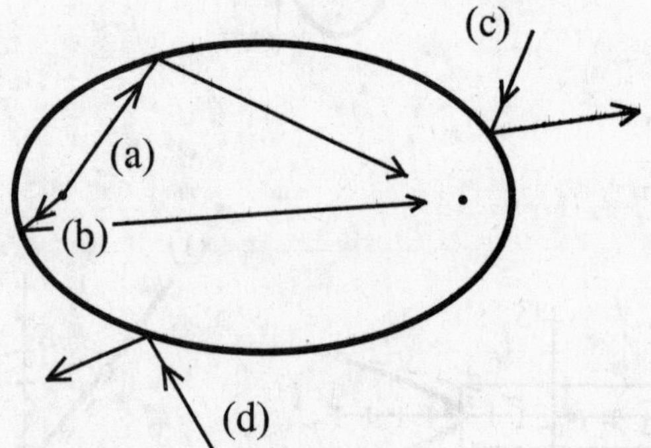

Fig. 19-26

8. See Fig. 19-27.

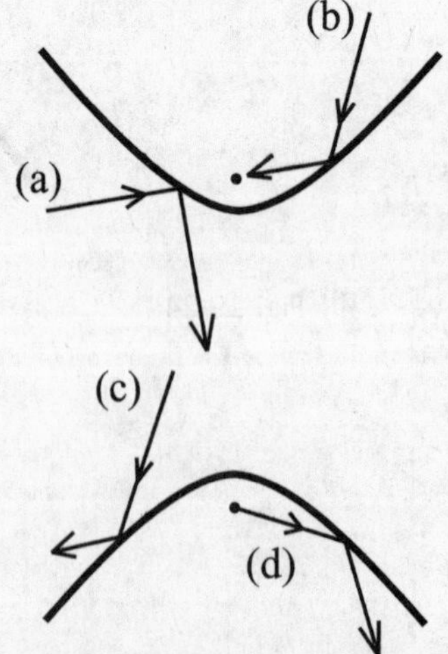

Fig. 19-27

Index